INTRODUCTION TO WORLD AGRISCIENCE AND TECHNOLOGY

INTRODUCTION TO WORLD AGRISCIENCE

JASPER S. LEE

Agricultural Educator
Demorest, Georgia

DIANA L. TURNER

Science Teacher
Starkville, Mississippi

AND TECHNOLOGY

IST AgriScience and Technology Series

Series Editor
Jasper S. Lee, Ph.D.

IPP Interstate Publishers, Inc.
Danville, Illinois

Cover Photo Credits:

Upper left—computer generated images of the earth courtesy of National Geophysical Data Center, U.S. Department of Commerce, Boulder, CO

Upper right—plant tissue analysis courtesy of Terra International, Inc., Sioux City, IA

Lower left—Tiger Swallowtail courtesy of Ross E. Hutchins Photograph Collection, Mississippi Entomological Museum, Mississippi State, MS

Lower center—white flies on the leaf of a cotton plant courtesy of Progressive Farmer Magazine, Birmingham, AL

Lower right—Quarter Horse courtesy of the American Quarter Horse Association, Amarillo, TX

Order from

Interstate Publishers, Inc.
510 North Vermilion Street
P.O. Box 50
Danville, IL 61834-0050
Phone: (800) 843-4774
Fax: (217) 446-9706
Email: info-ipp@IPPINC.com

Library of Congress Catalog Card No. 96–75862

Textbook ISBN 0-8134-3000-3
Teacher's Manual ISBN 0-8134-2960-9
Instructional Resource Guide ISBN 0-8134-3107-7
Computer Test Bank ISBN 0-8134-2975-7
Animal Science Transparency Set ISBN 0-8134-3024-0
Introduction to Biotechnology Transparency Set ISBN 0-8134-3022-4
Plant Science Transparency Set ISBN 8134-3023-4
Activity Manual ISBN 0-8134-2961-7
Instructor's Guide for Activity Manual ISBN 0-8134-2974-9
Cell Biology Transparency Set ISBN 0-8134-3025-9
Laboratory Skills Transparency Set ISBN 0-8134-3026-7

1 2 3 4 5 6 7 8 9 10 02 01 00 99 98 97 96

Preface

Introduction to World AgriScience and Technology is about meeting the needs of people. These needs are carefully described as food, fiber, and shelter. None of these could be met without applying science and technology.

This book blends science and agriculture. It emphasizes biological, earth, and physical sciences. A lot of science background isn't needed, but some will help. Further, knowledge and background in agriculture is not needed. Again, some will help!

This book deals with exciting subjects! Agriculture and its related areas are "science in action." The content of this book always relates the science information to a use in agriculture. Some things that sound complicated in the science classroom are used every day in agriculture!

Every effort has been made to make this book "friendly" to students and teachers. The content is organized and sequenced for ease of use. Many illustrations are used. Much of the art is new, and was produced using new computer graphics procedures. Photographs have been used from throughout North America and are from farms, agribusinesses, agencies, and schools. The entire process of writing the book involved new procedures in publishing not used before in agriculture education.

The contents of the book are based on research and development work. Reports of research were reviewed. Curriculum guides from other states and many different schools were used. Requirements for using agriculture classes to meet science credit in high schools were used. The book also includes experiences from a number of schools that had implemented programs of agriscience. The content was carefully reviewed by agriculture and science teachers, agriscientists, and others who had strong backgrounds in the subjects.

Introduction to World AgriScience and Technology is for use in modern agriscience education. It is for people who are serious about the science and technology of agriculture.

Begin your study with page 1 or wherever you wish. There is a lot to learn!

About the Second Edition

Students and teachers across North America have expressed tremendous enthusiasm for the First Edition of *Introduction to World AgriScience and Technology*. This enthusiasm became the driving force behind preparation of the Second Edition. The comments from users were so positive that the work on the Second Edition was exciting and easy for the authors.

The authors have worked to make the Second Edition better organized, more informative, and increasingly user-friendly. Personal appeal has been enhanced with full-color printing, attractive no-nonsense design, and more photographs and drawings. New developments in agriscience have been included to keep the content up to date. Specific areas where new information has been added include:

- revised statistical information on consumption of products and population
- more details on biotechnology, especially cell structure and processes
- synthetic applications of science, including synthetic biology and synthetic foods
- expanded information on technology applications in agriculture
- increased definitions and explanations of scientific terms
- enhanced details on being a good consumer
- practical examples relevant to youth in AgriScience Connections
- use of global positioning and environmentally sound agriculture

The parts, chapter titles, and objectives for the chapters have remained virtually unchanged. This means that the Second Edition can be used in classes right along with the First Edition. It also means that the Teacher's Manual, Activity Manual, and Instructor's Guide to the Activity Manual are unchanged.

As with the first edition, the authors believe that agriscience is first and foremost about agriculture—science in action. With quality instruction and higher expectations for students, awarding science credit for an agriculture class where this book is used is most appropriate. In fact, students will likely have a better understanding of the use of science in the world around them from this book than from many books used in science classes. Involvement with the National Science Teachers Association and using the *Scope, Sequence, and Coordination of Secondary School Science Guide for Curriculum Designers* has enhanced the science base.

Introduction to World AgriScience and Technology continues as the corner stone of Interstate's Science and Technology (IST) Series. It holds expectations for learners and creates excitement for the teacher. Go ahead and join the fun of agriscience learning!

About the Authors

Jasper S. Lee

Jasper S. Lee is an agricultural educator currently living in Georgia. He has been a high school teacher and a teacher educator and now operates an agricultural education consulting business known as Lee and Associates. He is a native of Clinton, Mississippi, and grew up with strong interests in agricultural education. FFA activities were most important during his high school years, with him holding chapter, federation, and state offices. His education beyond high school was at Hinds Community College, Mississippi State University, and the University of Illinois. His former students are in leadership positions throughout the United States and many other nations.

Lee has authored or co-authored a number of textbooks used in agricultural education and areas of applied science. Among the titles are *Aquaculture — An Introduction, Environmental Science and Technology, AgriMarketing Technology*, and *The Earth and AgriScience*. He serves as Editorial Consultant with Interstate Publishers, Inc., and works to help authors in writing as well as provide inservice education for teachers.

Diana L. Turner

Diana L. Truner is a high school science teacher in Starkville, Mississippi, with major interests in chemistry, physics, advanced biology, and environmental science. She is a graduate of Mississippi State University and widely recognized for excellence in science education in the state and the nation. She directs an outstanding science program that is often recognized as the top program in the state. In 1993, she was named the Mississippi Science Teacher of the Year. A year later she was named to the Education Hall of Fame by the Chamber of Commerce. She has been named the Star Teacher at her school on several occasions and is a leader among science teachers in the state. She is a strong advocate of quality science education. Her students often develop outstanding entries for science fair competition. Her former students are now in a wide range of graudate and professional schools in science.

Turner has authored several widely used textbooks and accompanying manuals. She had extensive involvement in the *Environmental Science and Technology* textbook and activity manual. Turner is popular as a leader of inservice education workshops for agriculture and science teachers. She often conducts national workshops as part of the Professional Development Program of Interstate Publishers, Inc.

USER-FRIENDLY FEATURES

- ➪ Uses teacher and student approved layout and design
- ➪ Reading level appropriate to intended grade of use
- ➪ Content is systematic
- ➪ Emphasizes biological, earth, and physical sciences
- ➪ Validated concepts and skills in mathematics
- ➪ Brings biotechnology to life through agriculture
- ➪ Enhanced details on being a good consumer
- ➪ Suggested activities beyond the classroom

CHAPTER ORGANIZATION

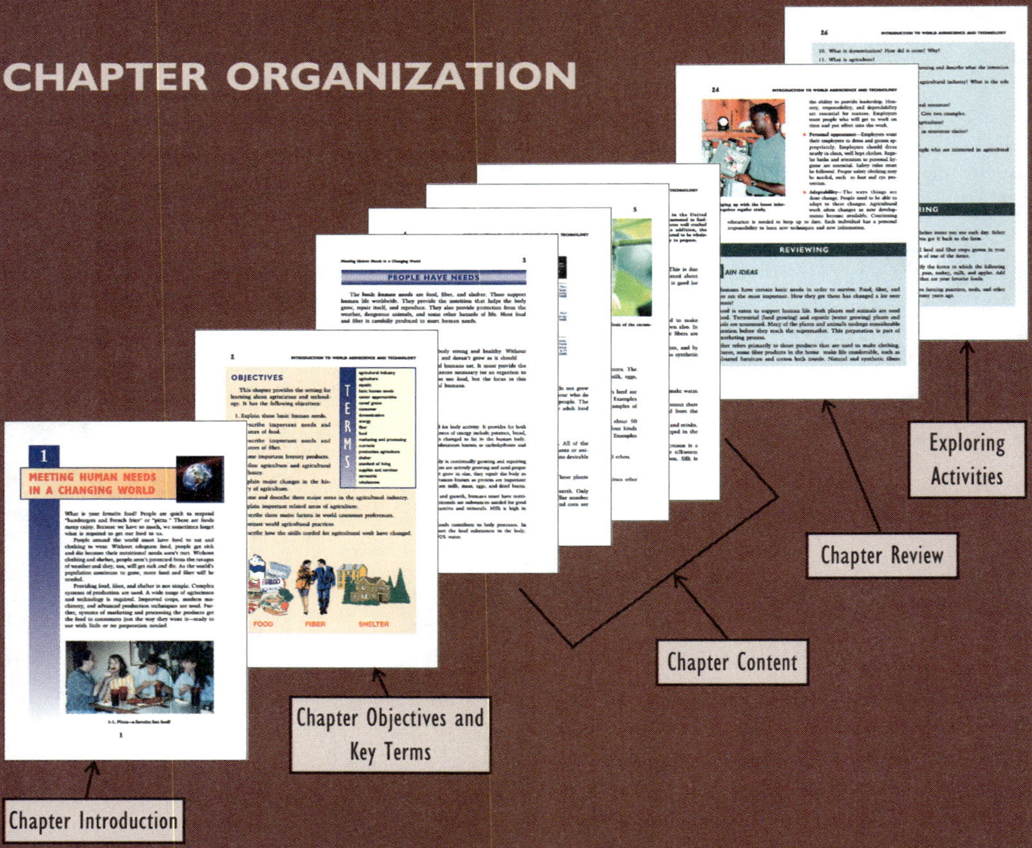

Introduction—each chapter starts with a short introduction to set the stage for learning the material by showing relevance to agriscience.

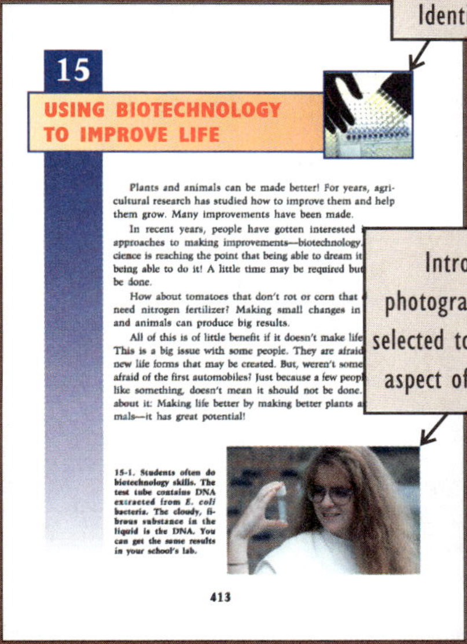

15

USING BIOTECHNOLOGY TO IMPROVE LIFE

Plants and animals can be made better! For years, agricultural research has studied how to improve them and help them grow. Many improvements have been made.

In recent years, people have gotten interested approaches to making improvements—biotechnology. cience is reaching the point that being able to dream it being able to do it! A little time may be required but be done.

How about tomatoes that don't rot or corn that need nitrogen fertilizer? Making small changes in and animals can produce big results.

All of this is of little benefit if it doesn't make life This is a big issue with some people. They are afraid new life forms that may be created. But, weren't some afraid of the first automobiles? Just because a few peop like something, doesn't mean it should not be done. about it: Making life better by making better plants a mals—it has great potential!

15-1. Students often do biotechnology skills. The test tube contains DNA extracted from *E. coli* bacteria. The cloudy, fibrous substance in the liquid is the DNA. You can get the same results in your school's lab.

413

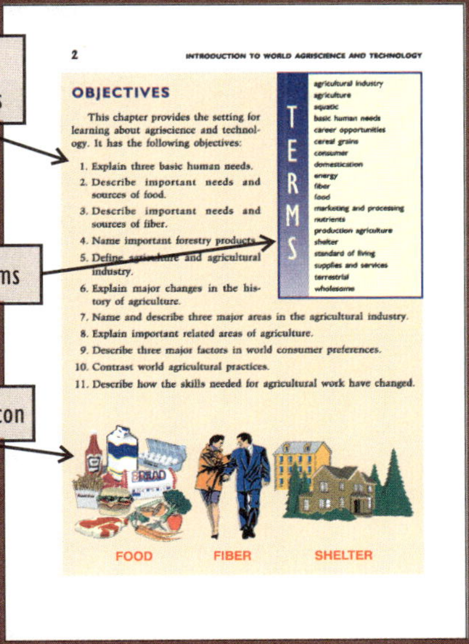

2 INTRODUCTION TO WORLD AGRISCIENCE AND TECHNOLOGY

OBJECTIVES

This chapter provides the setting for learning about agriscience and technology. It has the following objectives:

1. Explain three basic human needs.
2. Describe important needs and sources of food.
3. Describe important needs and sources of fiber.
4. Name important forestry products.
5. Define agriculture and agricultural industry.
6. Explain major changes in the history of agriculture.
7. Name and describe three major areas in the agricultural industry.
8. Explain important related areas of agriculture.
9. Describe three major factors in world consumer preferences.
10. Contrast world agricultural practices.
11. Describe how the skills needed for agricultural work have changed.

TERMS

agricultural industry
agriculture
aquatic
basic human needs
career opportunities
cereal grains
consumer
domestication
energy
fiber
food
marketing and processing
nutrients
production agriculture
shelter
standard of living
supplies and services
terrestrial
wholesome

Objectives and Terms—the second page of each chapter has Chapter Objectives written in specific terms but broad in scope. The terms represent the most important concepts in the chapter. In the chapter content where they are defined in context, the terms appear in a bold italic typeface. All terms are also defined in an extensive Glossary.

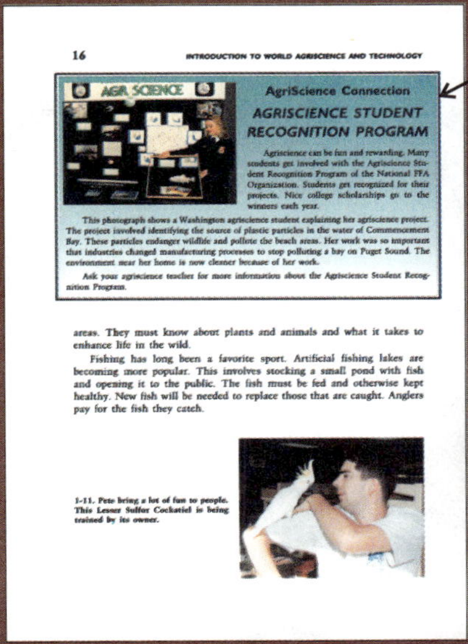

16 INTRODUCTION TO WORLD AGRISCIENCE AND TECHNOLOGY

AgriScience Connection

AGRISCIENCE STUDENT RECOGNITION PROGRAM

Agriscience can be fun and rewarding. Many students get involved with the Agriscience Student Recognition Program of the National FFA Organization. Students get recognized for their projects. Nice college scholarships go to the winners each year.

This photograph shows a Washington agriscience student explaining her agriscience project. The project involved identifying the source of plastic particles in the water of Commencement Bay. These particles endanger wildlife and pollute the beach areas. Her work was so important that industries changed manufacturing processes to stop polluting a bay on Puget Sound. The environment near her home is now cleaner because of her work.

Ask your agriscience teacher for more information about the Agriscience Student Recognition Program.

areas. They must know about plants and animals and what it takes to enhance life in the wild.

Fishing has long been a favorite sport. Artificial fishing lakes are becoming more popular. This involves stocking a small pond with fish and opening it to the public. The fish must be fed and otherwise kept healthy. New fish will be needed to replace those that are caught. Anglers pay for the fish they catch.

1-11. Pete bring a lot of fun to people. This Lesser Sulfur Cockatiel is being trained by its owner.

AgriScience Connection—each chapter has one or more AgriScience Connections which are practical examples relevant to youth in real life that bring science and agriculture together.

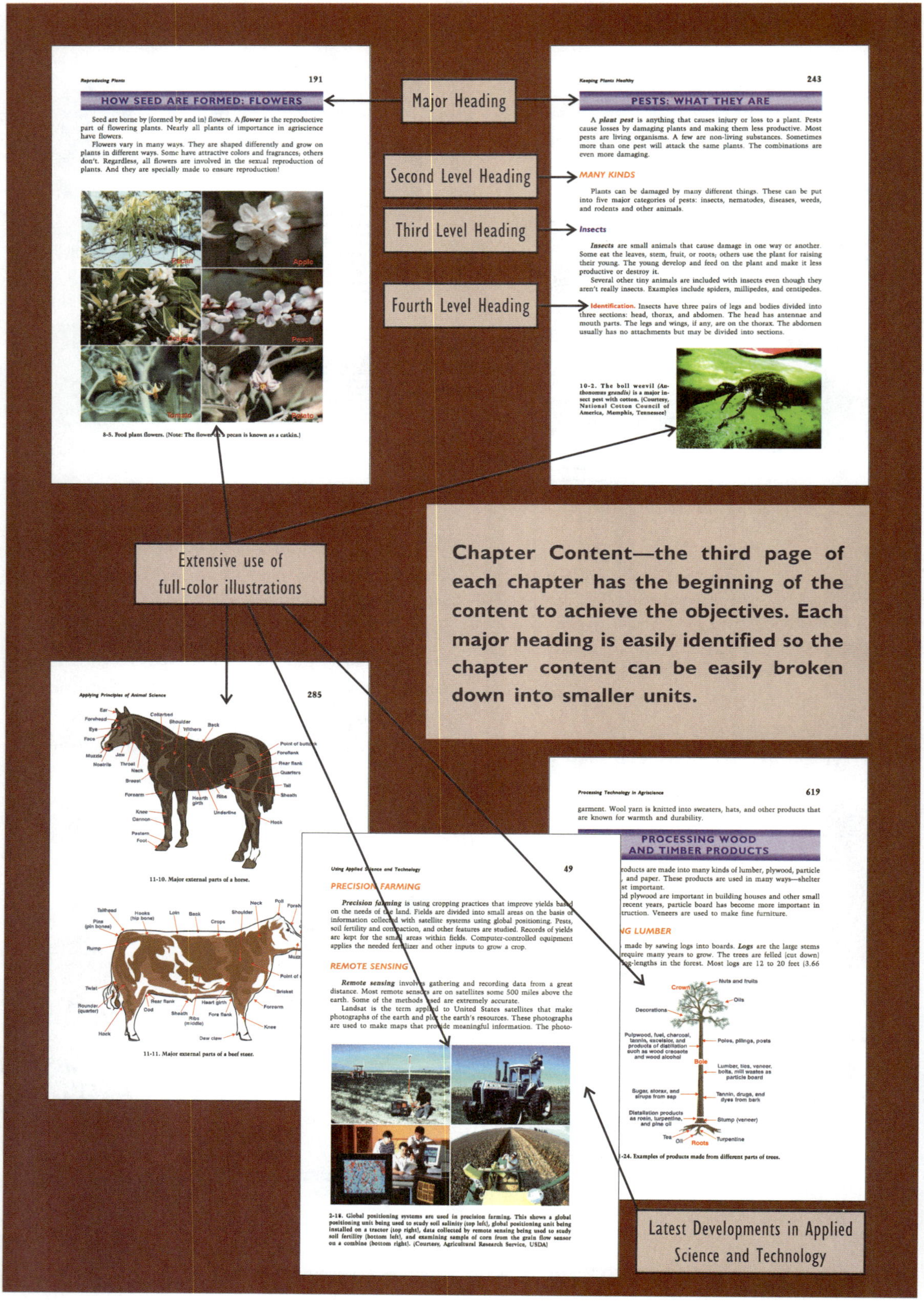
Major Heading
Second Level Heading
Third Level Heading
Fourth Level Heading
Extensive use of full-color illustrations
Chapter Content—the third page of each chapter has the beginning of the content to achieve the objectives. Each major heading is easily identified so the chapter content can be easily broken down into smaller units.
Latest Developments in Applied Science and Technology

Reviewing—each chapter ends with a Reviewing section that contains Main Ideas and Questions.

24 INTRODUCTION TO WORLD AGRISCIENCE AND TECHNOLOGY

1-15. Keeping up with the latest information requires regular study.

the ability to provide leadership. Honesty, responsibility, and dependability are essential for success. Employers want people who will get to work on time and put effort into the work.

- **Personal appearance**—Employers want their employees to dress and groom appropriately. Employees should dress neatly in clean, well kept clothes. Regular baths and attention to personal hygiene are essential. Safety rules must be followed. Proper safety clothing may be needed, such as foot and eye protection.
- **Adaptability**—The ways things are done change. People need to be able to adapt to these changes. Agricultural work often changes as new developments become available. Continuing education is needed to keep up to date. Each individual has a personal responsibility to learn new techniques and new information.

REVIEWING

MAIN IDEAS

Humans have certain basic needs in order t... shelter are the most important. How they get th... the years!

Food is eaten to support human life. Both p... for food. Terrestrial (land growing) and aquatic... animals are consumed. Many of the plants and ar... preparation before they reach the supermarket. ... the marketing process.

Fiber refers primarily to those products that... Of course, some fiber products in the home mai... upholstered furniture and cotton bath towels. N...

"Main Ideas" summarizes the topics covered in the chapter.

Meeting Human Needs in a Changing World 25

are used. The emphasis in agriscience is on cotton, wool, silk, fur, and a few others of less importance.

Shelter is used to protect the human and maintain health. Forestry is a major source of materials for shelter. Lumber, plywood, veneer, composition board, and paper are the major wood products.

Agriculture is the science of producing plants and animals. How it is done has changed considerably in the last 200 years. The agricultural industry has emerged and is made of three areas: supplies and services, production agriculture (often called farming), and marketing and processing.

Consumers are people who use food, fiber, and shelter. They have likes and dislikes, and these must be considered. Variations occur among people in different areas of the world. Climate, customs, and standard of living have a lot to do with what people eat, how they dress, and the kind of home they have.

Agriculture provides many opportunities for young people. Careers vary widely. A person may have an office in the city or work outside in the country. Education, experience, commitment, and personal traits are important to success.

QUESTIONS

Answer the following questions using complete sentences and correct spelling.

1. What are three needs of people? Why are they important?
2. What are the human food needs (nutrients)?
3. What are the two main sources of food?
4. List the parts of a plant that may be eaten by humans. Give an example of a plant for each part.
5. Distinguish between terrestrial and aquatic animals. Give an example of each that is used for human food.
6. What is fiber?
7. What are the common animal fibers? Give an exam...
8. What are the common plant fibers? Give an examp...
9. What is forestry? What are five products from fore...

The Questions can be used as part of the reviewing process by enabling the reader to evaluate their understanding of the chapter content. This helps in determining any area that may need further study or re-teaching.

26 INTRODUCTION TO WORLD AGRISCIENCE AND TECHNOLOGY

10. What is domestication? How did it occur? Why?
11. What is agriculture?
12. Name several inventions that changed farming and describe what the invention did.
13. What are the three major areas of the agricultural industry? What is the role of each in providing for human needs?
14. What is ornamental horticulture?
15. What is the meaning of renewable natural resources?
16. How is recreation a part of agriculture? Give two examples.
17. How is consumer choice important in agriculture?
18. What three major factors are important in consumer choice?
19. How do agricultural practices vary?
20. What important areas should young people who are interested in agricultural careers consider?

EXPLORING

1. Make a list of the food, fiber, and shelter items you use each day. Select one of the items and trace it from where you got it back to the farm.

2. Make a list of the plant and animal food and fiber crops grown in your area. Visit a farm to observe the production of one of the items.

3. Visit a local supermarket and identify the forms in which the following foods are available to the consumer: green peas, turkey, milk, and apples. Add other products to the list, especially those that are your favorite foods.

4. Tour an agriculture museum. Observe farming practices, tools, and other items that depict agricultural practices of many years ago.

Exploring—each chapter concludes with several activities to provide application of the content beyond the classroom.

Acknowledgements

Many people made important contributions to *Introduction to World AgriScience and Technology*. Without their help, the work in writing, illustrating and otherwise producing the book would have been impossible. Agriscience teachers, agriscientists, science educators, industry officials, government leaders and many others are thanked for their help.

Some of the important individuals are listed below.

AgriScience/agriculture teachers who helped with reviewing materials are recognized as follows:

- Tom Hurst, Michigan
- Bryan Starr, Oregon
- Larry Redding, Pennsylvania
- James Barbee, Texas
- Tom Clayton, Mississippi
- Lathen Walton, Mississippi

Government officials are acknowledged as follows:

- Commissioner Rick Perry, Texas Dept. of Agriculture
- Commissioner George Dunsmore, Vermont Dept. of Agriculture

Agricultural associations are acknowledged as follows:

- Rick Maloney, Hampshire Swine Registry, Illinois
- Wes Ishmael, North American Limousine Foundation, Colorado
- American Quarter Horse Association, Texas
- Ed Bible, American Polled Hereford Association, Missouri
- Donna Gonzalez, American Soybean Association, Missouri
- Linda Annis, Minnesota Soybean Growers Association, Minnesota
- Bill Agerton, Potash and Phosphate Institute, Georgia
- T. C. Nelson, National Cotton Council of America, Tennessee
- Sprice Trade Association, New Jersey

Agribusiness and industry contributions are recognized as follows:

- Earl Manning, Progressive Farmer Magazine, Tennessee
- Robert Pritchard, DEKALB Plant Genetics, Illinois
- Michael J. Montague, Monsanto Company, Missouri
- Ann P. Bruch, New York Cotton Exchange, New York
- Lee Cooper, Hach Company, Colorado
- Russell Fowler, KMN Modern Farm Equipment, Inc., Arkansas
- Charles E. Sukup, Sukup Manufacturing Company Iowa
- Kathy S. Ehmcke, Terra International, Inc., Iowa
- Adri G. Boudewy, California Milk Advisory Board, California

- McCarty Foods, Mississippi
- Marie Rindsig, American Breeders Service, Wisconsin

Education agency, agricultural agency and university personnel are acknowledged as follows:

- Jim Floyd, Extension Veterinarian, Auburn University
- Jackie Mullen, Plant Pathologist, Auburn University
- George Hurst, Professor of Wildlife and Fisheries, Mississippi State University
- Tim McCabe, Soil Conservation Service, USDA, Washington
- Beverly R. Howell, Extension Specialist, Mississippi Cooperative Extension Service
- Ray Waggoner, Agricultural Conservation and Stabilization Service, USDA, Washington
- James Jackson, Research Farm Supervisor, Mississippi Agricultural and Forestry Experiment Station
- Marshall Stewart, North Carolina Department of Education
- Marco Nicovich, Photography Manager, Mississippi Cooperative Extension Service
- Elizabeth Morgan, College of Veterinary Medicine, Texas A & M University
- Ronnie White, Alabama Cooperative Extension Service
- David Mitchell, Alabama Cooperative Extension Service
- Glen Anderson, Virginia Department of Education
- John T. Denmark, Jr., Regional Supervisor, Florida Department of Education
- Dave Doerfert, National FFA Foundation
- Marion D. Fletcher, Arkansas Department of Education

Other individuals are acknowledged for their valuable contributions to this work, as follows: Cindy Smith, science teacher at Starkville Academy, Starkville, Mississippi; Matthew Kanemoto, Kahuku School, Hawaii; Clay Walker, Forestry Suppliers, Inc., Jackson, Mississippi; Charles Wax, Climatologist, Mississippi State University; Jimmy Butler, Agri-Math Project Associate and now at the University of Tennessee at Martin; Sharon Hunter, Lela Ladner, Lou Boland, and DeShannon Davis, LEE AND ASSOCIATES; Lynn Wagner, administrator with Leake County Schools, Mississippi; Corinne Mounier-Lee, Sevier Pharmaceuticals, Orleans, France; Stephen J. Lee, Emory University, Atlanta; Ed Faurot, forestry instructor, Winston–Louisville Schools; and Susan E. Lee, Millsaps College, Jackson, Mississippi

All models for photographs are acknowledged for their assistance.

Special appreciation is expressed to Vernie Thomas, President of Interstate Publishers, Inc., for his support and innovative approaches to the development and publishing process for this book. Thomas has expressed a sincere dedication to meeting the needs of students and teachers through agricultural publishing.

Jasper S. Lee
Diana L. Turner

Contents

PART THREE
Plant Science and Technology

PART FOUR
Animal Science and Technology

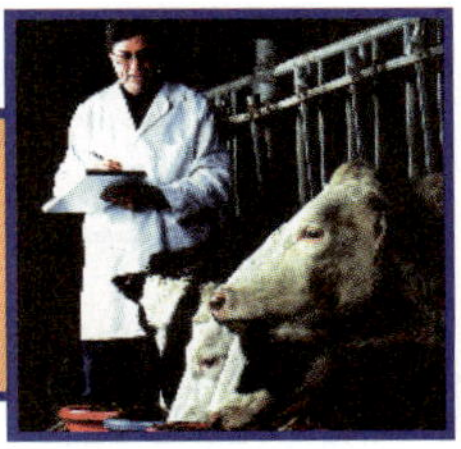

PART FIVE
Creating Improved Plants and Animals

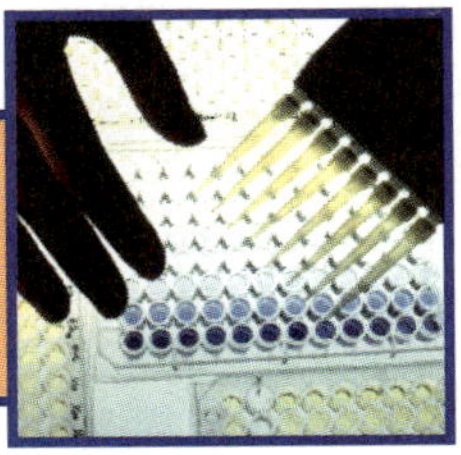

PART SIX
Earth Science and Technology

PART SEVEN
Physical Science and Technology

PART EIGHT
Food and Fiber Technology

Agriculture—Science in Action

Photographs courtesy, Agricultural Research Service, USDA and National FFA Organization

Photographs courtesy, Agricultural Research Service, USDA and National FFA Organization

1

MEETING HUMAN NEEDS IN A CHANGING WORLD

What is your favorite food? People are quick to respond "hamburgers and French fries" or "pizza." These are foods many enjoy. Because we have so much, we sometimes forget what is required to get our food to us.

People around the world must have food to eat and clothing to wear. Without adequate food, people get sick and die because their nutritional needs aren't met. Without clothing and shelter, people aren't protected from the ravages of weather and they, too, will get sick and die. As the world's population continues to grow, more food and fiber will be needed.

Providing food, fiber, and shelter is not simple. Complex systems of production are used. A wide range of agriscience and technology is required. Improved crops, modern machinery, and advanced production techniques are used. Further, systems of marketing and processing the products get the food to consumers just the way they want it—ready to use with little or no preparation needed.

1-1. Pizza—a favorite fast food!

OBJECTIVES

This chapter provides the setting for learning about agriscience and technology. It has the following objectives:

1. Explain three basic human needs.
2. Describe important needs and sources of food.
3. Describe important needs and sources of fiber.
4. Name important forestry products.
5. Define agriculture and agricultural industry.
6. Explain major changes in the history of agriculture.
7. Name and describe three major areas in the agricultural industry.
8. Explain important related areas of agriculture.
9. Describe three major factors in world consumer preferences.
10. Contrast world agricultural practices.
11. Describe how the skills needed for agricultural work have changed.

TERMS

agricultural industry
agriculture
aquatic
basic human needs
career opportunities
cereal grains
consumer
domestication
energy
fiber
food
marketing and processing
nutrients
production agriculture
shelter
standard of living
supplies and services
terrestrial
wholesome

FOOD FIBER SHELTER

PEOPLE HAVE NEEDS

The ***basic human needs*** are food, fiber, and shelter. These support human life worldwide. They provide the nutrition that helps the body grow, repair itself, and reproduce. They also provide protection from the weather, dangerous animals, and some other hazards of life. Most food and fiber is carefully produced to meet human needs.

FOOD

Humans need food. It keeps the body strong and healthy. Without adequate food, the body becomes weak and doesn't grow as it should.

Food is the solid and liquid material humans eat. It must provide the essential nutrients. ***Nutrients*** are substances necessary for an organism to live and grow. Animals and plants also use food, but the focus in this chapter is on providing for the needs of humans.

Human Food Needs

The nutrient needs of humans are:

- **Energy**—***Energy*** is a nutrient needed for body activity. It provides for both internal and external movement. Sources of energy include potatoes, bread, desserts, and meat. Excess energy is changed to fat in the human body. Energy primarily comes from food substances known as carbohydrates and fats.
- **Growth and repair**—The human body is continually growing and repairing itself. Infants, children, and adolescents are actively growing and need proper nutrition. Even though adults do not grow in size, they repair the body as it wears. Foods that are high in substances known as protein are important for growth. Examples of such foods are milk, meat, eggs, and dried beans.
- **Good health**—In addition to energy and growth, humans must have nutrients to stay healthy. Vitamins and minerals are substances needed for good health. Many vegetables contain vitamins and minerals. Milk is high in minerals.
- **Body processes**—All of the above foods contribute to body processes. In addition, water is needed to transport the food substances in the body. Blood is the primary liquid and is 92% water.

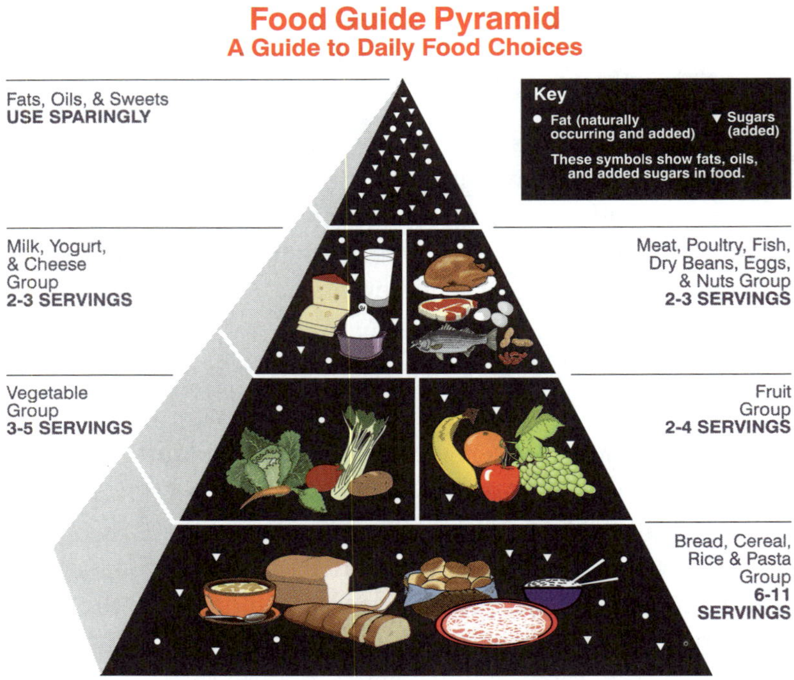

1-2. The food guide pyramid suggests foods for a healthful life.

Humans who do not get proper food may get sick. They do not grow and develop as they should. Some will not become adults. Those who do may be unable to work, to raise a family, and to help other people. The Food Guide Pyramid in Figure 1-2 shows recommended daily adult food choices.

Sources of Food

Food comes from two main sources: plants and animals. All of the thousands of food products in a supermarket originated as plants or animals. Modern processing has changed the plants and animals into desirable products.

Plant Sources. Many kinds of plants are used for food. These plants are often carefully grown to provide high-quality food.

Scientists have found about 250,000 kinds of plants on earth. Only about 600 are used to provide food for humans. An even smaller number provides most of the food. The cereal grains of wheat, rice, and corn are the major food items around the world.

Only certain parts of plants may be used as human food. Here are a few examples:

- **Leaves**—Lettuce, cabbage, and spinach; other parts of the plants discarded.
- **Seeds**—Beans, wheat, corn, rye, and nuts; other parts are discarded.
- **Roots**—Potatoes, carrots, onions, and ginger; some of these are structures that grow on roots, such as the potato which is a tuber.
- **Fruits**—Apples, strawberries, pears, and oranges.
- **Flowers**—Cauliflower and broccoli; the plants are discarded.
- **Stems**—Celery, rhubarb, and asparagus; other parts of the plants are not eaten; some plants are destroyed when harvested.
- **Multiple parts**—Turnips (leaves and roots may be eaten).

1-3. Cucumbers are fruit of the cucumber plant.

Animal Sources. Humans eat foods from several animal sources. The meat of some animals is important. Other animals produce milk, eggs, and related products that are eaten.

Both land and water animals are eaten. Those that grow on land are known as ***terrestrial,*** while those in water are known as ***aquatic.*** Examples of terrestrial animals are chickens, hogs, sheep, and cattle. Examples of aquatic animals are fish, shrimp, and clams.

Some two million species of animals live on earth. Only about 50 species are used to any extent for human food. Of these only four kinds are raised in large numbers: cattle, hogs, chickens, and sheep. Examples of foods from animals are:

- **Meat**—Muscles, organs, etc., of hogs, cattle, chickens, fish, and others.
- **Eggs**—Chicken, fish, and a few other animals.
- **Milk**—Cheese, ice cream, and liquid milk of cattle and sometimes other animals.

1-4. People in the United States are accustomed to finding grocery stores well stocked with food. In addition, the foods are expected to be wholesome and easy to prepare.

In recent years, more attention has been on raising fish. This is due to fewer wild fish in streams and oceans. More is being learned about how to raise water animals. Many people feel that eating fish is good for health.

FIBER (CLOTHING)

Fiber is not a food. It primarily includes materials used to make clothing and shelter. Just as food is grown, natural fiber is grown also. In addition to natural fibers, there are synthetic fibers. Synthetic fibers are made in mills from various products, such as petroleum.

Fiber is produced in three main ways: by animals, by plants, and by manufacturing. Fibers produced by manufacturing are known as synthetic or human-made.

Animal Fibers

Animal fibers include wool, fur, and silk. They are used to make warm and attractive clothing.

Wool is produced by sheep. Sheep naturally grow wool to protect their bodies from the weather. In the spring, the wool is removed from the sheep by shearing or other means.

Furs are obtained from a number of animals, such as rabbits and minks. Some of these animals are raised for their fur; others are trapped in the wild. The fur must be properly tanned before it is used.

Silk is produced by silkworms when making cocoons. A cocoon is a case of tiny threads that protects the developing stage of the silkworm moth. Manufacturing silk cloth involves unwinding the cocoon. Silk is said to be the strongest animal fiber.

Plant Fibers

Several fibers are produced from plants: cotton; flax; and hemp, jute, and sisal.

Cotton is by far the most important fiber. It is used to make clothing, upholstery goods, and other products. Cotton has qualities that are highly desired, such as being soft and crisp.

Flax comes from the stems of flax plants. It is used to make linen, which is a high quality cloth used for clothing, table napkins, and other items.

Hemp, jute, and sisal are coarse fibers that come from plants of the same name. All three are used to make twine, cords, and ropes.

1-5. Cotton fibers make comfortable and attractive clothing. (Courtesy National Cotton Council of America, Memphis, Tennessee)

Manufactured and Mineral Fibers

Some of these are made from wood; others are made from chemicals. A substance in wood known as cellulose is the major wood fiber. Examples of synthetic fibers include nylon and polyester. Mineral fiber includes asbestos, which is no longer widely used because of its possible hazards to health.

SHELTER (FORESTRY AND WOOD PRODUCTS)

Humans must be protected from the weather and other harm. They use shelter to defend against cold, heat, rain, and snow as well as wild animals and other pests.

Shelter is the housing humans use, such as apartments and houses. Many of the products used in building shelter are produced by forestry.

Forestry involves the growing of trees. Scientific processes are used to get trees to grow rapidly and economically. Some would say that forestry is a form of fiber production, and that is what it is! Since forestry is large and specialized, it is presented separately from plant fiber production in this chapter.

Here are a few examples of forest products:

1-6. Lumber and plywood are being used to construct a building.

- **Lumber**—Lumber is made by sawing logs into boards. It is used to construct houses, furniture, and other products for humans.
- **Plywood**—Plywood is made by cutting logs into large, thin sheets of wood that are glued together in layers. The wood product is strong and has improved qualities over lumber for certain uses.
- **Veneer**—Much like plywood, veneer is made by gluing a very thin layer of an expensive, beautiful wood over a lower cost wood. For example, a thin layer of valuable walnut wood can be glued over a thicker, less expensive wood. The resulting product has a wood finish with the beauty of a fine wood but at a much lower cost.
- **Composition board**—Several types of this product are made. It is produced in sheets similar to plywood or in boards like lumber. Wood chips or particles of wood are compressed together to make composition board.
- **Paper**—Paper is made from wood using a detailed chemical process. The wood is cut into chips and "cooked" until it turns to pulp. The pulp is rolled into thin sheets of paper. A wide range of paper products are created.

PROVIDING FOR THE NEEDS OF PEOPLE

Humans haven't always produced food, fiber, and shelter. Primitive humans living thousands of years ago searched and hunted for food. Hunting was difficult because of the lack of weapons. It was nearly impossible to capture some large animals. Snails, clams, oysters, and crawfish were probably among the first foods. Wild berries, melons, beans, and other fruits and vegetables were also eaten. Early humans used leaves and animal skins for clothing. They sought shelter in caves and under big rocks or trees.

Domestication is getting control over wild plant and animal species. It likely began about 7000 B.C. in Asia with donkeys, cattle, and dogs. The animals were brought to live near humans who cared for and protected them. Crops were being grown in China and Sweden in 1300 B.C. A few wealthy people had domesticated fish by 3000 B.C. Fish culture was well underway in China in 400 B.C. Agriculture in America was still primitive as late as 1500 A.D. A lot has changed since Christopher Columbus landed in America in 1492!

AGRICULTURE

Agriculture is the science of growing crops and raising livestock. The crops and livestock are used for human food, fiber, and shelter. Some people call this farming

Early agriculture in America focused on combining wild foods with those that could be grown. In the late 1700s, 90% of the people lived "on the land." Nearly everyone farmed. Their farming was largely subsistent, meaning that they barely produced enough for themselves. However, a few grew extra crops for sale to those who didn't farm.

As new developments made farming easier, farmers produced crops and livestock in increasing numbers. Fewer farm workers were needed. New developments provided more food and fiber than ever before.

1-7. Communication is important to the forester.

AgriScience and Technology Emerge

Many changes have occurred in the last 300 years. New insight into nature has helped improve crops and livestock. Inventions allowed people to get more work done with less effort. Agricultural inventions made it possible for more people to work off the farm. Factories were established to produce things for the farm as well as the home.

Inventions in agriculture changed farming. This gave rise to agriscience and technology. A few examples in agriculture are listed here.

- Charles Townshend developed the first crop rotation systems in the early 1700s. These involved using legume crops that added nitrogen to the soil. This work was done in England.
- Jethro Tull developed a seed planting machine in the early 1700s. This was the first machine that had internal working parts. Today's high-speed planters grew from this early effort.
- Eli Whitney developed the cotton gin in 1793. This invention greatly reduced the labor required to remove seeds from the cotton fiber. Large mills exist today thanks to Whitney's work.
- Cyrus McCormick patented a harvesting machine (reaper) in 1834. This machine greatly reduced the amount of hand labor required to harvest grain.
- John Deere, an Illinois blacksmith, developed the steel moldboard plow in 1837. This plow made it possible to till the heavy soils found in many parts of the United States.
- Gregor Mendel, an Austrian botanist, discovered basic principles of heredity. This work has been important in improving many crops and livestock.
- In the early 1900s, improved varieties of corn were developed in the United States. These were known as hybrid corn. By the mid-1990s, nearly all corn planted in the United States was hybrid seed.
- In the mid-1900s, new chemicals were developed to control insects, diseases and weeds. More advanced equipment became common on farms.
- In the late 1900s, genetic engineering and other advanced methods of improving crops and livestock were being developed.

1-8. The modern combine grew out of the reaper developed by Cyrus McCormick. This photograph shows six combines harvesting in one rice field in Texas. (Courtesy, Texas Department of Agriculture.)

Agricultural Industry Replaces Farming

Agriscience has resulted in many new ways to produce food and fiber. The use of these findings is known as technology. As the new science and technology of agriculture came about, producing food and fiber grew from farming into the agricultural industry.

Many changes occurred in agriculture. Each worker became much more productive and fewer people were needed on farms. Farm activities began to occur off the farm. More people were needed to provide for the needs of farmers and process what was produced.

Areas of Agricultural Industry

Agricultural industry includes all of the processes involved in getting food, fiber, and shelter to the consumer. It has three major areas: supplies and services, production agriculture, and marketing and processing.

Supplies and Services. Today, farming requires many things that are manufactured off the farm. These are known as ***supplies and services***. They are the inputs (materials, equipment, etc.) used to grow crops and raise livestock.

Major areas of supplies and services are:

- **Machinery and equipment**—A large industry exists to provide the tractors and implements needed on farms and in forestry. Agriscientists design equipment and test it in the field. People are needed to manufacture, sell, and service the equipment. Much of today's equipment involves sophisticated computers and laser systems.
- **Seed**—Seed companies have developed improved types of seed and produce them for the farmer. Agriscientists are involved in developing and producing the seeds. Seeds are cleaned and treated to assure that they are good. Proper bagging and storing are used to keep the seeds in top quality.
- **Chemicals**—Chemicals have many good uses in agriculture. Some ideas people have about chemicals are wrong. Chemicals aren't bad if used properly. They help control weeds, diseases, and insect pests. Fertilizers are chemicals that are used to help crops grow better. Agriscientists are involved in developing and testing chemicals. People who use them often need to be certified to do so.
- **Feed and animal medicines**—Most livestock producers get their feed from mills off the farm. They also use medicines to keep their livestock healthy.

Knowledge of scientific principles is important in feeding and maintaining the health of livestock.

Services of people who are highly trained and educated are often needed on farms. People who provide services include veterinarians, soil conservation technicians, and insect specialists.

Without supplies and services, today's agriculture would be much like it was 200 years ago. Farmers would have to save their own seed, make their own machinery, and provide for other needs. This approach is definitely out of date!

Production Agriculture. ***Production agriculture*** is the farming part of the agricultural industry. It involves the production of crops, livestock, and livestock products.

Today's farms are very specialized. Most produce only one or two crops or products. For example, a dairy farm will likely produce milk and no other crops to sell. A corn farm may grow corn and nothing else. The corn is sold to a mill or a farmer who has livestock to feed. They are said to be commercial farms. This is because they are producing for a specific market.

Some individuals also include tree farming (forestry) as a part of production agriculture. Trees are farmed much as any other crop, but they take longer to grow and require different practices.

Here are some examples of crops in production agriculture and why they are grown:

- **Rice**—for grain
- **Cotton**—for fiber and seed

1-9. Tomatoes being harvested for a specific market. (Courtesy, Texas Department of Agriculture)

- **Wheat**—for grain
- **Corn**—for grain
- **Soybeans**—for grain and hay
- **Sorghum**—for grain and feed

- **Vegetables**—for food; examples: tomatoes, potatoes, squash, snap beans, onions, lettuce, celery, and corn

- **Fruit**—for food; examples: apples, oranges, grapefruit, and cherries
- **Berries**—for food; examples: strawberries, blueberries, and raspberries
- **Nuts**—for food; examples: peanuts, pecans, walnuts, and almonds
- **Trees**—for timber and ornamental use, such as Christmas trees

Here are some examples of livestock in production agriculture:

- **Hogs**—for meat, known as pork
- **Sheep**—for meat and wool (the meat of young sheep is known as lamb, the meat of older sheep is mutton)
- **Cattle**—for meat, known as beef, and dairy (milk) cattle
- **Poultry**—chicken for meat and eggs and turkeys for meat
- **Horses**—for work and pleasure
- **Fish**—for food and recreation; examples: catfish, trout, tillapia, hybrid stripped bass, and crawfish

Production agriculture has come a long way from the subsistence farming of 200 years ago. Many farmers sell all that they grow and buy their food at the local supermarket. They do not even have a garden!

Successful farmers apply agriscience. They understand science principles and use them to produce crops and livestock.

Marketing and Processing. ***Marketing and processing*** is the part of the agricultural industry that moves crops and livestock from the farm to the people who use them. The crops and livestock are changed into forms that people want. Here is what might be involved with a fruit or vegetable:

1. The harvested crop is delivered by the grower to a packing shed.
2. At the packing shed, the crop is graded and packed into big crates for shipping to a processing plant.

3. In the processing plant, the crop is washed and otherwise prepared for canning or freezing.
4. As processing is finished, the produce is packaged and labeled for storage and delivery to supermarkets.
5. Specially equipped trucks haul the processed food to the supermarket.

All steps have procedures to keep the food safe. The food is carefully handled to maintain high levels of nutrition. Inspection is frequently made to assure the consumer that the food is ***wholesome*** (good to eat).

People also want food that is ready to cook and easy to use. They do not want to spend a lot of time preparing meals. Marketing and processing takes a lot of the work out of food preparation. This is why supermarkets have so many items. Some people like dinners that are ready to heat in the microwave oven. Others like to spend a little time preparing original dishes. However, very few people would think of slaughtering their own hog or chicken. Times have changed, and so has agriculture.

MAKING LIFE BETTER

Life is more than food, fiber, and shelter. People want to have fun and enjoy themselves. The agricultural industry has several areas that make life better. Three areas are presented here: ornamental horticulture, renewable natural resources, and recreation.

ORNAMENTAL HORTICULTURE

Ornamental horticulture is growing and using plants for their beauty.

Floriculture is the production and use of flowers and plants with attractive leaves. The plants may be grown in greenhouses, fields, or outside and inside buildings for beauty. The flowers grown in greenhouses are used to decorate homes, businesses, and other places. They are often made into artistic arrangements by floral designers. Local flower shops take orders and deliver floral arrangements to customers.

Landscape horticulture is using plants to improve our outdoor environment. The plants usually are used to make buildings, parks, and other areas more attractive. Shrubs (small woody plants) are grown in nurseries.

1-10. Plants and other natural objects can be used to add beauty to homes and businesses. This scene in Gatlinburg, Tennessee, makes use of shrubs, rocks, and mulch ground cover.

At the right size, they are moved and placed in landscapes. Carefully prepared landscape plans are often prepared. The plans describe the kind of shrub to use and where it should be located. A good knowledge of plants is needed to make and follow these plans. If the characteristics of plants aren't known, the wrong shrubs might be chosen. Of course, different shrubs have different growth requirements.

RENEWABLE NATURAL RESOURCES

Renewable natural resources include soil, water, air, and wildlife. These also relate to a quality living environment for people.

Quality soil, water, and air can be maintained. Practices can be used that prevent the loss of soil through erosion. Water and air can be protected from pollution.

Wildlife is important to maintaining a balance in the environment. People who enjoy hunting and fishing should follow all laws when involved in these sports. Licenses may be required.

RECREATION

Several areas of agriculture are connected to recreation. These range from hiking through mountain trails to fishing, hunting, and riding horses.

Nearly everyone enjoys being outdoors. Taking walks and hiking to see the wonders of nature can be refreshing. People who know principles of agriscience are necessary for establishing and maintaining parks and

AgriScience Connection

AGRISCIENCE STUDENT RECOGNITION PROGRAM

Agriscience can be fun and rewarding. Many students get involved with the Agriscience Student Recognition Program of the National FFA Organization. Students get recognized for their projects. Nice college scholarships go to the winners each year.

This photograph shows a Washington agriscience student explaining her agriscience project. The project involved identifying the source of plastic particles in the water of Commencement Bay. These particles endanger wildlife and pollute the beach areas. Her work was so important that industries changed manufacturing processes to stop polluting a bay on Puget Sound. The environment near her home is now cleaner because of her work.

Ask your agriscience teacher for more information about the Agriscience Student Recognition Program.

wildlife areas. They must know about plants and animals and what it takes to enhance life in the wild.

Fishing has long been a favorite sport. Artificial fishing lakes are becoming more popular. This involves stocking a small pond with fish and opening it to the public. The fish must be fed and otherwise kept healthy. New fish will be needed to replace those that are caught. Anglers pay for the fish they catch.

1-11. Pets bring a lot of fun to people. This Lesser Sulfur Cockatiel is being trained by its owner.

Hunting preserves are becoming more common. These preserves are needed because much of the wild game has been killed off. Due to better regulations, more gun safety is practiced.

Horses and horseback riding are growing in popularity. Previously, horses were kept for transportation and power. Today, they are kept for pleasure.

Rodeoing is a unique sport attracting increasing numbers of people. But there is more to it than people riding animals. Someone must have skills in how to care for the stock. This includes properly feeding, handling, and protecting the animals from injury. People have become more concerned about the humane treatment of rodeo animals in recent years.

1-12. Rare and exotic animals are kept in the zoo for people to learn about and enjoy. The giraffe is always interesting. It is the tallest four-legged animal.

Animals have served as companions to people for a long time. Many people have dogs, cats, fish, and other animals for pets. Zoos maintain exotic animals to educate people about animals found in nature. Some people depend on animals for work—blind people often use seeing-eye dogs.

PEOPLE MAKE CHOICES

In many parts of the world, people can make choices about food, fiber, and shelter. They may have strong likes and dislikes, known as preferences. It helps the agriscientist to know about preferences.

Food and fiber are produced to satisfy human needs. If consumers don't want what is produced, there is no need to produce it. The use of agriscience and technology can help make food and fiber more appealing.

PRODUCING FOR THE CONSUMER

Consumers are people—all of us! Consumers eat food, wear clothing, and live in housing. They have certain needs that must be met. They also have opinions about what they will eat and wear. These opinions dictate what is produced.

With the world population at 5.4 billion people and the United States with 260 million people, preferences can be very powerful. Consumers tend to reflect their backgrounds by what they eat and wear. Knowing where major populations are found can help meet demands of groups.

Demand is important. It is the willingness of a consumer to buy something. Of course, the price of the item and amount of money the consumer has are important. If prices are too high, consumers won't buy much of the item.

Agriculture is concerned with meeting these needs and demands. Crops and livestock should be produced only if there is a demand for them or their products. Every producer should study the market situation before beginning production. If no market exists, a person should not spend money on production because the money will likely be wasted.

What is produced must make a profit. Even if it can be sold, it must be sold for more than it cost to produce. If not, the producer will lose money.

CONSUMER PREFERENCES

A number of factors influence what people eat and use as clothing. There is a tendency for people in various regions of the world to be much alike in their preferences.

The major factors in world consumer choices are climate, customs, and standard of living.

Climate

Climate is the nature of the weather. Temperature, precipitation (rainfall), and direction and speed of wind are major factors in climate. Of course, soil fertility and drainage, as well as lay of the land, determine its use. Soil that has few nutrients won't produce much. Land that is too wet or too hilly can't be farmed.

The foods commonly eaten by people in a country are those that grow best in the climate. People in cool climates eat more cereal grains, potatoes, and meat. Those in warmer climates eat more fruit and vegetables, such as papayas and bananas. Some areas of the world have poor climate conditions. Animals and plants won't grow well. A good example is with the desert areas of Africa. There isn't enough rain to meet the needs of crops.

Table 1-1
Per Person Yearly Consumption of Selected Foods in the United States

Commodity	Pounds Consumed Annually in U.S.
Red meats (beef, veal, lamb & mutton, and pork)	163.0*
Poultry (chicken and turkey)	61.1
Fish and shellfish (wild catch)	14.9
Eggs	30.1
Dairy products and milk	572.2
Fresh fruit and melon	124.3
Vegetables (fresh, frozen, and canned)	243.7
Potatoes	135.7
Rice, flour and cereal products	189.2

Source: U.S. Department of Agriculture, 1994.
*Dressed weight

Customs

Customs are the long established ways of doing things. People often grow up eating certain foods and continue to prefer those foods as adults. Changes in customs occur slowly.

The world is divided into seven continents. Six have large numbers of people. Antarctica has no native people. The continents and their populations are shown in Figure 1-11.

Customs are identified with people from various continents and countries as well as within countries. In the United States, people in the north may eat more potatoes and wheat bread. Those in the south may eat more chicken, okra, and corn bread. People near rivers and oceans eat more fish.

Continents of the world have certain customs. The customs are often related to climate and native food plants and animals. Asia is associated with rice, tea, and chicken. Australia and New Zealand are known for wheat, beef, mutton, and dairy products. Africa is known for corn, sorghum, cassava (a root crop), and sweet potatoes. Europe is associated with potatoes, hogs, cereal grains, and dairy products. South America is known for cassava, dry beans, corn, coffee, and potatoes.

North America (United States and Canada) has a lot of variety. What is produced is related to the climate and the heritage of the people who

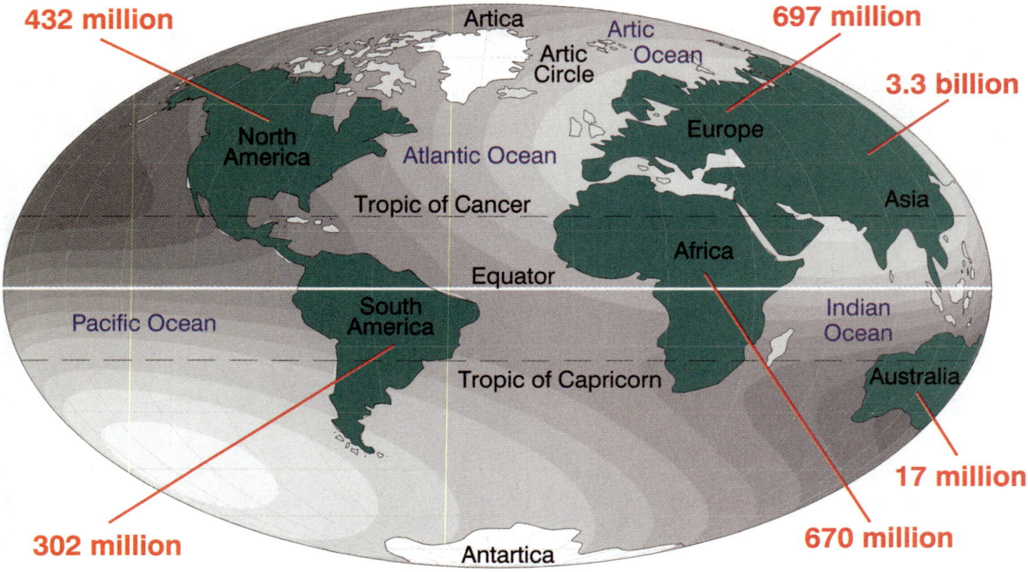

1-13. The location and population of the seven continents are shown here. Each has differences in climate, culture, and standards of living.

settled these areas. The southern United States grows vegetables, cotton, fish, and citrus crops, while the Plains produce wheat, soybeans, corn, and cattle. The northern United States produces wheat, potatoes, and dairy products. The West Coast produces a number of different crops, particularly valuable fruit and vegetables.

Standard of Living

Standard of living refers to the income of the people and the products available to them. People with more money often choose different foods. What they choose is also related to their housing and how food is prepared.

As income increases, food choices change. For example, people with higher incomes eat more beef, seafood, and other higher priced foods. Lower income people eat more rice, potatoes, and cassava. Some nations have many people who are living in poverty. Since they have low incomes, they buy low-cost foods.

Education is often associated with standard of living. Individuals and nations with better education have higher incomes. Educated people are often more willing to try new foods. People in countries with emphasis on technology are interested in the background of their foods. They choose those where technology was involved in production and processing.

AGRICULTURAL PRACTICES VARY

How people get food and clothing varies considerably around the world. Some have advanced food systems, such as Canada, the United States, Japan, and Europe. These are often referred to as developed nations. Other nations have very primitive agriculture. These are often known as the developing countries.

Nations with advanced food systems utilize science and technology. Those that are primitive use few or no improved practices. The contrast between nations is great. A few areas of contrast are:

- **Mechanization**—Developed countries use tractors, power equipment, and low levels of hand labor. Countries that are developing use considerable hand labor and little mechanization. Animals may also be used for power.
- **Improved crops**—Developed countries use crops that have been carefully selected to produce high yields. Developing countries may have few improved crops.
- **Commerce**—Developed countries have systems for buying and selling food and fiber. Developing countries may have poor transportation, few storage facilities, and no way to easily buy and sell.
- **Education of people**—People in the developed countries have higher levels of education. These countries have well developed systems of agricultural education and know about agriscience. The developing countries tend to have lower levels of education and very little education in agriculture. They know very little about agriscience.
- **Soil and climate**—Developed nations tend to have soils and climates that will support the production of food and fiber. Developing countries may have poor soils and extremely dry or wet climates.

BEING SUCCESSFUL IN AGRICULTURE

Outstanding ***career opportunities*** are available in agriculture. Career opportunities are good jobs that provide food, fiber, and shelter for people. Agriscience provides many types of jobs and ways to earn money. Even though the number of people who work on farms is down, the number in non-farm agricultural occupations has increased.

Today in the United States over 2 million people produce food and fiber. These are the people who work on farms. The number of people

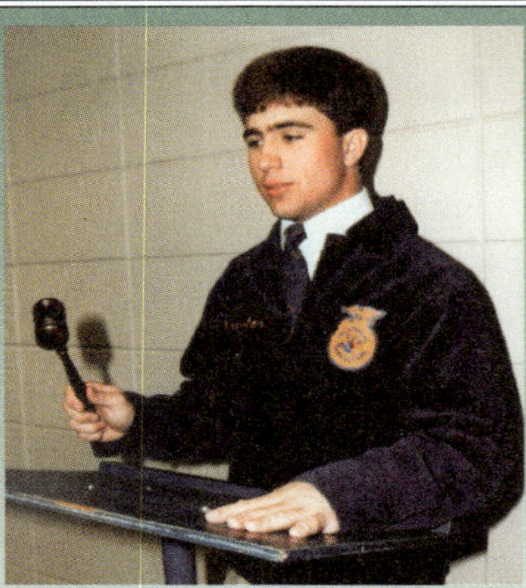

AgriScience Connection

BECOMING A LEADER

All people have great potential. Some people realize their potential; others don't. Students have exciting opportunities to begin realizing their potential, including important leadership skills.

Students in agriscience classes practice important skills through the FFA. The FFA is the national organization that provides many exciting opportunities for students. Members have a lot of fun traveling, meeting other students, and being involved. The FFA helps students learn in many ways, such as this Mississippi student who is presiding over a meeting at his school.

The FFA has plenty of activities for students who live in cities as well as in rural areas. Talk to your agriscience teacher about the FFA. Get involved in your local school.

who work off the farm in providing food and fiber is ten times greater than the number of farm workers, or about 20 million people. Many of these off-farm workers need considerable education in agriscience.

NATURE OF THE WORK

Agriculture has changed since the old days. Today, people use agriscience and technology to produce food, fiber, and shelter. They use modern machinery, computers, lasers, and remote sensing. They need to know mathematics, science, and communication skills. People in agriculture travel to work each day just like those who are in non-agricultural jobs. Soon, the notion of farm drudgery should be gone!

Agriculture has a wide variety of occupations. Some people can work with machines. Others work with plants and animals or in offices. Some people dress casually, while others wear business suits. Fields, forests, laboratories, offices, factories, and many other places are the settings of jobs in agriculture.

People in agriculture are paid well, although rates of pay vary. Employees get paychecks and important job benefits such as paid vacations and retirement programs. They have high standards of living with time for recreation and time to enjoy life. Of course, hard work and dedication are important to being successful. A young person can be as successful as he or she wants to be by taking advantage of the opportunities in agriculture.

SKILLS YOU NEED

Young people need to have skills so they can be productive. They need skills in agriscience and technology. They also need to be willing to put forth effort to achieve. Good skills help you do more work in less time and with less effort.

Five important skill areas for the success of young people in the exciting agricultural industry are:

- **Education**—Most agricultural jobs require training in how to do the work. Students can begin preparation in high school by taking courses in agriscience and technology. Some students combine the last two years of high school with two years at a community college to get what is known as a "tech prep" or 2+2 education. Other students want to get at least a four-year college degree in agriscience areas.

1-14. Computer skills are essential for success in agriscience.

- **Experience**—Practical experience in agricultural work is important. Students may begin while in high school with part-time and summer jobs. Some high schools have youth apprenticeship programs or supervised agricultural experience. Community colleges, state colleges, and universities often provide ways to gain practical experience.
- **Personal skills**—Many kinds of personal skills are needed. These include the ability to get along with people, courtesy in dealing with others, and

1-15. Keeping up with the latest information requires regular study.

the ability to provide leadership. Honesty, responsibility, and dependability are essential for success. Employers want people who will get to work on time and put effort into the work.

- **Personal appearance**—Employers want their employees to dress and groom appropriately. Employees should dress neatly in clean, well kept clothes. Regular baths and attention to personal hygiene are essential. Safety rules must be followed. Proper safety clothing may be needed, such as foot and eye protection.
- **Adaptability**—The ways things are done change. People need to be able to adapt to these changes. Agricultural work often changes as new developments become available. Continuing education is needed to keep up to date. Each individual has a personal responsibility to learn new techniques and new information.

REVIEWING

MAIN IDEAS

Humans have certain basic needs in order to survive. Food, fiber, and shelter are the most important. How they get these has changed a lot over the years!

Food is eaten to support human life. Both plants and animals are used for food. Terrestrial (land growing) and aquatic (water growing) plants and animals are consumed. Many of the plants and animals undergo considerable preparation before they reach the supermarket. This preparation is part of the marketing process.

Fiber refers primarily to those products that are used to make clothing. Of course, some fiber products in the home make life comfortable, such as upholstered furniture and cotton bath towels. Natural and synthetic fibers are used. The emphasis in agriscience is on cotton, wool, silk, fur, and a few others of less importance.

Shelter is used to protect the human and maintain health. Forestry is a major source of materials for shelter. Lumber, plywood, veneer, composition board, and paper are the major wood products.

Agriculture is the science of producing plants and animals. How it is done has changed considerably in the last 200 years. The agricultural industry has emerged and is made of three areas: supplies and services, production agriculture (often called farming), and marketing and processing.

Consumers are people who use food, fiber, and shelter. They have likes and dislikes, and these must be considered. Variations occur among people in different areas of the world. Climate, customs, and standard of living have a lot to do with what people eat, how they dress, and the kind of home they have.

Agriculture provides many opportunities for young people. Careers vary widely. A person may have an office in the city or work outside in the country. Education, experience, commitment, and personal traits are important to success.

QUESTIONS

Answer the following questions using complete sentences and correct spelling.

1. What are three needs of people? Why are they important?
2. What are the human food needs (nutrients)?
3. What are the two main sources of food?
4. List the parts of a plant that may be eaten by humans. Give an example of a plant for each part.
5. Distinguish between terrestrial and aquatic animals. Give an example of each that is used for human food.
6. What is fiber?
7. What are the common animal fibers? Give an example of each.
8. What are the common plant fibers? Give an example of each.
9. What is forestry? What are five products from forests?

10. What is domestication? How did it occur? Why?
11. What is agriculture?
12. Name several inventions that changed farming and describe what the invention did.
13. What are the three major areas of the agricultural industry? What is the role of each in providing for human needs?
14. What is ornamental horticulture?
15. What is the meaning of renewable natural resources?
16. How is recreation a part of agriculture? Give two examples.
17. How is consumer choice important in agriculture?
18. What three major factors are important in consumer choice?
19. How do agricultural practices vary?
20. What important areas should young people who are interested in agricultural careers consider?

EXPLORING

1. Make a list of the food, fiber, and shelter items you use each day. Select one of the items and trace it from where you got it back to the farm.

2. Make a list of the plant and animal food and fiber crops grown in your area. Visit a farm to observe the production of one of the items.

3. Visit a local supermarket and identify the forms in which the following foods are available to the consumer: green peas, turkey, milk, and apples. Add other products to the list, especially those that are your favorite foods.

4. Tour an agriculture museum. Observe farming practices, tools, and other items that depict agricultural practices of many years ago.

2 USING APPLIED SCIENCE AND TECHNOLOGY

People are learning more and more about the world they live in. Many questions about food, fiber, and shelter are being answered. Some answers tell how to do a better job in agriculture. Others tell how to control disease, prevent hunger, and make life better in general.

Years ago people believed in superstitions. These are beliefs based on ignorance, fear, and trust in magic. Sayings about black cats, signs of the moon, and walking under ladders have little to do with the real world. Today, people want to know if something actually exists; they want the facts! To help get the facts, people use detailed ways of looking at things. These give answers to questions.

2-1. Science and technology are used to grow fish. Here, an agriscience student checks on catfish being raised in the lab of his school in Hawaii.

Science is used to find answers in agriculture. Many answers have been found to common questions. In fact, agriscience is used to help learn more about our food, fiber, and shelter.

OBJECTIVES

This chapter provides basic information about agriscience. It has the following objectives:

1. Explain agriscience and technology.
2. Relate agriscience and technology to four areas of science.
3. Explain the scientific method.
4. Describe how the scientific method is used in agriculture.
5. Describe the role of technology in agriculture.
6. Identify new areas of agriscience.
7. List issues associated with agriscience and technology.

TERMS

agriscience
agronomy
animal science
aquaculture
biology
biotechnology
botany
computational science
earth science
entomology
environmental science
experiment
forestry
genetic engineering
horticulture
issue
laser
life science
mechanical technology
physical science
poultry science
radiation
remote sensing
research
science
scientific method
technology
zoology

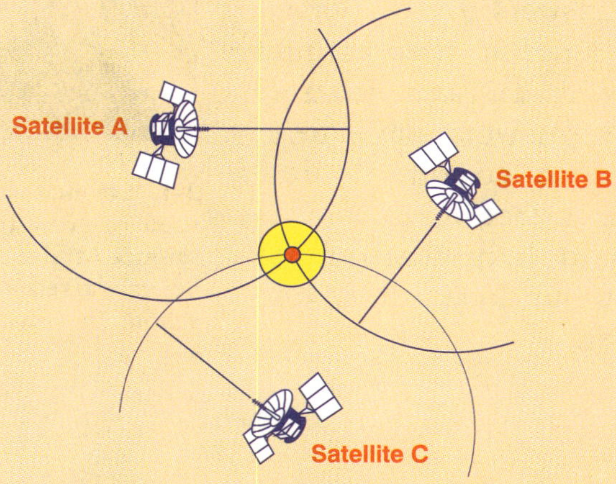

GLOBAL POSITIONING SYSTEM (GPS)

THE MEANING OF AGRISCIENCE AND TECHNOLOGY

Agriscience and technology go together. Both contribute to meeting important needs of people. Further, both are very important in producing plants and animals. Science is used in all areas of the agricultural industry.

AGRISCIENCE

Agriscience is the use of science in producing food, fiber, and shelter. All areas of science are involved. Emphasis is on those areas directly related to growing plants, raising animals and getting products to the consumer. There are two broad areas of agriscience: applied and basic.

Applied agriscience is the knowledge used in producing plants and animals. Scientific principles help agriculture in many ways. Applied agriscience answers the question: What should I do? For example, knowing what chemical to use to solve a crop problem is an example of applied agriscience.

Basic agriscience provides information for additional study. It is needed to gain an understanding of how a process works. The process may be of little practical value to agriculture but of great use in developing an applied agriscience practice. Basic agriscience answers the question: Why does it do what it does? For example, why and how a plant reacts to a chemical is basic agriscience.

People who work in agriscience may be known as agriscientists. They must often have considerable education in several areas of science.

TECHNOLOGY

Technology is the practical use of science. New methods of plant and animal production use technology. This makes possible increased production of food, fiber, and shelter.

In the definition of technology, emphasis is on the word *practical*. This means that the science can be put to good use. Benefits can be gained from using the technology.

Appropriate Technology

Technology can make big improvements in the ways things are done. In some cases, good judgment has to be made about what to use. Not all technology should be used. Some just isn't right for a given situation.

Appropriate technology is that which best fits the needs of people. Developed countries can use high levels of technology and make better use of more advanced technology. Lesser developed countries are different. They often benefit from simpler technology than that of the advanced countries.

Large machinery may be well suited to agriculture in the United States and other developed countries. The level of agriculture allows good use to be made of the latest machines.

Even in the United States, the use of some technology may be slow. This is because the technology may be viewed by some people as not appropriate (not right for the time). A good example is the dairy industry. Bovine Somatotropin (bST) was developed in the late 1980s. When used with dairy cows, the amount of milk given by a cow greatly increased. Fewer cows and dairy farms would be needed. This meant that some dairies will go out of business. A few leaders felt that it was better not to use the technology and keep the farmers than to use it and lose the farmers. After all, how would the farmers make a living? Today, fewer farms are producing more milk because of bST.

In developing countries, the latest technology is often too advanced. The people aren't ready for it. They don't know how to use it. Their farms are not set up for large machines. Big tractors could put a lot of people out of work.

People without jobs make other problems. For example, China and India have many people. Low levels of technology that involve a lot of hand labor help to keep more of them working. New technology with big machines might put a lot of them out of work. In these places, gradually changing to new technology is often best.

2-2. A mechanical cotton picker uses high levels of technology in harvesting valuable fiber from cotton plants. (Courtesy, Texas Department of Agriculture)

Leadership and Technology

Deciding when to use new technology is not easy. Agriscientists and government officials must often make decisions about what will be available. Government rules may keep some technology from being used. These rules are for the overall benefit of people.

Good information is needed to make decisions. Careful study of any new technology is a must. Leaders who have bad information can't make good decisions!

Once new technology is ready, information about it must be given to possible users. Educational programs may be needed. Farmers can't adopt practices they don't know about.

HOW AGRISCIENCE AND TECHNOLOGY RELATE TO SCIENCE

Science is knowledge about the world around us. It has many general truths or laws. Developments in science have often been tied to new ways of producing food and fiber. Early science was used to help understand plants and animals.

Observation is a fundamental scientific process. This means watching an event and recording what is seen. The events must be carefully carried out and good measurements must be made. For example, a new growth chemical may be used on a crop. The crop is carefully observed for change. The observations are compared to other crops of the same kind where the chemical wasn't used.

Science deals with observed facts and relationships among them. Procedures are carefully followed in making observations. The observations usually result in general principles that help explain the world around us. These principles serve as the bases for technology.

AREAS OF SCIENCE

The four major areas of science are mathematics, physical science, life science, and social science. Each has many divisions. Differences between them are not always clearly defined. Most areas of science overlap.

Mathematics

Mathematics is important in agriscience because it involves making and using observations. It is the science of numbers. Mathematics deals with how numbers can be measured and changed. Using mathematics, people can make exact statements and predictions. Here are two examples using the customary (English) system of measurements:

- **Weight**—A ton of feed is 2,000 pounds to both buyer and seller. This is a uniform measure that both can agree on. Mathematics allows for exact and uniform measurements.
- **Dimensions**—An acre of land is the same size in all cases: 43,560 square feet. When people say "acre" they are accurate. There is no lack of agreement on the amount of land.

Mathematics has several branches. Arithmetic and geometry are used by people in agriscience every day. Logic is used by researchers and others when they solve problems. Other areas of mathematics include algebra, calculus, probability, and statistics.

- **Arithmetic**—This is the study of numbers and methods of calculating. These are the simple things people do when counting cattle or determining the worth of a truck load of corn.
- **Geometry**—This is a practical area of mathematics. It involves showing mathematical relationships of points, lines, angles, surfaces, and solids. Geometry has many applications in agriscience, including building construction, land surveying, and making water drainage systems.

2-3. Considerable mathematics is used in constructing buildings.

- **Logic**—Logic involves reasoning. It is used to infer from one situation to another, such as one agriscience experiment to another. To infer is to make a reasoned judgment. Many things in our lives involve common inferences. For example, if the clouds are heavy we infer that it might rain or if there is smoke there is fire.

Deductive logic is reasoning from known principles to reach a conclusion.

Inductive logic is reasoning from repeated experiments to reach a conclusion. Principles may not be known.

Physical Science

Physical science is the study of nonliving matter around us. Physical science has three important areas for agriscience: earth science, chemistry, and physics.

Earth Science. Study of the environment in which plants and animals grow is ***earth science***. This includes soil, water, and the atmosphere.

Earth science is a large area of physical science. It has two major areas and three other areas grouped together, as follows:

- **Geology**—This is the study of the earth's composition, structure, and history. It includes movement in the earth by earthquakes and volcanic eruptions. Study of the soil is a part of geology that is important in agriscience. Geology can help maintain the soil and prevent erosion. This is often known as soil science or soil conservation science.

- **Meteorology**—This is the study of the earth's atmosphere. It includes studying the weather and making predictions. Good weather information is important in many agricultural activities: cutting hay, harvesting wheat, picking cotton, plowing land, protecting crops from freezing, and regulating irrigation systems.

- **Other Areas**—Hydrology, oceanography, and astronomy are three additional areas of earth science. All have use in agriscience.

  Hydrology and oceanography involve studying water. Both are used in agriculture. Hydrology is concerned with studying water on land. Its focus is on how water moves on and through the ground and ways to keep good quality water. It provides valuable information to farmers who use irrigation wells. Oceanography involves studying the water in the oceans. Wild and cultured aquatic crops may be effected by changes in ocean water.

Astronomy involves studying the planets, sun, and stars. The locations and processes of heavenly bodies are a part of astronomy.

Chemistry. Chemistry is concerned with the make up of materials or matter. Matter is anything that occupies space and has mass (weight or quantity). Chemists often refer to matter as substances. A substance is made of only one kind of matter. If a substance can be changed into two or more simpler substances it is said to be a compound.

Agriscience involves a lot of chemistry. Here are a few examples:

- **Fertilizers**—Fertilizer is used to enrich the soil. Many kinds are available. Chemistry tests tell which nutrients are present and needed in the soil. Using test results allows selection of the best fertilizer.
- **Pesticides and herbicides**—Pesticides are substances used to control insects and other pests. Herbicides control unwanted plants or weeds. Some pesticides and herbicides are elaborate chemical compounds. Agriscientists study complex materials to develop these products.
- **Rusting equipment**—Much of the equipment used in agriculture is made from iron. If not protected, iron will rust. Rusting is a chemical process. It occurs when iron is not protected from oxygen in the presence of moist air. The iron chemically combines with oxygen. Rusty iron is weak and may break. Painting iron and protecting it from moisture will prevent rusting.

Physics. Physics deals with the physical nature of objects. Areas of physics include the study of heat, light, electricity, and mechanics.

2-4. Using chemicals to control weeds improves lawns and gardens.

Mechanics is the area that gets the most attention in agriscience. It deals with objects and their motions. Most agricultural machinery and equipment applies many principles of physics. A large machine may be made up of many small, simple machines. (Note: The term mechanics has another meaning in agriscience. It refers to building and maintaining agricultural equipment.)

Here are a few examples of physics in agriscience:

- **Loading ramps**—Gradually sloping ramps make it easy to lift heavy objects. A fertilizer truck will back up to a ramp to be loaded or unloaded. A loading ramp uses a simple machine known as an inclined plane.
- **Wheels and axles**—Wheels and axles are often found on agricultural equipment. They are used to move the equipment from one place to another as well as to transfer power. In physics, power refers to the amount of work that is done. Just think of all the tractors and equipment we wouldn't have if there were no wheels and axles!

Life Science

Life science is the study of living things. It is sometimes called biological science or biology. ***Biology*** is another name for life science.

Two major areas in life science are: botany and zoology.

Botany. ***Botany*** is about plants. Since plants are very important in agriscience, there is a lot of emphasis on botany.

Plants have unique abilities to convert energy from the sun into food. The food is used by the plants themselves and by animals when they eat the plant.

2-5. Studying the leaves of plants helps understand how they grow.

Agriscientists study ways to help plants grow better. They study what plants need for growth and how to provide for these needs. A good understanding of botany is a must in agriscience.

Zoology. ***Zoology*** is about animals. Several kinds of animals are important in agriscience. Some of the animals are beneficial; others cause problems.

Beneficial animals are those that provide products people need. Food, clothing, medicines, soap, and other products may come from animals. Agriscientists study ways to improve animals and make them grow better.

Animals can cause harm in several ways. Small animals known as insects can damage crops. Large animals can attack and destroy sheep or cattle. These animals are often called pests.

Overlapping Areas. Botany and zoology have several common areas of study which overlap. These apply to both plants and animals. The overlapping areas are morphology, physiology, biochemistry, genetics, and molecular biology.

Morphology is the internal and external structure of living things. Anatomy is the study of the arrangement of the parts of plants and animals. Studying structure can provide information on how to improve a plant or animal.

Physiology is a study of how living things function. Life processes are studied. When these processes fail, the plant or animal will cease to grow and live.

Biochemistry is a study of the chemical processes in plants and animals. Both biology and chemistry are needed in biochemistry. A lot of biochemical

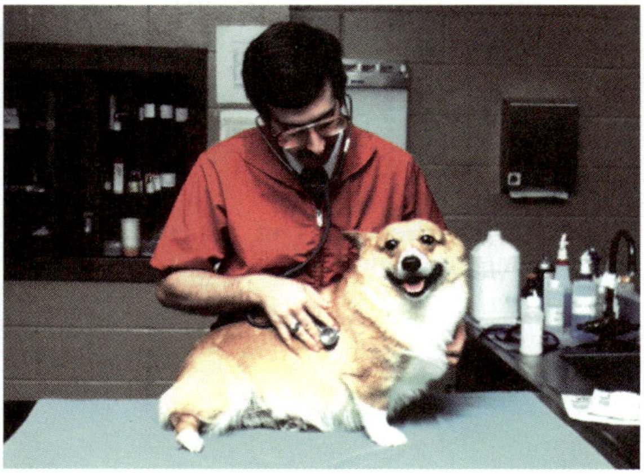

2-6. In-depth knowledge of animals is needed in treating a dog. (Courtesy, College of Veterinary Medicine, Texas A&M University)

research has been done using plants and animals together. For example, the composition of feed and how an animal uses it may be studied.

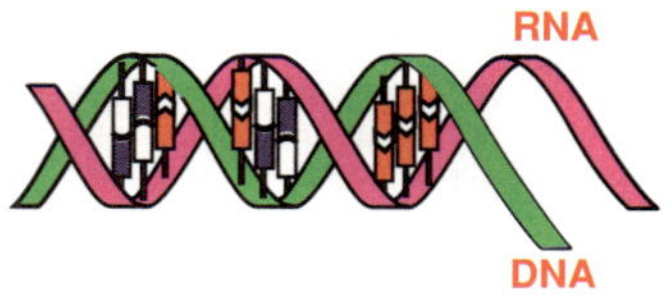

Genetics is a study of how plants and animals pass characteristics to their offspring. Agriscientists often work to improve plants and animals using various genetic principles.

Molecular biology deals with the organization of living matter. The substances studied are very tiny and must be enlarged. Knowing about these substances is often important in understanding disease, how feed is used, and other areas.

Social Science

Social science deals with human society. It is sometimes known as behavioral science. Anthropology, psychology, and sociology are areas of social science.

Many different characteristics of human behavior are studied. Some of social science relates to the interests of people (in groups or as individuals) in certain foods, clothing, or shelter.

The social science areas have indirect relationships with much of agriscience. Understanding them can help explain why people respond as they do to food and other needs of human life.

AREAS OF AGRISCIENCE AND TECHNOLOGY

Agriscience is made up of several areas of applied science. These relate to the organization of the agricultural industry and the crops and livestock that are produced. Some of the areas deal with plants; others with animals. In most cases, plant and animal areas overlap. Several important areas of agriscience are included here.

AGRONOMY

Agronomy deals with the study of plants and how they relate to the soil. Its purpose is to improve crop production and conserve natural resources. New kinds of crops may be developed and better ways to grow crops are studied.

Emphasis in agronomy is on crops grown for human food and animal feed. These are often known as agronomic or field crops. These crops

2-7. A good field of soybeans is the result of using the right cultural practices. (Courtesy, American Soybean Association, St. Louis, Missouri)

produce large amounts of food and fiber. Examples of agronomic crops are: wheat, corn, soybeans, rice, cotton, grain sorghum, and sugar beets.

Agronomy also includes the production of grass and other forage crops. A forage crop is a plant that animals eat in the pasture. Some animals graze (bite off and eat) the plants without them being harvested. Ornamental grasses (known as turf) are included. These grow on golf courses, football fields, and other places. Turf is a specialized area of agronomy.

Soil science is an area of agronomy that deals with studying the soil. Emphasis is on improving and conserving the soil. The soil may be classified and tested for nutrients. (Nutrients are the elements that plants need to grow, such as nitrogen and potassium.)

HORTICULTURE

Horticulture is the science of growing plants for food, comfort, and beauty. It includes ornamental horticulture and many plants grown for food. Much of horticulture involves skills similar to those in agronomy.

2-8. Plants are often started and grown in a greenhouse. This photo shows a glass greenhouse that protects the plants from cold weather.

2-9. A terrarium is a small enclosed container used inside for growing small ornamental plants.

Ornamental Horticulture

Ornamental horticulture is growing and using plants for their beauty. Some are inside of homes and buildings; others grow outside. Ornamental horticulture includes floriculture, landscape horticulture, and interiorscaping.

Floriculture deals with flowers and foliage plants. Flowers are the blossoms of plants that have special appeal. Foliage is the leaves of plants that are particularly attractive. Examples of flowers are the rose, marigold, carnation, and zinnia.

Some flowering and foliage plants are kept inside homes or businesses. Known as interiorscaping, the plants create pleasing areas inside of buildings. Examples of interiorscaping plants are the rubber tree, weeping fig, and philodendron.

Landscape horticulture is producing and using plants to make outdoor areas more attractive. Shrubs, vines, many kinds of flowering plants, and ornamental trees are used. Lawn and turf maintenance is a part of landscape horticulture. Examples of shrubs are the boxwood, holly, and gardenia. Grasses commonly used include Bermuda, Kentucky bluegrass, fescue, and centipede.

Food Crop Production

Many horticulture plants are important as food. Two main areas are involved: olericulture and pomology.

Olericulture deals with growing vegetables, such as snap beans and tomatoes. Many important food crops are included. Vegetables are grown in all areas of North America; however, some locations have unique climate advantages.

Pomology includes fruit and nut production, harvesting, and marketing. All areas of North America have some type of pomology. Common fruits include the apple, blueberry, orange, pear, and plum. Common nuts include the pecan, walnut, and peanut.

FORESTRY

Forestry is the science of growing and using forests. Large tracts of land are usually involved.

Many products are made from trees. Most trees are used for lumber, paper, and similar products. Some trees provide syrup (maple), rosin (pine), turpentine (pine), oil (tung), and decoration (Christmas tree).

Forests are of two main types: cultured and native. The difference is in how the forest is cared for.

Cultured forests are carefully developed, cared for, and cultivated. The trees are selected because of improved characteristics. Trees are planted as seed, or small seedlings are set out. All of the trees are usually of the same kind. These forests are sometimes known as tree farms.

Native forests are those that naturally grow. The trees grow wild and haven't been domesticated. A mixture of different kinds of trees may grow together. Native forests receive less care than those that are cultured.

Once trees are grown, they are manufactured into various products, known as forest products. Those who raise forest products focus on how to improve the use of wood. New ways of making furniture, building homes and other uses are a part of the science of forest products.

2-10. Two crops of pine trees are being grown on this tree farm. Trees in the background are about 12 years old. Those in the foreground are 3 to 4 years old.

ENTOMOLOGY

2-11. Traps are often used to determine the kinds and numbers of insects present in a field. Traps like the one above help determine when to use insecticides (insect poisons).

Entomology is the branch of zoology that deals with insects and related small animals. The science is often carried out in conjunction with agronomy, horticulture, forestry, and other areas of agriscience.

Entomology includes identifying both beneficial and harmful insects. Beneficial insects are those that do good things, such as the honeybee. Harmful insects cause damage to plants, animals, people, and structures. They include termites, corn borers, and white flies. Methods to control harmful insects are developed. Ways to encourage the growth of beneficial insects are studied.

In locations where crops are grown, entomologists work closely with farmers and others to identify and control pests.

AgriScience Connection

ENTOMOLOGY: THE STUDY OF INSECTS

Entomology is the study of insects and related small animals. The work is often exciting and requires problem solving skills. After all, how do you capture bees so that they aren't injured and you aren't stung?

In this photograph, a laboratory technician is getting a sample of wild bees from a dead saguaro cactus. A special type of vacuuming system is being used. This protects the bees from injury. Using a long tube allows the technician to keep some distance away and avoid being stung.

Just as collecting a sample of bees presents a challenge, many other areas of entomology have challenges. With a little investigation, you will likely find an area that is interesting to you. (Courtesy, Agricultural Research Service, USDA)

2-12. A knowledge of animal science helps these young people care for an orphaned lamb.

ANIMAL SCIENCE

Animal science is the area of agriscience dealing with the production of animals for food. Some animals are also used for fiber, recreation, and power.

Beef cattle are primarily produced for their meat. Some by-products are produced, such as leather, soap, and medical items. The meat produced is known as beef.

Dairy cattle are used for milk production. The milk may be manufactured into cheese, ice cream, yogurt, and other products.

Hogs are used for meat. Some by-products are made from the internal parts of hogs. The meat produced is known as pork.

Sheep are raised for meat and wool. The meat of young sheep is known as lamb, while that from older sheep is mutton. Sheep are sheared once a year for wool. The wool is made into cloth that makes warm clothing.

Horses are used for pleasure and power. Their role in providing power has decreased since gasoline engines were invented. Emphasis today is on horses for pleasure. Horseback riding, rodeoing, and cutting horse competitions are examples of the ways horses are used.

Other animals important in agriscience include goats, bison (buffalo), and llamas.

POULTRY SCIENCE

Poultry science is concerned with raising chickens and related fowl (birds). Chickens are by far the most important poultry. Other kinds of poultry include turkeys, ducks, geese, pheasants, quail, guinea fowl, and the ratites (ostrich and emu are the better known ratites). Poultry are produced for meat, eggs, and other products.

Poultry science seeks improved ways of growing chickens. This includes developing improved kinds of chicken. Providing the proper feed and environment is essential to get good growth.

AQUACULTURE

2-13. Aquaculture is an expanding area of agriculture. This photograph shows mature catfish being seined (captured) on a Texas farm. (Courtesy, Texas Department of Agriculture)

Aquaculture is the science of water farming. It is one of the newest areas of agriscience. Aquaculture deals with growing both aquatic animals and plants.

Aquatic animal crops are more popular in the United States than plant aquacrops. These include fish, such as trout and catfish; shellfish, such as oysters and clams; and crustaceans, such as shrimp and crawfish. Frogs, eels, and alligators are other examples of aquacrops.

Plant aquacrops include water chestnuts and water cress. Seaweed is popular in some places, particularly as a human food in Pacific Ocean island locations.

Important areas of animal aquaculture are water management, reproducing crops, controlling disease, feeding, and harvesting.

ENVIRONMENTAL SCIENCE

Everything external to a plant or animal makes up its environment. Temperature, light, water, air, soil, plants, animals, and other surroundings are parts of the environment.

Environmental science involves ways of wisely using and protecting the things around us. Humans use the resources of the earth to produce goods and make life better. They also assume a role in seeing that a good environment is available for future generations.

2-14. Many people enjoy the environment of a tropical beach in the winter.

Emphasis in environmental science is on natural resources. These include the air, soil, water, wildlife, native plants, and minerals. Major practices are to conserve and improve natural resources.

MECHANICAL TECHNOLOGY

Mechanical technology is the use of machines and equipment to do work. Gasoline-powered engines have replaced animal power. Sophisticated equipment has replaced hand methods in agriculture.

Mechanical technology involves considerable physics. Basic concepts of mechanics are important. Major areas include heat, light, sound, electricity, and magnetism.

Once equipment has been constructed, people are needed to keep it in good condition. We often call these people mechanics. They must have considerable knowledge of mechanical technology.

Important areas of mechanical technology in agriculture are:

- Agricultural power machinery and equipment, such as tractors, plows, planters, and harvesters
- Agriculture buildings, such as greenhouses and poultry houses
- Earth structures, such as ponds, drainage ditches, and terraces

Mechanical technology involves a number of applied skills. These deal with constructing and maintaining facilities and equipment. Examples of applied skills include the following:

- Fabricating metal which involves measuring, cutting, and welding skills
- Constructing with wood which involves measuring, cutting, fitting, and fastening, often with nails
- Wiring for electrical service which involves locating electrical service needs, installing electrical devices, running electrical wire, and fastening wires and devices

2-15. The tractor is just one kind of power technology used in agriculture. This photograph shows a 460-horsepower tractor used on a North Dakota farm. (Courtesy, Mollee Thomas, Danville, Illinois)

METHODS: THINKING LIKE AN AGRISCIENTIST

Agriscience involves using the ***scientific method,*** which is asking questions and seeking answers in an organized way. It is also known as scientific problem solving.

Questions often come up in agriscience. These may be the "what" or "why" of a particular plant or animal. All problem solving involves answering a question.

Thinking like an agriscientist also involves precise thinking. The thinking must be exact, accurate, and free of error. The experiences of others can also be valuable in solving problems, especially new problems.

THE SCIENTIFIC METHOD

The scientific method has five important steps. Sometimes the order of the steps may change or the number of steps used may vary, but a logical process must be used.

Step One: Identify the Problem

A problem can't be solved until it is defined. To help with this, most problems are first stated as questions.

In some cases, people only see symptoms of problems. Often more exploration is needed to accurately determine the problem. If hogs refuse to eat, failure to eat is a symptom; it is not the problem. Studying a symptom will not usually result in solving a problem.

Agriscience is often used to find improved methods of crop or livestock production.

Step Two: Get Information

This involves getting facts about the problem. Sometimes it is called data collection. The information is what is observed about the problem. Everything people see, hear, touch, taste, or smell can be a part of the observation.

Careful measurements of observations are a must. The measurements may be of several kinds. In agriscience, measurements are often made of increased growth, quality, or amount produced.

Step Three: Suggest an Answer

Agriscientists suggest answers to problems. These suggestions are known as hypotheses. Forming the hypothesis involves making a possible explanation of the problem.

Hypotheses are statements about the problem that can be tested. Hypotheses may be proven true or false. The findings are important either way.

Step Four: Experiment

This step involves performing experiments to test the hypotheses. An ***experiment*** is a trial or test. The conditions are carefully controlled except for the one that is being studied. The condition that is changed is known as a variable. In some experiments, several variables may be used. The part of an experiment that does not change is called the control or controlled variable.

Observations of changes are needed. Careful measurements must be made. Inaccuracy results in poor quality experiments. The solution to a problem is no better than the accuracy of the information.

Most agriscience experiments use control and trial plots. For example, experiments to measure the effects of increased fertilizer on yield must include plots that do not get the increased fertilizer. All plots in an experiment should be identical. If not, any differences that might be observed might not have been caused by the test that was carried out.

Step Five: Form Conclusions

A conclusion is a judgment formed after an experiment. It involves a process called inferring. The data are carefully looked at for meaning. Facts are used to infer traits to the problem being solved.

Sometimes conclusions need more study. This is known as replication (repeating). Most new agriscience developments need replication. This assures that the findings are accurate.

RESEARCH USE

Research is the careful and diligent search for answers to agriscience problems. The scientific method is the primary tool of researchers.

Agriscience research is used to find ways to improve many areas of agriculture. Experimental plots (small fields) are often used. Trials are carried out to determine the effects of something. Small plots of crops or groups of livestock may receive certain treatments (tests).

Agriscience research is carried out by land-grant universities, agricultural agencies, agribusinesses and other groups. A few farms do research. For example, many of the cotton varieties planted around the world were developed by Delta and Pine Land Company, a large farm near Scott, Mississippi. This large plantation employed agriscientists with advanced levels of education in areas of agronomy and genetics.

2-16. The Morrow Plots at the University of Illinois make up the oldest agricultural research land in the United States.

PRACTICAL USE

The scientific method is used for many purposes. Practical, everyday problems in agriculture can often be solved with the method. Here are a few examples:

- If a tractor engine stops running, the mechanic can use these steps to get it going again.
- If a plant shows signs of disease, the scientific problem-solving process can be used to determine how to control the disease.
- If the fish in a pond refuse to eat, the aquaculturist can use the method to solve the problem.
- If flowers wilt for some unknown reason, the problem solving process should help provide an answer.

NEW AREAS OF AGRISCIENCE

Exciting developments are underway in agriscience. These will result in new ways of producing food, fiber, and shelter. These new areas can provide answers to questions more rapidly than traditional methods.

BIOTECHNOLOGY

Biotechnology involves using biology to develop new products or processes. The work is carried out in laboratories where conditions can be controlled. Some biotechnology has been around for a long time. More advanced forms are now being used.

2-17. New technology is used to control pests. This photograph shows corn plants being inoculated with corn borer larva. (Courtesy, DEKALB Plant Genetics, DeKalb, Illinois)

Much of the work in biotechnology involves using yeast, bacteria, enzymes, and other natural substances from living things. Some of these have been used for many years. For example, yeast was used in making bread some 4,000 years ago in Egypt. Today more than 95% of the bread sold in the United States is made using yeast.

New uses of biotechnology hold a lot of promise. These deal with increasing production and protecting the environment. New vaccines are being developed to protect animals. Methods to get more lean meat from animals result in improved efficiency. The environment is being protected by new "waste eating" bacteria.

GENETIC ENGINEERING

Genetic engineering is an advanced form of biotechnology. It involves changing the nature of living organisms. The changes are intended to be very small and only of desired traits.

Genetic engineering involves "cutting and pasting" genes. Genes carry information about a living thing. They are found as a part of the chromosomes in the tiny building blocks known as cells. The gene that has an undesirable trait can be cut out and another one put in that has the desired trait. New techniques involve using tiny, high velocity guns to shoot genes into plant cells.

Genetic engineering has already produced important new approaches in agriscience. Crops have been developed that resist insect damage and spoilage. Some are being developed that kill off the weeds around them. (Black walnut trees have a natural substance that kills off certain plants that grow near them. The identification of the gene in the black walnut that has this trait is the first step. Once it is isolated, it can be moved to other plants. The other plants will kill off weeds that grow near their roots. This would cut down on the use of chemical weed killers near crops and help protect the environment.)

PRECISION FARMING

Precision farming is using cropping practices that improve yields based on the needs of the land. Fields are divided into small areas on the basis of information collected with satellite systems using global positioning. Pests, soil fertility and compaction, and other features are studied. Records of yields are kept for the small areas within fields. Computer-controlled equipment applies the needed fertilizer and other inputs to grow a crop.

REMOTE SENSING

Remote sensing involves gathering and recording data from a great distance. Most remote sensors are on satellites some 500 miles above the earth. Some of the methods used are extremely accurate.

Landsat is the term applied to United States satellites that make photographs of the earth and plot the earth's resources. These photographs are used to make maps that provide meaningful information. The photo-

2-18. Global positioning systems are used in precision farming. This shows a global positioning unit being used to study soil salinity (top left), global positioning unit being installed on a tractor (top right), data collected by remote sensing being used to study soil fertility (bottom left), and examining sample of corn from the grain flow sensor on a combine (bottom right). (Courtesy, Agricultural Research Service, USDA)

2-19. Black and white aerial photographs provide a lot of information about the land. This is a photo of part of the Shenandoah Valley in Virginia. (Courtesy, USDA, Washington, D.C.)

graphs aren't made with regular cameras that record light. They are made with remote sensors that view the earth in both visible and infrared light. Infrared light is a form of light that isn't visible to the human eye. Not only is the earth mapped but sea-floor (bottom) elevations are also mapped.

Aerial photographs taken from airplanes have been used for a number of years as sources of information.

Remote sensing in agriscience is beneficial in forecasting the weather, locating natural resources, detecting crop disease, and protecting the environment.

LASER TECHNOLOGY

A ***laser*** is a device that produces a special kind of light known as coherent light. Lasers are sometimes called "electric eyes." An intense, narrow beam of light is produced that is useful in many ways.

2-20. Tripod-mounted laser instrument used to guide land forming.

Laser technology has many uses in agriscience. It can be used in land forming to shape land to a certain contour. It is used in laying out rice fields, terraces, and fish ponds. The line of light does not bend and cause unwanted changes in elevation when moving the earth. The laser beam guides the tractor as it lays out the land.

Laser beams can be used to "watch" processes in manufacturing plants and other facilities. The light will detect changes in movement or other things indicating a problem. This saves humans from trying to watch high-speed processes in a factory.

Lasers can beam light to scan codes on merchandise, known as bar codes. Supermarkets frequently use scanners that help determine how much to charge the customer for groceries. The scanner is hooked to a computer that has all of the prices.

COMPUTATIONAL SCIENCE

Computational science involves using computers to do mathematics and science problems. Supercomputers with considerable capacity are needed. Individuals need not be at the computer site. People can access the supercomputer with a personal computer and a telephone modem (hook-up) from many miles away. Computational science has emerged since 1990 and offers great potential in certain areas.

Computational science is used in agriscience in designing facilities, developing products, and modeling the earth's resources.

RADIATION

Radiation is a kind of energy that travels as waves. There is natural as well as artificial radiation. Life depends on some kinds of radiation, such as that from the sun. It is known as cosmic radiation. Nuclear plants use radioactive materials to produce radiation power. Some people are very afraid of this type of radiation because of the potential danger it poses to humans.

Electromagnetic radiation, an artificial type, has a number of uses in agriscience. Foods can be exposed to radiation (irradiated) to prevent spoilage. (For example, meat products can be irradiated to kill bacteria that may be on or in them.)

Radiation can be used to control insects in food. No one wants to eat insect parts in their food! Wheat, corn, and other cereal grains often have weevils, worms, or other pests. By irradiating the stored grain, the pests are destroyed and the food is kept wholesome.

ISSUES ASSOCIATED WITH AGRISCIENCE AND TECHNOLOGY

An ***issue*** is something that has more than one side. When people argue, they are taking sides on an issue. Some areas of agriscience and technology involve issues of concern to society.

Sometimes science and technology have the potential of bad side effects. This means that some harm could result.

Many precautions are taken to prevent problems. New methods receive thorough study before being released for full-scale use. Laboratory analysis assures that new procedures are good and proper. Educational programs are held to keep people informed of how to use agriscience and technology safely.

Some of the issues facing people today are:

- **Food safety**—All people want to have food that is fit to eat. They do not want to eat food that could cause harm to the body. Foods need to be free of harmful substances. People do not want food that has traces of unwanted chemicals. They do not want food that has been contaminated with bacteria and other substances.
- **Pollution**—Maintaining a good environment should be the goal of all people involved with agriscience and technology. Fertilizer, pesticides, and other supplies should be properly used. Empty pesticide containers, used water, and other wastes should be properly disposed of.
- **Conservation of resources**—Making wise use of natural resources is important for all future people. Water needs to be used efficiently and not wasted. Soil erosion should be prevented. Fields should be managed to reduce water runoff. Tractors and equipment should be operated to conserve fuel.
- **Human rights**—Human beings are entitled to certain rights and should be treated fairly. The United States has a number of laws to protect people but some other nations do not. Poor quality food must not be sent to a nation where laws might not prevent its use. It is questionable whether food products should be imported from a nation where pesticides are used without regard to the environment and welfare of people.

AgriScience Connection

KEEPING CORN SAFE

People want safe food for themselves and their animals. Doing this requires study and careful handling in harvesting and storage. Science is used to help find solutions to problems that may arise.

Corn and other grains may contain a very toxic fungus known as *Asperigillus flavus*. This fungus causes aflatoxin in grain. Aflatoxin is a strong natural poison and possible cause of disease, such as cancer.

A biochemist is here using a photoacoustic infrared sensor to evaluate corn for fungal contaminants. The procedure has potential for use at grain elevators and food processing plants. When perfected, the procedure can be used to help reduce the natural toxins that get into food. (Courtesy, Agricultural Research Service, USDA)

REVIEWING

MAIN IDEAS

Agriscience is the use of science in producing food, fiber, and shelter. Science is knowledge of the world around us. Certain truths or laws have been developed about our world. Technology is the application of science. Many benefits are gained from the practical use of science.

Agriscience and technology are closely related to science. All areas of science have some application in agriculture. The major areas of science are: mathematics and logic, physical science, life science, and social science.

Physical science and life science are particularly important in agriscience. Physical science includes the study of soil, water, and related areas that influence the growth of plants and animals. Life science is sometimes known as biological science. It has two major areas: botany (plant science) and zoology (animal science).

Agriscience can be divided into a number of areas. The major areas are: agronomy, horticulture, forestry, entomology, poultry science, aquaculture, and mechanical technology.

Agriscience involves using the five-step scientific method.

Step One: Identify the problem (state question)
Step Two: Get information
Step Three: Suggest an answer (form hypothesis)
Step Four: Experiment (make trials)
Step Five: Form conclusions

New ways of producing plants and animals sometimes raise issues about their use. People want foods that are safe to eat. They want methods used to produce them that don't damage the environment and that protect resources for future generations. They also want the rights of people protected.

QUESTIONS

Answer the following questions using complete sentences and correct spelling.

1. What are agriscience and technology?
2. What is appropriate technology? Give an example.

3. What is science?
4. What are the four major areas of science? Define each.
5. What are the areas of physical science? How does each relate to agriscience?
6. What are the two major areas of life science? Why are these areas important in agriscience?
7. What are the areas of agriscience and technology? Briefly describe each area and give examples.
8. What is the scientific method? What steps are used? Briefly explain each step.
9. What is research? How is the scientific method related to agriscience research?
10. What are the new areas of agriscience? Briefly explain each. Give examples of how each can be used in agriscience.
11. What is an issue? Name and briefly explain four issues in agriscience and technology.
12. What procedures are followed to be sure new agriscience and technology is safe to use?

EXPLORING

1. Make a list of the areas of science that are taught at your school. Interview the teachers of the classes, asking about the general subjects that are covered and how they relate to agriscience.

2. If your school has a science fair, conduct an experiment related to agriscience and enter it into competition.

3. Make a tour to a local agricultural research station. Ask about the kind of research that is underway. Determine the exact processes used. Write a short paper that compares the processes with the scientific method.

4. Identify an agriscience problem (question) in the local community. Design a process for answering the question using the scientific method.

3 USING THE EARTH'S RESOURCES

The earth has many resources. They are to be used to help people meet their needs. How resources are used is important to all people. People must find a way to have enough to go around.

Some parts of the world do not have enough food. Soil, rain, and other resources will not support crop and livestock production. One in every three people in the world doesn't have enough to eat. About 12 million people die of starvation each year in the world. With the world's population growing at the rate of 1.7% a year, there could be 10 billion people by the year 2030. This represents nearly a doubling of the current 5.4 billion. How will we have enough food?

The goal in agriscience and technology is to use and protect the earth's resources. Some practices now being used make the resources better than they were before.

3-1. Information about the earth's resources helps us manage them better. This photograph shows cartographers (map scientists) preparing a map showing environmental features of the earth. (Courtesy, U.S. Department of Agriculture)

OBJECTIVES

This chapter gives information on the use and protection of resources in agriscience and technology. It has the following objectives.

1. Define environment and natural resources.
2. Explain ecosystems.
3. Explain and give examples of renewable natural resources.
4. Explain and give examples of non-renewable natural resources.
5. Explain pollution and identify sources of pollution.
6. Describe recycling and list examples.
7. Identify sources of pollution in agriculture and name rules to follow in preventing agriculture pollution.

BIOSPHERE—the area of earth that supports life extends from earth's crust into the thin layer of earth's atmosphere

TERMS

- abiotic factor
- agriculture pollution
- biodegradable
- biosphere
- biotic factor
- browse
- ecosystem
- effluent
- environment
- environmental pollution
- erosion
- food chain
- fuel
- game management
- habitat
- hydrosphere
- landfill
- natural resources
- niche
- nonpoint source pollution
- non-renewable natural resources
- particulate
- point source pollution
- recycling
- renewable natural resources
- soil
- sustainable agriculture
- symbiosis
- toxic products
- waste disposal
- water cycle
- water table

DETERMINING OUR ENVIRONMENT AND NATURAL RESOURCES

Environment is all of the factors that affect the life of a living thing. It is everything external to a plant or animal. Factors in the environment influence the growth and survival of plants and animals. A goal of agriscience is to maintain a good environment.

Natural resources are the naturally occurring things that support life, provide fuel, or are used in other ways by humans. People can use them to make life better. Examples include soil, water, air, minerals, plants, and animals. Some people include the beauty of nature as a natural resource. All of these are a part of the environment used by agriscientists.

A BIOLOGICAL VIEW OF THE ENVIRONMENT

Plants and animals can live in only certain places. The area of the earth that supports life is known as the ***biosphere***. The earth's biosphere is very thin. It extends only a few feet above the earth and into the oceans, depending on the kind of life.

Different plants and animals are suited to certain places or environments. Some live in water; others live on land. No one kind of plant or animal is suited to all places.

Identifying Ecosystems

All of the parts of a particular environment form an ***ecosystem***. Some parts are living; others are nonliving. The living things are constantly trying to have a balance in the system. The balance may involve nonliving things, such as water and food.

Two kinds of factors are present in an ecosystem. The living things in an ecosystem are known as biotic factors. The nonliving are known as abiotic factors.

Biotic Factors. The ***biotic factors*** are plants, animals, and other living things in an ecosystem. Living things compete for space, water, food, and other resources. The same kinds of plants and animals compete among themselves as well as with each other. For example, animals compete with their own kind and with others for food plants.

3-2. This Smoky Mountain stream shows an ecosystem of biotic and abiotic factors.

Biotic factors interact with each other. Some relate well together even though many kinds of living things may be present. They form a relationship that has benefits to at least one of them. This relationship is known as ***symbiosis***.

Some animals do not relate well together. One may kill another for its food. This is predation. Wolves may kill and eat sheep. Animals that do the killing are known as predators. Those that are killed are prey.

Abiotic Factors. The ***abiotic factors*** include water, temperature, sunlight, soil characteristics, and lay of the land, including the space available. They are the things that are not alive. These factors interact with each other and with the biotic factors.

Abiotic factors determine which plants and animals can live in an ecosystem. Plant growth is essential for animals to live. Biotic factors can't exist without the right abiotic factors.

How Ecosystems Work

A beautiful stream passing through a meadow has a working ecosystem. All of the biotic and abiotic factors interact together to form the ecosystem. The system may include: water free of pollution in the stream; fish and other life in the water; fertile soil surrounding the stream; plants growing on the surrounding land; insects in the plants; and birds in the air above.

People can sometimes get involved and mess up the system. Polluting the water, trashing the grounds, killing all of a certain kind of plant or animal, and anything else that changes the natural ecosystem can cause problems.

Maintaining ecosystems requires understanding how they work. Each of the biotic and abiotic factors has a unique part in the ecosystem.

Habitat. The area where a plant or animal lives under natural conditions is its ***habitat.*** Different living things have different habitats. Understanding the habitat of a plant or animal is important in helping it survive and grow.

A number of things make up a habitat: climate (temperature, rain, etc.); lay of the land; nature of the soil and water; and other plants and animals in the area.

Changes in a habitat effect all the ecosystem. For example, lack of rain can result in plants not growing. When plants fail to grow, animals may not be able to get food. The animals may starve or move to another location searching for food.

Practices in agriculture may sometimes effect wildlife habitat. For example, the habitat of some animals may be changed or done away with if large areas of land are plowed and put into a crop. The feeding and nesting area for birds, such as quail, may be destroyed. To solve this, strips of land around fields are often not plowed and left to grow wild plants to provide proper habitat.

Niche. Every plant and animal has its own special way of living. This is known as its ***niche*** or way of life.

A niche is the habitat of a plant or animal and determines how it responds in the habitat. This includes feed, shelter, raising young, and other ways of life.

An example is the beaver. Beavers live in small streams that pass through wooded areas. The beaver cuts down small trees, pulls them into the water, and makes dams that hold water year-round. The beaver lives in and around this water pool. Beavers will defend the area against other beaver.

3-3. The gecko (kind of lizard) has a niche on this wild ginger in Maui.

Food Chain. The living things in an ecosystem get their food in different ways. The sequence in which they get food is known as the ***food chain***. Plants and animals have different food chains.

Food begins with plants using energy from the sun to make food, which results in growth of the plant. Animals eat the plants, grow and die or are eaten by animals which later die. The remains of plants and animals rot and return fertility to the soil. The cycle begins

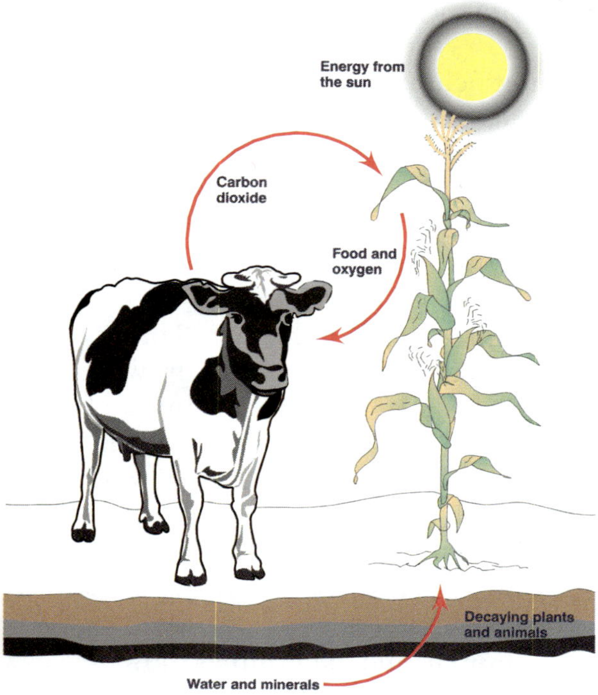

3-4. Cow and plant in the food chain.

again as new plants use the fertility for growth.

All kinds of living things participate in the food chain: plants of all types and animals, both large and small. Some are tiny and can only be seen when enlarged. These cause the rot of the dead plants and animals.

SUSTAINING THE ENVIRONMENT AND NATURAL RESOURCES

Understanding the environment helps use it wisely. Most natural resources can be used to the good of humans. It is important to use the resources in such a way that some will be available in the future.

Agriscientists use practices to sustain natural resources. This means that these resources are used in such a way that replacement occurs. For example, a crop such as corn removes fertility from the soil. After harvest, remains of the plant may be left on the field to rot.

The term ***sustainable agriculture*** is used to describe all of the things done in agriscience to maintain the ability to produce food, fiber, and shelter. Sustainable agriculture involves: protecting the soil, air, and water; maintaining wildlife habitat; using pesticides and fertilizers wisely; and disposing of wastes properly.

The goal of sustainable agriculture is to ensure food, fiber, and shelter for future generations. The long-term needs of people are considered as much as possible.

KINDS OF NATURAL RESOURCES

Natural resources can be classified into two major groups: those which are renewable and those which aren't.

RENEWABLE NATURAL RESOURCES

Renewable natural resources are those which can be replaced when they are used. When used up, more can be created. Agriscientists do things in their work that encourage natural resources to renew themselves.

Five main renewable natural resources are: soil, water, wildlife and fish, forests, and air.

Soil

Soil is the outer layer of the earth's surface. It is comprised of minerals, organic matter (rotting plants and animals), moisture and small amounts of other materials. Soil is capable of supporting plant life. Essential nutrients must be in the soil for plants to live and grow.

Top soil is the thin layer of the most fertile soil on the earth. In most places the top soil is about five inches thick. It can be deeper, or there can be none at all. If there is none, the soil will not grow much of anything.

The top soil is where seeds are planted and crops grow. Grass, trees, and other plants grow in the top soil. (Of course, big trees with deep roots go far below the top soil!) The top soil contains most of the fertility of soil. Many years are required for a one-inch layer of top soil to be formed. Top soil is made from rotting plants and animal remains.

Since top soil is loose and easily plowed, it is also easily lost. Water and wind can wash or blow it away. The washing or wearing away of soil is known as ***erosion***. Preventing erosion is a major goal in agriculture.

Most erosion occurs on freshly-plowed land and on construction sites for highways, houses and other buildings. Measures can be taken to protect the soil. Leaving plant materials on the surface and plowing as little as possible help keep the soil. One big rain on plowed land can wash away more top soil than can be formed in 20 or more years. Soil conservation measures are used to keep the soil in place. These same practices also encourage the development of new top soil.

Water

Water is often said to be the most important natural resource. Crops require water for growth. Animals need water to live. Humans need water for drinking, bathing, swimming, and other uses. Industries also use water in many ways.

3-5. Plowing the top soil is an important step in growing many crops. (Courtesy, KMN Modern Farm Equipment, Inc., West Memphis, Arkansas)

Water is a resource that appears in three forms: solid, liquid, and vapor. It is the only natural resource that can be in these three forms within the temperature range of the earth. The solid form is known as ice. Water becomes ice at 32°F (0°C). Water changes to vapor at 212°F (100°C). Since the earth is the only planet to have liquid water, it is sometimes referred to as the "water planet."

More than 70% of the earth's surface is covered with water. Much of this water has too much salt for most agricultural uses. About 97% of the earth's water is saltwater. Of the 3% said to be fresh water, two-thirds is frozen in glaciers and ice caps at the North and South Poles. This leaves only 1% of the water on earth for human, agricultural, and other uses.

World agricultural production is closely tied to water. Some places get too much rain; others get virtually none. In the United States, the desert of Nevada may get only 7 inches a year (too little for crops) while some mountains in Hawaii get over 500 inches (too much for crops).

Water Cycle. All of the water on earth forms the ***hydrosphere***. This includes water in streams, lakes, oceans, underground, and other places. It includes vapor as well as ice.

Water circulates from one part of the hydrosphere to another in a process called the ***water cycle***. Scientists often refer to the water cycle as the hydrologic cycle. In the cycle, water is continuously moving. It is cycled from the atmosphere (air) as a vapor to oceans, streams, living things, and underground pools.

The water cycle works like this: Water is evaporated (drawn up) into the atmosphere. As a vapor, it rises high into the air. As it rises, it begins to cool. The vapor forms tiny droplets in clouds. When it rains, these droplets get larger and fall to the earth. Some of the water soaks into the soil; other runs into streams that flow to lakes and oceans. The cycle is repeated again and again.

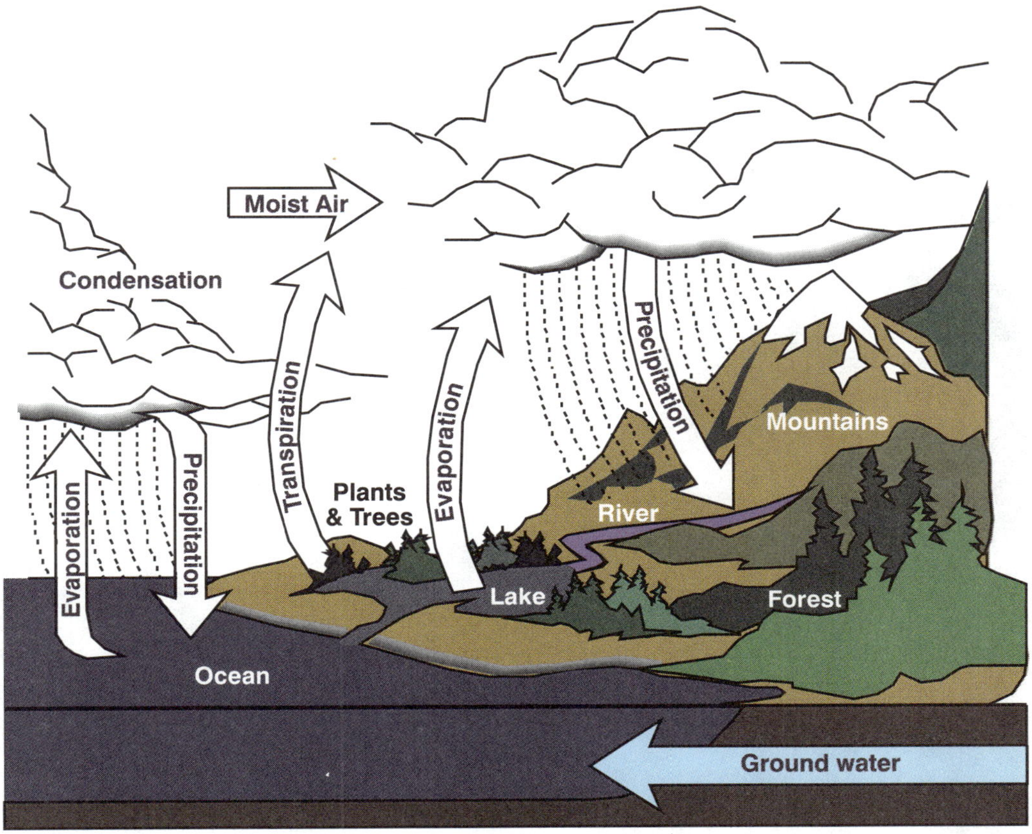

3-6. The water cycle.

Water Table. Water found below the surface of the earth is known as groundwater. It collects in spaces between rocks, sand, and soil. The water comes from precipitation and gradually soaks or seeps into the earth. The point in the earth where all of the spaces are filled and no more water can be held is the ***water table***.

People who drill wells carefully study the location of the water table. Sometimes they can get water by drilling 20 or 30 feet (6.1 or 9.1 m); other times they must go several hundred feet. Wells are often drilled to get water for human use, for factories, to irrigate crops, and for fish culture.

As water is pumped from wells, the water table goes down unless the water removed is naturally replaced by the water cycle. When the table lowers, wells must be dug deeper in order to provide water. In many places, too much water has been used and the water table has dropped. Some people say this means that we are running out of water!

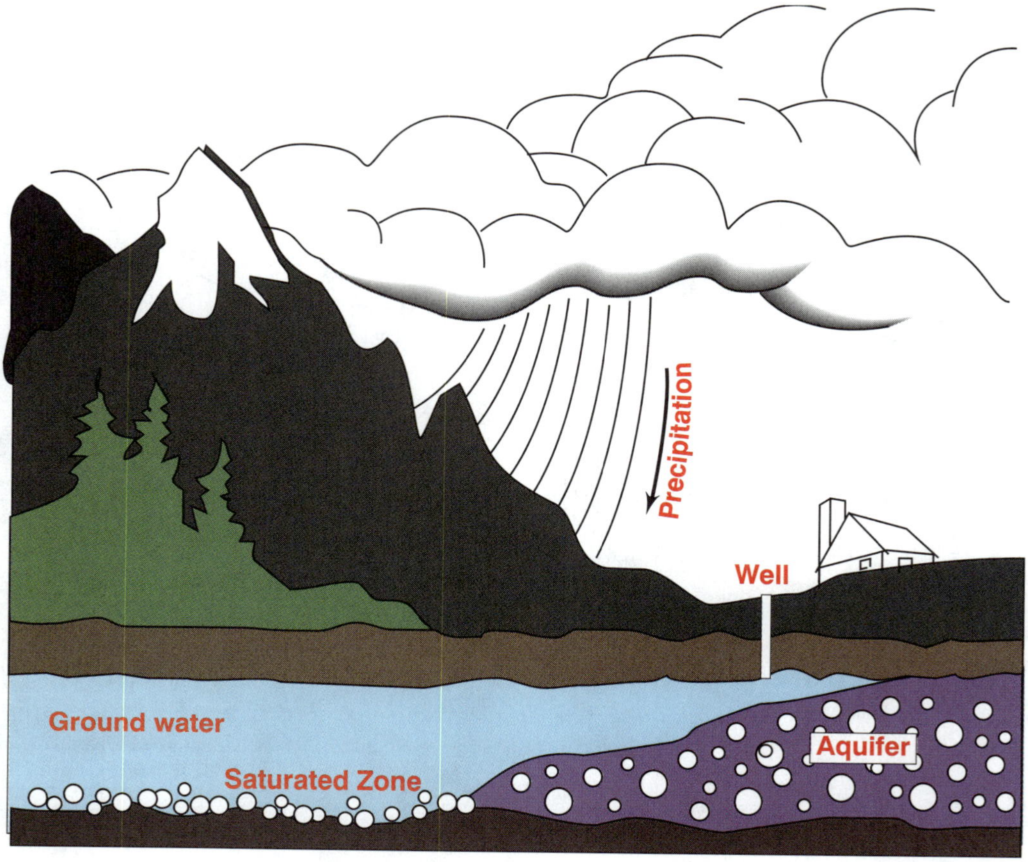

3-7. How the water table is restored and used up.

Renewing and Conserving Water. Steps can be taken to renew water supplies and conserve the use of water. The first step is to use no more water than necessary. Wasting water adds to the shortage of good water.

The supply of groundwater can be renewed by using structures to keep it from running off. Farm ponds, terraces, crop residue (stems, leaves, etc.), and other means can be used to hold back water. This allows it to soak into the earth and enter the groundwater.

Waste water can be treated for re-use. Sewage treatment plants can reclaim water for use in irrigating golf courses and other non-food production areas. Water given off by fish farms can be used to irrigate crops. Water from factories can be treated and used for other purposes, such as fish culture.

Renewing and conserving water involves using it wisely and trying not to waste any.

Wildlife and Fish

Wildlife and fish are plants and animals that haven't been domesticated. People often think of wildlife as the deer or rabbit they hunt or fish they catch from a stream or ocean. These provide recreation for many people.

Not all wildlife and fish have to be hunted to be enjoyed. People like to see the beauty of nature. Some like to watch birds or go hiking in wildlife areas. Those that are hunted are known as game animals.

Wildlife and fish includes: game birds (quail, turkey, pheasants, ducks, etc.), birds (red birds, robins, mockingbirds, etc.), butterflies and other insects, wild flower plants, rabbits, bears, elk, moose, deer, foxes, squirrels, raccoons, coyotes, beavers, alligators, frogs, and fish of many kinds.

Preserving Wildlife and Fish. Procedures can be followed to protect and improve wildlife and fish. These procedures are often known as ***game management***.

Game management begins with understanding the habitat requirements of wildlife and fish. The needed habitat can be provided and improved. In some cases, artificial stocking is used. This involves bringing in plants or animals from other locations. They are released or settled into the new environment. Only species that are suited to a habitat should be artificially stocked.

Streams and lakes are frequently restocked with fish. Young fish are raised to a few inches long in a hatchery and released into the wild. Trout, salmon, and catfish are three examples.

Many locations have laws that protect wildlife and fish. Over-fishing can remove all the fish so no new ones can grow. Over-hunting can kill all of the game so there won't be any for other people. Game laws are designed to protect animals from being killed off.

3-8. The ring-necked pheasant is a favorite among many wildlife enthusiasts.

3-9. Mule deer are found in some areas of North America.

Over Protection Problems. Some wildlife may have too much protection. The population may get so big that it causes problems. Laws which have been passed to protect some wildlife and fish may result in too many. The natural habitat may not be big enough to provide food for the wildlife. They may have to go to new places to get food. Sometimes they attack crops or livestock. Other times they don't get enough to eat, get sick, and die.

Many crops and livestock have been damaged by over-protected wildlife. The dollars lost may be great. Here are three examples:

- **Deer**—Protected deer populations may get too big for the natural ***browse*** (food). The deer will begin eating crops such as soybeans, peanuts, and sweet potatoes.
- **Coyotes**—Protected coyote numbers may grow so that food in the natural setting is low. They may attack and destroy sheep, baby calves, chickens, and other animals.
- **Beavers**—Beavers naturally dam streams to build their habitat. The dams may back water on to crop or pasture land. Use of the land may be lost.

Programs to protect animal wildlife must be reasonable and regularly reviewed. Once the wildlife population has grown to an adequate number, the laws should be changed. If not, crops may be lost to wildlife seeking food not available in their habitat.

Forests

Forests covered much of North America when it was discovered by Christopher Columbus in 1492. In the early 1600s, about half of the

United States was in forests. These were native forests, known as mature forests. As the United States developed, many of these forests were cut. Some were cut and used to build shelter and factories. Others were cut and burned to make room for crops.

Today, many areas have more trees than in 1700! Tree farms have been established much as crop or livestock farms. Tree farms use improved kinds of trees and follow approved procedures. Trees are planted and thinned. Diseases and insects are controlled. Unwanted fires are prevented and put out. Careful harvest practices are used.

Trees and the Environment. Trees make big contributions to the environment. They provide habitat for wildlife and help conserve the soil and water. They help maintain air quality by producing oxygen.

Trees add beauty to human life and provide recreation. Certainly, many people admire the giant sequoia trees of California, the tall pine trees of Mississippi, and the huge many-rooted banyan trees of Hawaii.

Cutting some large forests is said to be damaging the quality of the earth. This is an issue often associated with global deforestation, a practice of cutting trees without replanting them. Carbon dioxide content in the air has increased in recent years and is associated with tree cutting. Much of this is attributed to cutting the tropical rain forests in South America and Africa.

Commercial Forestry. Commercial forestry in the United States, Canada, and some other countries involves replacing trees as they are cut. Big wood and paper companies manage large amounts of land. As timber is cut, new trees are planted. The tree farms are carefully managed to get the most timber production and protect the environment.

Most commercial forestry operations use practices that sustain the forests. People need wood products as well as a healthy environment. The large companies need wood if they are to stay in business.

Air

Air is the earth's atmosphere. Air is evenly distributed over the earth and is essential for life. Good quality air is needed in the environment of all living things.

Air is made up of gases, moisture, and small solid particles, known as ***particulate***. The major gases are nitrogen and oxygen, which together

3-10. Air filter clogged with particulate found in a home. Humans and other living things breathe this air!

make up 99% of the air. Living things need oxygen for life processes. The air they get needs to be free of unwanted particulate.

Air Quality Problems. Air quality is often measured as dust count. An air quality index is found by measuring the number of particles in a cubic inch of air. Over large cities one cubic inch of air may contain as many as five million particles. Air from areas far removed from city life, such as over oceans, may have as few as 15,000 particles per cubic inch. The normal level is about 100,000 particles per cubic inch. Living things need air that is as free as possible of dust, smoke, and other particulate. Fires, factories, and engines of automobiles and trucks release substances into the air. Incinerators (places where trash is burned) may release dangerous materials. For example, the batteries used in flashlights and other small items are made using cadmium, a heavy metal that can damage human life. If batteries are burned, the cadmium may be released into the air.

Maintaining Air Quality. It is obvious that the way to maintain good air is not to release bad materials into it. This is impossible. Automobiles and trucks are needed to transport people and products. Factories are needed to produce things people want. Using air and releasing unwanted

3-11. Even a small fire produces smoke that damages the air.

wastes can't be stopped. In years past, air quality was sometimes ignored. Today, it isn't.

The goal is to avoid releasing anything that damages air quality. Devices have been installed on engines to reduce emissions (wastes into the air). Factories use procedures to "clean" the air before it is released. Open burning of waste products can be stopped in many cases.

All people have a role in keeping good quality air. Air can be maintained and renewed. Some cities have cleaned the air up by controlling smoke and cutting back other sources of pollution.

NON-RENEWABLE NATURAL RESOURCES

Some resources can't be renewed. Once used, they are gone. These are known as ***non-renewable natural resources***. Sometimes the word exhaustible is used to describe resources that can't be replaced.

People must use non-renewable natural resources wisely because there won't be any more! Conserving them is often not difficult.

The major non-renewable resources are those that provide fuel and certain minerals.

Fuel

Fuel refers to any material that provides energy. People use fuel to heat and cool their homes, cook food, power engines, make electricity, and operate machinery.

The major fuels are those from natural sources, such as coal, oil, and natural gas. Artificial fuels are sometimes made from corn and other products. These are based on alcohol. Gasohol is a mix of gasoline and alcohol.

A major energy crisis suddenly hit the world in the early 1970s. Costs of fuel jumped. U.S. motorists often waited in long lines to get a few gallons of gasoline. Some of the crisis went away when more fuel was produced and engines were made that used less fuel. People also were more careful in how they used it.

Natural fuels are of three major kinds: solid, liquid, and gas. More than 95% of all energy produced is from one of these.

Solid Fuel. Solid fuels include coal, peat, and wood. All of us are familiar with coal—the black substance formed over many years through the rotting of plants without air. Some people feel that coal is a dirty fuel. It is low in cost and is widely available throughout the world.

3-12. The gasoline used in an automobile is a non-renewable liquid fuel.

Wood is used for energy in cooking and heating. Some parts of the world have very little wood. Trees are taken and burned to cook food when they are quite small.

Liquid Fuel. Petroleum is the major liquid fuel. It is pumped from the earth as oil. A number of different products are made from petroleum, including gasoline, diesel fuel, engine oil, and lubricating grease. It is the major fuel used to power automobiles, farm tractors, and other engines.

Gas Fuel. Gas fuel is primarily natural gas, a form of methane. It is taken from the earth through wells, much like petroleum. Other gas fuels include butane and propane, which become liquids under pressure. Gas fuels are used to heat homes and businesses, manufacture products, and run some engines, such as those on harvesters.

Minerals

Minerals naturally occur in the earth. They are used in making many different products. We couldn't do without them!

Examples of minerals include iron, zinc, copper, lead, and magnesium. These are used to make many of the products we use in our daily lives.

Minerals are often dug (mined) from the earth. They must usually be refined before a final product is made. For example, iron must be smelted (melted) and cast (made into a certain shape). High temperatures are needed to melt iron. Other kinds of minerals may be mixed with iron to give it better qualities. For example, stainless steel is made by adding about 12% chromium (another mineral) to molten iron. Iron tends to be brittle (not bendable), while steel has more flexibility.

Many minerals are re-used, and this cuts down on the rate of depletion. All of us are familiar with automobile junk yards. The old cars are eventually melted down and made into new products. This helps to conserve minerals.

3-13. People like to live in clean communities such as at Rainy Lake near International Falls, Minnesota. (Courtesy, Mollee Thomas, Danville, Illinois)

ENVIRONMENTAL POLLUTION

Environmental pollution is made up of all the ways that people contaminate their environment. When something is polluted it is unclean. Some people say that pollution is the most serious problem facing humans today! Who wants to breath dirty air or drink bad water?

Four major resources are polluted: air, water, soil, and aesthetic (natural landscape beauty). In addition, there are other pollutants which can be serious. These include acid rain, nuclear accidents that release dangerous radiation, noise, and odors.

Humans cause most pollution. They are destroying the world they live in. If this is to be stopped, people must take important steps that begin with knowing the major sources of pollution.

Pollution is often described as from two kinds of sources: point and nonpoint. ***Point source pollution*** is from readily identifiable sources, such as a factory. ***Nonpoint source pollution*** is from sources that cannot be readily identified, such as all engines that release exhaust into the air. Pollution comes from several sources. Five are included here: toxic products, soil loss, waste products, discharged water, and junk and litter.

TOXIC PRODUCTS

Toxic products are substances that contain poison or have the potential to poison animals and plants. They can ruin the air, water, and soil.

Though not made to be a poison, many chemical products have characteristics that make them poisonous. Even helpful products can be poisons when not properly used.

Toxicology

The study of the effects of poisons on living things is toxicology. Toxicologists are people who study how these substances effect living things. All people are exposed to toxins because they are found in homes, workplaces, and outdoors. People are often unable to escape from toxins.

The effects of toxins are classified as acute or chronic. Acute effects appear immediately after exposure. They are associated with short-term exposure to rather high amounts of a toxin. The effects are often severe for a short time and then are gone. Chronic effects are long-term and may appear long after first exposure. Chronic exposure is often associated with human cancer, birth defects, and related problems.

Sources of Toxins

Any substance that has the potential of poisoning some part of the environment is a toxin source. Many products are used for non-toxic purposes but become toxic in the environment. Proper use of all materials is essential.

Major sources of toxins are:

- **Exhaust from engines**—Gasoline and diesel engines in cars and trucks burn fuel and release smoke into the air. This smoke contains harmful substances. Efforts to control engine exhaust include using emission control devices to capture particulate in the smoke and using fuel that doesn't have harmful substances, such as lead.

AgriScience Connection

DANGER UNDERGROUND

Two pipes coming out of the ground similar to those in this photograph are a signal of potential danger! They are connected to a fuel tank for a home heating system that is no longer used. One of the pipes vents the tank; the other is the pipe for filling the tank when it gets empty.

The fuel remaining in the tank, if any, is dangerous and is also a source of soil and water pollution. Tanks rust and leak over a number of years. What will happen to any fuel that has been left in the tank? Soil pollution as well as the danger of an explosion are major concerns.

Proper handling of the situation is needed. All fuel should be removed from the tank. The tank should be dug up and properly disposed of. The two pipes will no longer be in the ground nor in the way of mowing or other activities.

- **Burning materials**—Fires give off smoke. The smoke from fires often carries some of the substances from the materials that were burned into the air. Smoke can contain very toxic substances. One way of looking at smoke is that it is made of what is being burned. Open fires are not allowed in most places today. Incinerators (devices for burning materials) should be properly constructed to prevent the release of toxins into the air.
- **Agricultural chemicals**—Many kinds of chemicals are used in agriculture. Some have the potential of giving off toxins if not used properly. The important principle is to use agricultural chemicals according to directions and never use them when they aren't needed.

3-14. Citizens can help keep their communities clean by organizing litter removal programs.

Toxins Where People Work

Not only do toxins get into the air, water, soil, and elsewhere and affect people, they can be particular problems in workplaces. Many agricultural workers use or are present when others use chemicals. It is estimated that nearly four million workers in the United States are exposed directly or indirectly to pesticides every year.

The Environmental Protection Agency has developed guidelines, known as the Worker Protection Standards for Agricultural Pesticides (WPSAP). These guidelines are intended to reduce illness or injury from pesticides. Workers must receive information on the dangers of the pesticide they are applying. Protective clothing must be available. Training must be provided in safe pesticide use as well as how to handle emergencies.

Dangers are small when the pesticides are used properly. Individuals are protected and pollution is kept to a minimum.

SOIL LOSS

Wind and water can blow or wash soil away. In addition to the loss of fertility of the land, the soil particles get into the air, streams, lakes, and other water and cause pollution. The air may appear hazy and the water may have a muddy (brown) color.

Living things in the water may be damaged by muddy water. Natural processes in the water are disturbed. Oxygen levels may be too low to

3-15. Soil erosion results in land that is not productive. This photograph shows severe erosion due to water. The gully has been washed because of improper land management.

support aquatic life. Use of the water by industries or by cities may not be possible. The soil particles may float down streams to the ocean where they change the environment of sea life.

WASTE PRODUCTS

Waste products are of many kinds: garbage from homes or restaurants, left-over materials at factories, and crop remains and livestock wastes on farms. Livestock produce manure and other wastes, including an occasional dead animal to dispose of.

The major wastes made by people are paper, glass, aluminum and tin cans, food wastes, and yard wastes, such as grass clippings and leaves. The average person in the United States creates over 1,300 pounds (590 kg) of waste a year. Just think of all the paper left over from a lunch of hamburgers and french fries!

Agriculture creates some waste products. Empty feed sacks, plastic pesticide containers, used engine oil, and worn out tires need to be disposed of. Many of these produce toxic substances if not disposed of properly.

DISCHARGED WATER

Factories often use water in making products. Farms use water for irrigation, fish culture, cleaning barns, and other uses. In nearly all cases, some water may be discharged unused or after it has been used. This discharged water is known as ***effluent***.

Effluent can pollute the environment, particularly streams, lakes, and oceans. In the mid-1900s, many streams had a lot of water dumped into them. This water killed the natural life and changed the stream or lake.

Laws to prevent the release of dangerous effluent were passed. Today, many of the streams and lakes are naturally clearing up and losing the toxins.

JUNK AND LITTER

People discard all kinds of things: used cars, televisions, candy wrappers, soft drink cans, and old tires. Farms often have empty seed bags, worn out batteries, old equipment, and broken-down barns with scraps of lumber and roofing lying around. All of these pollute! They all damage the environment one way or another.

Litter along the roads and streets has been a target for years. No one likes to use a road littered with what someone else threw down. All people need to help keep the world an attractive place to live.

Old machinery and other junk left out in the weather is an eyesore. The machinery is no longer used. It detracts from the natural beauty and the value of the land is less. Old machinery and junk is often dangerous. The danger ranges from injuring a foot stepping on a nail to being hit by a falling piece of rusty machinery when walking by. Livestock can also be injured. A lot of things can be reused. Not only does it make the environment more attractive, reusing saves our natural resources.

DISPOSING OF WASTES

Waste disposal is the process of getting rid of wastes. In some form, nearly anything can be a waste. The uneaten bite of a hamburger and the last drops of cola that remain in a cup are wastes. We often don't view them that way, but we should when we leave them.

One way of reducing wastes is to use fewer things that are wasted. Rather than wrapping the food, putting it in a box, and then in a bag, just wrap or box it before putting it in a bag. Regardless, the modern way of life in much of the world involves producing a lot of waste.

FORMS OF WASTES

Wastes may be gases, liquids, or solids. Each form creates unique problems for disposal.

Some gases are very hazardous. Gases include carbon dioxide (given off by animals), carbon monoxide (produced by auto and tractor engines), and methane (given off as plant and animal matter rots).

Liquids are often produced in agricultural industry. Water runoff and effluent are two important examples. City sewage treatment facilities handle most residential and business liquid wastes.

Solid wastes include paper, plastic, food scraps, cans, manure, grass clippings, dead animals, and the unused stems and leaves of plants. These wastes can smell bad, promote the growth of rats and roaches, and harbor diseases. In addition, they are ugly to see along the road, in parks, and other places.

METHODS OF WASTE DISPOSAL

Disposing of wastes requires a big effort. People often don't realize how much waste is created and the cost of properly disposing of it.

Three methods are commonly used to dispose of wastes: landfills, incineration, and recycling.

Landfills

Landfills are large earthen pits for waste disposal. The are carefully designed to protect the environment from pollution. About 80% of the waste in the United States is disposed of in landfills.

Landfills are located on carefully selected sites. They are made by digging large holes in the earth. The earth should have a high clay content so that the wastes can't seep into the underground water. Some landfills put large plastic liners inside the earthen pit. The wastes are dumped on top of the liner or clay bottom, packed with a bulldozer, and covered with earth.

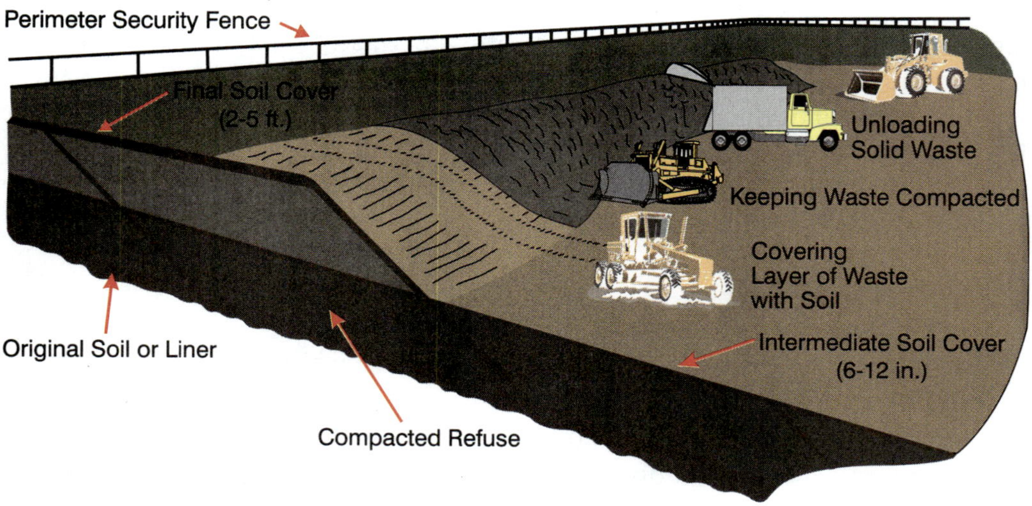

3-16. Diagram of a landfill.

Rotting (decomposition) takes a while. Some materials decompose easily; others break down slowly. Since landfills are packed and little air is present, materials are often slower to rot. Poultry farmers often use special pits dug in the ground for disposing of any chickens that die on their farms. These are known as composting pits. They speed rotting of the dead chickens.

The term ***biodegradable*** is used to describe materials that will break down from the action of bacteria. Some things are biodegradable, such as newspaper, while other things aren't, such as glass. New ways of manufacturing some products encourage rotting. For example, adding corn starch to some plastic materials will speed rotting.

Incineration

Some materials are burned in special structures known as incinerators. Smoke containing gases and particulate may be released into the air. These may settle to the earth and harm plants, animals and structures, such as the paint on farm buildings.

Most incinerators are equipped to prevent releasing harmful smoke into the air. Pollution control devices catch harmful materials. Only about 9% of the waste in the United States is incinerated.

Recycling

Recycling involves recovering and reusing materials instead of throwing them away. It conserves raw materials and reduces pollution. Interest in recycling has increased in recent years because of growing concern over protecting the environment. The hamburger you had today might have been in a recycled bag! (Look for the symbol that it was made of recycled paper.)

3-17. A recycling collection station. This portable recycling trailer is parked in a convenient location in Maryville, Tennessee, for citizens to sort and deposit materials for recycling.

3-18. The code on the bottom of plastic containers and on bags helps in sorting them for recycling. This shows the symbol and code on the bottom of a plastic milk container (top) and on a bag (bottom).

Common products that are recycled include paper, aluminum and tin cans, and glass. Other items recycled include motor oil, auto batteries, auto tires, plastic, and old cars. Farmers frequently save empty pesticide containers for recycling and proper disposal. Just about anything can be recycled in one way or another.

Many communities have recycling programs to help reduce pollution. Drop-off centers may be set up where people can take their trash. Some areas have curbside pickup of materials for recycling. Paper, plastic, glass, and cans must usually be sorted. Even different kinds of plastic need to be sorted. Mixed materials present big problems at recycling plants.

PREVENTING AGRICULTURE POLLUTION

Agriculture, more than any other area of human activity, uses the natural resources of the earth. In order to provide food, fiber, and shelter, agriscientists have developed new ways of doing things. Most of the time these developments have little bad effect on the environment. They do not pollute! In some cases and when they aren't used properly, they can cause pollution. Agriscientists want to prevent damage to our natural resources.

Agriculture pollution is any pollution caused by agriculture activities. It can be on farms, in greenhouses, at processing plants, or many other places.

SOURCES OF AGRICULTURE POLLUTION

Natural resources are threatened every time the soil is plowed, an animal is raised, and other work takes place. Knowing about possible sources of pollution can help prevent problems.

Several possible sources of agriculture pollution are:

- **Introducing pests**—Unwanted and damaging pests can be brought into a country or location on seed, plants, fruit, and other products. These pests can hide in the soil of a potted plant, on seed, in harvested crops, and other places. The producers of such things must rid them of any sources of problems. Many products have to be inspected before shipment. For example, papaya that have not been inspected cannot be brought from Hawaii to the U.S. mainland.
- **Exotic plants or animals**—People sometimes grow plants and animals that aren't native to a place. These plants and animals may be restricted. Their release into the wild is prohibited because of the damage they might cause. For example, fish and plants kept in aquariums as pets may damage the environment if released. Some natural lakes and streams have been damaged by people dumping pet plants and fish into them. They may take over the water and kill off the other plants and animals.
- **Chemicals**—Some people are quick to blame agricultural chemicals for a lot of pollution. This is unfortunate, because many of these chemicals cause little or no damage if used properly.
- **Waste water**—Excess water from irrigation, fish ponds, food processing plants, and other places can damage natural lakes and streams. Water should be treated before release if it could damage natural water in streams and lakes.
- **Processing wastes**—Preparing food products for people produces waste. Animal processing produces wastes that cannot be run into streams. Proper disposal of chicken feathers, internal parts of all animals, and other wastes is essential. Vegetables, fruit, timber, and other plants also produce various wastes. These include pods, hulls, sawdust, and tree bark.
- **New life forms**—Advanced science is producing new non-natural forms of plants and animals. These are considered threats by some people. They feel that farming them places the environment in danger. They fear that the plant or animal will "get loose" in nature and cause problems.

STEPS FOR PREVENTING AGRICULTURE POLLUTION

A few simple steps can prevent a lot of pollution. It is largely a matter of properly using agriculture technology.

Here are several important things to know:

- **Regulations**—Problems can be prevented by knowing the regulations that apply to using a chemical. It is essential to know how something is to be used. People must have training in how to use restricted materials, such as pesticides.
- **Follow rules**—Instructions for using most chemicals are printed on labels on the container for the material. Read the instructions and follow them!

3-19. Junk around a farm or agribusiness pollutes the environment and wastes non-renewable natural resources. Discarded items in this photograph could be recycled. Cleaning up would definitely make the scenery better!

- **Storage**—All materials that could cause problems should be properly handled and stored. Buy only the amount you need and keep it so that the containers aren't broken open.
- **Waste control**—Some waste occurs with about all chemical usage, such as when a sprayer is washed. How the sprayer is cleaned and what is done with the water is important. Never let wash water flow into streams, ponds or other places that could be damaged.
- **Use properly**—Use all materials in approved ways. Do not apply a chemical where it is prohibited or in rates greater than recommended. Do not let it get into water or drift into places where it shouldn't be.
- **Dispose of containers**—Chemicals often come in bags, cans, or plastic jugs. These should be disposed of properly. Some companies collect the containers their products were in back from farmers. Never throw containers into streams, woods, or pastures and never leave them in the field.
- **Safety**—Always practice personal safety. Dress properly. Use gloves, hats, eye protection, and air masks (respirators) as needed. Wash the skin if dangerous materials get on it. Know how to handle an emergency in case of an accident.

GOOD OUTWEIGHS BAD

Agriculture technology is good but its misuse can cause problems. Without new, improved ways of producing food, fiber, and shelter, we would not have a high standard of living. Crops and livestock would be lost to pests and disease. The amount produced would be far below what is needed. More people would go hungry and supermarkets would not have much food. People would not have many choices in clothing. Housing would likely be in short supply. The proper use of new methods of providing for the needs of people is essential. The technology isn't bad; its misuse is bad! (More information about the environment is in *Environmental Science and Technology* published by Interstate Publishers, Inc., Danville, Illinois.)

REVIEWING

MAIN IDEAS

The world has certain natural resources that are important to human life. These resources are a part of the environment, or all of the things that affect the life of living things.

Plants and animals live in the biosphere, or area of the earth that supports life. Within the biosphere, there are ecosystems or all of the parts that form an environment for a plant or animal. The ecosystem includes biotic (living) and abiotic (nonliving) factors. Plants and animals have habitats in which they live. Habitats vary for different plants and animals. Every living thing has a particular way of living, known as its niche. Plants and animals use different foods. They are all a part of the food chain.

Some resources are renewable; others aren't. The major renewable resources include soil, water, wildlife and fish, forests, and air. Non-renewable natural resources are gone once they are used. Examples include petroleum and minerals.

Sustaining natural resources has become a major goal. This involves using them in such a way that resources remain for future generations.

Environmental pollution is poisoning the environment. Many toxic substances are released by factories, homes, and farms. Using the materials according to recommendations will reduce threats to the environment.

Pollution can be reduced by properly disposing of wastes. Landfills, incinerators, and recycling are used to reduce pollution.

People have produced many inventions to help make life the way they want it. They have also produced things that harm the environment they live in. Steps must be taken to maintain the quality of the environment. Knowing and following safe practices will prevent much pollution.

QUESTIONS

Answer the following questions using complete sentences and correct spelling.

1. What is the environment? What are natural resources? How are they related?
2. What is the biosphere?

3. What is an ecosystem? Distinguish between factors in an ecosystem.
4. How does an ecosystem work? Describe habitat, niche, and food chain as a part of the ecosystem.
5. What is sustaining natural resources?
6. Distinguish between renewable and non-renewable natural resources.
7. What are the most important renewable natural resources?
8. What is the importance of the following renewable natural resources: soil, water, wildlife and fish, forests, and air?
9. What are the major kinds of non-renewable natural resources?
10. What is environmental pollution?
11. What are five important sources of pollution? Explain each.
12. What are the kinds of waste products?
13. What methods of waste disposal are used? Briefly describe each.
14. What are possible sources of agriculture pollution?
15. What can people do to reduce agriculture pollution?

EXPLORING

1. Survey your community to determine the waste disposal methods that are used. Contact the local solid waste disposal office, landfill, or government office for information.

2. How do people recycle in your community? Identify the different locations for depositing wastes to be recycled. Determine how the materials are sorted and prepared for recycling.

3. Make a survey of the water used by agriculture in your community. Determine sources of the water and its major uses, such as for drinking by livestock, cleaning barns, irrigating crops, and raising fish.

4. Make a list of the wildlife that are found in your local area. Determine if there are laws about hunting these animals. Talk to a local game conservation officer (game warden) to get the details.

4

USING THE SCIENCE OF COMPUTATION

Computation—a big, fancy word that is frightening! It shouldn't be frightening; it begins simple and gets more exciting.

As presented in Chapter 2, observations are important in agriscience. People must be able to observe and record accurately. The science of computation allows things to be measured and given values.

Suppose someone worked all day and there was no way to measure how long he or she worked. People want to know exactly how long they have worked so they know how much money they will be paid.

Or suppose a farmer produced a crop and hauled it to be sold and there was no way to measure it. How would its value be determined? The farmer wouldn't want to be paid less than it was worth.

Everything done in agriscience involves measuring and using numbers in one way or another. Some are simple and easy to do; others are more advanced.

4-1. Many measurements and computations are used in forestry.

OBJECTIVES

This chapter introduces the use of measures and numbers in agriscience. It has the following objectives:

1. Describe two important areas of computation in agriscience.
2. Explain the importance of standard measurements in agriculture.
3. Explain and give examples of measurement systems.
4. Name common measurements used in agriscience.
5. Define temperature and explain how it is measured.
6. Define linear measurement and explain how it is used.
7. Define area and explain how it is measured.
8. Define volume and explain how it is measured.
9. Define weight and explain how it is measured.
10. Use conversion formulas.
11. Calculate important measurement information.

TERMS

area
avoirdupois weight system
board foot
caliper
Celsius scale
chain
circle
cone
cord
cubic unit
customary measurement system
cylinder
Fahrenheit scale
linear measurement
measurement
measuring stick
metric system
parts per billion
parts per million
parts per thousand
parts per unit
rectangle
square
square unit
tape measure
temperature
thickness scale
trapezoid
triangle
volume
weighing scale
weight

COMPUTATION IN AGRISCIENCE

All areas of agriscience involve the use of computation methods. Some of the uses are simple and easy to do. Others are more involved. All typically focus on common problems facing the agriscientist.

Most agriscience computations involve two areas: measurement and mathematics. Numbers are used in both.

MEASUREMENT

Measurement involves determining the number of units in something. A unit is a single quantity that represents one complete amount of something. Units are the same size in all situations. For example, a foot is always 12 inches long and a bushel of shelled corn always weighs 56 pounds! (This example is based on the customary or English system traditionally used in the U.S.)

Different units of measurement are used. Units must often be converted from one to another. These are usually easily done. (Information in this chapter will explain many kinds of units and describe how common conversions are made.)

Measurement involves numbers. Units can be counted and divided into parts. Measurement has meaning only when a number is used with the unit. For example, we can relate to a board that is 12 feet long. Further, we know that it can be cut into two boards that are six feet long!

Uniform instruments are used to make measurements. They must provide the same value for the same quantity every time they are used. If they don't, they aren't good at measuring. All of us are familiar with scales that don't give an accurate weight every time we use them. They are no good!

MATHEMATICS

In agriscience, mathematics has two areas: skills and applications. The skills are needed to solve everyday problems. The problems involve the application of mathematics skills.

People often want to know which areas of mathematics are important. Research was used to get this information for agriscience. The findings of the research are given here.

The most important mathematics skills in agriscience are:

- **Adding numbers**—A simple example is: How many hogs do you have in two pens if one pen has 25 hogs and the other has 30? (Answer: 25 + 30 = 55 hogs in the two pens.) Addition is not always this easy, but it is done the same way.
- **Subtracting numbers**—Another simple example: If you had 15 horses and sold 6, how many would you have left? (Answer: 15 – 6 = 9 horses would be left.)
- **Multiplying numbers**—For example, if you had 12 calves and sold them for $200 each, how much money would you get? (Answer: 12 × $200 = $2,400.)
- **Dividing numbers**—For example, if you had 100 flowers and needed to make four equal arrangements, how many flowers would you put in each arrangement? (Answer: 100 ÷ 4 = 25 flowers.)

These skills are used in many areas of agriscience. However, their use often involves examples that aren't as simple as those used here.

The most important applications in agriscience are:

- **Calculating the area of squares, rectangles, circles, and other shapes**—This involves multiplying numbers.
- **Calculating the volume of cubes, cylinders, and other shapes**—This involves multiplying numbers.
- **Converting from one system of measurement to another**—This involves either dividing or multiplying, depending on what is being done.
- **Calculating board feet**—This involves both multiplying and dividing.
- **Calculating ratios and proportions**—This involves multiplying and dividing numbers.
- **Calculating profit and percentages**—These may involve using all four mathematics skills. They are more often connected with the business skills in agriculture.

4-2. Determining the amount of feed needed for the cattle in this California feedlot requires considerable mathematics.

Some of these applications sound complicated. And they can be! Regardless, all of them involve using the mathematics skills listed above. The examples show that the applications aren't that hard to do.

NEED FOR STANDARD MEASUREMENTS

Measurements need to be the same in all situations. They need to be standard. A standard is a model or example. Standards are set up as rules to follow in making measurements.

Here are three reasons why standard measurements are needed:

- Buyers and sellers must have the same standard when trading. For example, the scales used to weigh cattle that are being sold must always accurately weigh the cattle. If not, either the buyer or seller will be cheated.
- Persons mixing chemicals need to use the same standard that is given in the mixing instructions. If the standard measurements aren't followed, the chemical would be mixed wrong. This could result in a product that injures plants or animals.
- People who make products need to use the same standards. A good example is making lumber. All pieces need to be exactly the same size every time. Further, different mills must also use the same sizes.

MEASUREMENT SYSTEMS

Two kinds of measurement systems are used in agriscience: the customary or English system and the metric system. The customary system is widely used in the United States and Canada in non-technical areas of agriscience. The rest of the world primarily uses the metric system.

CUSTOMARY SYSTEM

The ***customary measurement system*** is used in every day life. Many of the measurements began in England hundreds of years ago. Agricultural measurements are often made in the customary system. However, measurements involving science are in the metric system.

Common measurements in the customary system are: inch, foot, yard, mile, ounce, pound, quart, pint, gallon, and acre.

The customary system is standardized so that a certain measurement is always the same amount. The standards grew out of the need for being consistent. For example, the inch was said to be equal to the length of three barleycorns when they are placed end-to-end. Another example is the length of one military marching step. A step is 30 inches.

Some of the measures used in agriculture aren't common in daily life; others are. Many crops are packaged in standard-sized containers. These containers are the size best suited to the particular crop. The U.S. Department of Agriculture publishes lists of container sizes. This makes it easy for people who work with the crops to keep track of the amount produced. Table 4-1 shows some of the most common measures used in agriculture.

Table 4-1
Measures and Weights of Selected Agricultural Commodities

Commodity	Unit	Customary Weight
		(pounds)
Alfalfa seed	bushel	60.00
Apples	northwest box	44.00
Apricots	lug	24.00
Asparagus	crate	30.00
Barley	bushel	48.00
Brussels sprouts	drums	25.00
Cabbage	open-mesh bag	50.00
Celery	crate	60.00
Coffee	bag	132.30
Corn		
ear (husked)	bushel	70.00
shelled	bushel	56.00
oil	gallon	7.70
Cotton (lint, no seed)	bale (gross)	500.00
Cream	gallon	8.38
Honey	gallon	11.84
Maple syrup	gallon	11.02
Milk	gallon	8.60
Oats	bushel	32.00
Onions (dry)	sack	50.00
Peanuts (unshelled)	bushel	17.00
Rice	bushel	45.00
Soybeans	bushel	58.00
Tobacco (Maryland)	hogshead	775.00
Tomatoes	crate	60.00
Wheat	bushel	61.00

Source: U.S. Department of Agriculture.

METRIC SYSTEM

4-3. A Haga Altimeter can be used to make height and slope measurements.

The ***metric system*** is used by all scientists around the world. Its official name is the International System of Units, often known as SI because of its French origin. The metric system is increasingly used in the United States. In 1975, Congress passed a law calling for the United States to change to the metric system. The change has been slow in agriculture.

Common measurements in the metric system are grams, meters, liters, and seconds.

The metric system is easy to use. It is based on a decimal system that increases or decreases numbers by 10s. In one sense, it is similar to the money system used in the United States and Canada.

The metric system uses prefixes to decrease or increase numbers. The prefix is added before the unit (measure) being used, such as "centi" is added to "meter" to form centimeter (one-hundredth of a meter). The common prefixes are shown in Table 4-2.

Table 4-2
Prefixes Used in the Metric System

Prefix	Symbol	Factor by which multiplied
Giga	G	1,000,000,000.0 (one billion)
Mega	M	1,000,000.0 (one million)
Kilo	k	1,000.0 (one thousand)
Hecto	h	100.0 (one hundred)
Deka	da	10.0 (ten)
Deci	d	0.1 (one-tenth)
Centi	c	0.01 (one-hundredth)
Milli	m	0.001 (one-thousandth)
Micro	µ	0.000001 (one-millionth)
Nana	u	0.000000001 (one-billionth)

MEASUREMENTS USED IN AGRISCIENCE

Several kinds of measurements are needed in agriscience. These may be made in both the customary and metric systems. More emphasis is being given to the metric system measurements, however. Common measurements are discussed here.

LINEAR

Linear measurement is the distance between two points. It is also known as length. For example, the length of a board is the distance from one end to the other. The board's length is a linear measurement.

Length measurements are often needed to determine other measures of an object, such as area and volume. Length measurements may be made in the customary or metric system. Table 4-3 shows common examples of linear measurements.

Length measurements can be converted from one system to another, as shown in Table 4-4.

AREA

Area is the measurement of surfaces. Examples include land, floor space in a greenhouse, and the top of a bench in an agriscience laboratory. Linear measurements must be made in order to determine area. It is reported in square units, such as square feet (ft^2) or square meters (m^2).

VOLUME

Volume is the total size of an object. Length and area are not enough. Volume is the amount of space something takes or holds. It is reported in cubic units, such as cubic feet ($ft.^3$) or cubic meters (m^3).

WEIGHT

Weight is the heaviness of something. It is related to gravity which is the pull of the earth. Weight is often measured using a scale, which is a machine for weighing. Weight may be measured using the customary system (pounds [lb]) or metric system (kilograms [kg]).

Table 4-3
Common Linear Measurements

Customary System:

1 foot (ft.)	=	12 inches (in.)
1 yard (yd.)	=	3 feet or 36 inches
1 mile (mi.)	=	8 furlongs or 1,760 yards or 5,280 feet

Metric System:
(Note: The meter is the basic unit of length. Measuring 39.37 inches long, the meter is slightly longer than a yard.)

1 millimeter (mm)	=	0.001 meter
1 centimeter (cm)	=	0.01 meter
1 kilometer (km)	=	1,000 meters

Table 4-4
Linear Measurement Equivalents and Conversions

Equivalents

Customary:		Metric:
1 inch	=	2.54 centimeters
1 foot	=	30.48 centimeters
1 yard	=	0.9144 meter
1 mile	=	1.609 kilometers or 1,609 meters

Converting Customary to Metric

If you know the length in:	Multiply by:	To find:
Inches	2.54	centimeters
Feet	30	centimeters
Yards	0.9	meters
Miles	1.6	kilometers

Converting Metric to Customary

If you know the length in:	Multiply by:	To find:
Centimeters	0.4	inches
Meters	1.1	yards
Kilometers	0.6	miles

AgriScience Connection

ACCURATE WEIGHT MEASUREMENT OF FEED

Large scales are used to weigh semi-trailer trucks loaded with feed. Here is how the weight of the feed is determined: The entire truck is driven onto the scales empty. The weight is recorded by a scale operator inside the office of the feed mill. After the truck has been loaded, it is weighed again. The weight of the empty truck is subtracted from the loaded weight to determine the weight of the feed.

If a truck and semi-trailer weigh 52,000 pounds when loaded and 12,000 pounds when empty, how many tons of feed does it haul? What would be the value of feed on the truck at $300 per ton? (Answers: The truck is loaded with 20 tons of feed valued at $6,000.)

TEMPERATURE

Temperature refers to how hot or cold something is. It is measured with a thermometer.

Two temperature scales are commonly used: ***Fahrenheit scale*** (F) and ***Celsius scale*** (C). The scales can be compared using water at sea level. Fahrenheit is the customary measurement of temperature, in which water boils at 212°F and freezes at 32°F. Celsius is the metric system of temperature measurement. On the Celsius scale, water boils at 100°C and freezes at 0°C. The Celsius scale is used in agriscience laboratories.

Temperature can be easily converted from one system to another, as shown in Figure 4-4.

TIME

Time was first measured in days, number of days between full moons, and the seasons of the year. Clocks were later developed that used hours and minutes. Each day is comprised of 24 hours, each hour made up of 60 minutes, and each minute of 60 seconds. This system is based on the rotation of the earth. A more precise measurement of time was developed in 1967 as part of SI. The newer system is not widely used in agriscience.

Converting Fahrenheit (F) to Celsius (C)

Formula: $C = {}^5/_9(F - 32)$

Steps:
1. Determine the F temperature.
2. Subtract 32 from the F temperature.
3. Multiply the difference found in step 2 by 5.
4. Divide the amount found in step 3 by 9.

(The result is the equivalent temperature in degrees C.)

Example: If the temperature is 100°F, what is the C temperature?

$100 - 32 = 68$
$68 \times 5 = 340$
$340 \div 9 = 37.7°C$

Converting Celsius (C) to Farenheit (F)

Formula: $F = {}^9/_5C + 32$

Steps:
1. Determine the C temperature.
2. Multiply the C temperature by 9.
3. Divide the amount found in step 2 by 5.
4. Add 32 to the amount found in step 3.

(The result is the equivalent temperature in degrees F.)

Example: If the temperature is 40°C, what is the F temperature?

$40 \times 9 = 360$
$360 \div 5 = 72$
$72 + 32 = 104°F$

4-4. Making temperature conversions involves using simple formulas.

DETERMINING LINEAR MEASUREMENTS

Making linear measurements is determining the number of units from one point to another. A standardized scale of units is used. The scale may be in the customary or metric system.

Care is needed in order to make good measurements. The instrument used to make the measurements must be accurate. The person using the instrument must use it properly or errors will result.

KINDS OF INSTRUMENTS

A variety of measuring devices can be used. The most common are included here.

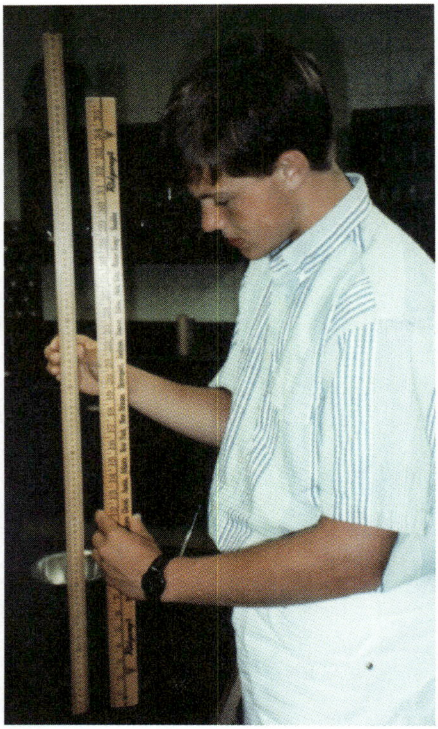

4-5. The meterstick is only about 3 inches longer than the yardstick.

Measuring Sticks

Graduated ***measuring sticks*** are typically made of wood, plastic, or metal. They are often the length of one unit and marked in smaller units. Common examples are the yardstick and meterstick. Yardsticks are 3 feet or 36 inches long. Metersticks are one meter long (39.37 inches or 100 cm).

In some cases, measuring sticks are 1 foot long and called rulers because they are used as straight edges when drawing lines.

One special kind of stick is the folding or zig-zag stick. It is used like a stick but folds for ease of use. These are typically used for measuring short distances of no more than a few feet or meters.

Accuracy requires being careful in placing and reading a measuring stick.

Tapes

Tape measures are flexible linear measuring devices made of metal or fabric. They may be in cases that the tape rolls into when not in use. Tapes may range in length from a few feet or meters up to 50 or more. Most tapes have a hook on the end to help hold it in place when a measurement is being made. Roll-up type tapes must be kept clean. Sand, grease or other substances can ruin the tape.

4-6. Tape being used to measure the circumference of a pine tree. Tree measurements are taken at 4.5 feet above the ground. This is known as *diameter breast high* or D.B.H.

Chains

Chains are typically used to measure land. They are made of lightweight chain or similar material. They are usually 66 feet long and may be marked in links, which are 7.92 inches long. Chains usually require two people to take measurements.

Calipers

Calipers are used for many important measurements. They have jaws or legs that are moved to determine the thickness, diameter, or distance between things.

Calipers are classified as inside or outside. Inside calipers measure the inside of tools or equipment. For example, measuring the diameter inside an engine cylinder involves inside calipers. Measuring the diameter of logs involves using outside calipers. Calipers can be adjusted to make very fine and accurate measurements.

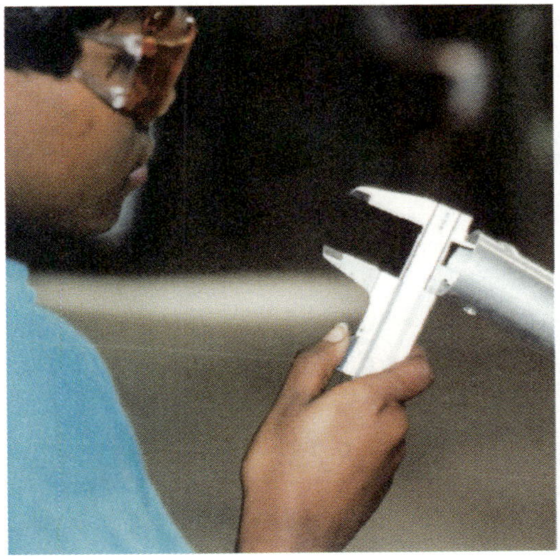

4-7. Caliper being used to measure the inside diameter of a pipe.

Remote Measurement

Remote measurement involves using aerial photographs or other information collected from a distance. With aerial photographs, planimeters are used to give measurements. A planimeter is a device that measures area by tracing the boundaries shown on the photograph. For example, the acreage in a field of corn can be determined using a planimeter.

Special Devices

Many scales for use in special situations have been made. One example is the ***thickness scale*** or gauge. It is used to measure very small distances and make settings. A common example is with engines to set the gap on spark plugs and in the points.

USE OF LENGTH MEASUREMENTS

Length is important in many areas of agriscience. People often measure land, trees, animals, and other items. Even the tiny fibers of cotton are measured for both length and thickness (diameter)! Here are four examples:

- **Marketing**—Measurements are used in marketing. Saw logs are sold after being measured for diameter and length. Grades of many foods depend on size. An example is the cucumber. Tiny, gherkin cucumbers are much more valuable than large ones that are sliced for dill pickles.
- **Planting rates**—Many crops must be properly spaced in the field. Fruit trees may need to be 20 to 40 feet apart. Seeds of corn may need to be planted 6 to 8 inches apart. Adjusting equipment for proper planting is important.
- **Chemical use**—Length measurements are often needed in figuring how much chemical to use. In some cases, the measurements are converted to area of surface to be treated.
- **Land measurements**—Land is often measured when it is sold or when size is important for other uses. Seeding rates and number of head of cattle that can be put in a pasture involve land measurement.

Length measurements are used to determine other measurements, such as area and volume.

DETERMINING AREA

Area is based on the length of the sides of an object or a space. Area is nothing more than the amount of space within linear measurements. For example, the area in a field of corn is based on the boundary of the field. Measurements are made of the corn field to get lengths and widths for use in calculating area.

The units used to measure a surface must be the same in order to determine the area. All sides should be measured in feet, inches, meters, or other units. In some cases, lengths can be converted to other units before the area is calculated.

Area is reported as square units. A ***square unit*** is one unit long and one unit wide of the linear measurement system that is being used. For example, if a flower bed is measured in feet, the area is reported as square feet.

The shape of the surface being measured determines how the area is calculated. Some surfaces are simple. They have regular length and width lines. Other surfaces are more difficult to calculate because the lengths and widths vary. The area of those with simple shapes is easier to determine.

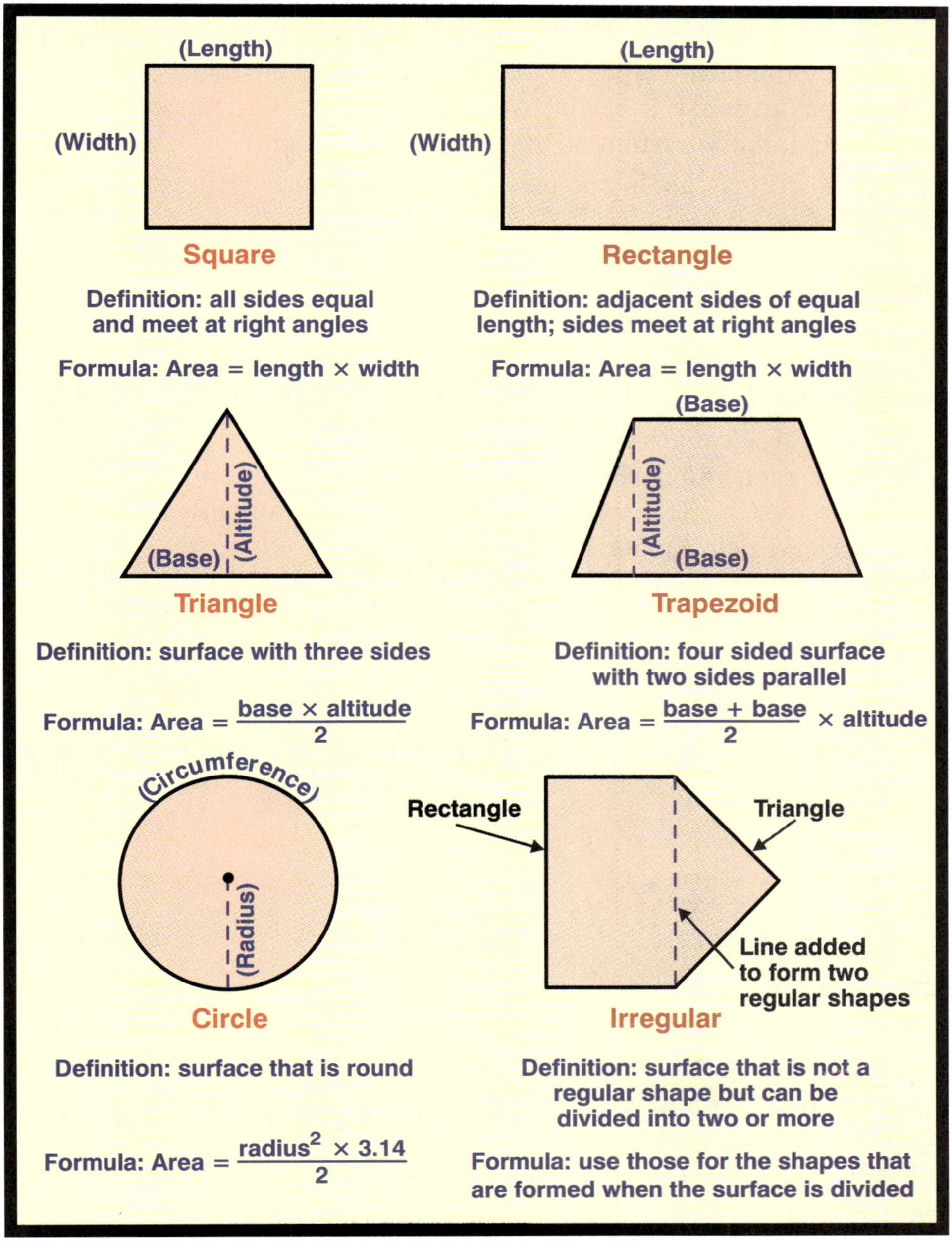

4-8. Common shapes and ways of determining area.

KINDS OF SHAPES

Agriscientists use areas with five shapes: square, rectangle, triangle, trapezoid, and circle. In some cases, surfaces must be divided into more than one shape in order to determine the area. Regardless of shape, area is the number of square units. A good example is an orchard that fits none of the standard shapes. The orchard, however, can be divided into several shapes to make it easier to figure the area. The measurements can be in the customary system or in the metric system.

The following symbols are used: a = area, l = length and w = width.

Square

A ***square*** is a surface in which all four sides are of equal length. The corners are "square" because they form right angles.

The area of a square is easily calculated by multiplying the length of two sides by each other. For example, the area of a sample of sod (lawn grass) that is 10 inches square is found by multiplying 10 inches × 10 inches, which equals 100 square inches.

Here is how it is done:

Formula: a = l × w

Example: The area of a square of sod that measures 10 in. is calculated as follows:

a = 10 in. × 10 in.

a = 100 in.^{2}

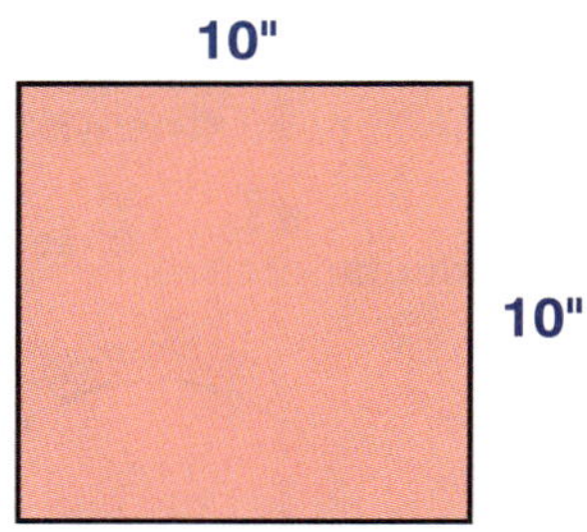

Square

Rectangle

A ***rectangle*** is a surface that is similar to a square but has an important difference: The two opposite sides are of equal length and the sides that touch are of unequal length. The longer sides are often referred to as *length* and the shorter sides as the *width*. The corners of rectangles meet at right angles.

The area of a rectangle is found by multiplying the length of a short side by the length of the long side. For example, a sheet of roofing that is 10 feet long and 3 feet wide is a rectangle. The area is as follows: 10 feet × 3 feet = 30 square feet.

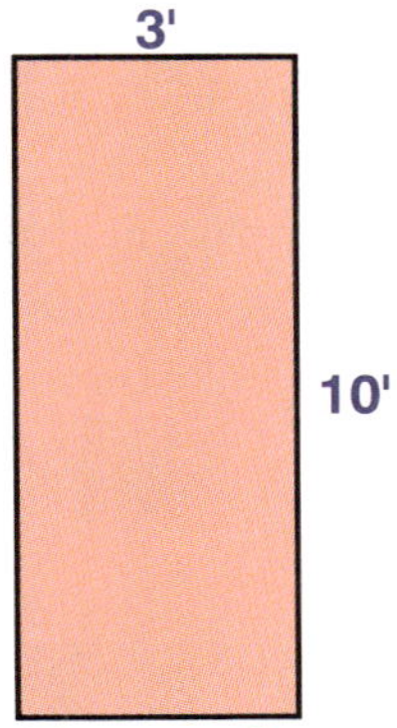

Rectangle

Here is how it is done:

Formula: $a = l \times w$

Example: The area of a rectangle that is 10 ft long and 3 ft wide:

$a = 10 \text{ ft} \times 3 \text{ ft}$

$a = 30 \text{ ft}^2$

Triangle

A ***triangle*** is a surface that has three sides. Though the sides may vary considerably, the area is figured the same way. One line of a triangle is known as the base (b). The point of greatest distance from the base is the height (h). Height is also known as altitude.

Figuring the area involves multiplying the length of the base by the height and dividing by 2.

Here is how it is done:

Formula: $a = \frac{b \times h}{2}$

Example: Find the area of a triangle with a base 20 cm and a height of 30 cm:

$a = \frac{20 \text{ cm} \times 30 \text{ cm}}{2} = \frac{600}{2} \text{cm}^2$

$a = 300 \text{ cm}^2$

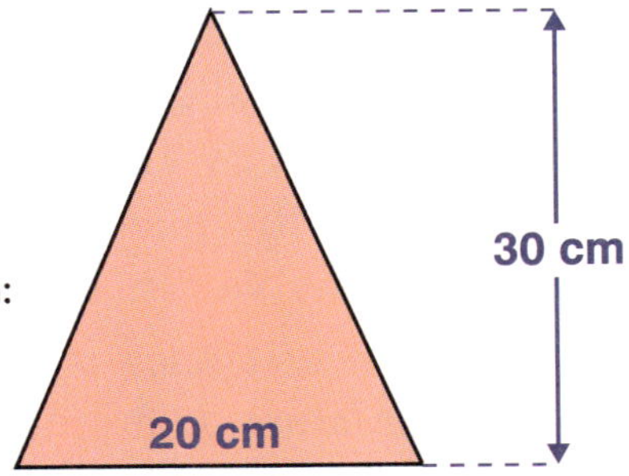

Triangle

Trapezoid

A ***trapezoid*** is a shape with four sides, two of which are parallel. The two parallel sides are known as bases. The distance between the sides is known as the altitude. The area of a trapezoid is found by multiplying the average length of the two bases by the height. For example, the area of a flower bed that is 4 feet wide and 10 feet long on the front and 14 feet long on the back would be 48 square feet. The first step is to find the average length of the bed by adding 10 and 14, which equals 24, and dividing by 2, which gives an average length of 12 feet. The average length is multiplied by 4.

Here is how it is done:

Formula: $a = \frac{b + b}{2} \times h$

Example: Find the area of a trapezoid that is 10 ft. long on one side, 14 ft. long on the other, and is 4 ft. wide:

$$a = \frac{10\,ft + 14\,ft}{2} \times 4\,ft$$

$$a = \frac{24\,ft}{2} \times 4\,ft$$

$$a = 12\,ft \times 4\,ft$$

$$a = 48\,ft^2$$

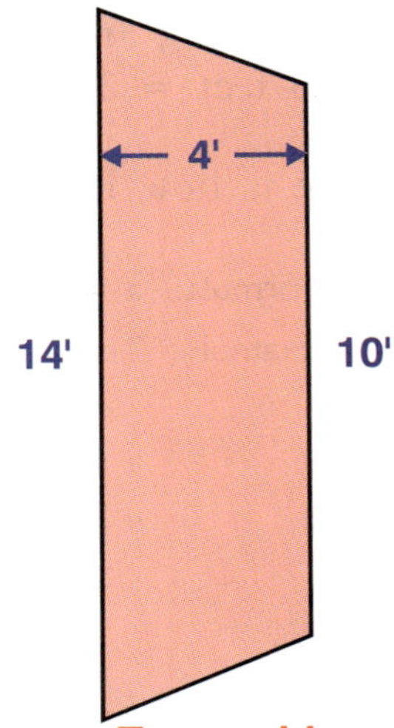

Trapezoid

Circle

Circles are round and can be measured in different ways. Sometimes the distance around a circle is measured. It is known as the circumference. Other times the distance from the center of the circle to the outside is measured. It is called the radius (r). Also, the distance across a circle may be measured. It is known as the diameter. In measuring diameter, it is necessary for the measurement to pass through the center point of the circle. In other words, the radius is equal to half of the diameter.

The area of circles is easy to determine if we use a constant of 3.14 in the calculation. (The constant of 3.14 is also known as pi.) For example, the area of a round flower bed is found by multiplying the radius squared (r^2) by 3.14. If a round (circular) flower bed is 10 feet in diameter, its radius is 5 feet. The area is 5 feet × 5 feet × 3.14 or 78.5 square feet.

Here is how it is done:

Formula: $a = r^2 \times 3.14$

Example: Find the area of a circle with a radius of 5 ft:

$$a = (5\,ft)^2 \times 3.14$$

$$a = 25\,ft^2 \times 3.14$$

$$a = 78.5\,ft^2$$

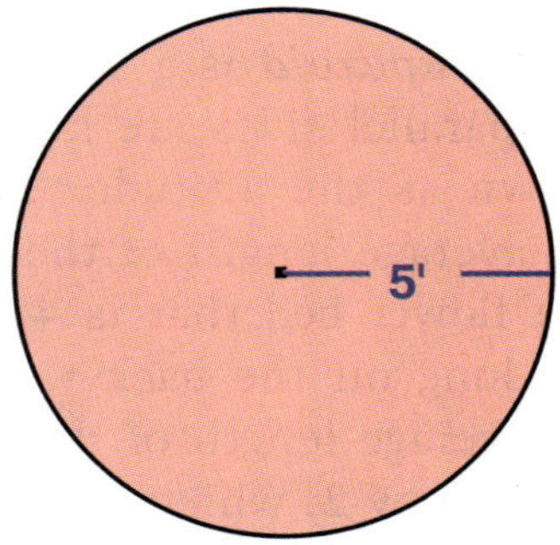

Circle

Irregular Shapes

Surfaces are not always perfect squares, rectangles, trapezoids and other shapes. The area can be figured by dividing the surface into regular shapes.

Most pastures, fields of crops, orchards, and forests are irregular. Determining the area will require forming regular shapes out of the irregular shape. A field may need to be made into several triangles and rectangles in order to determine its area.

LAND AREA

Land is measured in the customary system in the United States and the metric system elsewhere in the world. The customary unit is the acre. The metric unit for measuring land is the hectare.

Acres

Land measurement is important when selling the land, paying property taxes on it, and using it in other ways. The sizes of farms and ranches are often measured in acres. Residential and business lots may also be measured in acres or parts of acres. An acre is:

210 feet by 210 feet,
70 yards by 70 yards,
43,560 square feet,
4,840 square yards and
160 square rods.

Acres may be converted to hectares by dividing the number of acres by 2.47. For example, a farm with 247 acres would have 100 hectares of land.

Large areas of land are sometimes referred to as sections. A section contains 640 acres.

Hectares

Hectares are land measures based on the meter. A hectare is 10,000 square meters. One hectare may be described as follows in the customary system:

107,639 square feet,
11,960 square yards and
2.47 acres.

Hectares may be converted to acres by multiplying the number of hectares by 2.47. A farm of 100 hectares would have 247 acres of land.

DETERMINING VOLUME

Volume is the amount of space something occupies. It is measured as area times depth.

A good example is the cooler in a flower shop. Flowers are stored in coolers to keep them fresh until used. Coolers may be about 10 feet in length and width and as tall as the ceiling in the shop. The volume of a cooler is the amount of space from top to bottom and wall to wall inside the cooler. Some people refer to this as size. It is important to know the volume of the cooler when ordering flowers. You don't want to buy more than it will hold!

Volume is measured in two ways in agriscience: cubic units or standardized containers. Some containers are customary units, such as bushels and gallons. Other containers use the metric system, such as the liter. Most containers have been measured and their volumes converted to cubic units. For example, a gallon of milk is 231 cubic inches of milk.

Agriscience may use dry or fluid volume standards. The dry volume measures are used for non-liquid things such as corn or beans. The fluid measures are used for liquids such as milk and water. (Appendix A presents the common dry and fluid volume measures used in agriscience.)

A ***cubic unit*** is based on the cube. A cube has six equal sides. Cubes can be any size, depending on the unit being used. Cubic inches, cubic feet, cubic centimeters, and other measures are often used. Cubic centimeters (cc) may be used with small quantities of materials, such as animal medicines.

Cubic units are thought to be more accurate than some other ways of measuring and also easier to use.

SHAPES OF VOLUMES

Volumes have shapes that are based on area shapes, with an additional dimension known as height or altitude. The most common shapes of volumes are included here.

Square or Rectangular Containers

A square or rectangular container has length, width, and height (altitude). It is based on the square or rectangle used earlier in determining area. The walls of the container meet at right angles. Examples include boxes, rooms, warehouses, and covered truck beds or vans.

The volume of a square or rectangular container is found by multiplying length × width × height.

Here is how it is done:

Formula: $v = l \times w \times h$

Example: Find the volume of a container that is 10 m long, 9 m wide, and 8 m high:

$v = 10\text{ m} \times 9\text{ m} \times 8\text{ m}$

$v = 720\text{ m}^3$

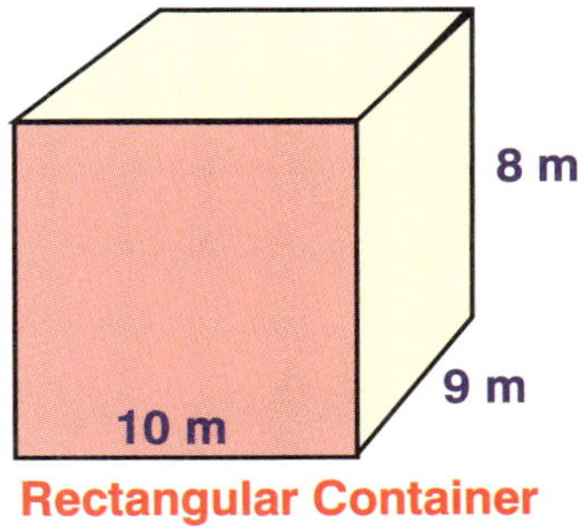

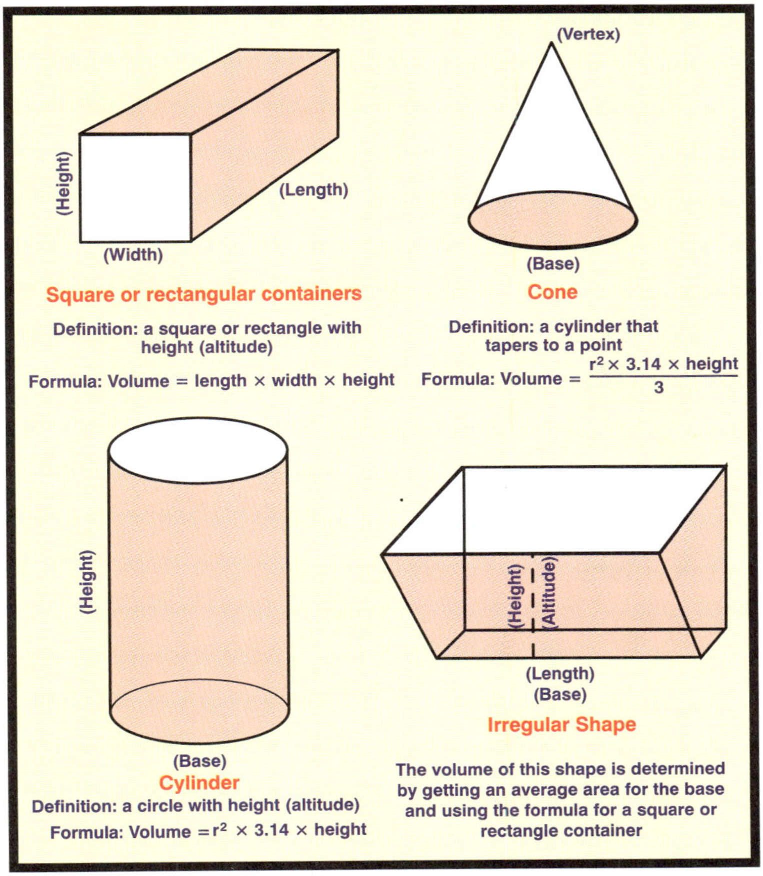

4-9. Common shapes and ways of determining volume.

Cylindrical Containers

A ***cylinder*** is a circle with height (altitude). The base and top of a cylinder are round and its sides are straight and parallel. Examples include round cans of food, silos where cattle feed is stored, grain bins, water tanks, and fuel tanks. People in agriculture often need to know how much can be held by bins or tanks. They want to know the volume of the bin or tank.

The volume of a cylinder is found by multiplying the area of the base by its height. A good example is a round grain storage bin.

Here is how it is done:

Formula: $v = r^2 \times 3.14 \times h$

Example: Find the volume of a grain storage bin (cylinder) that has a radius of 8 ft and is 20 ft high:

$v = (8 \text{ ft})^2 \times 3.14 \times 20 \text{ ft}$

$v = 64 \text{ ft}^2 \times 3.14 \times 20 \text{ ft}$

$v = 200.96 \text{ ft}^2 \times 20 \text{ ft}$

$v = 4{,}019.2 \text{ ft}^3$

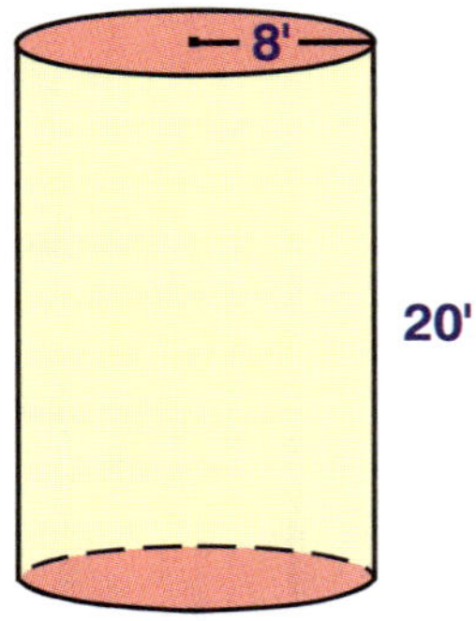

Cylindrical Container

The volume in cubic feet can be used to determine the number of bushels of grain the bin will hold. This is found by dividing 4,019.2 (cubic feet) by 1.244 (number of cubic feet in a bushel; see Table 4-5). The bin will hold 3,230 bushels of grain.

Conical Containers

A ***cone*** is a cylinder that tapers to a point at one end. Mathematicians may give more exact definitions but that is what a cone looks like. The large, round part of a cone is the base. The pointed tip is the vertex. Some people say a cone is like a pointed hat. The distance between the base and the vertex is the height or altitude.

Cones are found in many places in agriculture. Most tanks, grain storage bins, and similar things are made with cone-shapes or cones combined with other shapes.

The volume of a cone is found by multiplying the area of the base by the height and dividing the result by three.

Here is how it is done:

Formula: $v = \frac{r^2 \times 3.14 \times h}{3}$

Example: Find the volume of a feed hopper (cone-shape) with a base radius of 4 ft and a distance of 10 ft from the base to the vertex:

$$v = \frac{(4\,ft)^2 \times 3.14 \times 10\,ft}{3}$$

$$v = \frac{16\,ft^2 \times 3.14 \times 10\,ft}{3}$$

$$v = \frac{502.4}{3}\,ft^3$$

$$v = 167.7\,ft^3$$

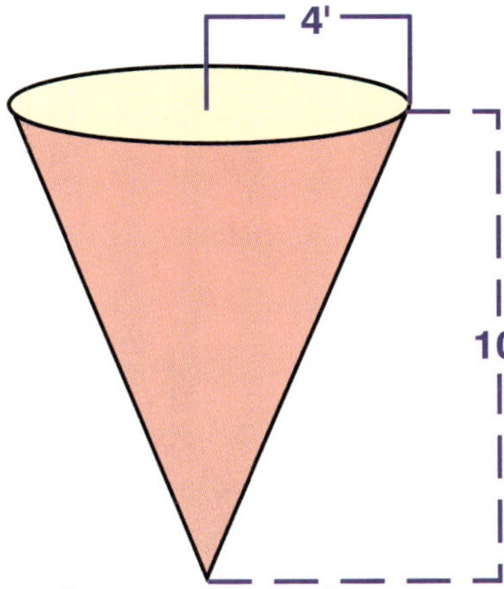

Conical Container

IRREGULAR-SHAPED VOLUMES

Many things in agriculture aren't exact squares, cylinders, and cones. These irregular structures can be subdivided into regular shapes. Two examples are used here.

- **Water trough example**—If a water trough is wider at the top than the bottom (base), the average area of the base can be determined. This is multiplied by the altitude to get the volume. A water trough that is 10 feet long, 1.5 feet in height (altitude) and 2 feet wide at the bottom and 3 feet wide at the top has a volume of 37.5 cubic feet.

Here is how it is done:

Formula: $v = \frac{b + b}{2} \times l \times h$

Example: water trough described above:

$$v = \frac{2\,ft + 3\,ft}{2} \times 10\,ft \times 1.5\,ft$$

$$v = 2.5\,ft \times 10\,ft \times 1.5\,ft$$

$$v = 37.5\,ft^3$$

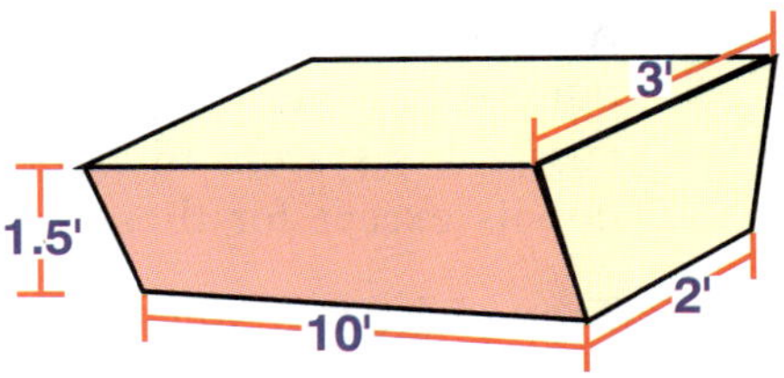

Irregular Container

- **Seed bin example**—If a seed bin is a cylinder with a cone at the bottom, the volume is figured using both the cylinder and cone formulas. First, the volume of the cone is figured and then the volume of the cylinder. The two amounts are added to get the total volume of the seed bin.

DETERMINING WEIGHT

4-10. Feed storage bin that has shapes of a cylinder and a cone. The volume is determined by forming the bin into the shapes of a cylinder and cone. Appropriate inside measurements are made of both shapes. The measurements are used in the formulas for cylinder and cone volume. Total volume is that of the cylinder plus that of the cone.

Weight is the heaviness of something. Scientists define weight as the measure of the force of gravity on an object.

Weight is sometimes referred to as mass, but mass and weight are not exactly the same. Mass is how much of a substance an object contains. Weight may change at different elevations, while mass is constant. The weight measured on top of a mountain may be less than that at sea level. This is because of the greater distance from the center of gravity. People in space may float in the air, as they have less weight than when on earth; however, they still contain the same amount of mass that they had before.

WEIGHT STANDARDS

Several standards for weighing things are used, including customary systems and the metric system.

Customary Weights

Three kinds of customary weights may be used: avoirdupois, troy, and apothecary.

- **Avoirdupois**—The ***avoirdupois weight system*** is based on the pound of 16 ounces and the ounce weighing 16 drams. It is the most widely used customary system in agriscience. (Note: All weights in this book are avoirdupois except for those in the metric system.)
- **Troy**—The troy weight system is based on the pound weighing 12 ounces and the ounce weighing 20 pennyweights.
- **Apothecary**—The apothecary weight system is chiefly used in weighing medical drugs. The weights are used to make drugs in very small amounts. Grains are the smallest measures in the apothecaries' weight.

Table 4-5 presents selected measures of weight in the customary systems. (More detailed measures of weight are presented in Appendix B.)

Table 4-5
Selected Measures of Weight

Customary Weights (avoirdupois)			Metric Weights		
437.5 grains (gr.)	=	1 ounce (oz.)	1,000 grams (g)	=	1 kilogram (kg)
16 oz.	=	1 pound (lb.)	1,000 kg	=	1 metric ton (t)
100 lb.	=	1 hundredweight (cwt)			
2,000 lb	=	1 ton (T)			

Metric Weights

The base weight unit in the metric system is the gram. As with linear units, the value can be increased or decreased using the prefixes shown in Table 4-2.

A gram is the mass or weight of 1 cubic centimeter of water, which is also the same as 1 milliliter of water. This makes it easy to convert from volume to weight and vice versa. This is often expressed as: 1 g = 1 cm^3 = 1 mL.

WEIGHING MACHINES

Machines that determine weight are known as ***weighing scales.*** Many kinds of scales are available. Accuracy is a must. Scales that make errors cost money and may cause harm to plants and animals through errors.

People need to know how to properly use and care for scales. Accurate scales are of little benefit if not properly used.

4-11. Using a triple beam balance to weight a mineral.

Types of Weighing Scales

Many kinds of scales are used. Some are designed for very small amounts of substances, such as animal medicines, and others are for large equipment, such as trucks loaded with grain.

Three major types of scales are used: balance, mechanical, and electronic.

4-12. Using a pan balance weighing scale.

Balance Scale. The balance weighing scale is the oldest, dating back to Egypt of 2,500 BC. There are two kinds of balance scales:

- **Equal-arm balance**—This involves suspending a bar over a sharp point. A pan is placed at each end of the bar. The item being weighed is placed in one pan. Objects of known quantity, known as weights, are placed in the other pan. Once the bar is balanced, the weights are added to determine the weight of the item in question.
- **Bar (steelyard) balance**—This type of scale involves using a bar that is anchored near one end so that a small weight can be used to balance the bar. The bar balance allows weighing heavy objects with small, light weights on the bar. Common examples in agriscience are the triple beam scale and pan balance scale.

Mechanical Scale. Mechanical scales were developed in the 1700s. They use levers that reduce the load to a smaller force. This means that weights equal to the weight of what is being measured aren't needed. Some of these scales involve beams, springs, and pendulum dials.

Electronic Scale. Developed in the 1950s, electronic scales are rapidly replacing other kinds of scales. They can weigh objects quickly—far faster than the other types of scales. Many electronic scales have built-in computer chips. These scales can do many things in addition to weighing, such as sorting and calculating price. Many meat markets, vegetable stands, processing plants, and other facilities use electronic scales.

4-13. Electronic balance (weighing machine) with beaker.

Calibration

Weighing scales need to be properly calibrated. This means that the weight they give is standardized for the unit being used. Objects of known weight are used to adjust scales that need to be calibrated.

SPECIAL MEASUREMENTS IN AGRISCIENCE

Many special kinds of measurements are used in agriscience. Two important areas are presented here: laboratory work and forestry and lumber.

LABORATORY WORK

Agriscience often involves working in laboratories where measurements are very important. Extremely small quantities may be measured. Accuracy and precision are essential.

Nearly all laboratory measurements are in the metric system. The metric system is easy to use because of the system of decimals and the relationships of the different units.

Conversions

Conversions from linear measures to volumes and weights are easy with the metric system. The original basis of measurement was with water, as follows:

1 cubic centimeter (cm^3 or cc, as it is often known) = 1 gram
1 cubic centimeter = 1 milliliter
1 gram = 1 milliliter

Table 4-6 presents the information most often needed when making conversions in agriscience. Other tables in this chapter also have useful conversion information.

Table 4-6
Most Needed Conversion Information in AgriScience

1 in.	=	2.54 cm
1 m	=	39.37 in.
1 kg	=	2.2 lb
1 L	=	1.06 qt

Measurements as Parts

Measurements are sometimes stated as ***parts per unit.*** This means that for a given number of units, another set of units is added. Parts per units are typically stated in three ways:

- ***parts per thousand*** (ppt)—number of parts in a thousand

- ***parts per million*** (ppm)—number of parts in a million
- ***parts per billion*** (ppb)—number of parts in a billion

Instructions for mixing materials may specify that a given number of parts be mixed with a quantity of another substance. For example, fish farmers must often mix therapeutants (fish drugs) in the water. If the rate to be used in the treatment is 5 ppm, 5 parts of the therapeutant should be mixed with 999,995 parts of water (5 + 999,995 = 1 million).

Determining how much to use involves these steps:

1. Measuring the vessel (such as a fish tank) containing the water.
2. Determining the volume of water in the vessel.
3. Calculating the amount of material that is to be mixed with the water. (For example, a dosage of 5 ppm would require 5.0 milligram per liter of water.)

Here are a few simple conversions that will help with measurements defined as parts:

1 ppm	=	2.72 pounds per acre-foot of water (acre-foot equals 1 acre of water 1 foot deep)
	=	0.283 grams per cubic foot of water
	=	8.34 pounds per million gallons of water
1 ppt	=	2,718 pounds per acre-foot
	=	3.80 grams per gallon of water
	=	28.3 grams per cubic foot of water

FORESTRY AND WOOD

Standing trees and wood products are measured with the customary system. These measurements often deal with the volume of timber or wood. Foresters undergo considerable training in how to make accurate measurements.

Timber and wood are measured in several ways: volume, weight, length, area, and board feet. The two most important ways are presented here.

Volume

The most common timber and wood volume measurements are:

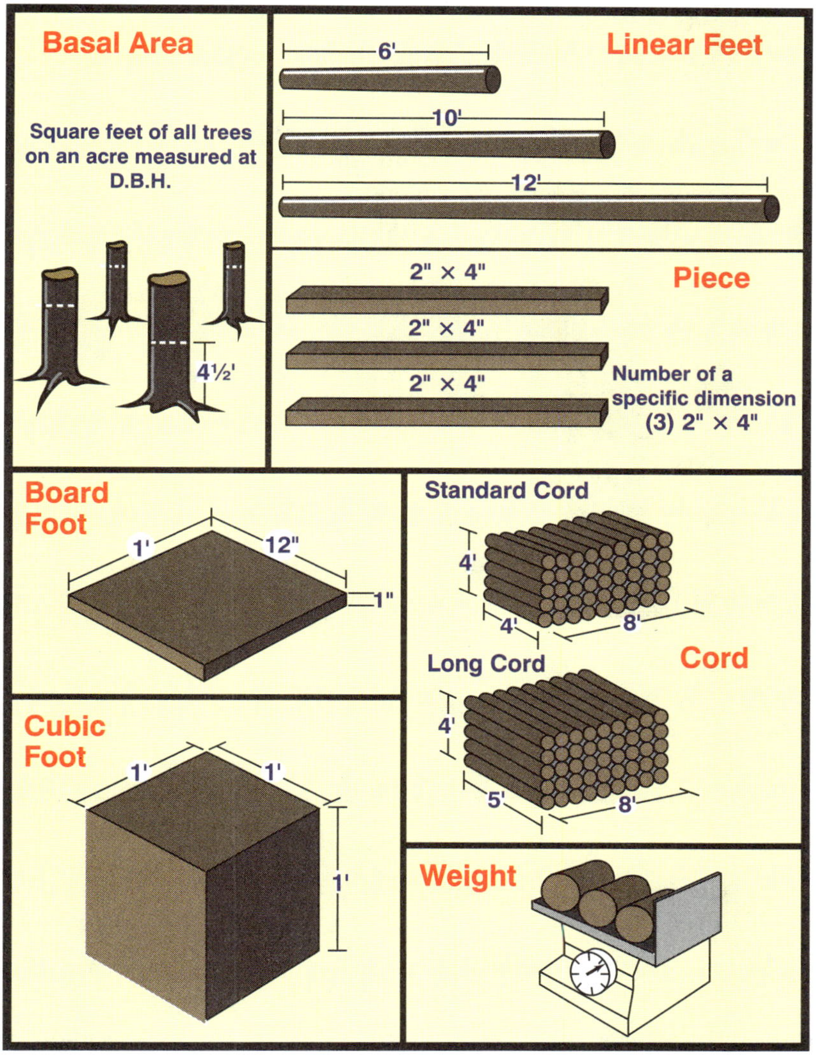

4-14. Common forestry and wood measurement units.

- **Board foot**—A ***board foot*** is a piece of wood that is 1 foot square and 1 inch thick. Trees are measured to determine the number of board feet of timber they contain. Finished lumber is slightly less than the dimensions used because some is trimmed away during planing (making smooth). Lumber dimensions are typically given in thickness, width and length. Most lumber is 1, 2, or 4 inches thick. Common widths are 2, 4, 6, 8, 10, and 12 inches. Lengths of lumber are measured in feet, with 8, 10, 12, 14, and 16 foot lengths being most common. Figure 4-16 shows how to calculate board feet.

4-15. Both boards contain one board foot!

- **Cord**—A ***cord*** is a stack of wood that measures 4 feet wide by 4 feet high by 8 feet long. The air spaces between the pieces of wood are counted in the measurement. A cord contains 128 cubic feet. Pulpwood, firewood, and fence posts are measured in cords.

Weight

Wood is increasingly being sold by weight. Loaded trucks are weighed as they arrive at a mill and weighed again as they leave after unloading. The difference in truck weight is the weight of the wood. Weight is a more accurate measure of wood than the cord because the air space has no weight. A cord of wood averages 5,300 pounds. About 12,000 pounds of logs are needed to produce 1,000 board feet of lumber.

Calculating Board Feet

Definition:	One board foot (bf) is a board that is 1 foot square and 1 inch thick. (The dimensions are 1' × 12" × 1".)
Formula:	bf = length × width × thickness divided by 12 (Length is measured in feet; width and thickness are measured in inches.)
Example 1:	The number of bf in a board that is 12 feet long, 6 inches wide, and 1 inch thick is determined as follows: $\frac{12 \times 6 \times 1}{12} = \frac{72}{12} = 6$ bf
Example 2:	The number of bf in a board that is 14 feet long, 4 inches wide, and 2 inches thick is determined as follows: $\frac{14 \times 4 \times 2}{12} = \frac{112}{12} = 9.33$ bf
Note:	To determine the bf in a stack of boards when all of the boards are of equal size, determine the bf in one board and multiply that by the number of boards.

4-16. Calculating board feet is easy to learn.

REVIEWING

MAIN IDEAS

Making observations and converting them to numbers is an essential skill in agriscience and technology. Measurement involves determining the number of standard units in something. It is essential for the same units to be standard for all uses.

Two measurement systems are used: customary and metric. The customary system is used in the United States and Canada in many areas of agriculture. The metric system is used in science applications around the world.

Common measurements are: linear (length), area, volume, weight (mass), temperature, and time.

Area measurements are based on linear measurements. They are reported as square units. The common area shapes are: square, rectangle, triangle, trapezoid, and circle.

Volume is the amount of space that something uses. It is measured in cubic units. Three regular volume shapes are used: square or rectangular container, cylinder, and cone.

Weight represents the heaviness of something, or the pull of gravity on something. Weights are determined with weighing machines, known as scales. With the metric system, some conversions can be made from volume to weight.

Laboratory work in agriscience and technology involves the metric system. Conversions within the system and to the customary system are fairly easy. An important thing to remember about conversions in the metric system is: 1 g = 1 cm^3 = 1 mL.

QUESTIONS

Answer the following questions using complete sentences and correct spelling.

1. What are the two areas involved in agriscience computations? Briefly describe each.
2. Why are standard measurements important?
3. What two measurement systems are used? Briefly explain each. Which is more widely used?
4. What are the six common measurements made in agriscience? Briefly explain each.

5. If someone tells you that the temperature is 90°F, what is the temperature on the Celsius scale?
6. What kinds of measuring instruments are used for linear measurements?
7. What is the area of the following:
   a. a square with sides 2 meters long?
   b. a rectangle that is 12 inches wide and 30 inches long?
   c. a triangle that has a base of 5 centimeters and altitude of 4 centimeters?
   d. a trapezoid with one base 3 meters long, the other base 6 meters long, and an altitude of 4 meters?
   e. a circle that is 10 inches in diameter?
8. What is the volume of the following:
   a. a rectangular container that is 2 feet by 3 feet by 5 feet?
   b. a cylinder that is 2 meters in diameter and 5 meters in height?
   c. a cone that is 10 feet in diameter at the base and has an altitude of 20 feet?
9. What are the following dimensions in the metric system:
   a. 1 foot is how many centimeters?
   b. 2 ounces is how many grams?
   c. 100 acres of land is how many hectares?
10. If you were instructed to use a chemical at the rate of 5 ppm and you had 1 million gallons of water, how much chemical would you use?
11. How many board feet are in a board that is two inches thick, 12 inches wide, and 10 feet long?
12. How many cords of firewood are in a stack of wood that is 4 feet wide, 4 feet high, and 16 feet long?

EXPLORING

1. Measure the size of your classroom using a yardstick or tape measure. Convert the size to meters.

2. Determine the number of square feet of floor space in your classroom.

3. Visit a farm supplies store and determine how chemicals, fertilizer, seed, and other supplies are measured. Write a brief report of your findings.

4. Keep a record of the temperature each morning when you get up for a week. Convert the thermometer readings from Fahrenheit to Celsius.

5

DETERMINING THE BASES OF LIFE

All living things have similarities. Whether large or small, living things share many common traits. Some are so small that they can be seen only when magnified. Others are very large and make a huge impression on the observer. Regardless, their similarities are greater than their differences. Isn't it interesting that most of the attention is focused on the differences!

Does this mean that horses are similar to the grass they eat when in the pasture? Or that dogs and cats are much more alike than different even though they may fight all the time? Or that cows are similar to fish? Yes! They are more alike than they are different. Certain structures and processes are present in all living things. Both plants and animals as well as other life forms are included.

Agriscience involves understanding similarities and differences. How plants and animals go about life involves important relationships. Some of the relationships are based on how life is different; others are based on how life is similar.

5-1. A cow guards her new-born calf that is only a few minutes old. The calf is not yet dry but soon will be in the warm sun.

OBJECTIVES

This chapter focuses on the common bases of all life. Some of the important differences are also included. It has the following objectives:

1. Explain important characteristics of organisms.
2. Describe the role of protoplasm.
3. Distinguish between living and non-living things.
4. Explain energy and its role in living organisms.
5. Explain life span and its stages.
6. Describe how organisms respond to their environment.
7. Describe the life processes in organisms.
8. Name the major parts of a cell and explain the function of each.
9. Explain cell specialization and the structures formed.
10. Explain heredity and genetics in agriscience.
11. Describe growth and cell division.
12. Describe ways organisms can be improved.

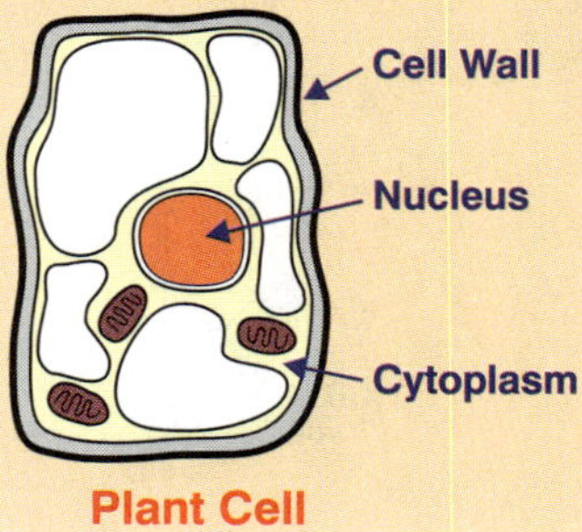

Plant Cell

TERMS

biogenesis
cell
cell specialization
cell structure
chromosome
circulation
circulatory system
controlled breeding
digestion
DNA
fertilization
gene
genetic code
genetic manipulation
genetics
growth
heredity
life span
living condition
locomotion
meiosis
mitosis
monogastric
multi-cellular organism
nucleus
organ
organism
organ system
ovum
oxidation
phloem
protoplasm
repair
reproduction
respiration
ruminant
sperm
tissue
vascular system
xylem

ORGANISMS AND AGRISCIENCE

Living things are ***organisms***! Regardless of size and shape, certain processes occur that are a part of life. When these processes stop, life ends. The aim in agriscience is to properly use living things to meet the needs of humans for food and fiber. By understanding organisms, people can produce better crops, livestock, and other commodities.

It is important to know about life processes to be good at growing plants and animals. Knowing some basic principles helps explain why plants and animals respond the way they do, including why they live and die.

Organisms have important characteristics. They are (1) chemically and structurally unique, (2) need energy, (3) have a life span, (4) are capable of growth and reproduction, and (5) respond to their environment.

ORGANISMS ARE UNIQUE

An organism is a complete and living thing. Organisms are made of substances that are carefully organized. Without the proper organization, there would be no life. How the substances are organized determines the kind of organism. Plants reflect a range of ways substances are organized; animals another. Even among plants and animals, there are many differences. Just think of the different plants and animals that exist in your local community!

Protoplasm—The Chemical Process

Life is based on certain chemical processes in an organism. These processes are very complex. Their unique chemical organization is ***protoplasm***. Protoplasm establishes the living condition in plants and animals.

Protoplasm is the only form of matter in which life exists. It is made of water and proteins. Complex chemical processes occur in protoplasm. The nature of the chemical activity determines the kind of life and nature of the organism.

Protoplasm varies considerably. It is different in all living things. It also varies from one part of a living organism to another. When protoplasm is properly organized, a living condition results.

Plants and animals need certain conditions to live and grow. In doing so, they may use fertilizer or feed and other items. What they are actually doing is helping protoplasm carry out more chemical activity. When this occurs, the organism grows better.

AgriScience Connection

GROWING CROPS BEGINS WITH IMPROVED SEED

Successful crop production requires good seed. Better yields are obtained from varieties of crops that have been improved. Improvements are made by controlled breeding and genetic manipulation.

Commercial seed companies sell seed that have been carefully improved and grow. A crop can be no better than the seed that are planted. Using poor quality seed wastes time, land, fertilizer, and other inputs. (Courtesy, Terra)

Cells—The Structural Unit

Organisms are made of cells. A ***cell*** is a tiny building block that has a definite structure. Some organisms are made of billions of cells, such as cattle and trees. They are said to be multi-cellular. Other organisms are made of only a few cells. Many one-celled organisms exist also.

Cells are produced only by living organisms. They are the basic units of living things. Of course, some parts of cells remain after an organism dies. The wood used to build houses or furniture is comprised of dead cell walls.

In addition to being made of cells, organisms also organize cells into forms that give structure. Distinguishing features of different animals or plants are based on how the cells are organized and different. (Cells will be discussed in more detail later in the chapter.)

Biogenesis

Biogenesis means that life comes from life. Nonliving things cannot produce life. Further, life tends to produce new life that resembles itself. No proof has been found showing that life has been produced other than by living things. Only living things can produce new life!

Scientists use synthetic biology attempting to create life-like processes in a laboratory. The work has not yet produced a living organism.

ORGANISMS NEED ENERGY

In order for life to occur, conditions must be right. Protoplasm activity requires energy. Energy is necessary for life processes in organisms. Energy must be available and the organism must be able to use it. All energy comes from the sun in one way or another.

Energy for plants and animals comes from food. Plants have the capacity to manufacture their food through a complex process known as photosynthesis. The food is broken down into energy that is used to support life processes. Life will continue only as long as energy is available and used by the organism. Animals often depend on plants as sources of food. Hence, the processes in plants are important to both plants and animals.

ORGANISMS HAVE A LIFE SPAN

The period of life is known as ***life span***. All organisms have a fairly definite life span. Activity must occur in protoplasm in the cells. When a condition favorable to the activity stops, the cells no longer live.

Some organisms live much longer than others. The average life span of humans in the United States is about 75 years. Cattle have life spans of about 12 to 15 years. Some plants live just one growing season. Insects may live only a few days.

Regardless of the life span, organisms go through five stages of life.

1. **Beginning**—All organisms have a beginning point. Beginnings vary between species, mostly depending on how the species reproduce.
2. **Growth**—This is a time when an organism grows rapidly. The growth stage occurs when an organism is young and immature. The rate of growth declines as maturity is reached.
3. **Maturity**—This a stage when the organism is said to be fully developed. Growth stops except for the repair and replacement of cells. The organism is usually capable of reproducing itself in this life stage.
4. **Decline**—This stage follows maturity. The organism loses its ability to maintain itself. Cells are no longer repaired and replaced. It is often associated with aging.

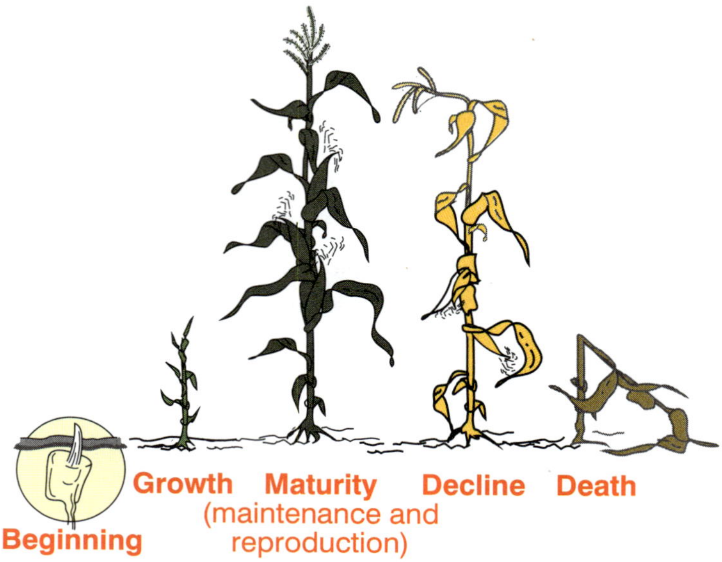

5-2. The life stages of a corn plant are shown here: beginning, growth, maturity, decline, and death.

5. **Death**—Death occurs as the organism is no longer able to replenish itself. Protoplasm ceases to carry out chemical activity and the organism no longer functions.

Plants and animals in agriculture go through these life stages. An example is the life stages of a horse. New life begins when the female's egg is fertilized by sperm from a male horse. The fetus (developing embryo or unborn horse) grows rapidly in the mare's (female's) reproductive tract. After birth, the foal (baby horse) continues to grow rapidly. Growth of the young horse slows as maturity is reached in about 2 to 3 years. The mature stage may last 10 to 12 years. Decline begins when the horse begins to lose teeth, develop bone problems, and otherwise deteriorate. Death follows decline.

When Life Stops

All things may be classified as living or nonliving. Some nonliving things were formerly living, but for some reason life stopped. Life ends when a system stops doing what it is supposed to do. The activity of the protoplasm no longer results in life. Both animals and plants may be killed when not handled properly. Growing crops and raising animals depends on encouraging life processes.

Nonliving things are classified as formerly living or never living. A tree that was once healthy and growing may die. This is an example of a living thing that has ceased to live. There is some reason for the death of the tree. In some cases, the death of the tree could have been prevented. Life and death are a part of the cycle of nature.

Some nonliving things were never living. Minerals, such as clay and iron ore, were never alive. They never had protoplasm to carry out complex chemical activities.

In some cases, nonliving things have to be carefully examined to determine if they were ever alive. Limestone rock may contain fossils, such as the shells of sea animals, which were once alive. Many of the fuels, such as coal and oil, are now nonliving but contain materials which were once alive.

Agriscience involves trying to encourage life. It helps understand how to provide for the needs of plants and animals. Controlling disease, furnishing nutrients, and providing protection from harm helps plants and animals grow.

ORGANISMS GROW AND REPRODUCE

Organisms will not last long if they to do not grow. This doesn't mean that they always increase in physical size. It means making new cells that may be needed to replace those which wear out in a mature organism.

Growth is when an organism increases in size by adding cells or by the cells getting larger. This is different from replacement, which involves growing new cells to replace those which have been destroyed. Mature

5-3. Sixty days are needed in a hatchery for fertile salmon eggs to be artificially hatched. Dark red eggs with eyes (black dots) have properly developing embryos. The light-color eggs do not have an embryo.

organisms must constantly replace older cells. When cell replacement slows, the organism begins decline. The activities in cells are similar during growth and replacement.

Reproduction is the process of organisms giving rise to new individuals of the same kind. Not all organisms reproduce but most are capable of reproduction at some stage of their lives.

Both growth and reproduction are important in agriscience. Crops and livestock must grow because new crops and livestock are constantly needed. The agriscientist must be skilled in reproducing and growing crops and livestock.

ORGANISMS RESPOND TO THEIR ENVIRONMENT

The surroundings of an organism comprise its environment. Many organisms are best suited to particular environments.

A number of factors make up environment. Organisms grow best when in environments to which they are adapted. In agriscience, the focus is on providing the best possible environment for a plant or animal. Here are a few examples.

5-4. Pineapple plants need sunlight, nutrients from the soil, and water. It took 18 months for this pineapple to bear its first fruit.

- **Rice**—Since rice grows in a warm climate and requires considerable water, a farmer would attempt to grow rice in warm climates and provide the needed water. Fertilizer may be used to help the rice grow better. Insects and weeds may be controlled to improve yields.
- **Bananas**—Tropical climates are required for bananas. Growing them in cold climates would not be practical.
- **Wheat**—Winter wheat is cold tolerant and can be planted in the fall for harvest in the late spring. Fertilizer may be applied to the crop to encourage growth. Weeds and insects may be controlled to keep down damage to the wheat.
- **Trout**—Trout thrive in cool, fresh water. They would not do well in warm saltwater. Best growth occurs if they are fed proper diets.
- **Beef cattle**—Cattle will thrive in a range of climates, although some breeds are especially

suited to warm or cold climates. Diseases should be controlled. Improved pastures, mineral supplements, and other nutritional needs may be provided.

Organisms have the ability to make adjustments to changes in their environment. Small changes are more easily adapted to than large changes. Major changes will result in loss of growth and production, and possibly death.

Plants sometimes need assistance in order to adapt. This is one of the roles of research. Researchers often focus on improving the ability of organisms to adapt to their environment. Breeding programs are used to develop new varieties. For example, several varieties of fruit trees have been developed that are known as "dwarf trees." This means that the tree size remains small, making it easier to harvest the fruit.

LIFE PROCESSES OF ORGANISMS

Organisms carry out several life processes. All of these processes are essential for something known as a ***living condition***. These are the characteristics that define life.

Eight life processes are used in agriscience. All living organisms, both plants and animals, have the processes in one form or another. The processes are important in growing crops and raising livestock.

GETTING AND USING FOOD

All organisms must have food. Without it, other life processes would stop. Food provides energy needed for activity. Since the life processes of a living organism never stop, the need for food doesn't stop either. Food is used for growth, repair, and movement within the organism even while it is resting. Food also provides minerals, vitamins, and other substances needed for other life processes to occur. Some organisms require food in ready-to-use forms. Others can take in nutrients and manufacture food.

Animals

For animals, getting and using food involves four processes: ingestion, digestion, absorption, and elimination. Taking in food is known as ingestion. Food is eaten (ingested) and then digested. ***Digestion*** is the process that changes food into simpler forms that can be absorbed by the animal.

5-5. The hogs being fed here are receiving a diet made of grain and other highly nutritious ingredients.

Digestion takes place in the digestive system. Various chemical processes occur that allow the nutrients to be absorbed. Absorption also occurs in the digestive system. It involves nutrients passing into the bloodstream of the animal. Once the food has been digested and nutrients absorbed, certain wastes remain in the digestive system. These are expelled from the animal through elimination (also known as excretion). Both liquids (urine) and solids (feces) are expelled.

Animals vary in their ability to use foods. Some must have food in forms that are ready to use. These animals have digestive systems that require highly-digestible foods. Others have digestive systems that can convert lower quality foods into more nutritious forms. They have digestive systems characterized by stomachs with more than one compartment.

Animals that digest food without making improvements to it are ***monogastric***. This means that they have a stomach with one compartment. They are not usually capable of further refining what they ingest. Feed that is higher in nutrients and ready to digest must be given to these animals. A good example of a monogastric animal is the hog. Diets of hogs typically include grain, grain by products, and other ingredients of high nutrient level. These feeds are said to be highly digestible.

Some animals are capable of further refining what they ingest. Examples are cattle and goats. A ***ruminant*** has a stomach divided into three or four chambers. The animal ingests the food into one chamber, which is later regurgitated and chewed. After chewing, the food goes to a different chamber where action by bacteria converts it into more nutritious food. This allows animals to ingest low quality forage (grass, leaves, etc.) and convert them to higher forms. This is why cattle graze in the pasture and later lie under a tree on a warm, sunny day. They are chewing their cuds! (More information is presented on animal nutrition in Chapter 12.)

Plants

Plants take in nutrients which are converted to food by photosynthesis. Most nutrients enter plants through their roots. Nutrient needs of plants vary with the species of plant. Some require more nutrients of certain kinds than others. Through the process of photosynthesis, the nutrients are converted into food for the plant. The food is typically a simple form of sugar.

Culturing plants requires knowledge of the nutrients that they need. Nitrogen is a nutrient needed by young plants for fast growth. Phosphorus and potassium are also important nutrients. Calcium, iron, magnesium, zinc, and other nutrients are required by plants, but in smaller amounts. Of course, plants must have water and sunlight in order to manufacture food from the nutrients. (More information is presented on plant nutrition in Chapter 9.)

MOVEMENT

Animals and plants move in different ways. Some movement is needed for life processes to occur within the organism. Internal movements are a part of using food, circulation, respiration, and carrying out other life processes. Some organisms are relatively stable in their environment; they don't move about. Others move about in their environment, and this is known as locomotion. Energy is required for movement.

5-6. Horses have long legs which allow them to run fast.

Locomotion is important to many animals. They must move about to get food and protect themselves from danger. A complex body system of nerves, bones, and muscles must be present for movement to occur. Animals vary in locomotion. Dairy cattle may graze in the pasture and walk to the barn for milking and additional feeding. Oysters move very little, as they ingest food by filtering it from the water that passes through their shell.

Movement in plants is usually related to growth and other life processes. Roots grow into the soil searching for nutrients and anchoring the plant.

The nutrients move through the roots and stems to the leaves. Leaves and stems grow and change positions to more effectively use the energy from the sun. Flower buds develop and open. Seeds develop in various ways, depending on the species of plant. A good example of plant movement is when leaves turn toward the sun. A house plant on the window sill will grow toward the sunlight.

CIRCULATION

The moving of food nutrients, digested food, and other materials throughout an organism is known as ***circulation***. Circulation varies between plants and animals. It also varies with the complexity of the plant or animal. The types of circulation described here are for many of the common agricultural plants and animals.

Animals

Circulation in most animals is carried out by a complex ***circulatory system***. The system contains a fluid known as blood. A heart and blood vessels keep the blood constantly flowing throughout the body. The heart is a hollow muscular organ that causes blood to move through the vessels. This results when the heart contracts (squeezes) and then releases the contractions. Heart movement is sometimes referred to as the heartbeat. The blood carries digested food, oxygen, and other materials to all parts of the body. As it flows, the blood picks up cellular wastes and carbon dioxide. These wastes are carried to organs where they are eliminated from the organism.

Injuries to animals can result in loss of blood or bleeding. If too much blood is lost, animals may be weakened or die. Some livestock management practices involve making cuts, such as dehorning cattle and docking lambs (removing the tail). These should be carefully done to reduce blood loss.

Plants

The circulatory system of plants is known as the ***vascular system***. It is often referred to as vascular tissue. Two kinds of vascular tissue are found: xylem and phloem. Of course, the way the vascular system works varies with the complexity of the plant.

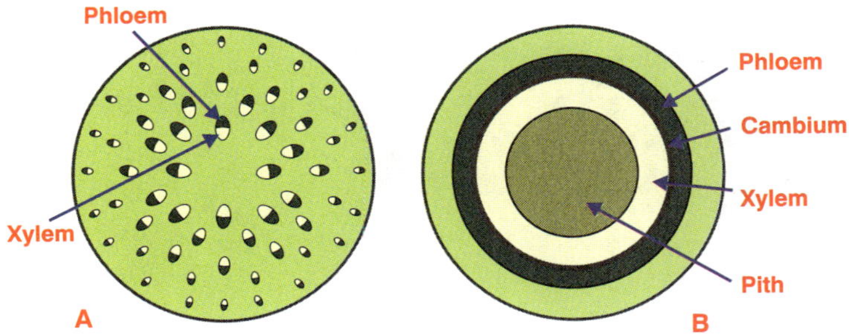

5-7. Internal stem structures of a monocot (A) and dicot (B).

Xylem conducts water and nutrients throughout the plant. This watery solution is frequently called sap. Xylem is found in roots, stems, and leaves. Water typically enters the roots and moves through the xylem to the leaves. Xylem is comprised of dead cell walls in woody plants.

Phloem moves food from where it is manufactured (usually in the leaves) to other parts of the plant. Phloem cells are living cells with perforations (small holes) in the cell walls. The perforations allow protoplasm to connect from one cell to another. This connection results in a continuous tube that conducts food (sugar and protein) down the stem.

Plants can "bleed" just as animals do. Cuts or breaks in stems will result in sap flowing out of the plant. When a plant loses sap, the effect is similar to an animal losing blood. Some cultural practices cause cuts or breaks. When plants are pruned (unwanted stems cut away), open cuts cause the loss of sap. This can be damaging to plants. A good example is when grapes are pruned too late in the winter so that excessive sap is lost and the production of the vine is reduced.

RESPIRATION

Respiration is the process an organism uses to provide its cells with oxygen so that energy can be released from digested food. Respiration occurs all of the time in living cells. Both plants and animals need oxygen to use food.

Animals

In animals, the blood carries nutrients, various compounds and oxygen throughout the body. A chemical process known as ***oxidation*** occurs between the oxygen and the digested food. This process produces chemical

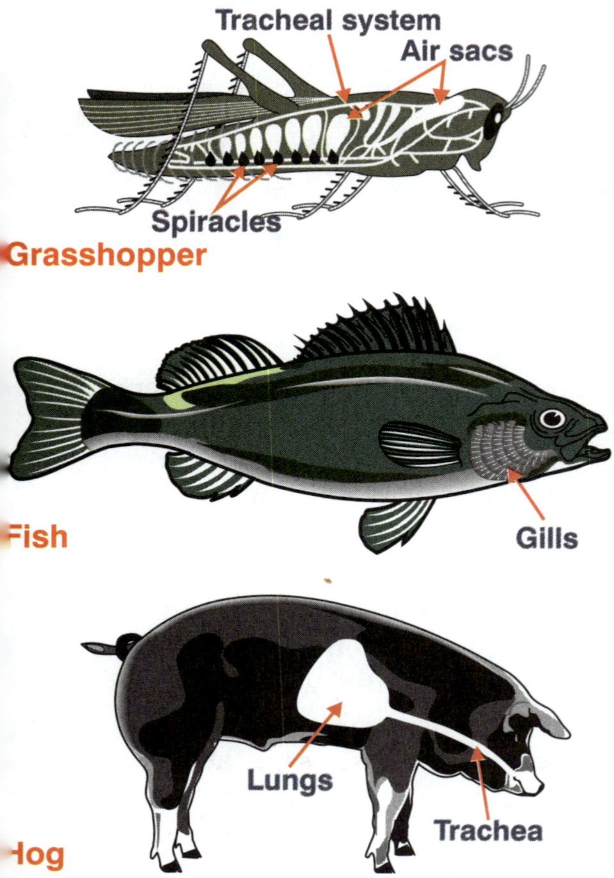

-8. Respiratory structures vary greatly in the grasshopper, fish, and hog.

energy and carbon dioxide, which is a by-product that must be removed from the organism.

Animals have various respiratory structures to obtain oxygen and release carbon dioxide. Cattle, hogs, and chickens have lungs. Air containing oxygen is inhaled through nostrils and temporarily held in the lungs. Blood vessels in the lungs remove oxygen and release carbon dioxide. Some cultured animals do not have lungs. One example is fish. Fish have gills that remove oxygen from the water and release carbon dioxide into the water. Fish farmers must monitor water and add oxygen if it gets low. Insects have tiny openings in their bodies called spiracles. Air enters through the spiracles and goes throughout the body in a network of tubes known as tracheae.

Plants

Unlike animals, plants use both carbon dioxide and oxygen. The carbon dioxide is used in photosynthesis as food is manufactured. The food is a sugar compound known as glucose. Photosynthesis also produces oxygen in the plant. This oxygen reacts with the glucose to produce chemical energy, much as in the oxidation process in animals. Plants use oxygen day and night in respiration. Carbon dioxide is used during photosynthesis which occurs primarily in the presence of light.

Respiration involves food being oxidized in the plant. The glucose reacts with the oxygen resulting in the production of carbon dioxide. The carbon dioxide is given off through stoma (tiny pores) in leaves. Water may also be given off by the stoma.

GROWTH AND REPAIR

Organisms undergo growth and/or repair from the beginning of life until death. The process of growth and repair is known as assimilation. It involves changing food substances into new living material. Since many foods are living or formerly living, the food substances are reorganized to form new cells. As presented earlier in the discussion of life stages, the rate of growth and repair is rapid in young organisms, slows during maturity, and virtually stops during decline.

Growth involves an increase in the size of the organism. It is obvious that young plants and animals grow rapidly. Mature forms (adults) also grow, but the nature of the growth is different. For example, the hooves on mature horses and the horns on cattle continue to grow. The mature animals may not change size to any great extent but growth of specific areas continues.

Repair occurs when parts of the organism wear or are damaged. This is sometimes referred to as self-repair. Injuries are healed; worn cells may be replaced; and in some animals, parts may be re-grown (known as regeneration). For example, the lizard's tail that is lost in an accident may re-grow.

SECRETION

Life in an organism depends on substances being produced and available for the organism. These are known as secretions. Specialized structures known as glands produce secretions that are needed for a wide range of functions. Secretions are juices or watery substances in the organism. These secretions make it possible for various functions to occur. All internal and external body fluids except urine are secretions.

A good example in animals is saliva. Saliva softens and lubricates food so that it can be swallowed. It also begins the digestive process by breaking down certain foods, such as starch. Other examples include tears (secreted by tear glands near the eyes to cleanse and protect the eye), bile (secreted by the liver into the first part of the small intestine to aid in the breakdown of fat), and gastric juice (secreted by gastric glands in the wall of the stomach to aid in digesting food).

SENSATION

Sensation allows an organism to respond to its environment. Complex systems of nerves may be involved in communication within an organism. Detecting changes in the environment is known as sensitivity. Both plants and animals respond to sensation.

The responses of animals to sensation are more obvious than plants. Larger animals have five types of receptors (often known as senses): touch, taste, smell, vision (sight), and hearing. The senses are receptive to stimuli, which are things that excite or stimulate. Most stimuli cause reactions from the organism. For example, if a cow touches an electric fence the stimulus shocks the animal causing her to move away. (She will not leave the pasture!) A lamb can see a person bringing feed and head for the feed trough.

Plants are particularly sensitive to light. Leaves may turn and stems grow toward sources of light. Plants are also sensitive to heat, lack of water, and other conditions in their environment.

REPRODUCTION

Reproduction processes vary among plants and animals and between species. In general, cells from the parents' bodies are contributed, such as a sex cell or living cells that will grow into a plant or animal. This results in the growth of a new individual, which will reach maturity and reproduce.

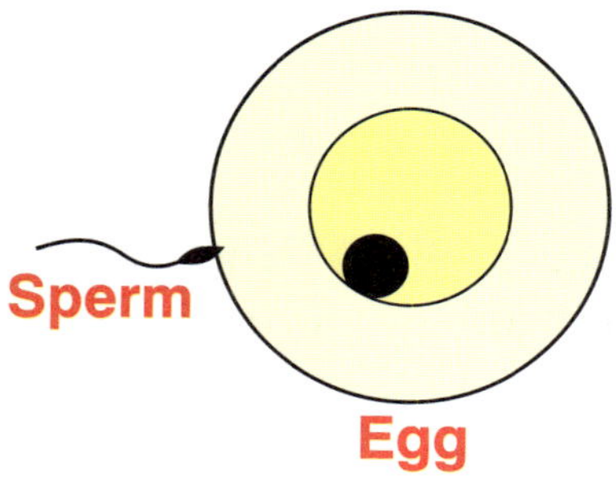

5-9. Sexual reproduction involves union of a sperm and egg.

Reproduction is not necessary for an organism to stay alive. However, it is necessary for new organisms to develop. Reproduction is necessary for particular plants and animals to remain in existence. Each develops reproductive organs or reproductive structures during the growth stage. These organs are capable of creating new organisms during the mature stage. Reproduction is the process by which new organisms are produced.

Two forms of reproduction are commonly used in agriscience: sexual and asexual.

In nature, most plants and animals reproduce through sexual means. Life begins when the ***sperm*** (male sex cell) unites with the ***ovum*** (egg) of

5-10. Sugar cane can be reproduced sexually or asexually. Mature stalks have flowers and bear seed. Most sugar cane is reproduced by planting small sections of the stalk. Buds at the joints in the stalk will grow into new plants.

the female. The union of sperm and ovum is known as ***fertilization***. The process of fertilization varies between plants and animals and within each group. Sexual reproduction involves two parents. (Later chapters will include sexual reproduction of plants and animals.)

Asexual reproduction does not involve the union of cells. Only one parent is needed. It typically occurs when existing organisms are divided into one or more organisms independent of each other. The offspring is identical to the parent in all respects. Asexual reproduction is most common in plants which have certain desired characteristics. Fruit and nut production often involves asexual reproduction.

Various forms of grafting or cuttings are used to create new plants. For example, a pecan bud from a tree that has the desired qualities can be grafted (joined) to a tree that is not desirable. The grafted part will grow and produce the desired pecans. Apples, peaches, plums, pears, and other fruit trees are often grafted.

Cuttings are used by taking a portion of a stem and placing it in a growing medium. Roots may develop on the cutting and a new plant will grow. Ornamental shrubs are often reproduced by cuttings.

STRUCTURAL BASIS OF LIFE

Cells are the structural bases of life. They contain protoplasm—the liquid, jelly-like substance that establishes life. Cells are involved in all of the life processes, as presented earlier in the chapter. Cells also bond together and specialize to form organisms.

CELL STRUCTURE

Cells tend to have similar structures regardless of the organism. Yes, the cells in both plants and animals are much alike. Close examination, however, will reveal that there are many differences. A microscope is required to see the structure. ***Cell structure*** is the general pattern of organization and relationship in a cell.

Cells have three major structural parts: cell membrane, nucleus, and cytoplasm. Each has specific functions in a cell.

- **Cell membrane or wall**—The cell membrane surrounds the cell and controls the movement of materials into and out of the cell. The membrane is comprised of several types of substances. Its arrangement varies in plants and animals.

  In plant cells, a wall outside the cell membrane provides rigidity and protects the cell. The cell wall also contains materials that make foods appealing and tasty. An example is pectin. Pectin is found in fruit and berries. It is important in preparing certain foods, such as jelly and preserves. Cooking fruit releases pectin so that a jelly is formed when it cools.

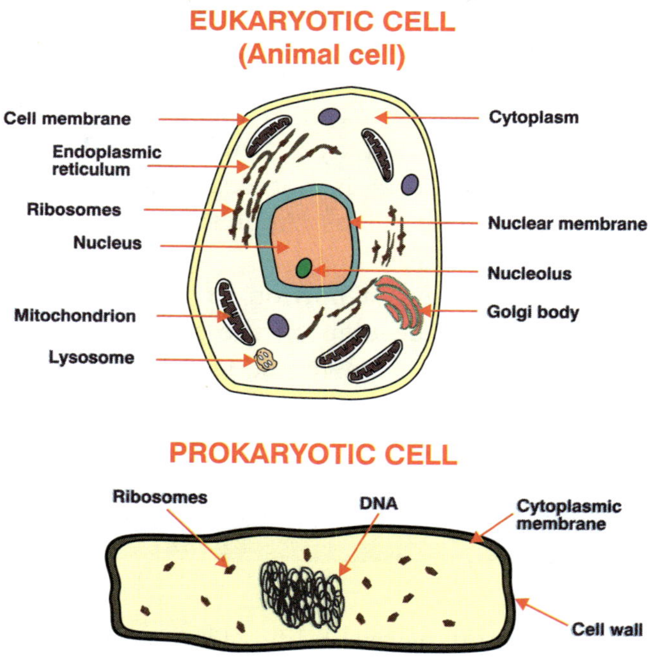

5-11. A comparison of eukaryotic and prokaryotic cells is shown here. (Prokaryotic cells do not have a separate membrane enclosing the DNA or other internal parts. Eukaryotic cells have at least one membrane that encloses the nucleus. Plants and animals have eukaryotic cells; bacteria are prokaryotes.)

Animal cells don't have walls. This gives the animal organism less rigidity. In order to provide body structure, many animals have skeletons. Skeletons give rigidity and body shape. Since their cell membranes are soft and pliable, animals are able to move about more easily than plants.

- **Nucleus**—Usually located near the center of a cell, the ***nucleus*** controls cell activity. The nucleus is made up of several materials, including a type of protoplasm known as nucleoplasm and fine strands of a substance known as chromatin. Chromatin makes up chromosomes.

  The activity of a cell is controlled by its ***chromosomes***. Chromosomes are made up of proteins and nucleic acid. The nucleic acid (DNA or ***deoxyribonucleic acid***) is the control substance in the cell nucleus. DNA has a spiral, double helix shape and controls all cell functions. DNA is the only chemical known that replicates itself. DNA molecules are further divided into segments, known as genes.

- **Cytoplasm**—Cytoplasm is found within the cell membrane and surrounds the nucleus. Cytoplasm is a thick solution that often flows in a cell. It contains several structures including organelles, which means "little organ."

  Several kinds of organelles are found in a cell: endoplasmic reticulum, mitochondria, lysosomes, Golgi apparatus, plastids, and vacuoles. The endoplasmic reticulum contains ribosomes, which control protein production. Mitochondria are the source of nearly all of the energy in a cell. Mitochondria also control the amount of calcium, water, and other substances in a cell. Plastids are found in plant cells. Chloroplasts, an important plastid for plants, contain the green pigment known as chlorophyll used in photosynthesis. Pigments that may give red, yellow, or other colors are found in plastids in certain plants. These plastids make the beautiful foliage on many ornamental plants!

CELL DIVISION

The two types of cell division are mitosis and meiosis. Both are important to organisms.

Division involves one cell splitting into two cells. The original cell is known as the parent cell. The new cells are daughter cells. The genetic character of daughter cells varies with mitosis and meiosis.

Mitosis is cell division for growth and repair. Each parent cell produces two daughter cells with identical genetic material. The number of cells in an organism increases through mitosis, and the consequence is a larger organism or cells that replace the damaged cells. In agriscience, emphasis is on helping organisms grow. Nutrients are provided to plants and animals so that mitosis can take place quickly and more often.

Meiosis is cell division in the sexual reproduction of organisms—both plants and animals. Each parent cell produces four daughter cells. Each daughter cell has half the number of chromosomes of the parent cell. The process of fertilization restores the number of chromosomes to that of the species of organism that is produced. Meiosis is sometimes known as reduction division.

More information on cell division is presented in Chapters 9 and 15.

MULTI-CELLULAR ORGANISMS

Organisms may be comprised of one or more cells. Most agriscience organisms, such as cattle and corn, are ***multi-cellular organisms***—have many cells. And the cells are carefully organized to form other structures, known as tissues, organs, and organ systems. Specialized kinds of cells are needed.

Cell Specialization

Cells vary because organisms require cells that perform specific functions. ***Cell specialization*** is when differences in cells exist so that the cells can perform unique activities for the organism. If all cells were alike, organisms couldn't exist! The grouping of specialized cells and organs forms a unique organism like the elephant.

Cells specialize to form all parts of the body. Eyes, ears, legs, tongue, and all other parts of the elephant represent cell specialization. The specialized cells work together to form an organism.

5-12. Elephants have many cells that form tissues, organs, and organ systems.

Tissues

Groups of cells that are alike in activity and structure form ***tissues***. The cells in tissues tend to be specialized. This means that the cells in tissues are suited to specific kinds of activities. In animals, tissues specialize as muscles, skin, nerves, bones, and in other ways. In plants, examples of tissues include melon rinds, potato skins, and flower petals.

Tissues have specific jobs to do in an organism. For example, the cells in muscle tissues provide motion. They are said to be specialized. Further, cells that are specialized for a certain job don't perform other jobs. Yet all cells must carry out the necessary chemical activity to keep them alive, such as taking in food and oxygen and producing energy through respiration.

Organs

An ***organ*** is a collection of tissues that work together to perform certain functions. The tissues may be different in the job they do yet each contributes to the functions of the organs. Examples of organs in animals are the heart, lungs, stomach, and liver. In plants, organs include roots, leaves, and stems.

Organ Systems

Higher forms of organisms have organ systems. An ***organ system*** is made of several organs which work together to perform an activity. These are the major systems of the bodies of many animals. For example, the digestive system is made up of several parts: mouth, stomach, small intestine, large intestine, and others. Organ systems are typically found in complex animal organisms.

HEREDITY AND GENETICS

Heredity is the passing of the traits of parents to offspring. New generations have characteristics of their parents. Heredity varies depending on whether reproduction is asexual or sexual.

In asexual reproduction, the offspring are identical to the parent (asexual involves only one parent). It is obvious that plants reproduced by cuttings or buds will have the same traits as the stock from which the cutting or buds were taken. A rose cutting from a bush that has red blossoms will

5-13. Offspring usually have a close resemblance to their parents. Here, a black cow has a red calf. What color is the calf's sire (father)? Most likely red!

produce red blossoms when it matures. This consistency of traits is very important in agriscience.

In sexual reproduction, the offspring are products of two parents. Further, each parent has different heredity. Heredity is more complicated in sexual reproduction. The offspring will exhibit traits of both parents and may have some that aren't obvious in either parent. These differences are known as variations. Variations arise because of new combinations of genetic factors.

GENETICS

Genetics is the study of heredity in plants and animals. Several genetic methods are used to improve plants and animals. Each applies fundamentals of heredity.

Gene Combinations

Sexual reproduction involves two sex cells—one from each parent. Fertilization occurs when the sex cells unite. Since each sex cell has one-half the number of chromosomes of the parent, the fertilized egg has a complete set of chromosomes. DNA is the substance in chromosomes that contains ***genes***. The genes are the specific determiners of heredity. Genes contain the units of heredity known as the ***genetic code***. Some genes are dominant and show up more often; others are recessive and show up less often. The developing new organism will be a combination of the two sets of genes.

Genetic engineering is used to alter the natural arrangement of genes in chromosomes. A process of gene splicing is used to give organisms certain desired traits. Genes of one type can be removed and those with desired traits added. For example, gene splicing is being used to develop cotton and soybean plants that are resistant to certain diseases and chemicals. This will lead to improved varieties that are easier for farmers to grow.

In effect, gene splicing creates new life forms. Some individuals are concerned about the creation of non-natural life. They fear that these new forms may have harmful effects on the environment and the quality of human life.

Improving Organisms

Two major approaches are used in agriscience to improve organisms (plants and animals): controlled breeding and genetic manipulation.

Controlled Breeding. ***Controlled breeding*** is breeding plants or animals to achieve certain characteristics in the offspring. Ranchers want cattle that will grow faster, resist disease, and provide more valuable beef. Wheat farmers want varieties that will produce more grain without increased fertilizer and other inputs.

Controlled breeding is considered to be a natural method of improving plants and animals. Agriscientists control the natural breeding processes. Three kinds of controlled breeding are used.

- **Selection**—Selection involves choosing a few parents with the desired traits. These are bred with the intent of increasing the amount of desired qualities in the offspring. A good example is the milk production capacity of dairy cattle. Dairy farmers want cows to produce 20,000 pounds or more of milk a year. The offspring of cows that produce large amounts of milk could be selected and mated. This mating should produce offspring with even higher milk production capacity.
- **Inbreeding**—Used for both plants and animals, inbreeding involves breeding offspring of the same parents to each other. The practice is somewhat different in plants than animals. Regardless, the purpose is to produce a new generation without the introduction of any new genes. For example, a plant is observed to have certain desired characteristics. The agriscientist may wish for all of the plants of that species to have the characteristic. The plant is protected so that pollen from other plants cannot reach it. The physical heredity of most of the new plants will be exactly like the parent. No new genes are a part of the heredity. Inbreeding also has problems. The weak or undesirable traits are also reproduced. In some cases, these

undesirable traits result in problems that make pursuing the desired trait impractical.

- **Hybridization**—This method of controlled breeding involves the crossing of two different but closely related plants or animals. Hybridization combines the traits of different parent strains. Strains are plants or animals that have a common heredity. With hybridization, the strains vary considerably. When bred, certain traits predominate in the offspring. Hybridization has been widely used by agriscientists. Examples include hybrid corn, hybrid striped bass, and mules. Hybridization has some problems, however. Many hybrids do not reproduce or if they do the offspring do not have the desired traits. Regardless, much of agriculture has been improved through the use of hybrids.

Genetic Manipulation. Artificial means of producing desirable traits in plants and animals are sometimes used. These means are known as ***genetic manipulation.*** Many of these efforts are in the experimental stage. Due to the nature of the work, some people feel that more information is needed before it is used on a large scale. Genetic manipulation involves possible hazards that must be overcome.

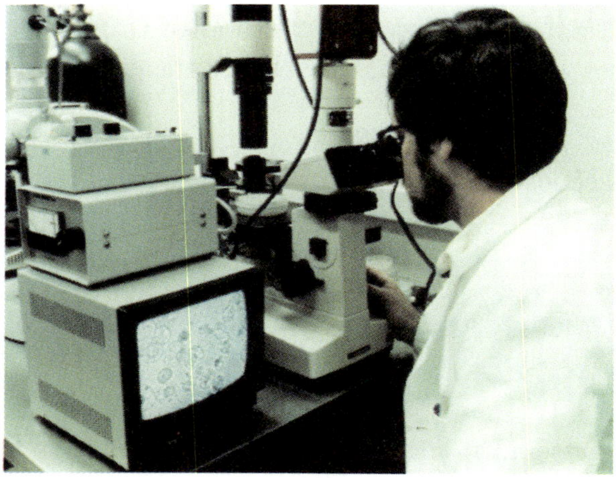

5-14. A scientist in a research laboratory is viewing plant cells that are being changed to improve the plant. (Courtesy, DEKALB Plant Genetics, Dekalb, Illinois)

Genetic manipulation requires considerable skill. Hereditary characteristics are moved from one species to another. In effect, agriscientists create new life forms. Gene splicing (as mentioned earlier) is a major part of their work. Genes are transferred from one plant or animal to another. The species involved may be unrelated to each other. In some cases, genes have been exchanged between plants and animals. For example, the luminescent (light) trait of a firefly has been spliced into tobacco plants. The tobacco plants that were produced gave off a low level of light. Though of little practical value, these efforts are used to develop new information about genetic manipulation. Several crops that are resistant to certain chemicals have been developed through genetic manipulation.

REVIEWING

MAIN IDEAS

Agriscience is concerned with promoting life processes. The efficient production of plants and animals depends on understanding fundamentals of these processes. The life processes of organisms are (1) getting and using food, (2) movement, (3) circulation, (4) respiration, (5) growth and repair, (6) secretion, (7) sensation, and (8) reproduction.

All living things are comprised of cells. The cells are much like building blocks. They have cell membranes, a nucleus, and cytoplasm. The cell contains protoplasm—the substance which establishes the living condition of organisms. Cells are grouped into tissues, organs, and organ systems.

Life forms reproduce themselves. Certain traits are passed by parents to offspring. Agriscientists study heredity (known as genetics) in order to improve life forms. Controlled breeding is used to improve plants and animals. Three kinds of controlled breeding are used: (1) selection, (2) inbreeding, and (3) hybridization.

QUESTIONS

Answer the following questions using complete sentences and correct spelling.

1. What is an organism?
2. Name and describe five important characteristics of organisms.
3. What is protoplasm? What is its role?
4. What is the difference between living and nonliving things?
5. What is the role of energy in living things?

6. Name and define the five stages of life.
7. How do organisms respond to their environment? List examples of plants and animals and their preferred environment.
8. What are the life processes of living organisms? List and explain each.
9. What is a cell? What are the major parts of a cell? Illustrate the major parts in a drawing.
10. What is cell specialization?
11. How are tissues, organs, and organ systems related?
12. What is heredity? Genetics?
13. What is the difference between controlled breeding and genetic manipulation?
14. What kinds of controlled breeding are used? Briefly explain each.
15. What is biogenesis?

EXPLORING

1. Make a survey of your local community to determine the kinds of plants and animals that are produced. Interview different producers to determine if improved varieties or species are being used. Prepare a report on your interviews.

2. Visit a local farm supplies store. Study the labels on seed to determine those which are improved varieties. Prepare a report on your observations.

3. Review seed catalogs for varieties that have been improved. If possible, determine if the varieties were improved by selection, inbreeding, or hybridization.

4. Arrange to tour a research facility where crops and livestock are studied. Determine the nature of the research. Prepare a report on your findings and show a relationship to the content of this chapter.

6

CLASSIFYING AND NAMING LIVING THINGS

All people have names. Their names come from the names of their families and those selected for them by their parents.

Culture (how people live) plays an important role in the names people are given. In the United States, the last names of people indicate the name of their families. Many family names originated because of the place ancestors lived or their occupation.

People also have first and middle names. First names (also known as given names) may or may not be associated with their family. However, the names parents use for their children are often the names of other family members.

Just as people have names reflecting their heritage, all living things have names. These names allow people to classify them and how they relate to each other. Without names, we would be unable to describe and compare living things to each other.

6-1. This young couple carefully selected a name for their daughter. Her last name is the last name of her father and mother. Her first and middle names were chosen from names in the mother's and father's family.

OBJECTIVES

This chapter provides an introduction to classifying and naming living things. It has the following objectives:

1. Explain taxonomy.
2. Describe a classification system for living things.
3. Name and distinguish between the five kingdoms.
4. Explain scientific names.
5. Match scientific names with the common names of agriculture species.
6. Describe how classification systems are useful in agriscience and technology.

TERMS

bacteria
binomial nomenclature
breed
common name
cultivar
dicot
fungi
kelp
kingdom
monocot
nori
phytoplankton
scientific name
seaweed
species
taxonomy
variety
wakame

Taxonomy of Farm Animals

Common Name	Hogs	Cattle	Sheep	Horses
Kingdom	Animalia	Animalia	Animalia	Animalia
Phylum	Chordata	Chordata	Chordata	Chordata
Class	Mammalia	Mammalia	Mammalia	Mammalia
Order	Artiodactyla	Artiodactyla	Artiodactyla	Perissodactyla
Family	Suidae	Bovidae	Bovidae	Equidae
Genus	*Sus*	*Bos*	*Ovis*	*Equus*
Species	*scrofa*	*taurus,* or *indicus*	*aries*	*caballus*

CLASSIFYING LIVING THINGS

The classification of living things is a science known as ***taxonomy*** or systematics. The first efforts to show how living things related to each other began some 2,000 years ago in Greece. Aristotle developed a simple system of placing plants and animals in separate groups. Simple subdivisions were made that are useful to this day.

Taxonomy involves using Latin—and sometimes Greek—words in the names. The common name of the ornamental astor is the Greek word meaning "star." Every known organism has a two-part ***scientific name.*** The names identify them much the same way as the names of people tell us their identity.

Common names are the names used by non-scientists. These names often vary for the same species from one location to another. Common names are given in everyday use for plants and animals.

A SCIENTIFIC CLASSIFICATION SYSTEM

Modern classification involves seven divisions or stages in the classification system. These show the relationships and differences between various organisms.

The modern classification system includes:

Kingdom
Phylum or Division
Class
Order
Family
Genus
Species

With this seven-division system, ***kingdom*** is the first group into which organisms are classified. The divisions get more specific with each move to the next division. The phylum is more specific than the kingdom, which is followed by class, order, family, genus, and species. All organisms in a species are very similar. They are more similar than those grouped together in the higher divisions, such as genus and family. ***Species*** is the most specific name used for an organism.

Organisms are sometimes further divided into subphylums or superfamilies. Agriscientists often work with varieties, cultivars, and breeds. These are used to further classify species.

A ***variety*** is a group of related organisms in a species. The difference is not enough to merit another species. Producers often plant certain varieties of crops, such as the seed of improved varieties of rice or wheat.

6-2. The scientific name for papaya is *Carica papaya*. People in all parts of the earth use the same scientific name. The above papaya trees are growing on the tropical island of Maui in the Pacific Ocean.

A ***cultivar*** is a crop variety that is cultivated and retains its features when reproduced. Many crop varieties have been developed that retain their characteristics from one generation to the next. However, the seed of hybrid crops do not retain their features when reproduced. They are often said to "not breed true."

Breeds are animals of the same species that have definite identifying characteristics and a common origin. Livestock production often involves breeds of beef and dairy cattle, hogs, sheep, and horses. For example, common breeds of beef cattle are Angus, Hereford, Santa Gertrudis, and Charolais.

Importance of Scientific Classification

Three important reasons to use scientific names are listed here:

- People in all parts of world use the same name for an organism. For example, a house fly has the scientific name of *Musca domestica*. There are many kinds of flies. Using the scientific name assures that all people will be talking about the same species of fly.
- Relationships among organisms are evident in their names. For example, all cattle have the same genus name of *Bos*. The species name will vary with the location where it originated and whether or not it has a hump.
- Differences between organisms are obvious from their names. For example, corn, oats, and many other plants are in the grass family, *Gramineae*. Strawberries and apples are in the rose family, *Rosaceae*. By knowing the family name, differences in organisms are easy to determine.

CLASSIFICATION KINGDOMS

All organisms can be classified into one of five kingdoms. A kingdom is the first division in the scientific classification system. The kingdoms are: monera, protista, fungi, plantae, and animalia.

In addition to the kingdoms, there is a pre-cellular group known as viruses. They are very tiny particles and are not complete organisms. Viruses do contain genetic material and protein, similar to living organisms. As incomplete cells, they cannot reproduce themselves. They invade other cells and use the enzymes and other materials. Viruses are associated with a number of disease problems of plants and animals. Viruses were discovered several years ago when electron microscopes were developed. These microscopes enlarge objects that are too small for regular microscopes.

MONERA

The monera kingdom is made up of simple, one-celled organisms. They are often called monerans. These organisms do not have complete cells, as do other organisms. Their cells lack a nucleus and membrane structures known as organelles.

The monerans have four phyla, as follows:

Phylum: **Schizophyta**—This is the largest phylum of the monera. It contains one-celled organisms commonly known as ***bacteria***. Bacteria are found just about everywhere. They are useful but also cause disease.

Phylum: **Archaebacteria**—This is a phylum with primitive organisms that live in harsh climates. It is thought that they are similar to the kind that existed when the earth was being formed.

Phylum: **Cyanophyta**—These organisms are known as blue-green bacteria. They are sometimes mistakenly called algae. The blue-green bacteria grow in water, such as ponds and streams. They may grow in large numbers and make the water appear green in color. The growth of blue-green bacteria is important in aquaculture ponds. Too much growth, known as bloom, as well as death of the bacteria can cause water problems. Most commonly the blue-green bacteria tie up oxygen and cause fish to die.

Phylum: **Prochlorophyta**—These are organisms which live in marine environments. They aren't widespread.

About 5,000 different species of monerans have been identified. Since they are very small, there are more monerans than any other kingdom on the earth.

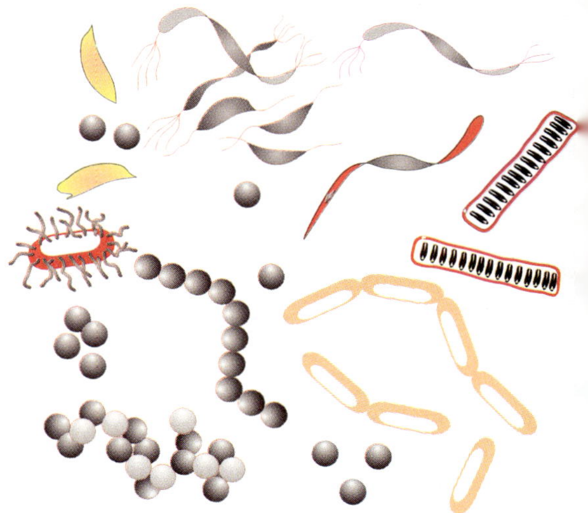

6-3. Shapes of bacteria vary widely. (Some appea[r] round while others may be rod-shaped, spiraled or as tiny filaments. Some bacteria form colonie[s] and others have flagella for movement.)

6-4. Making Swiss cheese involves the use of bacteria. Action of the bacteria create big bubbles of gas as the cheese is being made. The bubbles appear as holes in cheese slices.

Bacteria are present all around us. One of the most common is the *Escherichia coli*, the organism found in the intestines of humans. The *E. coli* live on the unabsorbed nutrients and help the body obtain vitamins. Bacteria have many uses although they can cause disease.

Here are a few examples of ways bacteria are used in producing food, fiber, and shelter:

- **Decay agents**—Bacteria act on materials to cause decay. This involves breaking down dead plants and animals into forms that can be reused. Bacteria are essential in natural recycling.
- **Food production**—Bacterial action helps produce many important food products. Cheese, vinegar, and yogurt are three examples. Bacteria are needed to change the original form of the food into a new product. Of course, only selected species of bacteria can be used for these purposes.
- **Medicine production**—Certain species of bacteria are used to make antibiotics, artificial insulin, and other medicinal products.
- **Food source**—Bacteria in ocean, lake, and stream water may provide food for animals. Several small aquatic organisms, including oysters and shrimp, feed on bacteria in the water.
- **Clean water**—Some forms of bacteria help to keep water clean. They convert materials in water into useful forms or forms that aren't dangerous. In fish ponds, bacteria convert a substance known as ammonia into a form that isn't harmful. These bacteria are added to filter systems in tanks where fish are grown to help keep the water clean.
- **Provide nutrients**—In rice fields, bacteria convert nitrogen from the air into a form that the rice plants can use. This saves the farmer the cost of buying fertilizer.

PROTISTA

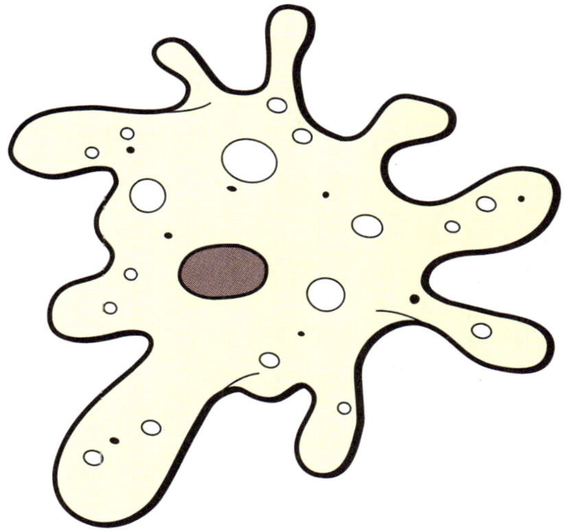

6-5. A greatly enlarged drawing of an ameba moving by extending part of the cell in one direction, with the remainder of the cell flowing into it.

The protists, as the protista are known, are one-celled or form collections of cells that may look like plants. Protists sometimes resemble plants and, at other times, animals. Their cell structure is more advanced than that of the monera. More than 65,000 organisms have been classified into the protista kingdom.

The phyla are:

Phylum: **Sarcodina**—The members of this phylum include the amebas. These organisms have cell membranes that are soft. Ameba move when part of their cell reaches out, somewhat like a foot, and the rest of the cell flows into it.

Phylum: **Ciliophora**—These organisms have cilia or tiny hairs that help them move about. The most common Ciliophora is the paramecium.

Phylum: **Zoomastigina**—These organisms have one long whip-like structure known as a flagellum, which they use to move about. The tiny euglena is an example. The members of this phylum often cause disease, such as human sleeping sickness. The organisms are carried by insects. When the insects bite, the organisms are transferred into the blood where they live.

Phylum: **Sporozoa**—These organisms have no means of moving about. They are carried in the bodies of host animals. Diseases of humans and other animals are caused by sporozoa. A deadly disease in poultry known as *coccidiosis* is caused by sporozoa.

Phylum: **Chlorophyta, Phaeophyta, Rhodophyta, Chrysophyta, and Pyrrophyta**—These are the five phylums of algae. Unlike some other protista, algae can make their own food. Algae are plant-like organisms that live as single cells or in groups.

Some scientists put the red, brown, and green algae in the plant kingdom. This is because they make food and live as plants.

Colonial algae are groups of algae cells that act as a unit. They share functions needed for life. Filamentous algae grow in string-like forms or rows of cells. Thalloid algae have cells which divide in all directions. They do not form definite structures. The thallus forms are known as seaweed.

Here are examples of ways the protists effect life on earth:

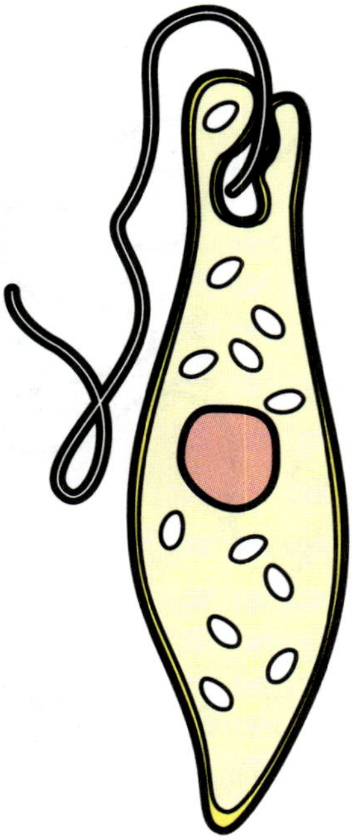

6-6. This greatly enlarged drawing of a euglena shows the structure of the flagellum.

- **Cause disease**—Disease of humans and other organisms may be caused by protista. A common example is malaria. It is transferred to humans by the bites of the anopheles mosquito. In the human bloodstream, the protista genus known as *Plasmodium*, a form of sporozoan, produces a toxin. The toxin causes chills and fever.
- **Use in polish or grind stones**—Diatoms, one form of protista, have extremely hard cell walls. They are mined and used to make materials to polish or grind metal and other materials.
- **Food**—Brown, red, and green algae are cultivated for food in some parts of the world. They are often known as ***seaweed***. The brown form is known as ***wakame*** in Japan and ***kelp*** in Europe and North America. The red algae is known as laver in Europe and North America and ***nori*** in Japan. Green algae is not grown as much as the brown and red. Of course, algae is not a popular food in North America. It is, however, often used to make food additives, such as that used in low-fat hamburger.
- **Add oxygen to water**—Forms of algae produce oxygen in the water of fish ponds and streams. These are called ***phytoplankton***. The oxygen is released in the water when the algae make food. Fish take the oxygen from the water and use it in their life processes.

FUNGI

Fungi tend to be more complex than the monerans and protists. Fungi range from tiny yeasts to large mushrooms. About 100,000 species of fungi have been identified. Four phyla are listed:

Phylum: **Zygomycota**—This type of organism includes black bread mold.

Phylum: **Ascomycota**—These include yeasts, truffles (food that grows below the surface of the soil), and *Penicillium* which is used to make penicillin.

Phylum: **Basidiomycota**—This group includes mushrooms.

Phylum: **Deuteromycota**—This includes imperfect fungi which cause ringworm and are a source of penicillin.

Here are some examples of ways fungi effect life:

- **Cause food spoilage**—Some fungi attack bread and other foods and cause them to be unfit to eat. This is easily seen on bread, and is known as mold. Fungi may also attack animal foods and ruin the feed. Proper storage can slow the growth of mold.

- **Make food products better**—Yeast is used as a leaven in baking bread. A leaven causes fermentation in the bread. Small amounts of yeast organisms are added to the mix as it is being prepared for baking. Warmth causes the yeast to grow. When baked, the holes left in the bread are due to gas bubbles caused by the fermentation.

6-7. Mushrooms are fungi that are frequently used as food.

- **Food source**—Mushrooms and truffles are desired as food. Mushrooms are preferred in the United States and Canada, Japan, and other countries. Truffles are preferred in areas of Europe. (Since truffles grow under the soil, trained hogs and dogs are used to sniff-out and dig up the mature truffle.)
- **Provide medicine**—Penicillin is an important medicine made from the *Penicillium* genus of fungi. It is used to treat many different kinds of disease in humans, livestock, pets, and exotic animals.
- **Cause plant disease**—Some plants are subject to mildew. This is a disease easily seen as a white coating on the affected plants. Control involves spraying the plant with a fungicide (fungus killer).
- **Damage property**—Some species of fungi grow on surfaces during damp weather. The stains or off-color on the trim and walls of houses is mildew. It must be removed before another coat of paint is applied. Many people use a chlorine and water solution to wash it off. Mildew is a problem in warm, humid climates.
- **Cause disease**—Some fungi are parasites. They live in or on other plants and animals. Ringworm is a common form that attacks humans, pets, and other animals. Athlete's foot is also a fungus.

PLANTS

The plantae kingdom is a large, important kingdom. Plants provide food, fiber, and shelter for humans. They provide feed for livestock, poultry, and other animals. Some 350,000 plants have been identified.

Plants are made of many cells and make their own food through a process known as photosynthesis. Some plants are quite large, such as the General Sherman sequoia tree in California—the earth's largest living organism. Others are very small plants, such as mosses and ferns.

The plantae kingdom has eight phyla or divisions, as follows:

Phylum: **Bryophyta**—Mosses and liverworts are the simplest land plants. They are generally found in damp, shady places such as in woods or near water springs.

Phylum: **Lycophyta**—Club mosses

Phylum: **Spenophyta**—Horsetails

Phylum: **Pterophyta**—Many species of ferns were thought to exist in pre-historic times. Today a small number of species remain.

Phylum: **Coniferophyta**—Conifers are trees or shrubs that bear cones, such as the pine tree. Most remain green the year round; therefore, they have the name of evergreen. Conifers are used in many ways, such as in making paper, lumber, plywood, and other wood products. Some produce rosin which is used to make turpentine and other products.

Phylum: **Cycadophyta**—Palm-like plants that grow in tropical areas.

Phylum: **Ginkgophyta**—Ginkgo plants grow fan-shaped leaves. Some are used for landscaping.

Phylum: **Anthophyta—**Flowering plants

The flowering plants phylum is the most important in agriscience. This phylum is divided into two classes:

Class: **Dicotyledonae (Dicots)**—***Dicots*** are plants with two seed leaves. Most fruits, flowers, trees, and vegetables are dicots.

Class: **Monocotyledonae (Monocots)**—***Monocots*** are plants with one seed leaf. Examples are onions, corn, wheat, oats, and sorghum.

6-8. The banana *(Musa paradisiaca)* is a large tropical plant that produces a popular fruit. This shows a large reddish bud with petals that open to reveal developing bananas.

Here are some examples of ways plants are used:

- **Food**—Many plants are used for human food. These include: fruit, such as the apple; grain, such as wheat; leafy vegetables, such as lettuce; roots, such as the carrot and beet; and flowers, such as the cauliflower.
- **Clothing**—Clothing is made from cotton, flax, and a few other plant fibers. Cotton is by far the largest provider of clothing material.

AgriScience Connection

LEARNING IN AN ARBORETUM

An arboretum is a type of outdoor botanical garden where trees, shrubs, and other plants grow in a natural setting. Some people refer to an arboretum as a plant identification laboratory. People go to the garden to study the plants and learn proper identification.

Most specimens in an arboretum are properly labeled with common name and scientific name. In some cases, other information is given such as where the plant originated.

Many universities have arboretums that also include trial sections for new varieties of new plants. Colorful flowering plants are often included in horticulture gardens.

- **Shelter**—Trees provide lumber and other materials that are used to build houses, make furniture, and other things people need.
- **Paper**—A number of different kinds of trees are used to make paper. In addition, other plants are also being grown for making paper, such as kenaf. (Kenaf is a kind of hibiscus, *Hibiscus cannabinus*, and also called Indian hemp.) More information is given about plants later in this book.

6-9. The star fruit *(Averrhoa carambola)* is a relatively new tropical fruit crop in south Florida.

ANIMALS

The animalia kingdom is made up of organisms that (1) have many cells, (2) can move about, and (3) get their food by ingestion (eating).

Over a million different kinds of animals are known to exist. These range from spiders, butterflies, and ants to elephants, horses, cattle, and chickens.

Animals are divided into 14 phyla, as follows:

Phylum: **Porifera**—These are the sponges that grow in the sea.

Phylum: **Cnidaria**—These are the jellyfish, coral, and other sea animals.

Phylum: **Platyhelminthes**—These are the flatworms.

6-10. The tom (male) turkey *(Meleagris gallopavo)* has the ability to strut, or make the feathers stand erect around his body.

Phylum: **Nematoda**—These are roundworms that are both useful and cause problems in agriculture. Nematodes in the soil can destroy crop roots. Roundworms in livestock cause animals to grow poorly.

Phylum: **Rotifera**—These are tiny round or worm-like animals that are found in the soil and in stagnant water.

Phylum: **Bryozoa**—These are animals that resemble moss but aren't plants.

Phylum: **Brachiopoda**—These are referred to as the lampshells.

Phylum: **Phoronida**—The phoronida are tube worms.

Phylum: **Annelida**—These are the segmented worms, such as earthworms and leeches.

Phylum: **Mollusca**—These are soft-bodied animals without segments in their bodies. Some have hard shells to protect their bodies, such as the oyster and clam. Examples of those without shells include slugs and squid.

Phylum: **Arthropoda**—The arthropods have bodies divided into segments and hard outer body covers known as external skeletons. The legs are jointed. Examples include lobsters, shrimp, crawfish, insects, mites, and ticks.

Phylum: **Echinodermata**—These are the starfish and other sea animals.

Phylum: **Hemichordata**—These include animals known as acorn worms.

Phylum: **Chordata**—This is by far the most important phylum of animals. Of these, the subphylum of vertebrates is widely grown in agriculture. Vertebrates are animals with internal skeletons and backbones. Examples include horses, pigs, cattle, sheep, birds, and most fish.

Here are some ways animals are used:

- **Food**—Many animals provide food for humans. Some are raised, such as hogs and chickens; others are captured in the wild, such as lobster and fish. Today, more fish are being grown on farms with the emergence of aquaculture (water farming). Various animal products are eaten, including meat, eggs, and milk.
- **Clothing**—Animals provide fur, wool, leather, and other products used to make clothing.

- **Medicine**—Various medicines to treat human disease come from animals. Insulin is made from cattle by-products and is used to treat diabetes. The intestines make sutures (threads) used to repair wounds or cuts in the human body.
- **Power**—Though the use of animals as sources of power has declined, they are still used for this purpose. Some countries rely heavily on animal power for agriculture. Other countries use animals in unique situations where they are best suited to provide power, such as to harvest maple sap in the United States. Animals used for power include the ox, horse, and water buffalo.

Many other examples could be listed, but more information about animals will be presented later in the book.

SCIENTIFIC NAMES

Scientific names are the two-word names of species. They are written in Latin and are always in italics or underlined.

Scientists call the names ***binomial nomenclature***. This means that all living things have two names made up of first, the genus, and second, the species. The binomial name of a species is its scientific name. After the genus has been given, the first letter of the genus name may be used instead of the entire word. For example, *Bos taurus* might be written as *B. taurus*.

The scientific names of common agriculture species are presented in Tables 6-1 through 6-6.

Table 6-1
Common and Scientific Names of Selected Terrestrial Farm Animals

Common Name	Scientific Name (genus and species)
Cattle (those with humps)	*Bos indicus*
Cattle (those from Europe)	*Bos taurus*
Chicken	*Gallus domesticus*
Goat (domestic goat)	*Capra hircus*
Hog	*Sus scrufa*
Horse	*Equus caballus*
Sheep	*Ovis aries*
Turkey	*Meleagris gallopavo*

Table 6-2
Common and Scientific Names of Selected Aquatic Animals

Common Name	Scientific Name (genus and species)
Bullfrogs	*Rana catesbiana*
Channel catfish	*Ictalurus punctatus*
Common carp	*Cyprinus carpio*
Crawfish	*Procambarus clarkii*
Goldfish	*Carassius auratus*
Oyster (American)	*Crassostrea virginica*
Rainbow trout	*Salmo gairdneri*
Shrimp (brown)	*Penaeus aztecus*
Tilapia (blue)	*Tilapia aurea*

Table 6-3
Common and Scientific Names of Selected Food and Feed Crops

Common Name	Scientific Name (genus and species)
Corn	*Zea mays*
Cotton (upland, as grown in North America)	*Gossypium hirsutum*
Cotton (long staple, as grown in Egypt and island locations)	*Gossypium barbadense*
Oats	*Avena sativa*
Rice	*Oryza sativa*
Soybeans	*Glycine max*
Wheat	*Triticum aestivum*

Table 6-4
Common and Scientific Names of Selected Ornamental Plants

Common Name	Scientific Name (genus and species)
Astor	*Astor novae-angliae*
Carnation	*Dianthus caryophyllus*
Gardenia	*Gardenia jasminoides*
Geranium	*Pelargonium graveolens*
Ginger	*Zingiber officinale*
Petunia	*Petunia x hybrida*[1]
Tulip	*Tulip gesneriana*
Zinnia	*Zinnia elegans*

[1]This scientific name designates a hybrid cross. Most ornamental plants have crosses as well as numerous varieties.

Table 6-5
Common and Scientific Names of Selected Trees

Common Name	Scientific Name (genus and species)
Cedar:	
eastern red	*Juniperus virginiana*
western red	*Thaja plicata*
Fir:	
Douglas	*Pseudotsuga taxifolia*
Maple:	
sugar	*Acer saccharum*
Oak:	
live	*Quercus virginiana*
Rocky Mountain white	*Quercus utahensis*
Pine:	
loblolly	*Pinus taeda*
longleaf	*Pinus palustris*
northern white	*Pinus strobus*
Spruce:	
blue	*Picea pungens*

Table 6-6
Common and Scientific Names of Selected Food Plants

Common Name	Scientific Name (genus and species)
Apple	*Malus domestica*
Banana	*Musa paradisiaca*
Coffee	*Coffee arabica*
Orange	*Citrus sinensis*
Papaya	*Carica papaya*
Pineapple	*Ananas comosus*
Asparagus	*Asparagus officinalis*
Blueberry	*Vaccinium virgatum*
Cauliflower	*Brassica oleracea*
Lima bean	*Phaseolus limensis*
Strawberry	*Frageria chiloensis*
Sugar cane	*Saccharum officinarum*
Tomato	*Lycopersicon esculentum*

HOW SCIENTIFIC NAMES ARE IMPORTANT IN AGRICULTURE

In agriculture, plants and animals go by their common names. A rancher hauls a load of beef cattle to market—not a load of *Bos taurus*! In fact, a cattle buyer would be surprised if the rancher said that the *Bos taurus* were for sale.

When are scientific names used? They are used when agriscientists need to communicate accurately about plants and animals. Since many people in agriculture use science, they need to know about scientific names.

Scientific names are universal. An organism has the same scientific name in all parts of the world. *Zea mays* is known as corn in North America and maize in other parts of the world.

CULTURAL PRACTICES

Producers of plants and animals often use information that was prepared for use around the world. Products used to kill weeds or treat sick animals may use scientific names. The product itself as well as the plants and animals it is to be used on may have scientific names.

Labels or instructions that use scientific names have little meaning unless the reader knows about the names. People only need to remember a few names. For the others, they need to know how to locate information on names.

LABORATORY WORK

Scientists nearly always use scientific names when doing research. They do so because other scientists around the world want to know what they found. The findings of research are reported using scientific names unless

6-11. A hybrid striped bass involves crossing the male white bass *(Morone chrysops)* with the female striped bass *(Morone szxatillis)*. The hybrid striped bass has excellent qualities for aquaculture.

the report is for a local area, such as the Shenandoah Valley of Virginia. People in this area would typically use the same common names. There would be no confusion about the name and which organism was involved.

When agriscientists talk with the general public they may use both the scientific and common names. The common name may be used followed by the scientific name. In some places, only the common name may be used. This is because all people would know the exact organism being discussed.

6-12. Many species of plants can be identified by studying leaf shape and other plant characteristics.

REVIEWING

MAIN IDEAS

Classification systems are used to show relationships and differences in organisms. The most common system used today is based on seven divisions:

Kingdom
Phylum or Division
Class
Order
Family
Genus
Species

In addition, breeds, varieties, and cultivars of crops and livestock may be used. These are based on variations in species. They retain their unique characteristics when they reproduce.

Scientific classification is important because it helps people in all parts of the world communicate. All people look at organisms the same way. Taxonomy is the study of classification of organisms into groups in the classification system.

Scientific names are based on scientific classification. All organisms have a two-part scientific name, known as binomial nomenclature. The name is made up of the genus and species. Scientific names are usually in Latin and are written in italics or are underlined.

QUESTIONS

Answer the following questions using complete sentences and correct spelling.

1. What is taxonomy?
2. What are the seven divisions in the modern classification system? What is the sequence of the divisions from broad groups to more specific groups?
3. What five kingdoms are used to classify organisms? Briefly describe each kingdom.
4. How do members of the five kingdoms provide for the needs of people? Give two examples for each kingdom.
5. What are the scientific names for the following animals: horse, sheep, goat, hog, chicken, humped cattle, and turkey?
6. What are the scientific names for the following crop plants: wheat, corn, cotton, and soybeans?
7. What are the scientific names for the following trees: live oak, sweet gum, and shortleaf pine?
8. Why is the use of scientific names important?

EXPLORING

1. Make a list of the common names of the major trees that grow in your community. Determine the scientific names of the trees by using reference materials or the assistance of a forester.

2. Review the labels on weed killers (herbicides) or insect killers (insecticides) and list the scientific and common names that are found. You may need to go to a garden center or farm supplies store to find labels.

3. Make a list of the common pets that people have in your community. Use reference materials or the assistance of an agriculture teacher to determine their scientific names. Study the classification of the pets to determine their relationships and differences.

4. Using a world atlas, geography book or other references, determine the common names that are used in different countries of the world for the following crops: *Oryza sativa*, *Capra hircus*, and *Zea mays*. (In North America, these are known, respectively, as rice, goat, and corn.)

7

APPLYING PRINCIPLES OF PLANT SCIENCE

All life depends on plants. Both humans and animals depend on plants to help meet their needs. Plants produce oxygen that is breathed by animals and other plants. Plants grow parts that are used for animal food. Even the animals that humans eat grew on plant materials—the cow in the pasture is eating plants!

Of the 350,000 plants on earth, only about 1,000 are used for food. Of these, a few serve as the major providers of food. The cereal grains, such as wheat and rice, are major human foods all over the world.

When conditions are poor for plant growth, famines develop. A famine is a long, widespread shortage of food. People may starve for lack of food. Domestic and wild animals die.

Understanding the basic principles of plant science can help when growing plants. Knowing what to do to make plants grow helps reduce the chance of famine. It helps people to live a better life.

7-1. Agriscientists are constantly studying new ways of producing plants. Plants in a research project are being watered. (Courtesy, Michael J. Montague, Monsanto Company, St. Louis, Missouri)

OBJECTIVES

This chapter provides information on the principles of plant science. It has the following objectives:

1. Distinguish plants from animals and explain how they are alike
2. Describe how plants are adapted to climate.
3. Explain plant life cycle and give examples.
4. Identify the major parts of plants.
5. Describe the functions of vegetative plant parts.
6. Distinguish between plant root systems.
7. Explain how roots absorb water and nutrients.
8. Explain the kinds of tropism and why they are important.

TERMS

aerial stem
annual
biennial
bulb
compound leaf
corm
fibrous root system
geotropism
growing season
herbaceous stems
leaf
life cycle
osmosis
perennial
phototropism
primary root
rhizome
root hairs
root cap
root
secondary root
simple leaf
single woody trunks
stem
subterranean stem
taproot
thigmotropism
transpiration
tropism
tuber
woody stems

PLANT LIFE

Plants, just like animals, are living organisms. Plants are a part of the ecosystem on earth. They maintain balance in nature and provide essential support for other life. Plants share the earth's space with other living things and with non-living things.

HOW PLANTS AND ANIMALS DIFFER

Plants are different from animals in important ways:

- Plants take in nutrients dissolved in water and make their own food. Animals depend on plants for their food.
- Plants typically do not move about because they are anchored in the soil or other medium. Animals are usually capable of locomotion.
- Plants have rigid cell walls with cell membranes inside. Animals have soft, pliable cell membranes (no cell walls).
- Plants take in carbon dioxide and give off oxygen during photosynthesis. Animals take in oxygen and give off carbon dioxide during respiration. In effect, photosynthesis and respiration are opposites!

HOW PLANTS AND ANIMALS ARE ALIKE

Plants and animals are alike in important ways:

- They have life cycles. They begin, grow, mature, decline, and die.
- They carry on life processes, such as circulation, respiration, and growth.
- They are made of cells. Though the cells may vary in structure, the cells tend to be similar.
- They must have food.

IMPORTANT PLANTS

The important plants in agriscience are those that have flowers and bear seed. They are the more advanced forms of plants and have specialized tissues and organs. Because of this, they are often known as vascular plants. These are the major plants for food, fiber, wood, and ornamental uses.

All future reference to plants in this chapter refers to the flowering plant (anthophyta) phylum. (Refer to Chapter 6 for an explanation of phylum.)

7-2. Food and agricultural plants.

HOW PLANTS ARE ADAPTED TO CLIMATE

Plants have certain conditions under which they grow best. In nature, most plants are adapted or best suited to a specific climate. Climate refers to the weather that exists in an area over a period of time.

The major factors in climate are temperature and precipitation (rain, snow, etc.). Also related to these are the length of the seasons of the year and the climate to which plants are best suited.

Plants vary in their growth characteristics. Some plants require more water than others. Some plants take longer to grow and cannot tolerate cold weather.

TEMPERATURE

Some plants are cool season crops; others must be grown in warm weather. Plants must be adapted to the climate where they are to be grown in order to grow successfully. The growing season of a plant must be known.

Warm season crops need a warm growing season; cold weather is not tolerated. ***Growing season*** is the period of time after the last frost in the spring and before the first frost in the fall. In North America, the growing season is longer in the southern part of the United States and shorter in the northern part of the United States and Canada. Examples of warm season crops are bananas, papaya, oranges, tomatoes, cotton, corn, and soybeans.

In cool season crops, the life cycle of the plant begins in the fall and ends when summer begins. The plants cannot tolerate warm weather.

7-3. Coconut palm trees are adapted to climates that other trees may not be suited to. This photograph shows a tree with a large crop of coconuts. If moved to another climate the tree would not bear coconuts and would likely die.

Examples of cool season crops include wheat, rye, oats, and some varieties of vegetables. Some of these crops can be grown in climates that are cool in the summer.

PRECIPITATION

Plants vary in the amount of water they need to grow. Some require much more than others. Crops need to be grown in areas where they are best suited to the available water unless irrigation will be used to add water. Crops that don't get enough water won't grow. Sometimes they die without producing anything.

LIGHT

Plants vary in the amount of light they need. Cool season crops will begin to die when the days get longer. Warm season crops begin to die when the days get shorter even if the temperature is still acceptable.

Light is often referred to as the photoperiod. It is the length of the light period in a day.

Long-day plants change from the growth to the reproductive stages as the length of day changes. For example, the mustard plant, a cool season crop, "goes to seed" when the days get longer in the spring. It quits growing leaves and grows flowers and seed.

Day-neutral plants are not influenced by the length of the day. They may produce flowers and grow regardless of the day length.

Short-day plants change from the growth stage to the flowering stage as the length of exposure to light is shortened. Plants that grow in the summer and bloom in the fall are short-day plants.

PLANT ADAPTATIONS

Plants can be successfully grown in an environment to which they are adapted. Water requirements, temperature preferences, and length of growing season are important factors. Before trying to grow a crop, study its requirements in detail.

AgriScience Connection

MACHINES ADAPTED TO THE GROWTH STRUCTURES OF PLANTS

Mechanical harvesting requires specially designed machinery. The machinery must be able to select the desirable parts of plants and remove the waste. A good example is the potato. As tubers, potatoes grow in the soil on the roots of potato plants. Getting potatoes out of the ground is not an easy job!

Machines that harvest potatoes must remove them from the soil, sort out stems and trash, and move the potatoes to a truck or container. All of this must be done so that the potatoes are not damaged and so that no potatoes are left unharvested in the soil. The above photograph shows a high-tech potato harvester on a Minnesota farm. (Courtesy, R.D. Offutt Farms, Fargo, North Dakota)

LIFE CYCLE

The ***life cycle*** of plants refers to the length of the plant's life. The length of life is known as life span.

Life spans may last only one season or for many years. Seasons may be only a few months long and are sometimes referred to as growing seasons. Growing seasons are related to the climate. Plants require certain climates in order to grow.

Climate varies widely in North America. Some places don't get very cold in the winter; others do. Some places get a lot of rain; others get very little. Many plants are sensitive to weather changes and die when it changes too much. Other plants adjust to weather changes and live for years.

Plants can be classified into three groups on the basis of life cycle: annuals, biennials, and perennials.

ANNUALS

Annuals are plants that have a complete life cycle within one year. This means that they grow from a seed, mature, reproduce or produce seed, and die in one season. Some go through the life cycle in a few weeks; others take months. Radishes and mustard plants have a life cycle of six to eight weeks. Corn and wheat may take three months or more. Many of the cultured crops are annuals.

Annuals may be divided into those that grow in the winter and those that grow in the summer.

Winter Annuals

Winter annuals are planted in the fall, grow in the winter, and mature in the spring. They die after seed have been produced when summer arrives.

Climate influences life cycles of winter annuals. In the southern and midwestern part of North America, wheat is a winter annual. In the northern areas, wheat is a summer crop. Wheat cannot survive the winter in very cold weather. It also cannot live in hot weather. Other examples of winter annuals include oats, rye grass, vetch, and clovers.

7-4. Corn is a summer annual. This crop is dry and ready to harvest.

Summer Annuals

Summer annuals are planted in the spring and die in the fall. They grow, mature, produce seed, and die during the summer.

Summer annuals are usually sensitive to cold weather and are killed by frost. This means that their life cycle is related to the frost-free time of the year. The frost-free time begins after the last frost in the spring and ends at the time of the first frost in the fall.

Examples of summer annuals include tomatoes, beans, corn, squash, and rice.

7-5. Soybeans are summer annuals. These young beans will produce a crop that can be harvested in the fall.

BIENNIALS

Biennials are plants that live two seasons. During the first season the plant grows from its seed and reaches maturity.

Food is made and stored in the plant for the second season. During the second year, the biennial grows a little more and reproduces or makes seed. After the seed are produced, the plant dies. Examples of biennials include cabbage, beets, some kinds of clover, and carrots. Biennials are not grown as widely as annuals or perennials.

PERENNIALS

Perennials are plants that live for more than two seasons. Some live for many seasons and may be hundreds or thousands of years old. Bristlecone pine trees found in California may be as much as 4,000 years old! Perennials tend to be large plants.

Perennials may grow only vegetative parts (leaves, stems, and roots) the first years. Flowers and seed aren't produced until later. The exact way a perennial develops varies with the kind of plant. For example, an oak tree may grow for years before it produces acorns (seed).

Perennials may adjust differently to the changing seasons. This further classifies them as deciduous or evergreen.

Deciduous plants lose their leaves in the winter. This creates leaves that people rake in the fall! Examples of deciduous plants are some oak trees, sweet gum trees, and sycamore trees.

Evergreens retain their leaves the year round. Examples include pine trees, holly shrubs and trees, junipers, and boxwood shrubs.

NOT ALL PLANTS ARE WHAT THEY APPEAR TO BE

Some plants aren't what they appear. Their natural life cycles are actually different from how they are used in agriculture.

Crops may be planted and harvested without the plant going through its life cycle. A good example is the carrot. Carrots are planted and harvested

7-6. Cotton growing in this field is treated as an annual. The crop is being cultivated and a herbicide applied at the same time. (Courtesy, National Cotton Council of America, Memphis, Tennessee)

in the first season after the enlarged root has developed. Except on farms that produce seed, carrots are treated as annuals.

Another example is cotton. Cotton is very sensitive to cold weather. It is planted in the spring and harvested in the fall. Cold weather kills the plant. If grown in a greenhouse, a cotton plant can live for years. Cotton is a perennial, yet we grow it as an annual.

CHANGING WHEN PLANTS GROW

Agriscience can be used to grow plants at times when they don't naturally grow. This involves studying how plants grow and the conditions they need for growth. Various facilities can be made that will provide the conditions needed.

7-7. This pansy plant has been started in a greenhouse and will be transplanted to a garden or flower bed. The individual container makes it easy to set out.

Greenhouses are the most common facilities used to grow plants. Greenhouses are constructed of plastic or glass placed over pipe or wooden frames. They have heaters or coolers to make the temperature right for the plants. The plants are carefully watered as needed. Diseases and insects are controlled and fertilizer is used as needed.

The photoperiod may be made longer or shorter in the greenhouse. It can be made longer by using artificial light for a few hours at night. It can be shortened by covering the greenhouse with cloth or other material to keep light out.

Here are two examples of how environments are controlled to produce plants:

- Tomatoes are warm season plants. However, they can be grown in the middle of winter in greenhouses where the climate is controlled. The grower must provide the best possible climate for the tomato in the greenhouse.
- Bedding plants are annuals such as petunias and marigolds that are used to beautify homes and other places. They are set out in flower beds. These plants are often sensitive to cold weather. Some people use growing structures to get a head start on the growing season. This helps them to have flowering plants ready earlier in the season. For example, petunia plants may be started in a greenhouse when the weather is still cold and transplanted into the flower bed or garden after the weather warms.

7-8. Heated beds are sometimes used to start plants early in the season. These sweet potato beds get the new plants growing while the weather is too cold outside.

There are other ways of altering when plants grow. Some involve using genetic engineering to alter the habits of the plant. Substances that can be applied to plants to protect them from bad weather are being studied. Many of these are used only on a small scale or experimentally.

VEGETATIVE PARTS OF PLANTS

Plants are made of several important parts. These carry out specialized functions in the life cycle of the plant. The parts of plants are of two major kinds: vegetative and reproductive. (In this chapter, only the vegetative parts are covered. The reproductive parts are covered in Chapter 8.)

Plants have three vegetative parts: leaves, stems, and roots. These work together as organs to help a plant carry out life processes.

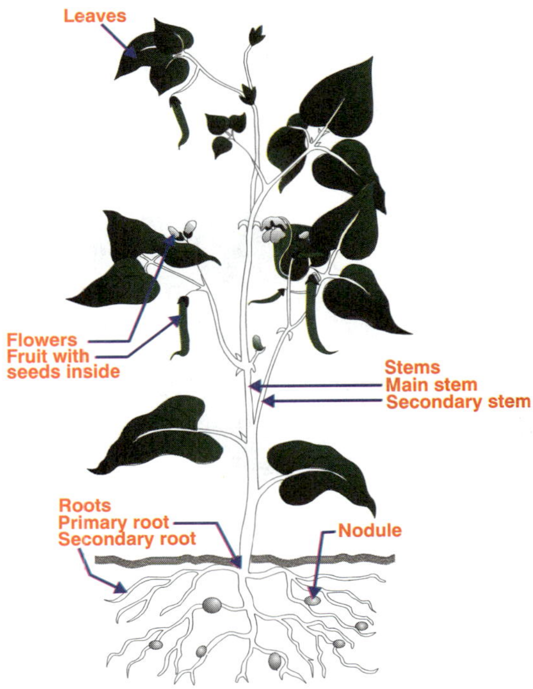

7-9. Parts of a legume plant are shown here. Legume plants form nodules of nitrogen on their roots.

LEAVES

Leaves vary in their appearance. Some are large, thin, and flat. Others are long and narrow like a needle. They are often green but may be of other colors. Regardless of shape and color, all leaves perform important functions for plants.

Leaves make food for the plant through a process known as photosynthesis. This involves the leaf converting water and other materials absorbed by the roots into food. Oxygen is given off by the process. Of course, there is a lot more to the process and this will be discussed in Chapter 9.

Parts of Leaves

Most leaves are made of several parts. All of the parts working together make it possible for the leaf to do what it is supposed to do. The leaves of dicots and monocots vary somewhat in arrangement.

The major parts of a dicot leaf are:

- **Blade**—The blade is the large, flat part where most of the photosynthesis occurs. Of course, some leaves aren't large and flat (such as the pine needle) but they function as the blade.
- **Veins**—The blade has veins which distribute water throughout the leaf. These are tiny tubes known as vascular bundles that form the framework of a leaf. How the veins are arranged is known as venation. Veins in monocots (grass family) are parallel (run side by side) while those in dicots have net or branching venation.
- **Epidermis**—The outer surface of a leaf is the epidermis. This holds the leaf parts together.
- **Stomata**—The stomata are tiny holes or openings in the epidermis. They are the openings through which water vapor, oxygen, and carbon dioxide pass into and out leaves.
- **Internal cells**—Various kinds of spongy and tall cells fill the space between the two layers of epidermis and veins. These cells carry out much of the photosynthesis.
- **Waxy coating**—The epidermis is coated with a waxy layer to protect the leaf. It prevents excessive evaporation of water.
- **Petiole**—The leaf is attached to the stem by the petiole.
- **Stipule**—Stipules are small structures that may grow where the petiole joins the stem.

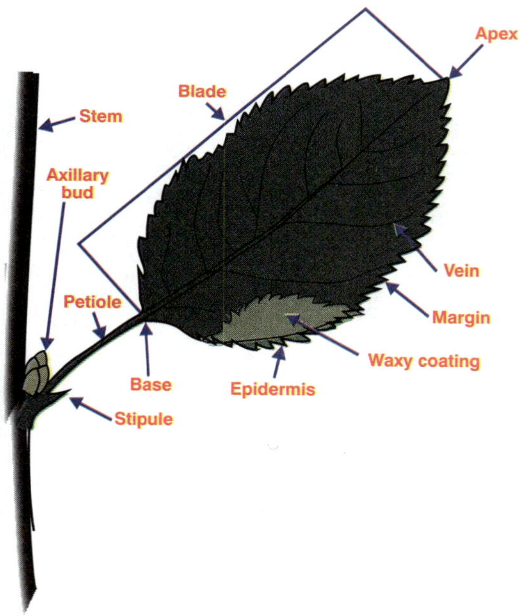

7-10. Parts of a simple dicot leaf.

7-11. Greatly enlarged drawing of open and closed stomata.

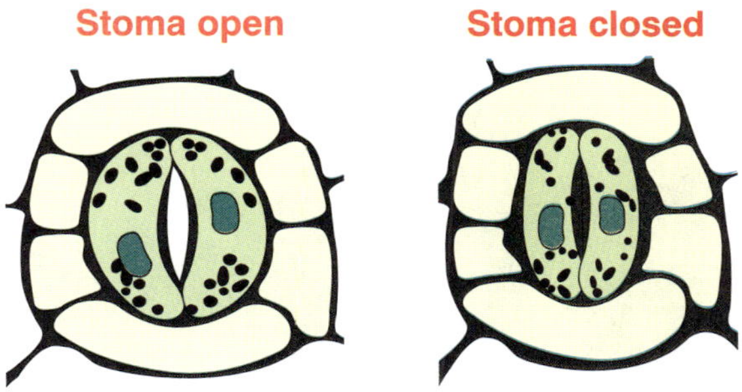

Monocot leaves are long, narrow, and thin with a point at one end. They are attached to the stem by means of a sheath. A sheath is a structure that wraps around the stem and holds the leaf in place. The middle vein of a monocot leaf is known as the midrib. It is a strong structure capable of supporting the entire leaf, such as on sugar cane plants.

Kinds of Leaves

There are two major kinds of leaves: simple and compound. A ***simple leaf*** has only one blade, while a ***compound leaf*** is divided into two or more leaflets. Plants with simple leaves include corn, oak and elm trees, and wheat. Plants with compound leaves include clovers, roses, and locust trees.

Shapes of leaves may vary from those with smooth edges to those with lobed edges and needles.

Leaf kind and shape is important in identifying plants. Many experienced agriscientists can look at a leaf and instantly know the kind of plant. Leaves also tell about the scientific classification of the plant. For

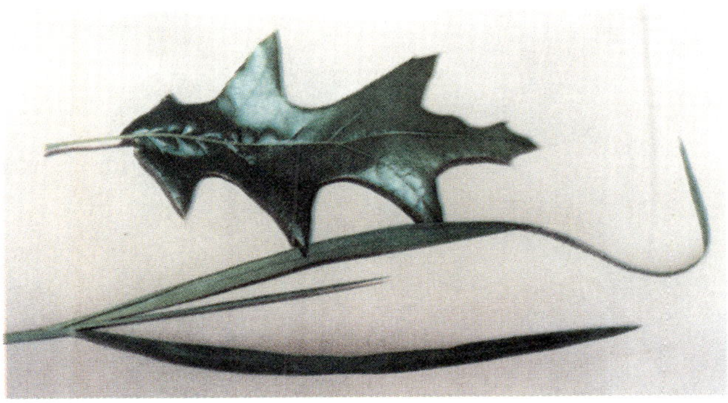

7-12. Differences in dicot and monocot leaves are shown here. The top dicot leaf (red oak tree) has net venation, while the bottom leaf (grass) has parallel veins and is a monocot.

7-13. Differences in simple and compound leaves are easy to see. The top leaf is a compound leaf from a locust tree. The bottom leaf is a simple leaf from a magnolia tree.

example, all members of the grass family have simple leaves with parallel venation.

Leaf Attachment

Leaves are attached to stems in certain ways. The leaves of plants of the same species are attached in specific and consistent patterns. Three basic patterns are found:

- **Alternate**—This is where only one leaf is located at each node on a stem. (A node is the place on a stem where leaves develop.) Grass plants have alternate leaf arrangement. The adjacent leaves point in different directions.
- **Opposite**—Opposite leaf arrangement is when two leaves are attached at a node opposite each other.
- **Whorled**—This involves three or more leaves being attached at each node.

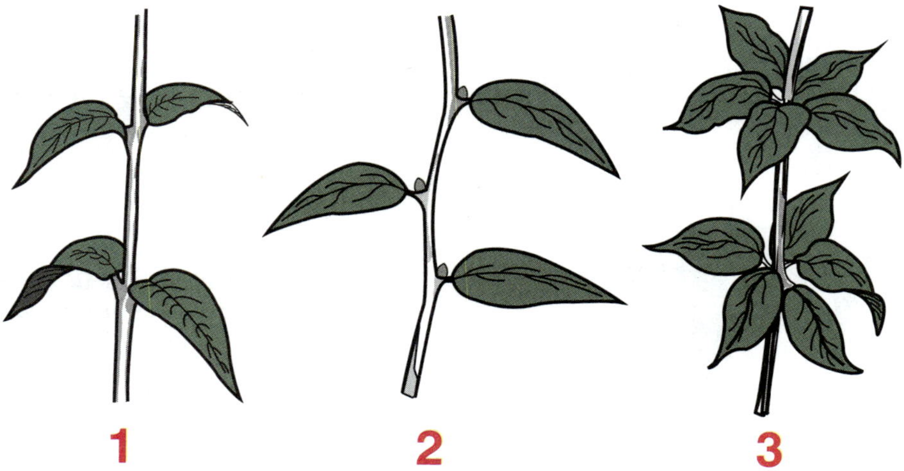

7-14. Three types of leaf arrangement are 1–opposite, 2–alternate, and 3–whorled.

STEMS

Stems support the leaves and vary a great deal in shape and size. Some are large and strong, such as a tree trunk. Others are short and small, such as that on a lettuce leaf.

Stems fulfill several important functions:

- Stems support the leaves and hold them so that they can get sunlight. Leaves must be where they can get as much sunlight as possible.
- Stems support flowers, fruit, and other structures that may be grown by the plant (including thorns on some species!).
- Stems transport water and other materials between the leaves and the roots.
- Stems grow. The tip of each stem has a bud. When the bud grows, the plant gets taller.
- Stems store food. A good example is the potato. Potatoes are special kinds of stems that grow below the ground. They store food manufactured by the leaves.

Kinds of Stems

The kinds of stems vary in two ways: where they are located and their structure.

Stem Location. There are two major kinds of stem locations: aerial and subterranean.

Aerial stems grow above the ground. They are the parts of plants that we typically call stems.

Subterranean stems grow below the ground. Examples include the following structures:

- **Tubers**—A ***tuber*** is an enlarged underground branch. A common example is the potato. Careful observation of a potato reveals buds, which are often

7-15. Potato with eyes beginning to grow. The eyes can be planted to start new potato plants.

7-16. An onion is a bulb. Note that the outer layer of leaves is dry and protects the inner, moist leaves.

called eyes. These eyes can be cut off and used to grow new potato plants. Potatoes store food for the plant. Of course, they are popular with people when mashed, fried, or baked.

- **Bulbs**—A ***bulb*** is an underground stem that has layers resembling leaves. A good example is the onion. Tulips, jonquils, and other ornamental plants grow from bulbs. Slicing a bulb in half reveals layers of leaf-like structure. The bulb stores food for the plant.
- **Corms**—***Corms*** are similar to bulbs but have thinner leaves and thicker stems than bulbs. Examples of plants with corms are taro, gladiolus, and garlic.
- **Rhizomes**—A ***rhizome*** is a long underground stem that sends up shoots to start new plants. Some pesky weeds have rhizomes, such as Johnson grass. Even when cut apart, a rhizome can grow into another plant.

Stem Structure. Stems are classified by the amount of woody material in the stem. They may be woody or nonwoody (nonwoody is also known as herbaceous).

Trees have ***single woody trunks***. The trunks have many cells with strong walls. Trees are used for lumber because of their cell structure.

Shrubs and many crop plants, such as cotton, have ***woody stems***. They have rigid tissue known as xylem. The stems do not grow nearly as large as tree trunks.

Herbaceous stems are soft and green and contain only a small amount of xylem. Many flowering, vegetable, and crop plants, such as corn, have herbaceous stems.

Parts of Stems

Stem parts vary by kind of plant and the woodiness of the stem. Stems with a lot of wood tend to have a more complicated structure.

All stems are divided into nodes. Leaves grow at the nodes. The space between nodes is known as the internode. The end of each stem has a terminal bud. This is where growth in length of the plant occurs. Terminal buds on trees usually show a spurt of growth at the start of every growing season.

The inside of a stem is designed to transport substances. Water and other substances go from the roots to the leaves. Food is moved from the leaves to the roots and other parts of the plant.

The cross-section of a woody stem has the following parts:

- **Bark**—The bark is the outer skin that protects the stem from injury and holds it together.
- **Cortex**—The cortex is the primary tissue in a stem. It is located between the bark and phloem.
- **Phloem**—The phloem transports sugar (manufactured food) throughout a plant.
- **Cambium**—The cambium is where growth occurs. Both phloem and xylem cells grow at the cambium.
- **Xylem**—The xylem conducts absorbed water and other substances from the roots throughout the plant.
- **Pith**—The pith is the middle part of a stem. It stores moisture and food.

Agriscientists refer to the xylem and phloem combined as the vascular system of plants. Vascular tissue is made of vessels and tubes that can convey fluids, known as sap.

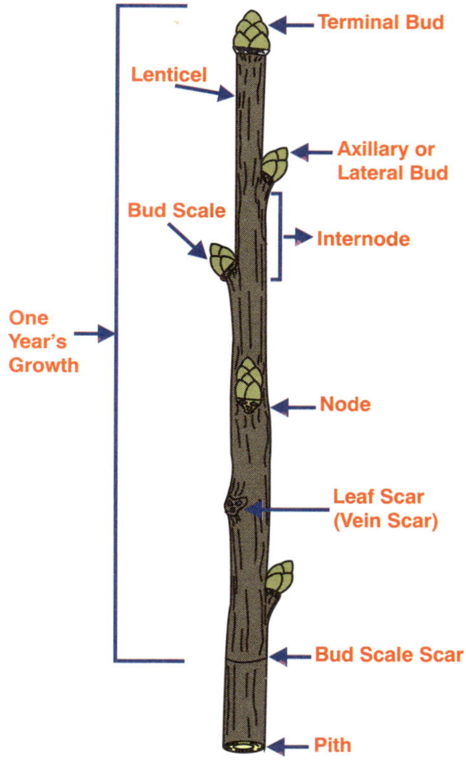

7-17. Typical parts of a stem.

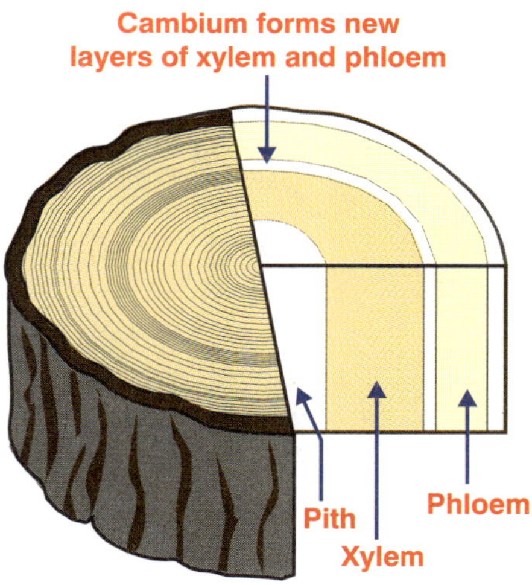

7-18. The structure of a woody stem has important tissue.

ROOTS

Roots are the parts of plants that usually grow underground. They do three important things for plants:

- **Anchor the plant**—Roots hold a plant in place and keep it from falling over onto the ground.

- **Take in water and minerals**—Roots absorb water and minerals from the soil. These materials are transported to the leaves for use in making food.
- **Store food**—Food made in the leaves may be moved to the roots for storage. Plants with large root structures store the most food, such as carrots and beets.

Kinds of Root Systems

There are two major kinds of root systems: tap and fibrous.

Plants with ***taproots*** have one main root that grows downward. It is a large root that may have a number of branches known as secondary roots. Pine and pecan trees, carrots, and beets have taproots. The taproots of trees are woody, while other taproots are fleshy, such as the carrot. Some weeds have well-developed taproots. A good example is the dandelion. This makes them difficult to kill.

Fibrous root systems are made of many small roots and spread out through the soil. Grasses have fibrous root systems. Wheat and corn are

7-19. The plant on the left has a fibrous root system, while the two on the right have taproots.

two common crop examples of this type of root system in the grass family. Other plants also have fibrous root systems, such as soybeans, tomatoes, and most ornamental bedding plants.

Some plants have modified root systems to serve their special needs. Corn has prop roots that grow out of the stem above the soil. When the plant becomes tall, the roots help to keep it from falling over. (These prop roots are sometimes known as adventitious roots.) Poison ivy and other vines, some lawn and pasture grasses, and strawberries also grow adventitious roots. These roots are used to anchor the plant and grow new plants. The roots on ivy anchor it to trees, buildings, and other structures.

Parts of a Root System

Root systems function to help the plant live and grow. They are connected to the lower end of the stem. Water and minerals are absorbed through the roots and passed to the stem.

Root systems have a primary root, secondary roots, and root hairs. The ***primary root*** grows from the seed and has branches known as ***secondary roots.*** Both primary and secondary roots have root hairs. The ***root hairs*** are very small and grow between soil particles. Every plant has many root hairs. The root hairs absorb water and nutrients from the soil, which are passed to the secondary and primary roots.

The tip of the primary root has a ***root cap***. The root cap protects the root as it grows into the soil.

On the inside, roots have vascular tissue. This is the tissue that conducts substances in the roots. It connects with the vascular tissue in the stem.

7-20. Two underground parts of plants; the carrot is a taproot and the jicama is a tuber.

How Roots Absorb Water

Water and nutrients are primarily absorbed by the root hairs. These roots have thin coverings that allow water to enter through a process known as osmosis.

Plant roots naturally have higher levels of certain nutrients than the soil and water surrounding them. The concentration of nutrients "draws" water. ***Osmosis*** is when water moves from a greater concentration to a lower concentration. The greater concentration is usually in the soil. Water moves into roots until the level of concentration is equal both within and outside the roots. Of course, dry conditions in the soil may not provide sufficient water for the plant.

After the water is absorbed by the roots, it is passed from cell to cell until it reaches the xylem. Upon reaching the xylem tubes in the roots, the water moves up to the stem and to the leaves. The petiole takes the water from the stem to the leaf veins which distribute it throughout the leaf.

Plant cells swell (get bigger) as they fill with water. They do not swell beyond the limits of the cell walls, however. As water goes into a cell, a pressure is formed. This is known as turgor pressure. The cell is rigid when filled with water. This helps the plant and leaves to stand erect.

If the conditions outside the plant are dry and the plant cell has more water than the plant's surroundings, the cell will lose water. This occurs when plants wilt. Many people let their house plants go without water and they wilt. If the plants get water soon, they will absorb it and again stand erect.

The leaves of plants give off some water through a process known as ***transpiration***. This creates a kind of upward pull from the leaves through the stem to the roots.

In addition to water, roots absorb a variety of other substances. These are known as inorganic nutrients. Plants need some of the nutrients in large amounts, such as nitrogen and potassium. Those needed in large amounts are known as macronutrients. Plants also need other nutrients in smaller amounts, known as micronutrients. Examples of micronutrients are iron, zinc, and sulfur.

HOW PLANTS MOVE

Most plants do not move about. Their roots anchor them to the soil or other material. Some movement is essential, however, if they are to get needed sunlight and respond to their environment.

Plant movement is known as ***tropism***. Various things in the environment of plants cause them to move. These things are known as stimuli. Tropisms are named for the various stimuli that influence plants.

7-21. Food and agricultural plants. (Note: Kenaf is a new crop used in making paper.)

KINDS OF TROPISM

The most common tropisms are:

- **Phototropism**—Plants respond to light. ***Phototropism*** involves plants growing or turning in the direction of light. A good example is when a plant sits a long time on the window sill and bends toward the open window. If the plant is turned around, the stems will again bend toward the window if given enough time. Phototropism is why stems grow up!
- **Geotropism**—Plants respond to gravity. ***Geotropism*** involves the roots growing downward. Other plant structures, such as tubers, also grow down.
- **Thigmotropism**—Plants respond to solid objects. ***Thigmotropism*** refers to how plants grow over and around objects such as rocks and sidewalks. A good example occurs when fence wire is attached to a tree. Over time, the tree will grow around the wire so that it winds up on the inside of the tree. Of course, the tree has been damaged and it may be dangerous to saw or cut where the wire is located.

HOW TROPISMS HELP AGRICULTURE

Tropisms help crop plants grow properly. Sometimes plants get twisted or knocked over, such as when a storm damages a field. Tropisms help them regain their normal position.

Crop plants may lodge in rain and wind. Lodging means that the plants have fallen over. It is particularly a problem with grains, such as corn, wheat, and oats.

With wheat, the heads of grain may be on the ground and will be damaged if they are there very long. The attraction of sunlight straightens the plants and gets them back into a normal position. Wheat that is lodged cannot be harvested very well. Much of the grain will be lost. Combines (grain harvesting machines) don't do a very good job of harvesting lodged wheat.

Tropisms cause seeds to sprout and grow normally. Seed that are planted upside down will grow as well as those planted upright. Geotropism attracts roots downward as they grow from the seed. Phototropism attracts stems, buds, and leaves upward. So regardless of how the seed is turned, the plant grows right!

Many of the ways crops are grown and harvested are based on the plants being in a normal growing position all of the time. Tropisms help to keep them there!

REVIEWING

MAIN IDEAS

Plants are essential for a balance in nature. Plants provide food, fiber, and shelter for humans. They also provide the food for nearly all animals, including cattle, hogs, sheep, horses, chicken, fish, and pets.

Life cycle is the length of a plant's life. Plants can be classified into three groups on the basis of life cycle: annuals (live one year and die), biennials (live two years and die), and perennials (live more than two years and may live hundreds of years, depending on the plant).

Through agriscience, we can grow plants at times of the year that aren't natural for them. For example, greenhouses can be used to produce summer annuals in the winter.

Plants have three main vegetative parts: leaves, stems, and roots. All three are needed in order for a plant to grow.

Leaves are usually thin, flat organs that make the food for the plant. They are positioned above the ground to get sunlight. Leaves may be simple (one blade) or compound (several blades or leaflets).

Stems support the leaves and connect the leaves with the roots of the plant. Stems conduct water and nutrients from the roots to the leaves. They take manufactured food from the leaves to the roots. Stems grow both above and below the ground. Specialized kinds of stems below the ground include tubers, bulbs, corms, and rhizomes.

Roots absorb water and nutrients from the soil. They are connected with the stem so that the water and nutrients flow upward. Leaves anchor plants and store food for the plant. Many of the roots that store food are eaten by humans, such as carrots and beets.

Water enters roots by osmosis. This process involves the movement of water from places of higher concentration to lower concentration. From the roots, the water moves through the stems to the leaves. Nutrients are also absorbed with the water. Some of the nutrients are essential to plant growth.

Tropisms influence the direction plants grow. Geotropism is the attraction of certain parts of plants to grow toward and into the ground. Phototropism is the attraction to plants to light.

QUESTIONS

Answer the following questions using complete sentences and correct spelling.

1. How are plants and animals alike? How do they differ?
2. What are the factors in the adaptation of plants to climate?
3. What are the three ways of classifying plants on the basis of life cycle? Distinguish between the life cycles.
4. How can the life cycles of plants be altered?
5. What is the major role of leaves?
6. What are the major parts of a leaf? (Sketch a leaf and label its parts.)
7. What are the kinds of leaves? Give an example of each.
8. How are leaves attached to plants? What patterns are found?
9. What is the role of the stem?
10. What are the two major kinds of stems? Distinguish between them.
11. What are the parts of a stem? (Sketch a cross section of a woody stem and label the parts.)
12. What does the root system do for a plant?
13. What are the two major kinds of root systems? Distinguish between them.
14. What are the parts of a root system? What is the role of these parts?
15. What is tropism? Name three kinds and explain how they affect plants.

EXPLORING

1. Determine the climate in the area where you live. Select five crops that are adapted to the climate. Use reference materials to help with this activity.

2. Make a list of the important crops and/or ornamental plants in your area. Classify them as to annuals, biennials, and perennials.

3. Make a tour of a greenhouse growing facility in your area. Determine the names of the plants being grown. Study how the environment is regulated to help the plants grow.

4. Make a collection of leaves of the common plants in your area. Classify the leaves as simple or compound. Use reference materials and identify the common and scientific names of the plants.

8

REPRODUCING PLANTS

New crops are grown every year. New animals are born every day. Tree seed sprout and grow every spring. It's a continuous cycle of new life.

A major area of agriscience is reproducing plants. Naturally, most plants reproduce on their own, but for agricultural crops, the process is regulated. It is important that crops reproduce when, where, and in a way that provides the desired products in a timely manner.

If this process is not understood, the seeds people plant will not come up (sprout) and plants won't grow. Knowing how plants reproduce helps improve crops and produces better seed. Many improvements have been made in crops as a result of using the processes involved in controlling reproduction.

Even the beautiful flowers of ornamental plants make lives happier! A vase of flowers on a dining table improves the meal. A corsage for the school prom helps make the event more festive. Attractive flower beds around homes and businesses make life more pleasant.

8-1. Flowers are important in reproducing plants. These are flowers of the hibiscus plant.

OBJECTIVES

This chapter provides basic information on reproducing plants. It has the following objectives:

1. Explain ways plants reproduce.
2. Describe the role of seed.
3. Identify and explain the parts of a seed.
4. Explain the role of flowers in reproducing plants.
5. Describe how seed are formed.
6. Describe germination and the conditions under which it occurs.
7. Explain the use of vegetative propagation.
8. Explain the importance of improved seed.

TERMS

- asexual reproduction
- budding
- clone
- complete flower
- cutting
- dry fruit
- embryo
- endosperm
- fleshy fruit
- flower
- fruit
- germination
- good seed
- grafting
- imperfect flower
- incomplete flower
- layering
- ovule
- percent germination
- perfect flower
- planting depth
- pollen
- pollination
- propagation
- scion
- seed
- seed certification
- seed coat
- seed-germ
- sexual reproduction
- tissue culture
- vegetative propagation
- viability

WAYS PLANTS REPRODUCE

Many of the things humans need and want involve growing plants. Just think of the number of wheat plants that have to be grown every year to provide the flour used to make bread!

Most plants reproduce naturally to keep their kind living. Reproduction is an important part of nature. Agriscientists study and use the processes that are involved. New ways of reproducing plants make it possible to have better plants.

PROPAGATION

Agriscientists often use the term ***propagation*** in place of reproduction. Propagation involves increasing the number of plants. It includes both natural plant reproduction as well as other means of getting more plants.

Plants are propagated in two major ways: sexual and asexual.

Sexual Reproduction

Sexual reproduction involves using seed to propagate plants. Flowers are important in this process because they contain the reproductive organs. These organs produce sex cells.

AgriScience Connection

WHEAT: THE MOST IMPORTANT FOOD CROP

Wheat is the earth's most important food crop! People rely on foods made from wheat as a major part of their daily meals. The seed (known as kernels) are used to make flour which is used in making bread, pasta, and breakfast cereal and as feed for animals.

The wheat plant is a grass-type plant that reaches maturity in one growing season. In warmer areas of the southern United States, wheat is planted in the fall. In Canada and the northern states, it is planted in the spring.

The major wheat growing area of North America is in the great plains of Texas, Oklahoma, Kansas, Nebraska, and Colorado. Nearly 96 million acres (39 million hectares) of land are planted to wheat in North America each year.

This shows a nearly mature wheat field. The insert shows a close-up of harvested wheat kernels.

Two kinds of sex cells are produced: female and male. The female sex cell is known as the egg or ***ovule***. The ovules of flowers are very small. Ovules are produced in the ovary of a flower.

The male sex cell is the ***pollen***. The pollen contains sperm that mature during pollination. Through the process of fertilization, the sperm in the pollen unites with the ovule in the ovary and a new seed is formed.

A seed contains the tiny embryo of a new plant. An ***embryo*** is an immature plant. It will grow into a mature plant if planted properly. Embryos are sometimes called ***seed-germ***.

Sexual reproduction involves the creation of a genetically new individual. Many, many combinations are possible, depending on the different parent plants involved. When agriscientists get involved, they sometimes produce improved plants in ways that nature can't.

Asexual Propagation

Asexual reproduction involves reproducing plants by using vegetative parts of the plant. There is no union of sperm and egg. Asexual reproduction is widely used with many important food and ornamental plants.

Some asexual reproduction occurs in nature, but most involves the work of agriscientists. In nature, some plants send out runners or develop shoots that grow into new plants. A good example is the strawberry. Mature strawberry plants grow several runners (stolons) that are a few inches long. New plants are formed along the runner. Each runner may start several new plants. The stolon can be cut between the plants so that each lives on its own.

8-2. Pineapples are reproduced asexually. This shows a field ready for setting the piles of slips through holes in the strips of plastic.

Asexual reproduction does not result in genetically different plants. Those that are produced are genetically like their parent. This is known as a ***clone***. In the example of the strawberry, the plant that develops from the runner is identical to the parent plant. The berries of the new plant will be very similar to those of its parent.

Asexual reproduction assures that the new plant will be like the old one. This is important in fruit and nut trees, ornamental plants, and many others.

Here is an example of asexual reproduction: Native pecan trees produce very small pecan nuts. Occasionally one will grow that produces larger pecans. When this occurs, pieces of the tree with the large pecans are cut off and grown as another tree. The new tree will produce pecans just like the old one. If we sexually reproduced the pecan by planting the large pecan as seed and letting it grow, the resulting pecan nut would most likely be small and not like the parent. Pecan orchards with many trees may actually be just one tree growing separately many times!

USING PROPAGATION

Successfully growing plants involves using methods of propagation. People in agriscience study the best ways and use them in their work. New plants won't develop and grow if the wrong practices are followed!

When properly planted, a seed will sprout (germinate) and produce a new plant. Knowing when and how to plant seeds is important. Seeds with living embryos can be planted wrong and plants will not grow. After the plants come up, the care (culture) needed for them to grow must be provided.

Asexual reproduction involves skill in knowing how and when to use it. Care is needed to assure that the new plant lives.

SEXUAL PLANT REPRODUCTION: SEED

Seed are containers of new plant life. They are formed in the ovaries of flowers, which become fruit. Seed are designed to allow new life to grow and become established as plants. Seed insure continuing life as well as provide food and other products.

In agriscience, plants must be able to produce a lot of seed. This requires understanding seed and how it develops.

KINDS OF SEED

Seed can be classified as two kinds: monocot and dicot. The important major difference in the seed is evident in the plants that are formed.

Monocots are plants that have seed with one seed leaf, known as a cotyledon. The embryo in the seed will have one leaf. As the embryo grows, the leaves develop with parallel venation. Corn, wheat, rice, and all grasses are monocots.

Dicots are plants that have seed with two cotyledons. The embryo in the seed has two seed leaves. In addition, the leaves of dicots have net venation. Examples include tomatoes, beans, petunias, carnations, and most all other plants with net venation in the leaves.

The words monocot and dicot are shortened from the longer words of monocotyledon and dicotyledon. The original, longer words more fully described that the difference was due to the number of cotyledons. "Mono" means one, while "di" means two. When used before the word, cotyledon, we know how many seed leaves are in the seed.

Many more dicot plants than monocots are found on earth. Scientists estimate that there are 185,000 species of dicots as compared to 90,000 species of monocots.

The seed of both monocots and dicots have the same basic parts. Seed vary considerably from different species of plants within dicots and monocots.

8-3. Two views of a corn kernel (monocot) and lima bean (dicot) illustrate outside differences in the seed.

PARTS OF A SEED

As a container of life, a seed must perform certain functions. It must keep the life viable (alive) and provide food for it to grow.

The structure of dicot and monocot seed varies even though the same basic parts are present.

Dicot Seed

Dicot seed vary in shape depending on the species of plant. Bean seed tend to be flat, while other seed may be round, such as the radish. In learning about seed, the seed of a lima bean is a good example to use.

Bean seed are often rounded and flat. They develop in a pod with several other seed. The pod (bean fruit) develops where the flower grew on the plant. Upon maturity, the pod comes open and the beans fall out. Each bean seed is capable of starting a new plant.

The outside (external) parts of a bean seed are:

- **Seed coat**—The seed coat is the outer covering of the seed. It protects the fragile embryo from injury and holds the seed together. The coat must also allow moisture to enter the seed when it is planted so that it can sprout (germinate). Seed coats are also known as testa.
- **Hilum**—The hilum is the place where the seed was attached in the pod. It is sometimes known as the seed scar.
- **Micropyle**—The micropyle is the tiny opening near the seed scar where the pollen entered the ovule to form the seed.

The inside (internal) parts of a bean seed are:

- **Cotyledons**—Bean seed have two large cotyledons. They break apart forming two symmetrical (alike) halves of the bean. The cotyledons are usually fleshy in form because they contain the food for the developing embryo. The food is high in protein and oil. This is why beans are important crops!
- **Radicle**—The radicle is the part that becomes the root system of the plant. It is sometimes known as the embryonic root.
- **Hypocotyl**—The hypocotyl is the part that connects the radicle and the cotyledons. It is shaped like a stem and connects the embryo with the food supply in the cotyledons.
- **Epicotyl**—The epicotyl forms the stem of the plant. It is attached to the hypocotyl on one end and has tiny, undeveloped leaves (embryonic leaves) on the other end.
- **Plumule**—The plumule is the part of the seed that develops the above-ground part of the plant. It includes the epicotyl and embryonic leaves in the seed.

Monocot Seed

Monocot seed contain an embryo and food supply, just as dicot seed. How the seed are made is different, however. A good example is corn.

A corn grain (kernel) is actually a fruit, much as the entire pod of the bean is a fruit. Under proper conditions, the corn seed will sprout and grow.

The external parts of a corn seed are:

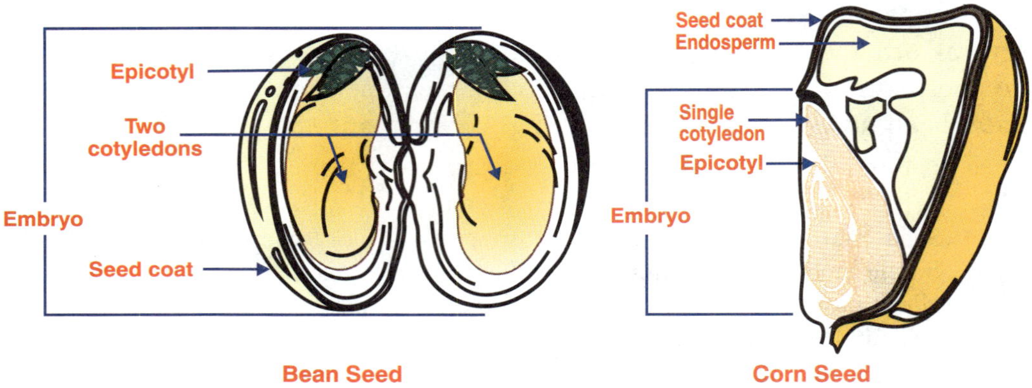

8-4. The major parts of dicot and monocot seed are shown here using a bean and corn kernel.

- **Seed coat**—The ***seed coat*** protects the embryo and holds the seed parts together.
- **Seed scar**—This where the corn grain was attached to the cob. A corn cob is a long, rounded structure formed inside of the corn ear. The ear is the part of the corn plant that holds the seed (grain). The ear is covered with husks (shucks) that protect the developing seed. Ears form from the fertilized female flowers of the corn plant (stalk).
- **Silk scar**—The silk scar is on the opposite side of the grain from the seed scar. The silk scar is where the silk (pollen tube) in the female flower was attached to the ovule. Silks extend outside the developing corn ear in the female flower to receive the pollen from the male flower.

The internal parts of a corn seed are:

- **Endosperm**—The ***endosperm*** is where the food is stored for the embryo to use when the seed sprouts. (It is also the part of the corn grain that provides much of the nutrition in food and feed made from corn.)
- **Radicle**—The radicle becomes the root system. It grows into the soil as the seed germinates and the plant develops.
- **Hypocotyl**—The hypocotyl connects the radicle with the food source. It does not grow in monocots as it does in dicots.
- **Epicotyl**—As with dicots, the epicotyl forms the stem of the corn plant.
- **Cotyledon**—The cotyledon does not store food, as in the bean plant. It absorbs food from the endosperm and moves it to the embryo.
- **Plumule**—The plumule is the part of the corn seed that develops into leaves on the growing plant. In the seed, it is only one leaf.

HOW SEED ARE FORMED: FLOWERS

Seed are borne by (formed by and in) flowers. A ***flower*** is the reproductive part of flowering plants. Nearly all plants of importance in agriscience have flowers.

Flowers vary in many ways. They are shaped differently and grow on plants in different ways. Some have attractive colors and fragrances; others don't. Regardless, all flowers are involved in the sexual reproduction of plants. And they are specially made to ensure reproduction!

8-5. Food plant flowers. (Note: The flower on a pecan is known as a catkin.)

TYPES OF FLOWERS

Flowers are categorized in two ways: complete or incomplete and perfect or imperfect.

Complete or Incomplete

Complete flowers have four principal parts:

- **Sepals**—The sepals are the outer parts of the flower. They are usually green but may have bright colors, such as the tulip. Sepals cover the bud and protect it as it develops. In open flowers, the sepals are found at the base of the petals. They provide support for the petals and help hold them open.
- **Petals**—Petals are located just inside the sepals. They are often brightly colored and showy. They attract insects which help with the fertilization process by carrying pollen. All of the petals on a plant form the corolla.
- **Stamens**—The stamens are the male part of the flower. They are located inside the ring of petals. Each stamen has an enlarged anther at the end that produces pollen or sperm. The stalk-like structure that supports the anther is the filament.
- **Pistil**—The pistil is the female part of the flower and is located in the center of the flower. The outer end of the pistil has an opening known as the stigma. The tube-like structure that connects the stigma with the ovary is the style. The ovary is the enlarged base where the ovules (egg cells) are located. Fertilization occurs when pollen grains come into contact with the stigma and are moved through the style to the ovules. The ovules develop into seed after being fertilized.

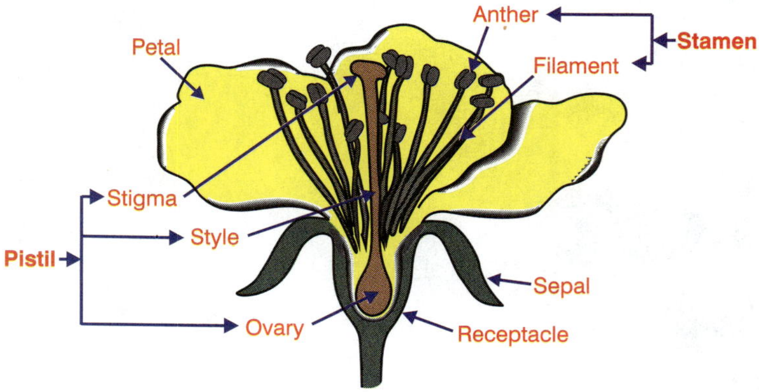

8-6. Parts of a perfect flower.

Incomplete flowers do not have all four principal parts. Some flowers do not have sepals and petals, such as the flowers on wheat and oats. The process of fertilization can take place without sepals and petals. Of course, flowers must have ovules and anthers so seed can be formed. In some cases, the male and female flower parts are separate on plants. Both are needed for a seed to be formed.

Perfect or Imperfect

Perfect flowers have the stamen and pistil in the same flower. Those flowers that lack either stamens or pistils are imperfect. Some imperfect flowers have the male sex organs; other imperfect flowers are female.

Plants may have both male and female ***imperfect flowers*** on them. When they do, they are known as monoecious. Corn is a monoecious plant.

On corn, the male flower is the tassel at the top of the plant. Pollen is produced by the tassel. The female flower forms where the ear grows on the plant. The female corn flower is observed as a small shoot with silk at the end. Usually two shoots develop on one corn stalk. They develop where leaves join the corn stalk but only a few leaves in the middle of the stalk are involved. Other monoecious plants are squash, melons, and pumpkins. Flowers on these plants are easily observed as male or female. The female bud and flower has a small structure that resembles a fruit. The male flower does not.

Some species of plants are dioecious. This means that some plants have male flowers and other plants have female flowers. Plants that are male do not bear fruit. This is the reason why dioecious plants are often propagated by vegetative means. People who grow crops must have plants

8-7. This research plot shows corn tassels (male flower) ripe with pollen. The shoots with silks are along the stalks. (Courtesy, DEKALB Plant Genetics, Dekalb, Illinois)

that bear fruit! Examples include strawberries and asparagus. To be sure that a strawberry plant will produce berries, they are propagated vegetatively using stolons (runners).

Combinations

Flowers can be combinations of complete and incomplete and perfect and imperfect. Two examples are (1) wheat, which is incomplete and perfect and (2) cotton, which is complete and perfect.

POLLINATION

Pollination is the transfer of pollen from an anther to a stigma of a flower of the same species. Pollen may be moved by wind, insects, birds, and other natural means. Insects and birds move pollen because it sticks to their bodies. As they move from flower to flower, some of the pollen rubs off. In some situations, agriscientists move the pollen to insure that the plant is properly fertilized.

8-8. This flower on a pine tree will produce an abundance of pollen.

Some plants have very attractive, brightly colored flowers. This is nature's way of attracting insects to assure that the flower is pollinated.

Flowers may be cross-pollinated or self-pollinated. Cross-pollination is when two different plants are involved. Pollen from an anther on one plant is moved to the stigma on another plant. Self-pollination involves flowers on the same plant. The pollen is moved from one flower to another on the plant.

A good example is the corn plant. A tassel may make as many as 15,000,000 pollen grains. Only one grain is needed for the formation of a kernel in an ear of corn. In a field of corn, the plants may self-pollinate and cross-pollinate. Scientists believe that no more than 5% of corn kernels are self-pollinated. In other words, most kernels on a corn plant are the result of two different parent plants.

FERTILIZATION

Fertilization is the union of the pollen cell with the ovule (egg). After getting on the stigma, the pollen grain grows a long tube cell through the style toward the ovule. A pollen grain develops two sperm. One of the sperm unites with the ovule in the ovary and forms an embryo. The other sperm forms tissue known as endosperm in the embryo sac. Some agri-scientists refer to this as double fertilization, which is unique to flowering plants.

SEED FORMATION

Fertilization initiates the growth of fruit and seed. A flower that is not fertilized will not produce seed.

An ovary that has been fertilized forms ***fruit***. In some cases, the ovary grows very large and has seed inside. Examples include watermelons, cucumbers, green peppers, and tomatoes.

Once an ovary has been fertilized, the flower is no longer useful. Flowers on many plants dry up and fall to the ground soon after fertilization. For example, the big, beautiful flower on a squash plant may last only one or two days. It quickly dries up and may be seen on the end of the developing squash as a dead flower.

The seed that are formed are a product of their parents. All parts of the seed must develop if the seed are to be planted and grow into other plants. They must remain on the parent plant until mature. If fruit are pulled too soon, the seed may not grow because they are immature.

8-9. The apple is a fleshy fruit with seed in the middle.

KINDS OF FRUIT

Different plants form different kinds of fruit. There are many different kinds of fruit. In general, all fruit can be classified into two types: fleshy and dry.

Fleshy fruit refers to those that have large fibrous structures surrounding the seed. Examples include apples, pears, blackberries, oranges, and grapes.

8-10. Corn is a dry, hull fruit. This ear of dry shoe peg corn illustrates arrangement of kernels on the cob.

Dry fruit refers to fruit that develops as a pod or in a hull. Pod fruits develop in structures that have lines or places where the fruit has definite seams. They can usually be easily divided into half. Beans, peas, peanuts, and cotton have pods. Hull fruits do not have definite seams in the shell of the fruit. Hull fruits include pecans, acorns, corn, oats, wheat, elm tree seed, and the dandelion.

The fruit that develops from the ovary of a plant is important as food and for other purposes. Agriscientists often study ways of producing more and larger fruit.

SEED DISPERSAL

Seed need to be dispersed (scattered) over the land to grow new plants. In nature, seed are scattered in many ways. Birds may remove a plum from a tree and carry it a few hundred yards before the seed is dropped. Wild animals may carry fruit from a tree some distance away for eating and discard the seed. Some seed have structures that help them float long distances in the air.

As plants were domesticated, humans developed systems of dispersing and planting seed. The methods used are designed to allow plants to grow and produce large quantities of fruit.

8-10. Snap beans are dry fruit. The beans form in pods with definite seams.

8-12. Examples of fruit.

HOW PLANTS ARE FORMED: GERMINATION

Germination is the sprouting of a seed to grow a new plant. Seed must be mature in order to germinate. Some seed will germinate soon after full size is reached in the fruit. Other seed must go through a time of rest, known as dormancy. Seed that are properly stored will last for years.

VIABILITY

Agriscientists want seed that will grow. This quality is known as ***viability***. When seed are viable, they come up and grow plants.

In order to germinate, seed must be in an environment that is appropriate for the species. Seed of different species sprout under different conditions. Some need warmer temperatures than others. Some need more moisture and must be protected from sunlight.

Seed Testing

Seed should be tested for viability. Seed that are sold usually have labels attached to the bags indicating the percent of seed that will sprout and grow. This is known as percent germination. If a label says that the seed have 90% germination, 90 out of every 100 seeds should sprout and grow. Of course, the seed must be properly planted!

Germination testing is done by sprouting a sample of seed. For example, 100 seed may be taken from a sack of corn and artificially sprouted. This may involve wrapping the seed in moist paper towels and keeping them at a warm temperature, such as 85°F for several days. After a few days, the number with sprouts is counted and divided by 100 (the number of seeds in the test). If very few seed sprout, the seed have low viability.

Percent germination is important because it influences the number of seed to plant. If you want 20,000 plants to grow on each acre of land, you need to plant more than 20,000 seeds. If percent germination of seed is 90%, 10% more seed should be planted. In this case, 22,223 seed would need to be planted on the acre. Of course, the seed must be properly planted and the weather must be good.

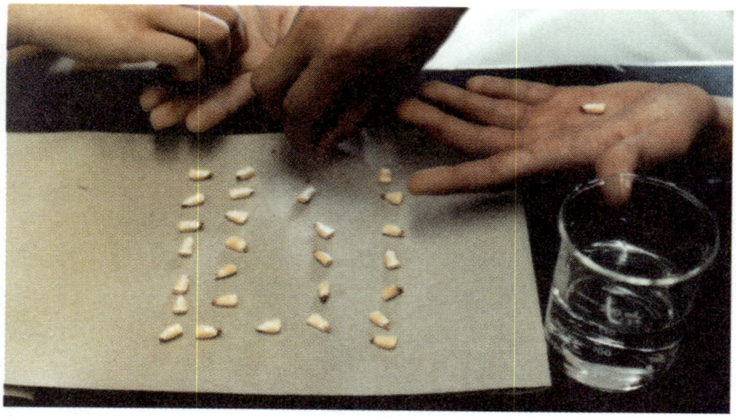

8-13. These corn seed are being prepared for viability testing. The seed will be wrapped in the moistened paper and kept in a warm place for several days. Those that sprout are viable.

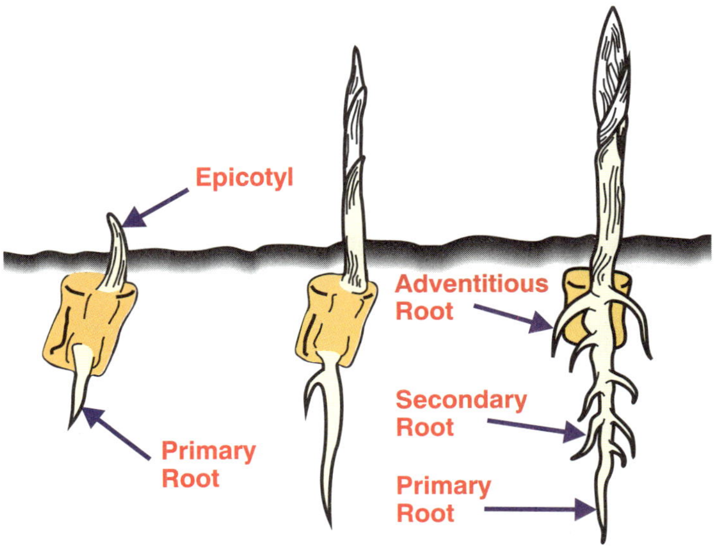

8-14. The sprouting and emergence of corn—a monocot plant.

Viability is related to the amount of food stored in the seed and how the seed are handled before planting. Seed that are fully developed as opposed to shriveled usually have enough food. Seed should be stored in cool, dry places and protected from injury by insects, rodents, and rough handling. Dropping a bag of seed injures the seed inside and may cause them not to grow.

Germination Needs

Seed are planted in some kind of medium that encourages germination and provides water and other nutrients. The medium must help seed sprout and grow.

Soil is by far the most widely used growing medium. Wheat, corn, pasture grasses, trees, ornamental plants, and others are planted in soil. Soil is an important natural resource.

Seed planted in greenhouses and other specialized growing structures may be planted in "artificial" media. The artificial media may be made of natural materials. Examples of media include vermiculite, peat, and sphagnum moss. These materials are sometimes mixed with soil or manure to form the growing medium best suited for a species.

Most seeds require three conditions to be within certain ranges before germination will occur. These conditions are moisture, temperature, and

8-15. Artificial media are often used to provide a good growing medium. This false aralia is being transplanted to a larger pot with a mixed medium of both artificial and natural material.

oxygen. Some seed require light in order to germinate. Most of those used in agriculture do not. In fact, light can damage some sprouts as they emerge.

Moisture is needed for germination. Seed usually absorb considerable water when planted. The water softens the seed coat and causes the seed to swell. Too much moisture can damage the seed. Bacteria, fungi, and other organisms can attack the seed. Wet seed will rot and not sprout, particularly in cold or cool weather.

Since most seed are planted in the soil, soil temperature is important. Seed will not sprout if the soil isn't warm enough for them. Some species require warmer soil than others. Corn needs soil that is at least 50°F. Many crops need a soil temperature of 70°F or higher. Seed planted in greenhouses may be placed in soil or other growing media that is heated to assure good germination.

Oxygen is needed for the embryo to live and the sprout to grow. Soil that is too wet may not have much oxygen present.

PLANTING DEPTH

Planting involves covering the seed with soil or other media. The layer of soil incubates the seed. Conditions should encourage the seed to sprout.

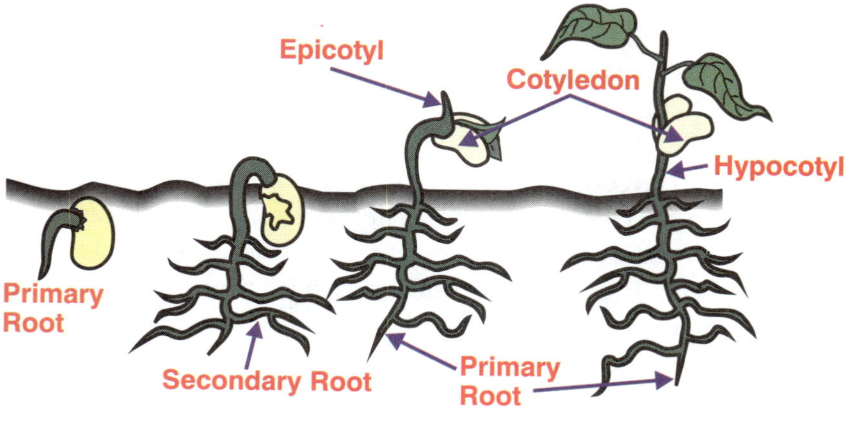

8-16. The sprouting and emergence of a bean—a dicot plant.

Temperature, moisture, and available oxygen vary by depth in the soil. ***Planting depth*** is how deep a seed is planted.

Seed of different species are planted at different depths. It is important to know the requirements of each species. Sometimes a general rule of thumb is used: plant seed at a depth that is eight times the thickness of the seed at its thinnest place. Very small seed may not be covered with any soil. Merely applying water will be sufficient for the seed to germinate. An example are the tiny seeds of the begonia.

Depth of planting is related to the kind of soil. Seed are planted a little deeper in sandy soil than in clay soil. Also, seed should not be planted quite as deep if a rain is expected. A heavy rain can compact the soil over a seed making it hard for the new plant to push its way through the soil. (Table 8-1 shows suggested depths for planting some crops.)

Table 8-1
Planting Depth of Selected Crops

Crop	Planting Depth
	(inches)
Beet	1
Carrot	½
Corn	2
Cucumber	1½
Peas	2
Potato	4
Radish	½
Snapbeans	2
Spinach	1
Tomato	½
Watermelon	1
Wheat	1½

Note: Seed should be planted a little deeper in sandy, dry soil and not quite as deep in heavy, moist soil.

ASEXUAL PLANT PROPAGATION

Asexual plant propagation is reproducing plants without seed. This is sometimes known as vegetative propagation. A piece of the parent is cut off and put in another place to grow. The plants are reproduced without an ovule being fertilized.

The plants that are produced are known as clones. A clone is an organism that has traits identical to its parent. It's a copy of its parent!

Vegetative propagation involves getting a part of a plant to grow into another plant. Leaves, stems, or roots are used to grow the new plant. In some cases, parts of desired plants are transferred to less desirable plants.

BENEFITS OF VEGETATIVE PROPAGATION

Vegetative propagation has several benefits. These help assure quality crops and ornamental plants.

Two major benefits are:

- **True traits of parents**—Seed produce plants that may vary from the parents. This is known as not breeding true. Even in the same species, variation occurs when sexual reproduction is used. Many fruit trees have been carefully selected for large, delicious fruit. If the seed are planted, the plant they grow might not produce the same high quality fruit. For example, the seed of an apple might grow into a tree that would bear fruit that would not be like that of its parent. Vegetative propagation assures that the new plant is identical to its parent.
- **No seed**—Some plants do not produce seed. An example is the seedless grape. Other plants are vegetatively propagated because they have seed that are very small or difficult to get.

METHODS OF VEGETATIVE PROPAGATION

How a plant is propagated varies with the nature of the plant. Some plants are easy to propagate with buds, stems or other parts; others are difficult. Some plants can't be vegetatively propagated with much success.

Plant parts located above and below the ground can be used.

AgriScience Connection

FUTURISTIC FRUIT

Many techniques are used to produce plants. Some require high levels of scientific skill. The dishes in this plant growing facility contain tiny peach and apple trees. Grown from lab-cultured cells, the experimental trees will be assessed for practical use in providing nutritious food.

Many foods of the future are being developed in laboratories today. Tissue culture and genetic alteration are often a part of developing the plants. Years of field trial and testing will be needed before the plants are released for production. (Courtesy, Agricultural Research Service, USDA)

Below-Ground Parts

8-15. Eye cut from a potato tuber that is ready to plant.

Some plants grow stems and root structures below the ground that are used for vegetative propagation.

In some cases, the parts with buds are cut into pieces. A good example is the potato. Potatoes are below-ground stems that store food and have buds. A potato can be cut into sections with buds (known as eyes) and planted. Each eye will develop into a potato plant. Sometimes entire potatoes are planted.

In other cases, the parts are used without cutting. Bulbs and corms are two examples. Some plants grow a number of bulbs. This is often known as multiplying. The parent plant can be dug and the new bulbs pulled apart and planted separately.

Above-Ground Parts

Many plants are propagated using buds, stems, and other structures that grow above the ground. The major methods of vegetative propagation are: layering; using cuttings, buds, and grafts; and using tissue culture.

Layering. ***Layering*** involves getting roots to grow from the stems of plants. It is done in several ways.

Soil may be pulled up around a stem to get it to grow roots. For example, some people layer gardenias using a low limb on a gardenia. The limb is bent down to the ground and partially covered with soil. The covered part will grow roots in a few weeks. After enough roots have developed, the limb can be cut from the parent plant. It will have enough roots to grow on its own and can be moved to another location where the gardenia will grow.

Air layering involves placing a growing medium around a stem that includes a bud or node. Sometimes a cut is made in the stem and it is covered with a medium known as sphagnum moss. A layer of plastic may be wrapped around the moss to hold moisture. After roots develop, the top part is cut off just below where the plant was layered to make the

8-18. Air layering is being used to reproduce this plant.

new plant. This new plant can be set out to grow on its own. It is now free of support from the parent plant.

Cuttings. A ***cutting*** is a short section of plant stem used for propagation. The section of stem is cut from the parent plant and put in a growing medium to sprout roots. Once the roots have developed, the plant can be set out and grown to maturity. Many plants can be grown from cuttings: begonias, gardenias, pineapples, and coleus are examples.

Budding. ***Budding*** involves taking a bud from one plant and moving it to another. The bud is removed from a parent that has desired traits. It is placed on another plant (stock) to grow. If properly done so that the layers of the stem match, the bud will grow. Of course, buds can only be placed on stocks that are compatible. Examples include fruit trees, roses, and nut trees.

Grafting. ***Grafting*** involves placing a section of a stem (known as ***scion***) of one plant onto another plant. Just as in budding, the scion and stock must be compatible.

The process involves attaching the scion so that it grows on the stock as if a natural part of the plant. The part that grows from the scion will

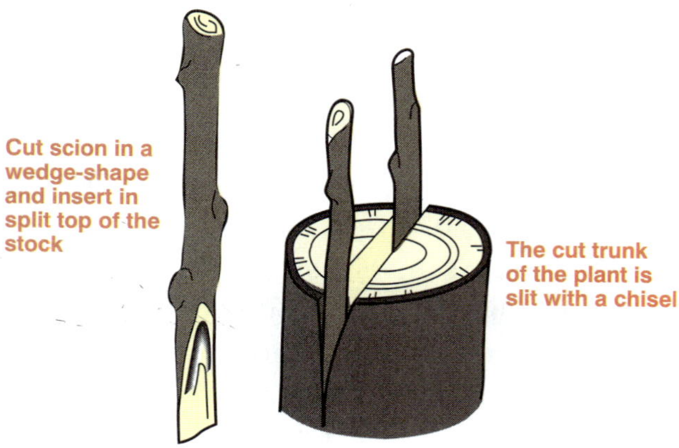

8-19. The cleft graft is a popular way of grafting young trees.

8-20. The process of making a bark graft involves several steps. The scion is prepared by shaping to fit the stock, the stock bark is split, the scion is placed so that cambium layers match with the stock, and the scion is secured in position to begin growing.

have traits of its parent. Any growth on the stock will be like the stock. On pecans, small sections of stem of the desired pecan are placed on small tress grown from seed, known as seedlings. The seedlings would not produce big, nice pecans but they are excellent stocks for the scions. Grafting requires matching the cambium (growing part of the stem) of the stock and scion. If done properly, the two parts will grow together as one plant.

Tissue Culture. ***Tissue culture*** is one of the newer methods of vegetative propagation. It involves taking a group of cells or a single cell and growing it into a plant. Many, many plants can be grown from one parent.

Specialized laboratory techniques are needed to assure success. Sanitation is most important. The cultures are placed in warm, well-lighted places in a liquid or semi-solid growing medium. Since the cultures start very small, some of the work may require magnification of the cells. Producers of ornamental plants, such as chrysanthemums and carnations, are finding tissue culture a good way to vegetatively propagate plants. (Tissue culture can also be used to produce new plants from underground parts.)

USING IMPROVED SEED

Many plants have been selected and improved to make better food or other products. The seed planted should be of the best-suited variety. Agriscientists have made improvements in many plants.

VARIETY SELECTION

Only varieties of plants suited to a particular location or use should be planted. People who are planning to grow crops should get information on the best varieties to plant. For example, varieties of corn that grow well in Canada might not do well in the United States.

8-21. Cotton seed tend to have lint still attached. They are often chemically cleaned to remove the lint left by the ginning process. Seed with lint don't work well in planting equipment.

Varieties suited to the climate should be selected. Most local seed stores sell only the varieties recommended for the area. Ordering seed from companies that are located far away might result in varieties that aren't suited to the area. If unsure about which variety to get, contact the agronomy specialist at a college or research station.

Varieties should be selected that will meet the desired purpose. Using corn as an example, some varieties are used for grain; others are for people to eat fresh as roasting ears. Varieties have been developed that have characteristics that make them best for particular purposes. Of course, people can eat corn being grown for livestock feed but the corn may not be as sweet and tasty.

Seed Certification

Many states in the United States have seed certification agencies or associations. Seed certification assures that the seed are of good quality. This process is used in the improvement of plant varieties and to assure genetic purity of the seed planted.

Four classes of seed are certified, as follows:

- **Breeders seed**—These are seed that are used to breed or maintain seed of a particular variety. Only plant breeders usually have breeders seed, as they aren't sold to the public. These seed are placed in containers labeled with a white tag and the words "Breeders Seed."
- **Foundation seed**—These are seed that are produced from breeders seed. The seed are carefully grown to maintain the genetic traits established by the breeder. Containers for these seed usually have white tags labeled "Foundation Seed." The tags are issued by the seed certification association.
- **Registered seed**—These seed are produced from foundation seed. They are carefully grown to assure purity of the variety. Containers have a purple tag that provides information about the seed.
- **Certified seed**—This is the most widely used class of seed. They are produced from registered seed. High standards are to be followed by all seed producers to assure good seed. Purple labels are used for certified seed.
- Seed certification is primarily used for the major crops, such as wheat, soybeans, and corn.

8-22. Quality seed are pure and free of weed seed, aren't cracked or broken, and are free of disease. The soybean seed shown here are of high quality. (Courtesy, American Soybean Association, St. Louis, Missouri)

GOOD QUALITY SEED

Good seed are essential. Good seed, when planted properly, will come up, grow, and become the intended plant. Some people grow plants and save their own seed. Other people buy the seed they plant. In either case, good quality is a must.

Several things to look for in good seed are:

- **Purity**—Seed should be pure (contain only the seed of the intended variety). Seed that have other varieties of seed mixed in are impure.
- **Contamination**—Seed should be free of trash, such as: pieces of leaves, pods, and stalk; wood pieces; and stones. Most seed are carefully harvested and cleaned to keep trash out.
- **Percent germination**—The seed should come up and grow. Only seed with 90–100% percent germination should be selected.
- **Uniformity**—Seed should be of uniform size and shape. Small seed or those that aren't shaped properly may not germinate. Buying them is a waste of money!
- **Damage**—Seed coats should not be cracked or broken. Seed that have been damaged may not come up. Rough handling in harvesting, cleaning, and moving seed causes damage. Broken seed are a waste of money.
- **Disease**—Seed should be free of disease. Seed that contain disease will not produce well and can infect an entire farm with a disease.
- **Treatment**—Some seed are treated to prevent disease when the seed are planted. These seed should be carefully handled and never fed to animals. The chemicals used to treat seed may cause injury to animals.
- **Name of grower and/or dealer**—Most seed growers and dealers are honest, reputable people. They produce only good quality seed. Avoid using seed from sources that are unknown or known to be dishonest.

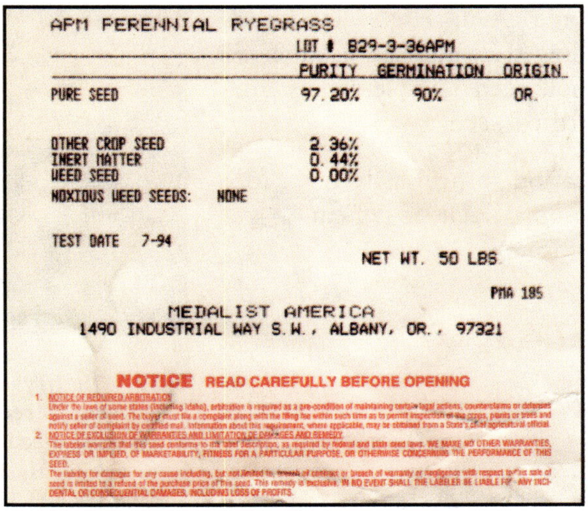

APM PERENNIAL RYEGRASS

LOT # 829-3-36APM

	PURITY	GERMINATION	ORIGIN
PURE SEED	97.20%	90%	OR.
OTHER CROP SEED	2.36%		
INERT MATTER	0.44%		
WEED SEED	0.00%		

NOXIOUS WEED SEEDS: NONE

TEST DATE 7-94

NET WT. 50 LBS

PMA 185

MEDALIST AMERICA
1490 INDUSTRIAL WAY S.W., ALBANY, OR., 97321

NOTICE READ CAREFULLY BEFORE OPENING

1. NOTICE OF REQUIRED ARBITRATION
[illegible]
2. NOTICE OF EXCLUSION OF WARRANTIES AND LIMITATION OF DAMAGES AND REMEDY
[illegible] WE MAKE NO OTHER WARRANTIES, EXPRESS OR IMPLIED, OF MARKETABILITY, FITNESS FOR A PARTICULAR PURPOSE, OR OTHERWISE CONCERNING THE PERFORMANCE OF THIS SEED.
[illegible] IN NO EVENT SHALL THE LABELER BE LIABLE F[illegible] ANY INCIDENTAL OR CONSEQUENTIAL DAMAGES, INCLUDING LOSS OF PROFITS.

8-23. Sample seed label.

REVIEWING

MAIN IDEAS

Plants are propagated (reproduced) sexually and asexually. Sexual propagation involves the union of a male and female sex cell. Asexual propagation is the use of vegetative parts of a plant to grow another plant.

Sexual reproduction involves the formation of flowers, pollination, fertilization, fruit and seed growth, and properly planting the seed. Flowers have male and female parts. In some cases, separate flowers are male and female. Plants produced by sexual means have genetic traits of both parents. Pollen is the male sex cell produced by the anther in a flower. Pollination involves the pollen contacting the stigma or female part of the flower. The pollen develops sperm which fertilize ovules in the ovary of the flower. The fertilized ovary develops into fruit which contain seed. Seed are containers with living plants inside. They must be carefully handled to assure their growth.

Germination occurs when seed sprout and grow. Seed must have the right conditions in order to germinate: moisture, temperature, and oxygen. Most seed are planted in growing media, usually soil, that provide the needed germination environment. Only seed that are pure and have a high percent germination should be planted.

Asexual propagation results in a plant identical to its parent, known as a clone. The genetic traits remain the same. Many plants can be produced from one parent plant. Both above-ground and below-ground plant parts are used in asexual propagation, depending on the species. Asexual propagation is done in several ways: layering, cuttings, budding, grafting, and tissue culturing. All methods produce a clone of the original plant. Tissue culture is being used with valuable plants that are difficult to grow by other means.

The success of a crop can be no better than the seed used. Growers should select seed of the best possible quality. Seed should be pure, free of disease and damage, and have a high percent germination. It is best to get seed only from reputable sources. All seed should be labeled with information about the quality of the seed.

QUESTIONS

Answer the following questions using complete sentences and correct spelling.

1. What is the purpose of plant propagation?
2. Distinguish between the two major ways plants are propagated. Give advantages and disadvantages of each.

3. What is a seed? Sketch a seed and label its parts.
4. What is the purpose of flowers?
5. What are the parts of a complete flower? Sketch a flower and label its parts.
6. What is pollination? How does it occur?
7. What is fertilization?
8. How are seed formed?
9. What kinds of fruit are produced by plants?
10. What is germination? Why is it important?
11. What does a seed need to germinate?
12. Why is vegetative propagation used?
13. What parts of plants can be vegetatively propagated?
14. What methods of vegetative propagation are used?
15. Why should quality seed be used? What determines seed quality?

EXPLORING

1. Visit a place where seed are sold (farm supplies store, garden center, or the display rack in a supermarket). Read the labels on the seed. Determine the stated percent germination, purity, and other characteristics.

2. Conduct a germination test of seed. Use corn, beans, peas, or another crop with large seed. Place the counted number of seed (usually 25) on a moist paper towel on a tray and cover it with three or four layers of moist towel. Place the towels with seed in a warm location. Keep the towels moist. Check the seed in 3–5 days and note any swelling or sprouting. The amount of time varies with the species of plant but a week is usually long enough for common seed. After a week count the number of seed that sprouted and use that number to get a percent germination for your seed test.

3. Use the eye on a potato to grow another potato plant. The eye should be planted two to four inches deep in soil and keep moist and warm. The potato eye will begin to grow in a few days and should come up in two to three weeks.

4. Arrange to tour a seed processing facility. Note how the seed are cleaned and protected from injury. Prepare a report on your observations.

9

PRODUCING PLANTS

Producing plants requires knowledge and skill. Sure, plants grow wild in nature—the forest is filled with wild trees! The plants in most forests are growing without much attention to spacing, season of the year, and how fast they are growing.

When growing crops, producers want them to grow quickly and produce heavily. A top-quality product is also preferred. People are impatient—most can't wait around for nature to provide food, fiber, and shelter.

Producing plants when, where, and how they are wanted is more involved than some people think. There must be knowledge of how they grow and what can be done to make them grow better. They must be kept healthy and free of damage!

Plant growth includes principles of science. If the way plants grow is understood, things to help them grow better can be done.

9-1. Careful examination of plants is often needed to prevent problems and help them grow. (Courtesy, Michael J. Montague, Monsanto Company, St. Louis, Missouri)

OBJECTIVES

This chapter provides important information on how plants grow. It has the following objectives:

1. Describe how plants grow.
2. Explain important factors in plant growth.
3. Explain photosynthesis and its importance.
4. Define transpiration and tell why it is important.
5. Name the nutrients plants need and describe how plants get them.
6. Describe nitrogen fixation in legumes.
7. Explain the use of fertilizer.

TERMS

annual growth rings
auxin
cell cycle
cell division
chlorophyll
dry fertilizer
fertilizer
fertilizer analysis
gibberellic acid
glucose
glycolysis
heartwood
legume
liquid fertilizer
major elements
mulch
nitrogen
no-till cropping
phosphorus
photosynthesis
plant hormone
potassium
primary growth
sapwood
secondary growth
side-dressing
soil analysis
soil sampling
tissue analysis
top-dressing
trace elements
wilting

HOW PLANTS GROW: MORE CELLS

The tiny new plant that emerges from a seed usually gets off to a good start. It survives on the food stored in the seed until it can get its roots established and send leaves through the soil's surface into air and sunlight. Once the plant reaches this stage, it can make its own food.

Plants are made up of many tiny cells. Growth involves increasing the number of cells in a plant. Size, shape, and functions of cells vary. The cells in a plant all work together to form tissues and organs. This makes it possible for the plant to live and grow.

CELL DIVISION

Plants are multicellular organisms. They are made of hundreds, thousands, millions, or billions of cells, depending on plant size.

The number of cells in a plant increases as the cells divide. This type of ***cell division*** is known as mitosis. Cells divide over and over. They go through a process, known as the ***cell cycle***. Cell cycle is the sequence of events that occur in a cell from one mitosis to the next. It is a continuous process.

The cell cycle can be divided into three parts:

- Part 1: Interphase—This is the normal phase of a cell. Often this phase of cell life is considered a part of mitosis because the DNA in the nucleus is replicated at the end of interphase. Because the cell is growing rapidly and increasing in size during the interphase, it is often called the "resting stage" of cell life and not considered a part of cell division. A good environment encourages cells to grow bigger. This includes plenty of water, nutrients, light, and the right temperature.
- Part 2: Mitosis—This is the part of cell life that includes the division of the nucleus. There are four phases of mitosis (five if interphase is included) during which two nuclei are formed. Once the nuclear material is divided, the cell prepares for division of the remainder of the cytoplasm.

9-2. This tiny tree seedling with seedcoat attached has a lot of growing to do in order to become a log.

9-3. The small terminal bud on the end of the stem of a pine tree will grow rapidly and increase the height of a tree several inches a year.

9-4. The tassel (male flower) on a corn plant illustrates primary growth. (Tassels form at the end of the main upright stem.)

- Part 3: Cytokinesis—This part of cell life occurs immediately after mitosis. A membrane grows across the cell that divides the cell material between the two nuclei. Each new cell, known as daughter cells, contains one of the new nuclei formed during mitosis and half the original cytoplasm.

KINDS OF GROWTH

Plants have two kinds of growth: primary and secondary. Both are important when growing plants that will produce good crops.

Primary Growth

Primary growth is when a plant gets taller. Stems and roots get longer and leaves and flowers grow. Some people refer to primary growth as linear growth.

New cells grow at the tips of stems and roots in areas known as apical meristem. The meristem is the plant tissue that is capable of cell division. The buds on the ends of stems are made of meristem tissue. Spurts of meristem activity occur in the spring. Cells elongate and divide rapidly. The tips of stems of trees may grow several inches in a few days.

Primary growth is important when plants are required to grow rapidly and bear fruit or to be used for products.

Secondary Growth

Secondary growth increases the diameter of stems and roots. Tree trunks increase in size due to secondary growth. Therefore, secondary growth is very important in producing logs, pulpwood, and many other plant products.

New cells grow in meristem tissue known as lateral meristem. Much of this growth is in the cambium layer of a plant. Some of these cells become xylem; others become phloem. All plants tend to grow alike. Of course, annuals do all of their growing in one year.

In trees, the phloem is near the bark and the cambium lies between it and the xylem. The xylem is toward the inside of the tree and is often called wood. Each year another layer of cells is grown.

The cells form rings that are easy to see when a tree is cut down. The rings are known as ***annual growth rings***. One ring is the amount of xylem a tree grows in one year.

9-5. The annual growth rings are easy to see in this fast growing tree. How old is the tree? (Count the number of dark circles.)

The width of rings tells how rapidly the tree was growing. Narrow rings indicate that the tree was growing slowly, while wide rings indicate fast growth. Pine trees may have layers of growth that are one-half inch wide each year under good conditions.

The number of rings can be counted to estimate the age of a tree. A tree usually grows one ring each year; therefore, a tree with 25 rings would be about 25 years old. The age of a tree can be estimated by using an increment borer to take a sample of the annual growth rings.

Rings have dark and light colors. The light-colored rings were grown in the spring when growth is rapid. Larger cells which have lighter colors grow in the spring. The darker color is a sign of smaller cells which grow in the summer.

9-6. A sample of the annual rings is being removed for study to determine rate of growth and age of the tree. The tool used to get the sample is known as an increment borer. The tree will soon heal where the sample was taken.

9-7. The wood sample shown here was taken with an increment borer from a fast-growing tree. The width of the rings indicates the rate of growth. The number of rings indicates the age of the tree.

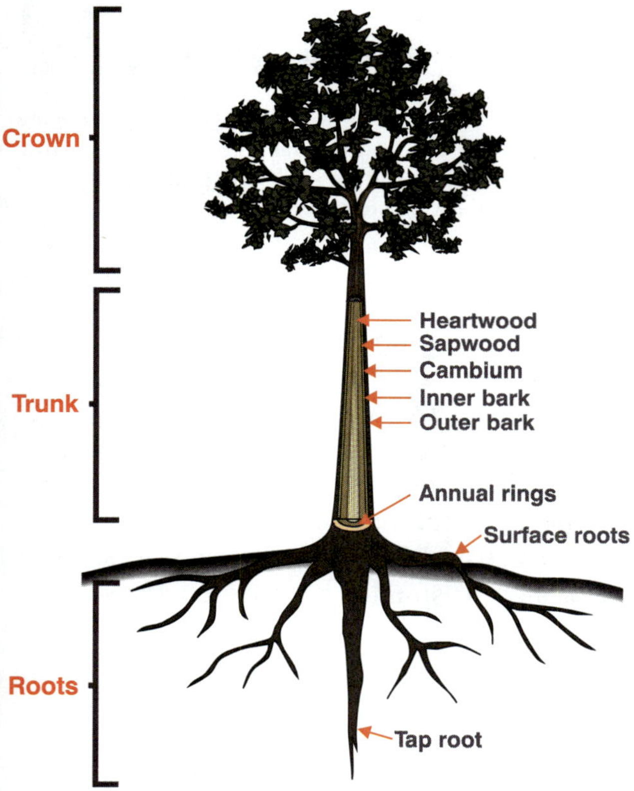

9-8. Parts of a tree.

The innermost xylem of a tree is the ***heartwood***. In pine trees, the heartwood has a lot of rosin and resists decay much better than the ***sapwood***. Sapwood is the xylem that is two to four years old. The color of sapwood is lighter or white, while heartwood is darker. Sapwood lumber is not very resistant to rot and should not be used to build structures that aren't protected from the weather. Heartwood lumber is much better at resisting rot.

PLANT GROWTH PROCESSES

As organisms, plants carry out a number of complex processes. These processes make it possible for cells to grow and divide. They keep the entire plant healthy and growing toward maturity. (These were presented in Chapter 5 and are listed here for review.)

The processes include: absorption of water and nutrients, moving water and other materials in the vascular system of the plant (circulation), making food through the process of photosynthesis, converting food to

AgriScience Connection

CANDLES ON PINE TREES: FESTIVAL TIME

Have you ever seen a candle on a tree? Rapidly growing young pine trees may give the appearance of having candles at the ends of stems. In fact, branching new growths often resemble candelabra—bunches of candles on giant candle holders. Spring time and the exciting new growth is a cause for celebration where forestry is important.

Communities often celebrate the new pine growth. Spring festivals bring citizens out to special events to commemorate the new growth.

The growth is important to people who work in forestry. It signals that the trees are getting larger and more valuable. More lumber, paper, and other products can be made from a tree.

tissue by the assimilation process, taking in and giving off carbon dioxide and oxygen through respiration, and adapting to the environment through transpiration and reproduction.

ANIMAL AND PLANT DIFFERENCES

Animals and plants both grow by cell division. How the cells divide and the way the tissues and organs work together are somewhat different.

Plants are indeterminate in their growth. This means that they often do not stop growing. They grow throughout their lifetimes. Animals are said to be determinant because they reach a certain size at maturity. Before a new animal is born or hatched, we have an image of the mature size of the new animal in the adult form.

WHY PLANTS GROW: HORMONES

Plants grow because of complex processes that occur in them. These processes are directly or indirectly caused by substances known as phyto-hormones or ***plant hormones***. Phyto-hormones are also known as growth regulators.

Plant hormones naturally occur in plants. They regulate the growth and development of plants. Even very small amounts can result in a change in plant growth. All of the stages of plant growth are started and controlled by hormones.

AUXIN

Auxin is a plant hormone that controls stem growth and regulates fruit development. Auxin also causes the spurt of growth that plants have in the spring.

Auxin promotes the elongation of primary growth cells in stems and roots. It causes cells to stretch and divide.

The growth cycle of plants is caused by auxin. Swelling plant buds send auxin to the cambium in the spring and cause it to produce xylem cells. In the fall, leaves fall from trees because they stop making auxin.

Auxin supplements are used in agriscience. Cuttings may be treated with auxin to encourage rooting. It is also used to regulate fruit setting on plants.

Auxin-based materials are used to produce weed killers, known as herbicides. These are made from artificial auxin.

GIBBERELLINS

Some 50 forms of gibberellins are formed by different plants. Gibberellin is sometimes called ***gibberellic acid***.

Gibberellins are as important in plant growth as auxin. They affect stem and leaf growth, fruit development, flowering, cell division, and other plant activities. Gibberellins are closely associated with organ enlargement. Stems, fruit, and leaves expand because of its action.

Gibberellins cause some species of plants to flower. Spraying gibberellic acid on some plants will cause them to produce flowers at times of the year when they wouldn't do so naturally. This use is important to some growers of ornamental flowers.

CYTOKINS

The cytokin hormones promote cell division. They also perform other functions in plants, just as the auxins and gibberellins.

Cytokins regulate cell expansion in roots, leaves, and stems. They are often used when asexually propagating plants, particularly with tissue culture. The use of cytokins causes the cells to grow more rapidly.

Cytokins are used with ornamental plants to encourage the development of buds and flowers. A good example is the ornamental plant, African violet. Buds form more quickly when cytokins are put on the plant.

ETHYLENE

Ethylene is a hormone that is a gas. It causes potato tubers to sprout, pineapples to flower, fruit to ripen, and leaves to fall off plants. It also regulates the height of small grains, such as oats and wheat.

As a gas, ethylene has been difficult to use artificially because the plants had to be put in some type of enclosed place where the gas couldn't escape. New methods of producing ethylene have been developed. Some newer approaches involve spraying plants with liquids that contain ethylene.

ABSCISIC ACID

Abscisic acid is named for its major role in plants: dropping of leaves and fruit. Leaves and fruit fall from a plant because a layer of cells develops over the stem where the leaf or fruit is attached. This layer of cells is the absciss layer.

This hormone helps regulate plant functions and promotes hardiness. Hardy plants can withstand changes in their environment. They can tolerate colder weather than they have had in a protected environment. Hardiness also refers to how easily a plant can be moved. Bedding plants may be damaged by the sun if suddenly moved from a protective greenhouse to a hot field. Plants that aren't hardy are known as tender plants. Abscisic acid helps plants develop hardiness. Gradually exposing plants to a change in their environment also helps develop hardiness.

PHOTOSYNTHESIS

Photosynthesis is the process by which plants make food. Though a fairly complex process, it is easily carried on by plants under the right

9-9. Photosynthesis is occurring rapidly in this young growth on a muscadine grape. Note the tiny grapes between the leaves.

conditions. The major product of photosynthesis, sugar, provides energy for plant functions.

THE PROCESS

Photosynthesis involves two major phases: energy gathering and sugar making. Energy is gathered from sunlight by the leaves of the plant. The energy is changed to a form known as chemical energy. Sugar is produced when the chemical energy rearranges the combinations of carbon dioxide and water in the plant. A green-colored substance known as ***chlorophyll*** must be present for rearrangement to take place.

The sugar made by photosynthesis is known as glucose, or simple sugar. Bonding more than one glucose sugar together may make other kinds of sugar. An example is sucrose (table sugar). Some of the sugar also changes to starch.

What Plants Need

Photosynthesis is carried out in live plants. The plants must also be in an environment that helps them with the process. Certain things must be present for photosynthesis to occur.

Plants need the following for photosynthesis:

- **Chlorophyll**—This is a green-colored substance in green plant leaves. The chlorophyll is located in bundles known as chloroplasts. Sugar is made in the chloroplasts.
- **Sunlight**—Leaves must get plenty of sunlight. Many leaves will turn toward the sun to help get the amount of light they need. Chlorophyll transforms the light energy of the sun into chemical energy.
- **Carbon dioxide**—Plants get carbon dioxide (CO_2) from the air. It enters the leaves through tiny holes known as stomata. Carbon dioxide is a major substance from which glucose is made. The carbon dioxide is split apart by the chemical energy in the plant.
- **Water**—Water (H_2O) is the other major ingredient needed for the plant to make sugar. It is absorbed by the roots and goes to the leaves through the vascular system. Water is split apart by the chemical energy and partially combined with the elements of carbon dioxide. Water also serves other roles, such as transporting materials throughout the plant.

AgriScience Connection

WHEN NATURE RUNS LOW

Sometimes the soil does not have the moisture that plants need to grow well. Producers help nature out with irrigation—the artificial application of water to the soil. Irrigation helps plants have the moisture they need.

How plants are irrigated varies with the kind of plant and how it is grown. Some fields are sprinkled, others flooded, and in others each plant has its own water emitter—drip irrigation. The later uses water more efficiently because only the area immediately around the roots gets irrigated.

Four methods of irrigation are shown here: center-pivot sprinkler, furrow flood, individual emitter, and sprinkler irrigation. Variations in these methods are used with different crops.

The Equation

Photosynthesis is often expressed as an equation. The equation illustrates the splitting apart and recombining of the ingredients. The ingredients begin as one kind of molecules and end up as different kinds of molecules. Molecules are the smallest sized particle that a substance can have. Molecules are often made up of two or more elements. For example, water (H_2O) is made of two atoms (basic unit of matter) of hydrogen (H) and one of oxygen (O).

The equation for photosynthesis is:

$$\text{light} + 6CO_2 + 6H_2O \xrightarrow{\text{chlorophyll}} C_6H_{12}O_6 + 6O_2$$

The equation can be explained as follows:

> Six molecules of carbon dioxide (CO_2) and six molecules of water (H_2O) are combined. Two substances are formed: sugar (glucose) and water. The sugar ($C_6H_{12}O_6$) remains in the plant as food. Six molecules of oxygen (O_2) are given off into the air through the stomata.

In the equation, chlorophyll is a catalyst (causes the process to take place). The chlorophyll is not changed in the process. It remains to continue promoting photosynthesis.

THE PRODUCT

Photosynthesis produces two products: oxygen (O_2) and ***glucose*** (sugar) ($C_6H_{12}O_6$). Plants use or dispose of the products.

Most of the oxygen is given off into the air. It replenishes the oxygen that is used by animals, combustion of engines, and other activities. Of course, the plant also uses some of the oxygen.

The sugar is transported to various parts of the plant and stored for future use. The form of the sugar may change. Much of it is stored as starch in the stems and roots. One example is the potato. The potato tuber is filled with starch made by the plant from sugar. Other examples are fruit and seed. Both have a lot of sugar and starch that are stored to help the young embryos live.

Agriscientists want plants to grow and carry out photosynthesis. They want plants to have many of the desired parts, such as fruit and seed. This is essential if products are to meet human needs. Culturing plants involves trying to provide the best possible conditions for photosynthesis.

RESPIRATION

Respiration uses oxygen. It occurs in living cells when energy is burned (oxidized). The energy comes from the sugar made by photosynthesis. Waste products are given off—carbon dioxide and water.

Energy is used to help plants grow and reproduce. The energy is released by a process known as ***glycolysis***. Glucose is broken down into pyruvic acid. The process occurs in the cytoplasm of cells. Oxygen is required for normal respiration to occur. When glycolysis occurs without oxygen, it is known as fermentation.

Respiration occurs 24 hours a day. Unlike photosynthesis which occurs mainly in light, respiration takes place in the dark. Plants use energy all of the time.

TRANSPIRATION

Transpiration is the loss of water from a plant. Most water is lost through the leaves. Transpiration is sometimes called water evaporation. A tree may transpire as much as 80 gallons of water in one day!

Plants must have enough water. Water is used in making food, transporting nutrients and food for storage, and to help the plant fit into its environment.

Most of the force that causes water to move up the stems of plants is due to transpiration. Plant roots must be able to absorb more water from the soil. This process is known as transpirational pull or turgor pressure. In addition, roots appear to have some upward pressure on the water. This is evident when a plant is cut off and water comes out of the cut.

Weather effects transpiration. Warm weather results in more transpiration. Also, transpiration increases in windy weather and when the relative humidity is low.

Water goes out of leaves through stomata. The stomata open to let water out when the surrounding air doesn't provide the needed environ-

9-10. A drop of water has formed on this cut stem. Pressure within the plant is forcing the water out.

9-11. Plants in homes and offices must be properly watered.

ment. If the air is humid and not too warm, the stomata don't let out much water.

Too much water loss damages plants. Every species of plant has different needs. Those that grow in deserts are adapted to the hot, dry climate. Those that grow in swamps are best suited there. Even when plants are being grown in their best environment, excessive water loss can occur.

WATER SHORTAGE

Shortages of water effect plants in different ways. The first sign may be ***wilting*** of the plant. The stems and leaves are no longer rigid and erect. Wilting may be due to excessive transpiration or other factors. All relate to a shortage of water in the plant.

Plants wilt for several reasons, as follows:

- **Dry soil**—Plants with shallow roots are especially hard hit by dry soil. Adding water to the soil will help the plant get more water. Farmers often use irrigation to add water for crops.
- **Damage to root system**—The root systems of plants may be damaged by plowing, while being transplanted (moved), or in other ways. Damaged roots cannot absorb water as usual. All plants should be given water when transplanted. Deep plowing or digging near a plant can cut the roots. When transplanting shrubs, the tops are usually cut back (pruned) so that there

9-12. Small trees shown here are on sale at a garden center. They have been grown in containers to keep both roots and stems healthy. This helps assure that the trees have good root systems when planted.

are fewer leaves for the roots to provide with water. All plants have a balance of roots and stems and leaves. When one is damaged, the others suffer.

- **Disease**—Wilting of plants may be due to disease. The disease may damage the ability of the plant to absorb water and carry on life processes. Some diseases live in the soil and interfere with the work of the roots.
- **Too much water**—The roots of plants need oxygen. When soil is too wet for a period of time, all of the oxygen is driven out. The roots suffer from oxygen depletion and cannot take up moisture.
- **Too much fertilizer**—Plants that get too much fertilizer have their roots burned. This is especially true if the fertilizer is applied too close to the roots of the plants. Concentrated fertilizer can draw water out of the roots! Plants should never be fertilized when the soil is very dry.

HELPING PLANTS

We can help plants avoid too much water loss. Here are several things to do:

- **Avoid damaging roots**—Plowing around plants can damage the root system. Do not plow (cultivate) too close to plants or too deeply. Improper cultivation does more harm than good. Cultivation is used to control weeds and create a soil mulch on the land. If not controlled, weeds damage crops. They compete for water, nutrients, and space.
- **Plant drought-tolerant species**—Some species of plants have been developed that will grow with less moisture. These should be selected if the growing conditions will be dry.
- **Fallowing**—This practice is used in places with small amounts of precipitation, such as western parts of North America. Crops are grown on the land every other year. The year following growth of the crop the land is kept clean of all plants. This allows the soil to store water for the next crop.
- **No-till culture**—***No-till cropping*** involves growing crops without plowing the soil. Water loss is greater when soil is plowed. Plowing exposes the soil to the sun and wind, which dry up the moisture. No-till cropping also reduces soil loss through erosion. The dead vegetation of crops stays on the top of the soil and holds it in place. The use of no-till has increased in recent years. Some farmers use minimum-till. This involves only partial breaking of the land to plant the crop.

9-13. Pine needle mulch has been used with these newly planted pansies. The mulch keeps moisture in the soil, prevents soil erosion and keeps weeds from growing.

- **Mulch**—A ***mulch*** is a cover on soil to hold the moisture. It also keeps down the growth of weeds. Common mulches are straw, tree bark and leaves, sheets of plastic, sawdust, and paper. The plants are set in the soil through the mulch. Many people use mulches in flower beds and around shrubs. Freshly plowed soil also serves as a mulch. Fields where crop remains from previous years cover the ground have a form of natural mulch.
- **Terracing**—The land can be formed into ridges or embankments to hold back water. These structures are known as terraces. Terraces keep rainfall from running off quickly. Slowing the rate of runoff allows more water to soak into the soil.
- **Irrigation**—Irrigation is adding water to the land. Some crops require a lot of water, such as rice. Other crops require less, such as wheat. Water can be pumped from wells or reservoirs and put on the fields. Reconditioned water from sewage plants is sometimes used to irrigate areas that aren't used to grow food crops. Of course, irrigation may deplete the supply of fresh water in the earth and reservoirs or streams. Crops are irrigated by sprinkling, flooding, or dripping the water around the plants. Large pivot irrigation systems are used in some places. These systems water many acres at one time.

AgriScience Connection

ROOTS NEED PLENTY OF ROOM

Plant roots need plenty of room to grow. Plants do not grow well and may die if their roots are restricted. People often have the problem of restricted roots with ornamental plants.

This shows the root system of an ornamental plant that died because its roots were restricted by the container in which it was growing. The condition is known as "root-bound." It is prevented by growing plants in pots large enough for the root system to continue growing.

Small plants must be transplanted to larger pots as they grow. Plants that live several years may need to be re-potted once or twice a year when they are young.

PLANT NUTRIENTS

Nutrients are essential for plant growth. Nutrients are the minerals (substances) in the soil which are absorbed with the water through the roots. Plants don't grow well if there aren't enough nutrients. The color and size of plants are related to nutrient supply.

Nutrients are chemical elements. Some are needed in large amounts; others in small amounts.

The dry stems and other parts of plants have been analyzed to find out what they contain. More than 95% of the total dry weight of plant material is made up of carbon, hydrogen, and oxygen. The remainder (approximately 5%) is made of other elements.

ESSENTIAL ELEMENTS

Research by agriscientists has shown that plants must have certain elements to grow. Those that are needed in large amounts are known as ***major elements*** or macronutrients. Nutrients that are needed in small amounts are known as ***trace elements*** or micronutrients. Both major and trace elements are needed for plant growth.

Seventeen elements have been found to be essential for plant growth. Three other elements are used by plants but research has yet to determine how important they are in growth.

Essential elements are often known by their chemical symbols. Chemists have developed symbols for all elements. For example, the symbol for nitrogen is N and the symbol for calcium is Ca. Table 9-1 lists the essential elements and their symbols.

Table 9-1
Essential Elements for Plant Growth

Macronutrients	Symbol	Micronutrients	Symbol
carbon	C	iron	Fe
hydrogen	H	zinc	Zn
oxygen	O	copper	Cu
nitrogen	N	boron	B
potassium	K	molybdenum	Mo
phosphorus	P	manganese	Mn
calcium	Ca	chlorine	Cl
magnesium	Mg	cobalt	Co
sulfur	S	vanadium	V
sodium	Na	silicon	Si

The three major elements are nitrogen, phosphorus, and potassium. These are needed in large amounts.

Nitrogen

Nitrogen is among the most important elements in the growth of plants. Nitrogen makes leaves healthy and green. Dark-green, fast-growing fields of corn require large quantities of nitrogen!

Nitrogen is the element that is most frequently deficient in crops. Even though the atmosphere is 78% nitrogen gas, plants cannot use the nitrogen in "air" form. It must be converted into other forms by bacteria or algae or by industrial processes.

Nitrogen is a vital part of cell protoplasm. It is found in chlorophyll and other substances in plants.

Plants that don't get enough nitrogen are stunted and have pale green or yellow leaves. Plants that get too much nitrogen are too green and grow too much weak, succulent (juicy) foliage.

Sources of Nitrogen. Nitrogen is produced by natural processes in the soil as well as added as a chemical fertilizer. Plants absorb nitrogen in nitrate (NO_3^-) or ammonium (NH_4^+) forms.

The decomposition of organic matter produces nitrogen. Bacteria form nitrogen when they act on stems, leaves, and other materials in the soil. The nitrogen becomes available when the bacteria die.

Crops may be fertilized with chemical fertilizers, the most effective sources of nitrogen. The major nitrogen fertilizers are:

- **Ammonium nitrate (NH_4NO_3)**—33% N; solid, round, white prills about ⅛ inch (0.32 cm) in diameter
- **Anhydrous ammonia (NH_3)**—82% N; a liquid under pressure
- **Urea [$(NH_2)_2CO$]**—45% N; applied as a solid or liquid, with the liquid suitable for spraying on the leaves
- **Sodium nitrate ($NaNO_3$)**—16% N; originally mined in Chile, now manufactured

Since nitrogen is readily lost from the soil, the amount applied should be no more than necessary. Recommended levels of fertilizer use have been established for many crops. These vary with the climate and crop variety. Corn, for example, can benefit from 120 pounds or more of nitrogen

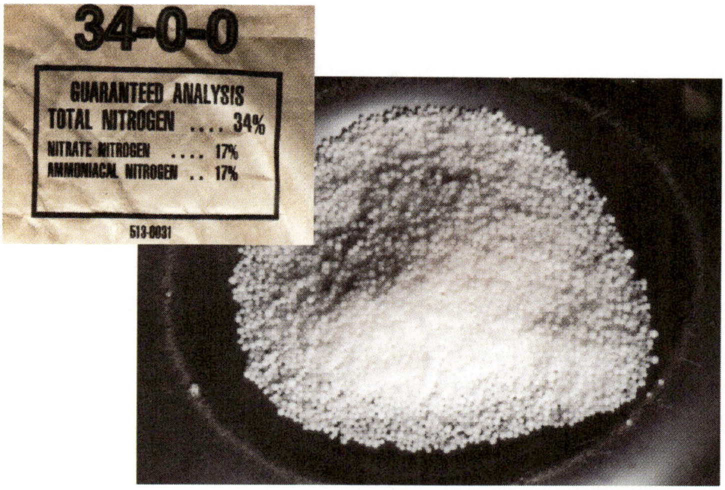

9-14. Ammonium nitrate is 34% nitrogen—an excellent source of nitrogen for many crops. It is manufactured as prills—small round particles.

fertilizer per acre, depending on the plant population (number of corn stalks). Many farmers use lower amounts. Applying more than the crop will use is a waste.

Legume Plants. A ***legume*** is a plant that has the ability to take nitrogen from the air and put it into nodules that grow on its roots. This process is known as fixation, or nitrogen fixation. The roots of legumes may have nodules—small, rough, and rounded attachments about the size of a pea.

Certain kinds of bacteria must be present for fixation to occur. The *Rhizobium* bacteria are most important. If the soil doesn't contain the bacteria, fixation will not occur.

Seed of legumes are sometimes inoculated with the nitrogen-fixing bacteria before they are planted. Packages of inoculant are available at farm supplies stores. The bacteria must be alive and capable of fixing nitrogen.

Legumes are plants that have seed pods. Examples of legumes are soybeans, peas, vetch, and locust trees.

When growing legume crops it is seldom necessary to apply nitrogen to the field. Sometimes legumes are grown with other crops and provide nitrogen for them also. Many of the so-called cover crops (often grown on land in the winter to protect it from erosion) are legumes that add nitrogen. Crops grown following the cover crops in the summer may not need nitrogen fertilizer. If they do, the requirement is lower than when no cover crops are grown.

Phosphorus

Phosphorus is a nutrient needed by plants to store and transfer energy and to grow. Plants that don't get enough phosphorus don't grow well. They appear stunted and the older leaves fall off. Some of the leaves may turn red or purple. Phosphorus is needed for seed germination. It stimulates flowering and aids the plant in seed formation.

Phosphorus is not plentiful in the environment like nitrogen. The soil may contain small to large amounts. Most phosphorus is in the phosphate form (PO_4^{-3}).

Phosphorus is often applied as superphosphate or mixed with other fertilizer elements. Superphosphate is made by treating rock phosphate with sulfuric acid. The phosphorus content in superphosphate is about 9%.

Potassium

Potassium is a nutrient needed for photosynthesis, moving sugar, and other functions. The opening and closing of stomata are regulated by potassium. Plants need potassium to be healthy and tolerant of changes in weather. Seed and grain need potassium to develop properly.

Plants that do not have enough potassium show signs in the older leaves first. The edges of the leaves lose their green color and may die. Dead leaf tips are often signs of potassium deficiency. Plants that don't have enough potassium will have small seed and fruit.

Potassium is usually added to soil as potash (K_2O) fertilizer. Muriate of potash (KCl) is mined in North America, Europe, and other places as a source of potash.

Other Essential Elements

Healthy plant growth depends on a number of other essential elements. These include calcium, sulfur, magnesium, iron, boron, zinc, manganese, copper, molybdenum, chlorine, and sodium. Of course, carbon, oxygen, and hydrogen are available from the atmosphere.

Research has shown that crops need varying amounts of essential elements. People should study the needs of crops when they grow them and provide the needed nutrients.

HOW PLANTS GET NUTRIENTS

Plants get nutrients from the soil and air. Most are from the soil and are absorbed by the roots. Water carries the nutrients throughout the plant.

Plants that do not get enough nutrients will not grow and be productive. Culturing plants requires the addition of the needed nutrients.

People who grow crops add nutrients to the soil or water, spray it on the foliage, or provide them in other ways. Some have provided nutrients by injecting them into trees, but these tend to be extreme cases.

9-15. The fibrous root system on this grass is about the same size as the stems and leaves of the plant. Plants must have good root systems to absorb water and nutrients.

USING FERTILIZER

Producing plants often requires the addition of ***fertilizer***. Fertilizer is any substance used to provide plant nutrients. It may be added to the soil or water or provided in other ways.

Some soils do not have the needed nutrients. If crops are to grow, the nutrients must be added.

KNOWING WHAT IS NEEDED

Crops use nutrients when they grow and bear fruit. The nutrients must be available if the crops are to have good yields. When crops grow, they take nutrients from the soil. The nutrients must be replaced if future crops are to do well.

The kind and amount of fertilizer to use depends on several major factors: the requirements of the crop that will be grown; the nutrients available in the soil; and soil pH, whether or not the soil is acidic or basic. (More is presented about this in another chapter.)

Crop Needs

Crops have varying nutrient needs. Some require more nitrogen than others; some require other elements in greater amounts. It is important to know the nutrient needs of crops.

Different crops remove different kinds and amounts of nutrients from the soil. For example, 100 bushels of corn require 90 pounds of nitrogen,

Table 9-2
Amounts of Selected Nutrients Used by Certain Crops

Crop	Amount of Yield[1]	Pounds of Nutrients Used		
		Nitrogen (N)	Phosphorus (P_2O_5)	Potassium (K_2O)
Cabbage	20 tons	130	35	130
Corn	100 bu.	90	35	25
Cotton (seed and lint)	1,500 lbs.	40	20	15
Potatoes	400 bu.	80	30	150
Soybeans	40 bu.	150	35	55
Sugarcane	30 tons	96	54	270
Tomatoes	15 tons	90	30	120
Wheat	40 bu.	50	25	15

[1]Yield is based on the production from one acre of land. Many producers get higher yields. In order to do so, they must add more nutrients to the soil. Plants can't produce without the needed nutrients!

35 pounds of phosphorus, 25 pounds of potassium, and 6 pounds of calcium. Other crops require varying amounts of nutrients. Table 9-2 shows the amounts of nutrients required by some crops.

Once we know the amounts of nutrients crops use, we need to know the amount to apply to fields. This varies with the climate, nature of the soil, and other factors.

The land-grant colleges in most states have specialists who establish recommended rates of fertilizer use. This information is available from agriculture teachers and other trained agriculture professionals.

Soil Testing

Tests are used to determine the nutrients present in soil. The process is known as soil testing or ***soil analysis.*** This information is used to determine what to add to the soil for a crop.

Soils change. Some nutrients wash away when it rains. Other nutrients are used up by crops. Soil should be tested annually or at least every three years.

Soil testing involves several important steps:

9-16. Soil analysis results can be no better than the sample. Proper procedures should be followed in collecting soil samples.

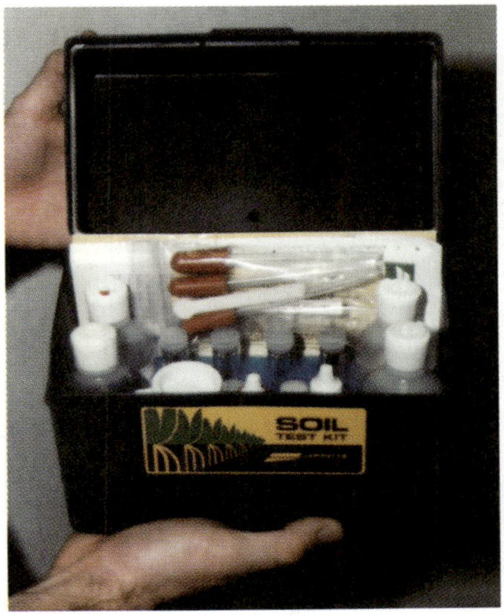

9-17. A small soil analysis kit that can be carried into a field.

- **Sampling**—***Soil sampling*** involves collecting samples of soil for analysis. These must accurately represent the soil in a field. The sample must be collected from several places in the field at a depth of 4-6 inches in the soil. Clean tools and containers must be used. If not, the sample will be contaminated and the results of the test will not be based on the soil that is actually in the field. Bad tests are not any better than no tests! Remember, the sample should be representative of all the soil in the field.
- **Sample analysis**—Soil samples are tested to determine the nutrients they contain. This involves using chemical processes to get the information. Samples are often sent to soil testing laboratories that have the equipment to do a good job. Some people test their own soil using various test kits and laboratory procedures. These can be accurate if done properly. The analysis involves determining the amounts of nutrients in the soil sample. On the basis of the analysis, recommendations can be made about the amount of fertilizer to use on a field for a given crop.
- **Interpreting results**—Once test results are known, the nutrient content of the soil is compared to the needs of crops. This tells how much fertilizer to add. For example, if a crop requires 100 pounds of nitrogen per acre and the soil only contains 10 pounds, 90 pounds must be added. If the right amount isn't added, the crop yield will be low.

Tissue Analysis

Tissue analysis involves testing samples of plant tissue. The analysis will give information about nutrient deficiencies in the plant. These deficiencies reflect the content of the soil in which the plant is growing. Fertilizer can be used to add the elements that the soil needs.

Tissue analysis normally involves testing samples of leaves. The leaves are dried, ground to a powder, and dissolved in a solution. The solution is tested for the presence of particular elements.

Tissue samples must be carefully taken and tested. They should be representative of the crop and not contaminated with other substances. For example, crop leaves that are placed in an empty fertilizer bag for hauling will get some remains of the fertilizer on them. This will give inaccurate test results.

Results of tissue testing can be used to tell the kind of fertilizer to use. Some people combine soil and tissue analysis. Agriscientists feel that using both tests gives more information for recommendations.

TYPES OF FERTILIZER

Many kinds and types of fertilizer are available. In general, fertilizer is one of two types: dry or liquid. However, other special types are available.

Dry Fertilizer

Dry fertilizer is produced as pellets or granules. These are small particles about the size of a BB. Dry fertilizers have been used for many years. They are often sold in bags or in bulk (unbagged) form. Since bags add to the cost of fertilizer, most people who use large quantities buy it in bulk. Bags are used by small farmers, homeowners for their lawns, and others who don't have equipment to handle large amounts.

Liquid Fertilizer

Fertilizer nutrients may be dissolved in water and applied by spraying on the soil. The use of ***liquid fertilizer*** has increased in recent years. Special tanks with pumps and spray rigs are needed to apply the fertilizer.

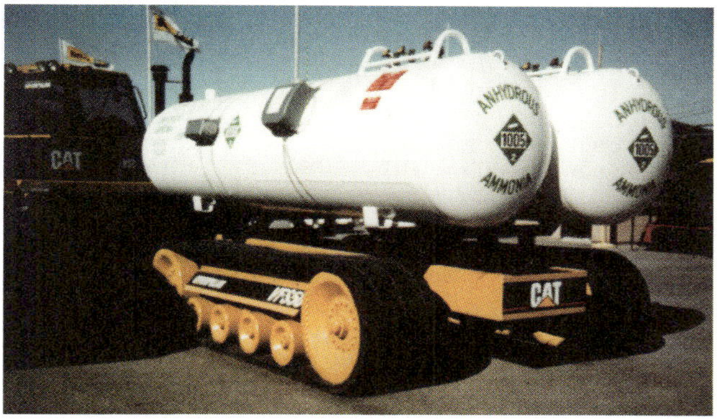

9-18. Special tanks are used to transport anhydrous ammonia.

Other Forms

Fertilizer is sometimes applied in other forms. Some of these are specialized uses.

Anhydrous ammonia, a source of nitrogen, is a liquid form that readily becomes a gas in the atmosphere. It is held in pressure tanks and injected into the soil. Any that is released in the air quickly becomes a vapor and is lost. Anhydrous ammonia has had widespread use with crops on some farms, but is decreasing in use.

Solids and spike forms of fertilizer are sometimes used in special situations. Some fertilizer may be made into nail-like spikes that are pushed into the soil. These are used in decorative pot plants and other special places. The fertilizer slowly dissolves and is released into the soil.

Fertilizer and other chemicals are sometimes combined. Lawn care products may include pellets of fertilizer and herbicide (weed killer) mixed together.

FERTILIZER ANALYSIS

Most fertilizer is made out of several different materials. Such products are known as blended fertilizer. The materials in it give it an analysis or grade. The ***fertilizer analysis*** tells us the nutrients that are in it.

A label on a fertilizer container usually has three important numbers as well as other information. These numbers are always used in the same sequence and have the same meaning. Here is what they mean:

- **First number**—The first number is the percent of nitrogen in the fertilizer.
- **Second number**—The second number tells about phosphorus. It gives the percent of the anhydride of phosphoric acid (P_2O_5) in the fertilizer. Phosphorus is not reported in the elemental form, or P.

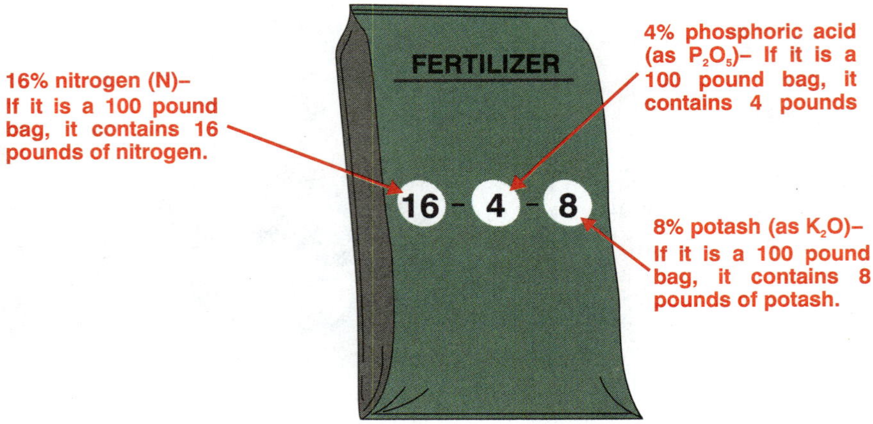

9-19. Interpreting fertilizer analysis usually involves three numbers, as shown above.

- **Third number**—This number tells about the potassium in the fertilizer. The number represents the percent of potash (K_2O) in the fertilizer. Potassium is not reported in the elemental form, or K.

When the three numbers are put together, they give the overall analysis of the major nutrients in the fertilizer. Here are two examples:

- **0-20-20**—This means that the fertilizer contains no nitrogen and 20% P_2O_5 and 20% K_2O. A 50-pound bag of fertilizer would have 10 pounds of P_2O_5 and 10 pounds of K_2O.
- **9-12-12**—This means that the fertilizer has 9% N, 12% P_2O_5, and 12% K_2O. A 50-pound bag would contain 4.5, 6, and 6 pounds, respectively, of each of the three nutrients.

Other information on the label will tell about trace elements and availability of the nutrients.

FERTILIZER PLACEMENT

Fertilizer must be put where plants can get it. Since most is applied to the soil, it must be placed properly.

Proper placement means that the fertilizer is close enough for the roots to get it but not so close that it damages the roots. Putting fertilizer too close damages the roots. Some dry fertilizer can also damage leaves and stems if it gets on them.

9-20. A six-row planter with fertilizer applicator. (The three large hoppers hold dry fertilizer. The six small hoppers hold the seed.) (Courtesy, James Lytle, Mississippi State University)

Pre-Planting Application

Fertilizer may be applied to a field before the crop is planted. It may be spread or broadcast evenly over the soil. The field may be tilled with a disk harrow or "do-all" (large tiller pulled by a tractor) to mix it into the soil.

Planting Application

Fertilizer may be applied at the same time as planting. Planting equipment may have fertilizer distributors attached. The fertilizer is often placed in narrow bands four to six inches from the seed. It should never be placed in the same drill (small trench) as the seed. Fertilizer can burn sprouts and keep seed from coming up.

Post-Planting Application

Fertilizer is sometimes applied to crops after they are up and growing. This may be known as ***top-dressing*** or side-dressing, depending on the location of the fertilizer.

Pastures, wheat, oats, and other crops are often top-dressed. This means that the fertilizer is applied over the ground where the crop is growing. Rainfall dissolves the fertilizer and the water soaks it into the soil.

Side-dressing is applying fertilizer along the sides of rows of growing crops. Corn, cotton, and many vegetable crops are side-dressed. The fertilizer is placed six inches from the root area. Sometimes the crop is cultivated to mix the fertilizer into the soil. In other situations, the fertilizer is placed on top of the soil and carried into the soil by rain or irrigation

9-21. Sugar beets have become an important source of sugar.

water. Care should be used to avoid damaging plant roots, stems, and leaves when side-dressing.

HANDLING FERTILIZER

Fertilizer is a chemical compound. If not used properly, it can cause damage to people, equipment, and the environment.

Here are a few rules to follow when using fertilizer:

- Use only the recommended kind and amount. Using the wrong kind or more than is needed wastes fertilizer and pollutes the environment. Excess fertilizer may wash into streams and rivers.
- Apply fertilizer only where it is needed. Wasting fertilizer on the ground at the ends of rows costs money and doesn't grow a crop.
- Store fertilizer in dry places where nutrients won't be lost. If in bags, they should not be broken open or cut while being moved. Since some fertilizer ingredients are explosive, keep fire away from storage areas. Fertilizer should not be stored near animal feed or water.
- Clean equipment after it is used. Fertilizer will chemically react with metal and other materials. The equipment can be damaged and not work properly the next time it is used.
- Wear protective clothing and wash after handling or using fertilizer. Some fertilizer can burn the human body or that of animals.

Remember, fertilizer is a chemical. It can cause reactions that are dangerous and damaging.

REVIEWING

MAIN IDEAS

Agriscience is concerned with getting plants to grow. Providing what they need for their best growth is the major goal. Knowledge of how plants grow helps in producing them.

Plants have two kinds of growth: primary and secondary. Primary growth is when plants grow longer. It occurs at the buds on the ends of stems in tissue known as meristem. Secondary growth is when the stems get thicker. It involves layers of cells expanding the size of the stem.

Several hormones are important in plant growth. These influence how the plant grows and the stages it goes through. Some hormones can be artificially added to get plants to grow in different ways.

Photosynthesis is making food. Plants use energy from sunlight, carbon dioxide from the air, and water from the soil to make a simple sugar known as glucose. Chlorophyll is needed for the process to occur. The sugar is moved to other locations in the plant and may change form, such as to starch. Some of the sugar is stored; the rest is used by the plant in respiration.

Respiration involves use of the food by the plant. It occurs in the cells and involves "burning" the sugar to release energy.

Transpiration is the release of water from the plant. This occurs as the plant adjusts to its environment. Wilting is a sign of not enough water. Plants need water so that they aren't damaged.

Plants need certain nutrients to grow. Twenty nutrients are essential. Three of the nutrients are needed in large amounts: nitrogen, phosphorus, and potassium. Plants can get some of the elements from the air, such as carbon dioxide and oxygen. Others are obtained from the soil as water or nutrients in the water.

Fertilizer is used to increase the nutrients available to plants. Most fertilizer is applied to the soil. Soil testing and tissue analysis are used to find out the kind of fertilizer that should be used. As with all chemicals, fertilizer should be carefully used. It can damage animals, crops, people, equipment, and the environment.

QUESTIONS

Answer the following questions using complete sentences and correct spelling.

1. How do plants grow?
2. What is cell division? What phases are involved?

3. What kinds of growth occur in plants? Distinguish between the two.
4. What kinds of hormones are found in plants? What does each do for the plant?
5. What is photosynthesis? What are the major phases of the process?
6. How is photosynthesis carried out? Write the equation.
7. What is respiration?
8. What is transpiration?
9. Why do plants wilt? What can be done to prevent wilting?
10. What are the essential nutrients for plants? List their names and chemical symbols.
11. What is the importance of nitrogen? What are the sources of nitrogen?
12. What is a legume?
13. What is the importance of phosphorus and potassium?
14. How is the fertilizer to use determined?
15. What types of fertilizer are used?
16. What does "fertilizer analysis" mean? Explain the following label on a fertilizer bag: 5-10-5.
17. How is fertilizer placed?
18. What are the important rules to follow in handling fertilizer?

EXPLORING

1. Visit a local farm supplies store. Ask to see the labels on fertilizer bags. Study the labels for their content. Make a list of the different kinds of fertilizer.

2. Make a soil analysis of the soil at your home or on the school grounds. Take the sample and send it to a lab or use a testing kit and do it yourself. The teacher will need to assist with this activity.

3. Design an experiment in the school laboratory or at your home. Use corn or bean plants. Vary the kinds of fertilizer you use. Note differences in the color of the leaves and rate of plant growth.

4. Arrange to tour a tissue analysis laboratory. Take a sample of plant leaves and have them analyzed. Prepare a report on your observations.

10

KEEPING PLANTS HEALTHY

Plants can get sick! They are subject to damage by pests. Healthy plants can be attacked and lost. Entire crops can be destroyed. Fortunately, most pests can be controlled and losses reduced or eliminated.

Human life is often affected by plant pests. The loss of major food crops can cause big problems. Two examples are potatoes and corn.

Potatoes have been a major food crop in some countries for many years. Ireland is a good example. In the mid-1840s, the potato crop in Ireland was lost several years in a row because of the potato blight. Thousands of people starved to death. Thousands left Ireland for other places—often the United States—because they were faced with starvation.

Corn has long been a major crop in North America. In 1970, corn leaf blight struck the United States. This resulted in 40 million fewer bushels of corn grown in 1970 than in 1969. This loss had huge impacts on human food and livestock feed.

10-1. A field of healthy corn plants. The plants are growing rapidly and will produce a good crop. (Courtesy, DEKALB Plant Genetics, Dekalb, Illinois)

OBJECTIVES

This chapter includes information on plant pests, including prevention and control methods. It has the following objectives:

1. Explain five major kinds of pests.
2. Explain three conditions needed for pest problems.
3. Describe how pests affect plants and cause losses.
4. Describe how pests are prevented.
5. Describe methods of pest control.
6. Explain integrated pest management.
7. List safety practices to follow in pest control.

PESTICIDE	PEST CONTROLLED
Insecticide	Insects
Miticide	Mites
Acaricide	Ticks and Spiders
Molluscicide	Snails and Slugs
Fungicide	Fungi
Avicide	Birds
Rodenticide	Rodents
Nematicide	Nematodes
Bactericide	Bacteria
Herbicide	Weeds
Piscicide	Fishes
Predacide	Predatory Animals

TERMS

- bacteria
- bactericide
- beneficial insect
- biological pest control
- chewing mouth
- contact poison
- cultural pest control practices
- environmental disease
- fumigant
- fungicide
- genetic pest control
- herbicide
- insect
- insecticide
- integrated pest management
- mechanical pest control
- metamorphosis
- nematocide
- nematode
- non-selective herbicide
- parasitic plant disease
- pesticide
- plant disease
- plant pest
- protective treatment
- rodent
- roguing
- seed-borne pest
- selective herbicide
- stomach poison
- sucking mouth
- surfactant
- systemic poison
- therapeutant
- translocated herbicide
- trap cropping
- weed

PESTS: WHAT THEY ARE

A ***plant pest*** is anything that causes injury or loss to a plant. Pests cause losses by damaging plants and making them less productive. Most pests are living organisms. A few are non-living substances. Sometimes more than one pest will attack the same plants. The combinations are even more damaging.

MANY KINDS

Plants can be damaged by many different things. These can be put into five major categories of pests: insects, nematodes, diseases, weeds, and rodents and other animals.

Insects

Insects are small animals that cause damage in one way or another. Some eat the leaves, stem, fruit, or roots; others use the plant for raising their young. The young develop and feed on the plant and make it less productive or destroy it.

Several other tiny animals are included with insects even though they aren't really insects. Examples include spiders, millipedes, and centipedes.

Identification. Insects have three pairs of legs and bodies divided into three sections: head, thorax, and abdomen. The head has antennae and mouth parts. The legs and wings, if any, are on the thorax. The abdomen usually has no attachments but may be divided into sections.

10-2. The boll weevil *(Anthonomus grandis)* is a major insect pest with cotton. (Courtesy, National Cotton Council of America, Memphis, Tennessee)

10-3. Body sections of an insect.

There are many different kinds of insects. Over 86,000 species of insects have been identified in North America alone. Entomologists study insects and their scientific classification. Various books are available with detailed keys to help identify insects.

Mouth Parts. Insects have different kinds of mouths. How the mouth is made determines the way an insect gets its food. How a plant is damaged is related to the mouth. Identifying their structure is important in trying to control insects.

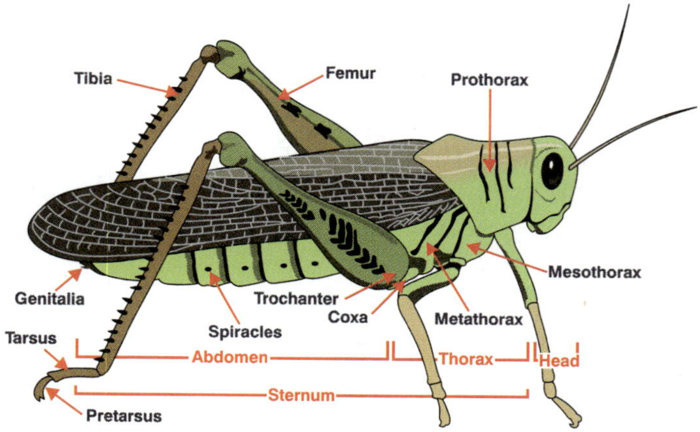

10-4. Major external parts of an insect.

Insects have two kinds of mouths, as follows:

- **Chewing**—Insects with a ***chewing mouth*** bite off, chew, and swallow parts of plants. They eat holes in leaves and other parts. Examples include cutworms, bean beetles, and armyworms.
- **Sucking**—Insects with a ***sucking mouth*** suck sap from a plant. They pierce the outer layer of the plant to get to the sap. They do not chew the plant. Examples include aphids (also known as plant lice), chinch bugs, thrips, squash bugs, and leafhoppers. (These insects are sometimes known as piercing–sucking insects.)

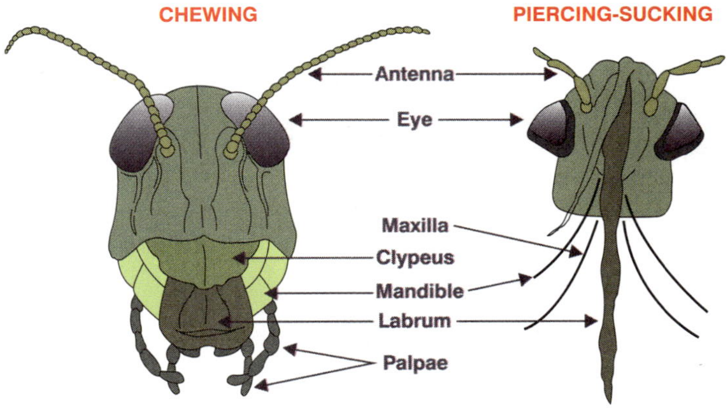

10-5. Mouth parts on insects.

Life Cycle. Insects go through definite life cycles. They begin life in one form and go through one or more other forms before maturity and death. The series of stages an insect goes through is known as ***metamorphosis***.

Some insects have complete metamorphosis; others have incomplete metamorphosis. Incomplete metamorphosis is also known as simple metamorphosis.

Complete metamorphosis involves four stages of development. The stages are different so that the adult form has little in common with the other forms. The stages of complete metamorphosis are:

1. egg
2. larva
3. pupa (also known as the resting stage)
4. adult

Examples of insects with complete metamorphosis include moths and butterflies, beetles, and boll weevils.

With incomplete metamorphosis, the young and adult forms resemble each other. The stages of incomplete metamorphosis are:

1. egg
2. nymph
3. adult

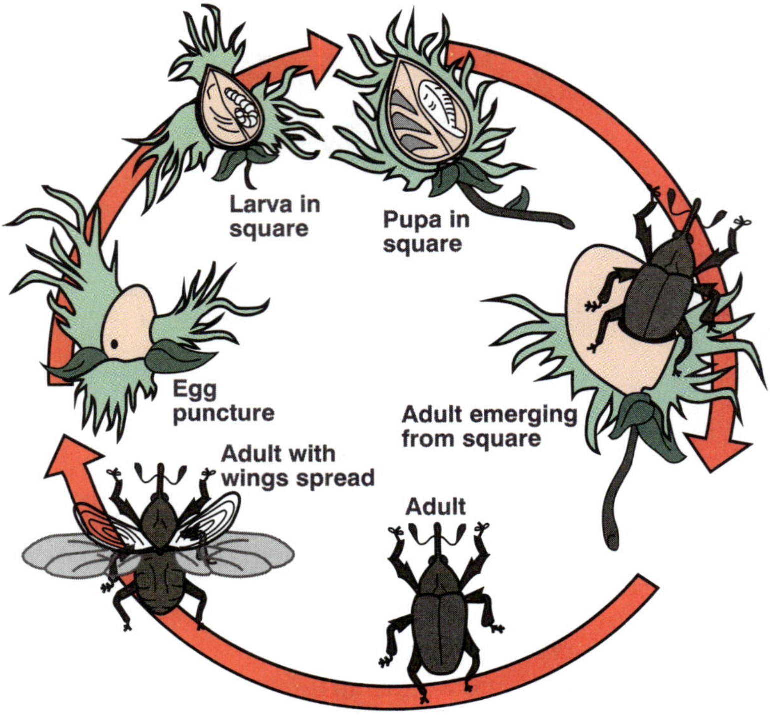

10-6. The life cycle of a boll weevil is a complete metamorphosis.

10-7. Tent caterpillars (larvae of the *Malacosoma americanum* moth) construct silk tents and can eat all of the leaves off of trees. They are often seen in apple, cherry, and pecan trees in the summer.

The nymph looks similar to the adult. The nymph is smaller, doesn't have well-developed wings, and has no reproductive organs. The grasshopper is a good example.

Life Stages and Damage. Insects cause damage at different life stages. Some cause damage by making holes in fruit to lay eggs. As the eggs hatch, the larva eat the fruit or surrounding leaves and stem.

A good example of damage in the larva stage is the tomato hornworm, also known as the tobacco hornworm (*Protoparce sexta*). The adult form is a moth that doesn't cause damage. The eggs are laid by the adult on plants that will provide food for the larva. As the eggs hatch a few days after laying, a worm two to three inches long with a horn-like structure at one end develops. This worm has a big appetite and can strip the leaves from tomato or tobacco plants in a few days. In the pupa stage, it is found "resting" in the soil and often dug up when the soil is plowed.

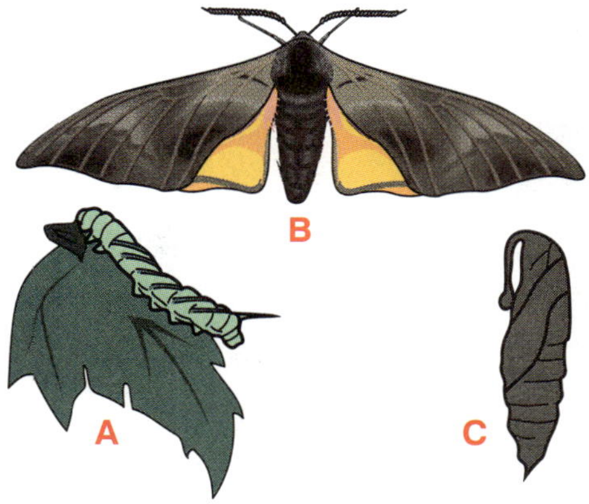

10-8. The tomato hornworm (*Protoparce sexfa*) does its damage in the larva stage (a). Following the larva stage, it goes into the pupa stage (c) and later the adult (b). The adult lays eggs on the plant and the life cycle is repeated.

Two other examples are the boll weevil (*Anthonomus grandis*) and the Hessian fly (*Mayetiola destructor*). Adult boll weevils feed on the bud or boll (fruit) of cotton and lay eggs in the feeding hole. The larva eats the inside of the bud or boll and destroys it. The larva change to the pupa stage and rest before emerging as an adult boll weevil. The Hessian fly lays eggs on wheat leaves. The eggs hatch into maggots in five days. The maggots suck juice from the stem.

10-9. Hessian fly (*Mayetiola destructor*) is only about 1/8 inch (3.2 mm) long. (Courtesy, James Lytle, Mississippi State University)

Agriscientists are better able to control insects if they know their life stages.

Beneficial Insects. Not all insects are bad. Some do a lot of good. They help maintain a good environment for plants.

Beneficial insects help crops grow. They may do so by improving the environment for the crop plants. Here are three ways insects are beneficial:

- **Improve the soil**—Some insects burrow into the soil. The tunnels they make allow air to enter the soil and improve the soil's physical condition. Insects also bury decaying parts of dead plants and this helps improve soil fertility. Examples of insects beneficial to the soil include ants, wild bees, and beetles.

10-10. The honeybee is a beneficial insect. It helps pollinate flowers and provides honey. Some people raise bees in hives. A bee farm is called an apiary.

- **Help pollinate plants**—Insects often go from flower to flower searching for food, for instance the juice in flowers known as nectar. As the insects move about, pollen sticks to their bodies. Some of it rubs off when they go to another flower. This helps with pollination. Examples of insects that are beneficial in pollination are bees, butterflies, wasps, and beetles.
- **Destroy insects**—Some insects attack and kill harmful insects. This helps to keep down the population of harmful insects. Examples include lady beetles, dragon flies, wasps, and ants. Lady beetles are especially useful in destroying the harmful aphids that may get on some plants.

AgriScience Connection

INSECTS CONTROLLING INSECTS: GOOD GUYS VERSUS THE BAD GUYS

Not all insects are bad! Many rid the environment of damaging insects. For example, some species of wasps attack forms of other insects and destroy them. Parasitic wasps usually only attack the targeted pests, such as the corn ear worm. (The corn ear worm is also known as the tomato fruitworm and cotton bollworm.)

This shows a wasp (*Microplitis croceipes*) laying an egg inside a caterpillar. The egg hatches in a few days and begins feeding on the caterpillar. The caterpillar is unable to grow to the pupa stage and dies. The wasp larva eats its way out of the caterpillar and becomes an adult in two weeks.

Using pesticides in a field kills the beneficial insects plus the harmful insects. Having 300 to 600 tiny wasps per acre of crop will usually provide adequate control of the targeted pest. (Courtesy, Agricultural Research Service, USDA)

Nematodes

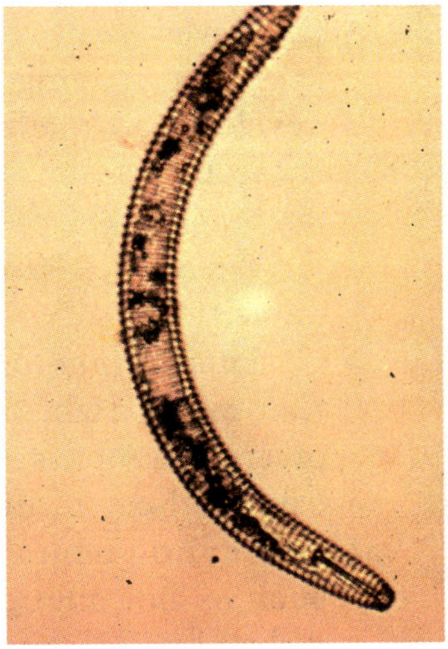

10-11. Nematodes attack the roots of plants and reduce growth. (This photo was made using a microscope to enlarge the tiny nematode. (Courtesy, Frank Killebrew, Plant Pathologist, Mississippi State University)

A ***nematode*** is a tiny worm-like organism that lives in the soil and attacks the roots and stems of plants. Their damage may go unnoticed for awhile. Nematodes sometimes cause knots on the roots of plants. The damage is not considered nearly as great as that of insects. (Nematodes are sometimes incorrectly listed as insects.)

Identification. Nematodes are very small animals only one millimeter long. Their identification often requires laboratory study by a trained agriscientist. They are often tentatively identified by the kind of damage they cause plants. Roots, stems, buds, and other parts of the plant show damage in different ways depending on the kind of nematode. Accurate identification requires laboratory testing.

Life Stages and Damage. Nematodes go through several life stages. The adults lay eggs which hatch into nymphs. The nymphs molt (shed old skin and grow new skin) four times. The adults have spear-like structures that pierce plants and suck out the juice. Nematodes are also carriers of certain diseases, such as ring-spot virus in apples and peaches. Nematodes attack a number of crops, including corn, soybeans, cotton, and vegetables.

Diseases

A ***Plant disease*** is an abnormal condition in living plants. The plant is different from normal and often appears stunted or is not growing properly. Functions in the plant may be interrupted. Plants sometimes die from disease. Diseases are caused by many different living organisms and non-living materials.

Identification. Diseases do not just happen; they result from a cause. Many things can cause disease. Careful laboratory study is often needed to accurately identify the disease.

Diseases are visible by the kind of damage (signs) we see. Signs of disease include the following: (1) rotting plant parts, particularly the fruit;

(2) leaves turning yellow or having an unnatural color; (3) plants wilting; (4) plants having twisted leaves or stems; (5) buds, flowers, or fruit not developing or falling off; and (6) dead plants. Various combinations of these may also be found.

Types of Disease. Diseases are of two major types: environmental and parasitic.

Environmental diseases are caused by factors in a plant's environment that aren't right for the plant. Since plants vary in needs, their tolerance of environments varies. Some plants can live with a condition that will cause others to die.

Environmental diseases include the following: (1) nutrient deficiencies, such as not enough potassium in the soil; (2) damage to plant parts, such as by plowing improperly; (3) chemical injuries, such as by using chemicals wrong; (4) pollution injuries, such as smoke from a factory; (5) weather, such as by storms or too much rain; and (6) naturally-occurring genetic abnormalities.

Tissue analysis and soil testing may be needed to help identify some kinds of environmental diseases.

A ***parasitic plant disease*** is caused by very small organisms that live in the plant and disrupt its normal life. A microscope is often needed to see the organisms. Laboratory study is used to make accurate identification.

Diseases are caused by several organisms:

- **Fungi—*Fungi*** cause more plant diseases than any other parasite. They are small, one-celled organisms that grow on and in plants. Fungi cause plant mildew, rusts, and smuts. Fruit rot is often caused by fungi. The Dutch elm disease is caused by fungi, and has killed many trees. Fungi are spread

10-12. This corn suffers storm damage—an environmental disease.

10-13. This tomato is rotting because of tomato blight. The disease is caused by the fungus *Sclerotium rolfsii.* This disease is most damaging during hot humid weather. (Courtesy, Jackie Mullen, Auburn University)

by wind, water, insects, and in many other ways. For example, diseases in ornamental shrubs can be spread by trimming the shrubs with shears that have been infected by a shrub with disease. However, just as there are beneficial insects, there are some beneficial fungi.

- **Bacteria**—***Bacteria*** are small, one-celled organisms. Some bacteria are beneficial and others cause plant disease. Bacteria often get into plants through cuts or breaks in the bark or epidermis. Some enter through flowers and natural openings in the stem and leaves. Apples and pears may get a bacterial disease known as fireblight. Rot in fruit and vegetables is often due to bacteria.
- **Viruses**—Viruses are special forms that don't have an organized nucleus. They are not considered to be living organisms. They are very small and are visible only with powerful electron microscopes. They can't be seen with regular microscopes. A number of diseases are caused by viruses. Examples include cucumber mosaic, citrus tristeza, and tomato ring spot. Viruses are spread by insects, equipment used to plow fields, and by vegetative propagation.

10-14. The circle area without clover has a fungus disease often called dollar spot. The clover has been killed. The spot will increase and destroy more clover.

10-15. The leaves on this shrub are discolored and have spots because of a viral disease.

Weeds

A ***weed*** is a plant that is growing where it isn't wanted. They compete with crop plants for nutrients and space. Weeds reduce crop yields and lower the quality of the crop that is produced.

Plants can be useful in one place and weeds in another. In effect, a weed is a plant that is out of place.

Kinds of Weeds. Weeds are plants. They may be classified much the same as crop and ornamental plants. Knowing the growth characteristics of weeds helps in selecting methods of control.

Weeds may be classified on the basis of life cycle, as follows:

- **Annuals**—Annual weeds go through a complete life cycle in one season. They sprout from seed, develop into mature plants, and die in one season. Annual weeds can also be classified as summer annuals and winter annuals.

10-16. Dandelions growing in a lawn are weeds. The seed in the round seed heads are ripe and ready for a wind to blow them away where they will grow more dandelion weeds.

Summer annual weeds are those that grow in the summer. Their seeds lie dormant in the soil through the winter. They sprout and grow in the spring at about the same time that crops are planted. Common summer annual weeds are morning glory, cocklebur, pigweed, and crabgrass. These weeds cause problems in crop plants that grow in the summer, such as corn and soybeans, vegetables, ornamental plants, and lawn grasses.

Winter annual weeds go through their life cycles in the winter. The seed sprout and begin growing in the fall and winter. The plants mature and seed are produced in the spring. The seed fall to the ground and lie dormant through the summer. Examples of winter annual weeds include chickweed and henbit. These weeds cause problems in wheat, oats, rye grass, and winter turf grass.

10-17. The poison ivy vine growing on the tree is a perennial weed.

- **Biennials**—Biennial weeds go through a complete life cycle in two years. They come up and grow roots, stems, and leaves one year. They grow flowers and fruit the second year and die. Examples include thistles and wild carrot.

- **Perennials**—Perennial weeds live more than two years. They are often difficult to control because of the root systems they grow. Some have fleshy roots, such as the dandelion and plantain. Others have rhizomes that grow below the ground or stolons that grow above the ground. Examples include Bermudagrass and Johnsongrass. Trees growing along the edges of fields may be weeds. If they shade out the sunlight or take water that the crop needs they are weeds. The trees are growing where they are damaging crop plants.

What Weeds Do. Weeds cause problems in producing plants. A few problems are:

- **Keep plants from growing**—Weeds compete with desired plants for nutrients, water, sunlight, and space. A few weeds may not cause much of a problem in a crop. A lot of weeds can cause crop plants not to grow.

- **Waste nutrients**—Since weeds use nutrients, fertilizer applied is wasted when it is used by weeds. Fertilizer used by weeds doesn't help the crop.

- **Lower quality of crop**—Weeds lower the quality of harvested crops. Weed seed in wheat must be removed. Trash from weed leaves that sticks in cotton must be cleaned out. Weed leaves in harvested spinach or other leafy crops must be removed. Weeds that grow in hay crops aren't usually removed and lower the quality of the hay. Some weeds could even be poisonous to cattle.
- **Make harvesting harder**—Harvesting is harder when there are many weeds. Harvesting machines must cut and discard the weed pieces. More fuel is needed to run the machines. The equipment wears out more quickly.
- **Hiding place for insects and disease**—Weeds can harbor (hold) insects and disease. These can attack the crop plants and destroy them.
- **Look bad**—Weeds aren't attractive in flower beds, lawns, and other decorative areas. Who likes a weedy lawn? Agriscientists often feel that weeds detract from the appearance of a field of crop plants.

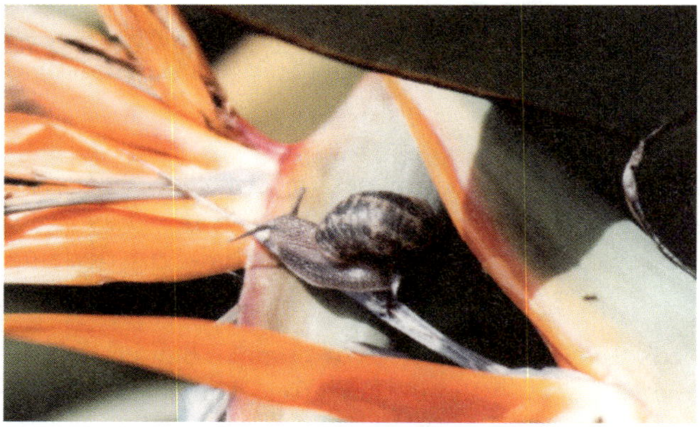

10-18. A snail feeding on a bird of paradise flower.

Rodents and Other Animals

Rodents and other animals can attack and destroy plants. A ***rodent*** is a small animal such as a rat or rabbit. Other animals that are pests include birds, beavers, deer, and raccoons. These pests eat leaves, stems, fruit, and roots of plants. The amount of loss caused by these animals is not as great as that caused by other pests. Individual farmers, however, may have large losses.

A few examples are: deer that eat soybean plants growing in a field, raccoon that climb corn stalks and get roasting ears, rabbits that bite off the tender leaves of new bean plants, birds that eat grain from sorghum, rats that eat fruit growing near the ground, and slugs and snails that eat fruit and leaves.

Preventing and controlling animal pests involves destroying animal habitat and getting rid of the animals. Cutting weeds and bushes around fields destroys habitat. However, some of the animals are protected by laws. For example, it may be illegal to kill birds that are causing damage to crops.

RIGHT CONDITIONS

Conditions must be right for pests to damage plants. Since most crops are grown crowded together, pests may attack more easily and cause greater problems.

Three conditions are needed for pest problems to develop:

- **A pest**—If a pest doesn't exist where plants are grown, it can't cause a problem. Culture practices can be used to keep pests away. For example, growing different crops on the same land in different years helps to keep pests away. This is known as rotation. Tomatoes are often rotated with other crops because of blight and other pests that may get into the soil. By not growing tomatoes every year, the pest loses a host that helps it survive. A host is a plant that provides a pest with food.
- **A susceptible plant**—Some plants are more likely to be attacked by pests than others. Pests that feed on one kind of plant can be present but don't cause problems for other kinds of plants. Once the affected plant is grown, the pest has a host to attack. Producers who know they have certain pest problems grow crops that aren't affected by the pest. Again the tomato is a good example. People who know that the cause of tomato blight is in their soil will not grow tomatoes. They grow other plants that the blight doesn't attack.

10-19. Determining the kind of disease may require careful laboratory examination. This photograph shows tissue from a diseased plant being prepared for growth in a laboratory. (Courtesy, Jackie Mullen, Auburn University)

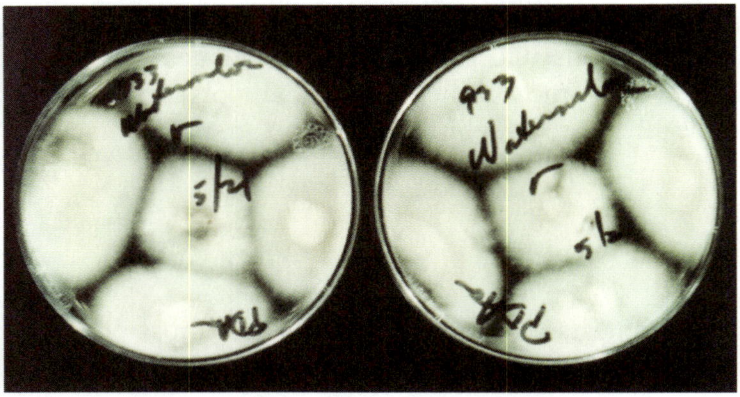

10-20. The Fusarium spores taken from the stem of a wilted watermelon plant are shown here. The spores were grown from plant tissue in a petri dish. (Courtesy, Jackie Mullen, Auburn University)

- **The right environment**—Pests and susceptible plants can be growing together without problems. This is because the environment is not right for the pest to cause problems. For example, if a plant that prefers warm, dry weather is in cool, damp weather and a pest is present, it may attack the plant. Growers can use practices that make a bad environment for the pest. In some cases, varieties of crops that are resistant to disease can be grown. Pesticides (pest killers) are used to make the environment deadly to pests.

PESTS CAUSE DAMAGE

Pests damage plants by interfering with their growth and reproduction. When this happens, the products that plants provide are damaged and/or the amount produced is less.

Pests cause damage in several ways.

HOW PESTS CAUSE DAMAGE

Plants and the crops produced by the plants are damaged by pests. The kind of damage varies with the kind of pest and the plant. Some plants aren't bothered by the presence of pests, while other plants may be severely damaged by the same pests.

Pests Chew Holes in Plants

Some pests make holes in plants and their products. Insects may eat holes in the leaves of plants. This makes the plant less productive. If the plant is a leafy food crop, such as cabbage, the plant is of little value if the cabbage heads are full of holes. This kind of damage is easy to see.

Pests Attack the Vascular System

Pests sometimes attack a plant on the inside. This kind of damage can go unseen for a long time. For example, borers can attack the vascular systems of trees. They may bore or tunnel around the tree and cut the vascular system apart. When this happens, the tree may be unable to conduct water to the leaves and food to the roots. The tree will die.

10-21. Worms have destroyed the productivity of this corn.

Pests Attack the Fruit

Some insects attack the fruit of plants. This either damages the fruit or stops the fruit from developing. Not only does this reduce fruit production, it also reduces the seed and reproduction of the plant. A boll weevil completely destroys a cotton bud or boll. No cotton or seed will be produced by that flower.

Pests Contaminate Products

Products from plants infested with pests are contaminated. Seed may be infested with a disease. Wherever they are planted, the disease will develop. Buds and scions used for vegetative propagation carry any disease of their parent with them. The new plant will also have the disease.

Pests Rob Plants of Food

Sucking pests take the juices out of plants. They take away what the plant needs. Food for the plant is reduced. The plant will not grow as well. Fruit will be smaller and there will be less of it.

Pests Damage Land

Some pests dig holes and build mounds. These damage land and the plants that grow on it. Rodents may dig tunnels under the ground. Crawfish and ants may build mounds above the ground. Ant mounds above the

10-22. A fire ant mound has damaged this ornamental shrub. Fire ants are difficult to control and infest thousands of acres in the Southeastern United States.

ground damage plants and equipment. The fire ant found in the southern United States and parts of Central America also makes a bad bite on people and animals, causing a festered sore.

HOW DAMAGE AFFECTS THE PRODUCER

People who produce plants are affected when diseases strike their crops. This damage reduces the income to producers. Costs of production are increased and the volume sold is reduced.

Reduced Yields

Pests cause plants to produce less. When insects eat the leaves off of a plant, it can't carry on photosynthesis to make food. Even minor damage cuts back on food production. Plant food is needed to produce large fruit. Fruit is lost when attacked by pests.

Lowered Quality

Crops damaged by pests produce fruit and other products that are damaged or smaller than usual. Apples with worms aren't the same quality as those without. Lettuce with holes in the leaves isn't fit to sell. Producers can't sell some damaged crops or, if they can, they get less money for them.

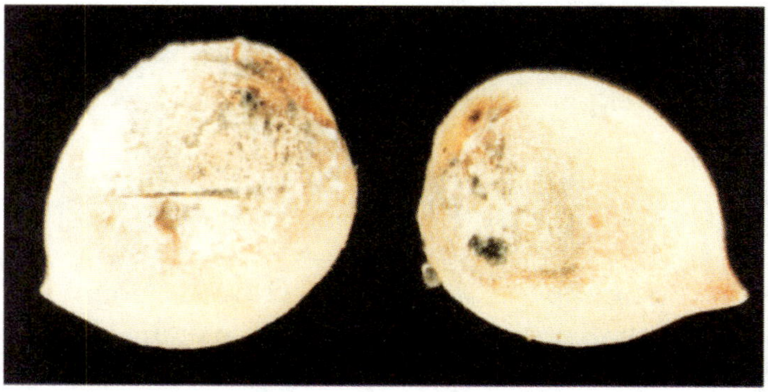

10-23. These peaches have brown rot caused by the fungus *Monilinia fructicola*. They have no value and must be thrown away. (Courtesy, Jackie Mullen, Auburn University)

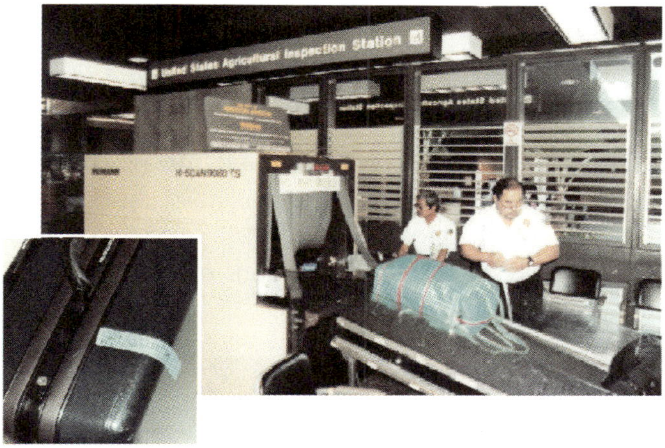

10-24. Inspection services are required by the U.S. Department of Agriculture for items being transported into the United States and to and from selected locations within the Nation. This shows luggage being x-rayed at the Honolulu Airport. Luggage that passes inspection receives a seal of inspection.

Increased Production Costs

Pests increase the cost of producing plants. Controlling them is costly. Pesticides (pest killers) may be bought and applied to the crop. Equipment to apply the pesticides and fuel to run the equipment are needed. In addition, labor is required to do all of the activities.

Hiding Place for Pests

Pests can hide in products from plants with disease. The folds in leaves, stem areas of fruit, and other characteristics make hiding places for pests. Worms on spinach leaves can be harvested right along with the leaves. The worms hide on the leaves. Of course, the worms must be removed before processing and cooking.

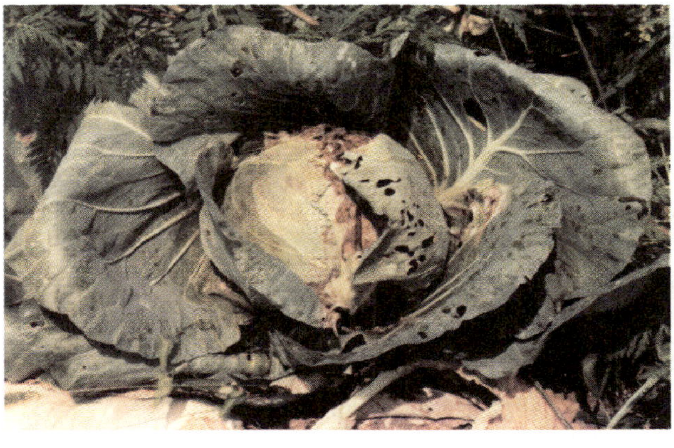

10-25. This cabbage has been damaged by insects and rot.

Restrict Marketing

Taking certain products from some places to other places is prohibited. This is because the source of the product has pest problems. For example, papaya can't be carried from Hawaii to California because of possible disease contamination of the papaya. Products that are possibly diseased might carry the disease with them if shipped. Inspection services are sometimes used to check for pests. If it is found to be free of pests, the product can be traded (sold).

PREVENTING PEST PROBLEMS

Many pest problems can be prevented. Good practices in producing plants will help to reduce the problems. The steps to take aren't that hard to follow.

Preventing pest problems is far better than treating them. If there is no problem, there is no cost to control it.

Here are a few suggestions for preventing pest problems:

- **Use good seed**—Some pests are ***seed-borne***. This means that they are in the seed and go where the seed go. Diseases are more likely to be in seed than other pests. Use seed from sources that are disease free. Seed companies with good reputations will likely have the best seed. Use seed and plants that have been appropriately inspected and certified as disease free.

10-26. Quality treated seed are being used to plant this field. (Courtesy, American Soybean Association, St. Louis, Missouri)

- **Destroy diseased plants**—Destroying diseased plants often destroys the pests that are in them. Cutting and moving trees with borers protects uninfected trees from the borer. Shrubs that have disease should be buried, burned, or destroyed in another way. By removing the plants that first show signs of pests, we may keep other plants from having the pest.
- **Use the right fertilizer**—Plants can suffer from nutritional diseases. Providing the needed fertilizer will overcome these problems. Soil testing may be needed to determine the kind of fertilizer that is required. Plants that don't have the needed nutrients are weaker and subject to attack by other pests.

- **Disinfect equipment**—Equipment can carry pests from one field to another. Equipment used in a field with pests should be washed before using it in another field. This even includes lawn mowers that may bring pests from an infected lawn to a healthy one!
- **Use good water**—Irrigation water can bring pests to a farm or field. Use water that is free of pests. Water that is polluted with chemicals can bring problems to crops. Overflow water from streams and creeks can bring pests onto the field. Water that is high in certain minerals can poison plants and the soil. For example, water treated with chlorine should not be used to water tomato plants. The plants are damaged by chlorine.
- **Control animal movement**—Animals can carry pests in their bodies, drop uneaten food, or take pests on their feet and bodies. It isn't easy to control some wild animals, but good fences can keep out larger animals.
- **Use chemicals properly**—Plants can be damaged by the improper use of chemicals. They can be stunted or killed or may not produce fruit if harmful chemicals get on them. Always use chemicals in approved ways.
- **Use tests to check for pests**—If a pest is suspected, have the plant tested. Take samples of leaves or other tissue to a lab for analysis by trained people. It is easier to control a pest if it is caught before it affects a large area.

AgriScience Connection

SCOUTING: ESSENTIAL IN PEST MANAGEMENT

Scouting is inspecting a field for insects, weeds, diseases, and other conditions that damage crops. This helps determine how and when to use various pest control measures. It prevents waste and helps protect the environment.

Scouts go into areas of a field and examine the crop for insect damage. They may use nets and capture insects using a sampling procedure. Their findings are used to decide what, if any, actions to take with crops such as the cotton shown here.

Scouting is a major part of integrated pest management (IPM). Without good information, only poor decisions can be made about managing the pests in a crop. (Courtesy, Marco Nicovich, Mississippi State University)

GETTING RID OF PESTS

10-27. A coffee plantation in Hawaii uses Macadamia nut shells for mulch. This keeps weeds down and reduces moisture loss.

If pests get into crops and other plants, they must be controlled or the crop may be lost. Several different methods of control are used. The method must be right for the crop and the pest. Also, laws regulate practices that can be used. Some pesticides have been declared unsafe. They are no longer approved for use on crops, trees, and ornamental plants.

Pests are controlled in the following ways: physical or mechanical methods, cultural practices, chemicals, biological methods, and genetic methods.

MECHANICAL METHODS

Mechanical pest control involves using tools or equipment for control. The pests are destroyed or removed. Here are examples of mechanical methods:

- **Plowing**—Plowing destroys some pests, particularly weeds. The pests are cut off or dug up. Conventional cultivation with hoes, harrows, and other equipment destroys weeds. This works best in hot, dry weather. Plants can be "set out" and grow again if a rain immediately follows plowing. They aren't dead!
- **Mowing**—Mowing cuts off weeds. Not only does it partially destroy the weeds, but it also destroys places where other pests can hide. Most weeds can sprout out after being mowed. Mowing may be needed several times in a growing season. Of course, mowing can't be used to control pests in some crops because it would destroy the crop.
- **Mulching**—Covering the ground with a layer of plastic, sawdust, or other material prevents weed growth. Other pests won't have the weeds as hiding places. Mulching is used with ornamental plants and certain vegetable crops.

CULTURAL PRACTICES

Cultural pest control practices are those things that can be done to keep pests out of a crop. Some are easy to do; others are more involved. Here are a few practices to control pests:

10-28. Cutting stalks after harvest is a good cultural practice to reduce insect populations.

- **Rotate crops**—Grow different crops on the land. Growing the same crop year after year allows pests that harm a particular crop to build up in the soil. Crop rotation involves planting one crop, such as soybeans, one year and another crop, such as corn, the next year. Without a host, the pest may die before a particular crop is planted again.
- **Roguing**—***Roguing*** is removing infected plants from a field, forest, or orchard. Diseased trees can be cut down and moved. This prevents the infection of other plants by the infected plant.
- **Trap cropping**—***Trap cropping*** involves planting a small plot of the crop near the field about two weeks before the main crop is planted. The pests attack the trap crop. Pests that reproduce in the earliest plants of the year go to the trap plots, where they can be killed before they damage the main crop.
- **Burning**—Setting dry fields on fire after they have been harvested destroys some pests. It also does away with plant remains that could serve as hiding places for pests. Of course, burning produces smoke which damages the air.

10-29. This area of weeds and bushes around a field is a good hiding place for some pests.

- **Use resistant varieties**—Some varieties of plants have been developed that resist pests. Growers should plant these varieties to help reduce pest problems.
- **Clean around fields**—Weeds and bushes around fields serve as hiding places for some pests. Mowing and spraying can be used to keep them down.
- **Cutting stalks**—Insect pests can hide in plant stalks. Cutting the stalks after harvest removes hiding places.

CHEMICALS

Chemicals used to control pests are known as ***pesticides***. Most of these chemicals are deadly—they are made to kill the pest. As "killers," they can also cause problems for other life. However, the advantages of many pesticides have far outweighed the disadvantages. They make it possible to produce more food!

Most chemicals are complex compounds. They are developed in laboratories and tested on trial plots. Their use must be approved by various government agencies, such as the Environmental Protection Agency. The chemicals can be hazardous to humans.

Chemicals are often mixed with other materials when they are used. Many chemicals are mixed with water. In some cases, substances to help the solution "work" are used. One example is a ***surfactant***. A surfactant is a material that helps solutions do what they are supposed to do. Most surfactants are added to help solutions spread evenly over the waxy surfaces of leaves.

There are many kinds of chemicals used for pest control. Most of them can be classified as those for insects, diseases, and weeds. They can be further classified by how they control the insect, disease, or weed.

Insect Chemicals

Chemicals that control insects are known as ***insecticides***. Some of the substances used are very poisonous. The material that actually does the killing is known as the active ingredient. Labels on containers describe the amount of active ingredient. Some are more powerful (concentrated) than others. Most insecticides are mixed with other materials for use.

Insecticides are made in several forms: dusts, granules (small solid pieces), insecticide-fertilizer mixtures, various kinds of powders, and solutions. Each works best under certain conditions. Most of the powders and solutions are mixed with water for spraying on the affected plants. The equipment must be properly adjusted and operated.

Insecticides are classified by how they get into the insect's body.

10-30. Airplane being used to apply insecticide to cotton. (Courtesy, Marco Nicovich, Mississippi State University)

- **Stomach poisons—*Stomach poisons*** are eaten by the insects. Layers of the insecticide must be on the plant leaves or other materials the pest eats. Stomach poisons work best on chewing insects, such as leafworms, armyworms, and bagworms.
- **Contact poisons**—These poisons are absorbed through the insect's skin. To work, the ***contact poison*** must come into contact with the insect. This means that the poison must be sprayed on the insect or on places where the insect will go. Contact poisons are best for controlling sucking insects, such as aphids.
- **Systemic poisons—*Systemic poisons*** are inside the plant. The poison is applied to the soil and taken up by the roots. Some systemics are sprayed on the leaves and stems and absorbed into the plant. The vascular system moves the poison all over the plant. When an insect bites the plant or sucks its juice, it gets poison.
- **Fumigants—*Fumigants*** are insecticides in gas form. The poison enters the insect's body through the respiratory system. Fumigants can be used only in enclosed places. People and other animals must stay out during the treatment. Fumigants are good for controlling insects in seed, such as wheat in a bin, or plants in greenhouses. Sometimes they are used to treat the soil. When this is done, the ground is covered with a layer of plastic to hold the gas in contact with the soil.

Care must be used in handling and applying insecticides. They are dangerous! Humans should wear protective clothing and masks. Always wash skin and clothes after using insecticides.

Nematode Chemicals

Chemicals used to control nematodes are known as ***nematocides***. Since nematodes are in the soil, treatments must be made that reach into the soil.

Fumigants are sometimes used. The soil is usually plowed and covered with a plastic sheet. The nematocide fumigant is sprayed under the plastic. In some cases, the nematocide is injected into the soil.

Tear gas (chloropicrin) is commonly used for fumigation. It kills nematodes, fungi, and weed seed.

Disease Chemicals

Many kinds of chemicals are used to control plant diseases. The kind used depends on the disease problem. Accurate diagnosis of the disease is a must. If this is not done, the wrong chemical might be used and the disease wouldn't be controlled. When used on plants that aren't diseased, they are known as ***protective treatments***. When used on plants with disease, they are known as therapeutic chemicals or ***therapeutants***.

Chemicals used to control disease caused by fungi are known as ***fungicides***. The best fungicides are systemic in action. They get into the vascular system of a plant and reach all parts. Fungi are killed all over the plant. Fungicides that kill by contact must be sprayed directly on the growing fungi.

Bactericides are chemicals used to control bacteria. The chemicals are also known as germicides. These solutions are frequently sprayed on plants. Caution should be used to avoid damaging the plant.

Weed Chemicals

Chemicals used to control weeds are known as ***herbicides***. Many kinds are available. The right kinds must be selected for the needed control.

Selective herbicides will kill only certain kinds of plants. They will usually kill either broadleaf weeds or grasses but not both. For example, cockleburs (broadleaf weeds) can be controlled in corn (grass-type leaf) with herbicides that can't be used on soybean or other broadleaf plant crops. When used properly, selective herbicides will kill weeds without damaging the crop plants.

Non-selective herbicides kill upon contact. They are sometimes known as contact herbicides. They kill all living cells whether weed or crop plant. Non-selective herbicides are used to kill all of the vegetation where they

10-31. The soil under the plastic has been fumigated. Turf grass seed will be planted here. Notice how the edges have been sealed to prevent loss of chemical by using soil.

are applied. For example, a non-selective herbicide can be used along the edge of walks in lawns to keep grass from growing over the walk.

Translocated herbicides are absorbed into the plant and moved all through its parts. They are also known as growth regulators. These herbicides upset the growth processes of plants. They interrupt the use of food. Results of use may take time. The plant has to move the herbicide throughout the organism before it is killed. These herbicides often cause plants to grow in twisted, unnatural ways the first few days after use. The plant usually turns yellow and brown as it dies within a couple of weeks. Translocated herbicides work best if the weather is warm and the plant is growing rapidly.

10-32. A small sprayer and liquid herbicide are used by homeowners to control weeds.

Spraying is the most common way to apply herbicides. They are also applied as granules mixed with fertilizer and applied to the soil. The action takes place slowly as the granule is dissolved. Herbicides may be applied before, during, or after planting.

Before-planting use is known as preplant application or preemergence. This keeps weeds down as the seeds germinate. Weed competition is reduced. The herbicide is sometimes incorporated (mixed) in the soil. Care must be taken so the herbicide does not damage the seed or tender plants as they begin to grow.

Some herbicides are applied at the time of planting. Sprayers are attached to the planter. A band 8- to 12-inches wide is sprayed above the drill after the seed are covered with soil.

Post-emerge herbicides are used after crop plants are up. They are sprayed on the ground and leaves of weeds. The herbicide used must be selective in action; otherwise, the crop plants will be killed.

BIOLOGICAL METHODS

Biological pest control usually involves using living organisms to control pests. These include using insects that eat other insects, using bacteria and fungi, and altering the reproductive processes of the pests.

Many insects are beneficial in controlling other insects. Pests have natural enemies in the environment. Lady beetles are notorious for their roles in controlling a range of insect pests. Predatory mites have been used to control

10-33. Chemical methods can be used on lawns to control winter weeds. This shows henbit in a dormant Bermuda grass lawn before (top) and two weeks after spraying with a herbicide (bottom).

other kinds of mites on apples. Of course, many gardeners like to have a few toad frogs around to eat insects! Caution: Insecticides should not be used on plants to control insects if predatory insects are present. The insecticide will kill the beneficial insects.

Forms of bacteria and fungi have been developed for release into the environment. These organisms attack the pests and destroy them. A good example is the bacterium *Bacillus thuringiensis*. When released in fields, the bacteria attack and kill various species of worms.

Several major insect pests have been controlled by altering the reproductive process. Since some insects mate only once, a female that mates with a sterile (no sperm) male lays eggs that won't hatch. The male insects are raised in laboratories and sterilized. They are released in areas with insect problems. The males from the laboratory mate with the females. The eggs produced are infertile (won't hatch). This method has been used to control some pests, including the pink bollworm which attacks cotton in California. Increased use will be made of biological methods in the future. Research is underway to develop new ways of using plants and animals to control pests.

GENETIC METHODS

Genetic pest control is developing crops that are resistant to pests. This means that the crop plants have some trait that repels the insects.

Science has had great success with plants that are resistant to disease. Wheat, tomatoes, potatoes, corn, soybeans, strawberries, and many other plants have been developed which resist at least one disease.

Some plants have been developed that are resistant to nematodes. Crops that resist nematode pests are more productive and economical to produce.

Pests are sometimes able to overcome the resistance of certain varieties. This means that new resistant plants must be continually developed.

More emphasis is being given to genetic methods of pest control. These are thought to be less damaging to the environment than other methods. However, some people fear that the new genetic forms may pose problems themselves.

INTEGRATED PEST MANAGEMENT

Integrated pest management (IPM) is a planned process for controlling pests. It involves using the control methods that have the best outcomes for the well-being of society. A combination of measures are used to reduce pest damage with least disruption to the environment.

IPM involves using a blend of pest control techniques in a planned program. Research has shown that no single control measure works consistently over a long period of time. One reason is because pests can develop resistance to pesticides. IPM is a combination of physical, chemical, and biological methods of control.

Emphasis is on reducing pest numbers to a level where they cause only some damage. IPM does not result in the elimination of all pests. It brings them under control so that they do not destroy too many leaves, buds, fruits, or other parts of plants. Fields must be checked regularly for insects, known as scouting.

A field of crops is managed as an ecosystem. This means that all organisms live together and share the space and its resources. IPM attempts to keep the pests to a level that doesn't do much damage to the plants. A little damage is usually expected.

IPM has fewer adverse (bad) effects on the environment and society. It is used to keep the environment clean and free of harmful materials.

SAFETY PRACTICES IN PEST CONTROL

Many of the methods used to control pests are dangerous. They can injure people and other animals. They can pollute the environment and contaminate water and food. Government laws have been passed to regulate the use of pesticides. Also, pesticides are approved only for certain uses. Here are a few safety rules to follow in pest control:

- **Use only approved pesticides**—Government regulations allow only certain pesticides to be used. Follow the laws!
- **Know the pesticide**—People who use pesticides should be informed about them. Labels on containers give a lot of information. Use the pesticide according to directions on the label. Do not use them where they aren't approved for use.
- **Use the pesticide with low toxicity**—Toxicity refers to how poisonous the pesticide is. Use the pesticide that will do what needs to be done but that is no stronger than needed.

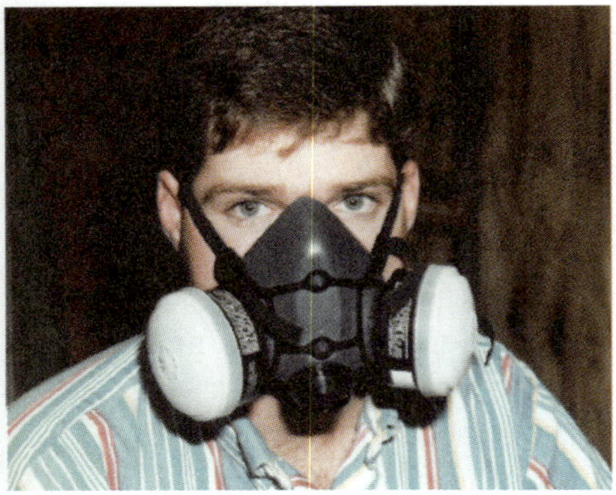

10-34. A properly fitting respirator may be needed when applying pesticides.

- **Use pesticides only when needed**—Pesticides should be used when pests are to be controlled. Using a pesticide without need damages the environment and wastes money.
- **Do not contaminate resources**—Pesticides can pollute the environment. They should never be dumped into streams or on the ground. Leftover pesticides should be properly disposed of when no longer needed. Keep pesticides away from food and water.
- **Wear protective clothing**—People need to protect themselves from pesticides when they are applying them. Protective clothing and respirators (masks) should be used. The clothing should be properly washed after it is worn. The Worker Protection Standard (WPS) law specifies how pesticides are to be used and labeled.
- **Wash the skin after contact**—Wash with soap and water immediately if any pesticide gets on the skin. Wear rubber gloves to keep pesticide off skin.

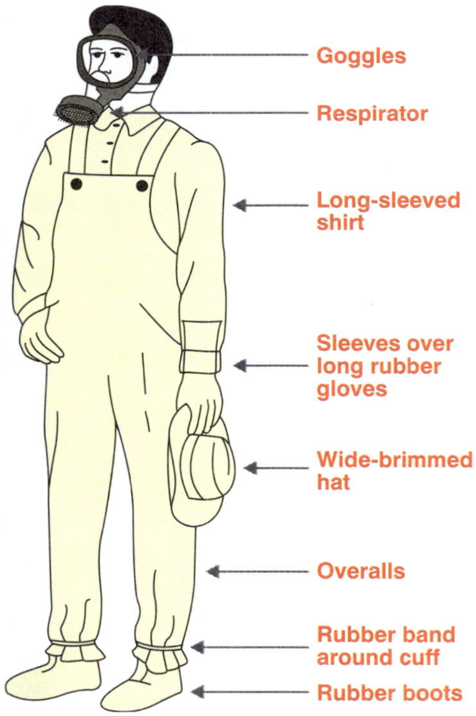

10-35. Proper protective clothing is needed when applying chemicals.

- **Dispose of empty containers properly**—Empty containers should never be thrown into creeks or gullies. Some manufacturers take empty containers back. Generally, empty containers should be rinsed out three times and returned for recycling or sent to an approved solid waste facility (dump).
- **Apply in good weather**—Pesticides should be used when they will have the most effect on pests. Wind blows sprays and dusts away. Sometimes other crops, water, or livestock can be damaged by drifting poisons.
- **Use the right equipment**—This includes funnels to help in pouring and measuring and mixing containers. Spraying equipment should be adjusted properly so it applies no more than is needed.
- **Know the emergency measures**—People who use pesticides should always know what to do in case of an accident. Local physicians know who to contact for help when people are poisoned. It is a good idea to have the emergency telephone number handy.

REVIEWING

MAIN IDEAS

Plant pests cause injury or damage to plants. They may be living organisms or non-living substances. The five major kinds of pests are: (1) insects, (2) nematodes, (3) diseases, (4) weeds, and (5) rodents and other animals.

Certain conditions must be present for a pest to be a problem. The three conditions are: a pest, a host plant, and the right environment. If one is missing, there will be no problems.

Pests damage plants in many ways. They generally cause reduced yields of crops even if they don't destroy plants.

It is far better to prevent pests than to try to control them after they get into a crop. The major ways to prevent pests are: (1) use good seed, (2) destroy diseased plants, (3) use the right fertilizer, (4) disinfect equipment, (5) use good irrigation water, (6) control wild animal movement, (7) use chemicals properly, and (8) use tests to check for pests.

Controlling pests can cost a lot of money. It may also cause other problems, such as damage to the environment. Methods of pest control should be selected that are most effective. Several methods of pest control are: (1) use mechanical means, such as hoeing and plowing; (2) use appropriate cultural practices, such as rotation; (3) use chemicals, such as insecticides and herbicides; (4) use biological methods, such as predatory insects; and (5) genetic methods, such as planting resistant varieties.

Integrated Pest Management (IPM) involves a combination of ways to control pests. It allows some pests, with control measures used when the pests reach a certain level. Regular scouting of the crop by trained people is needed.

Always follow safety procedures when using pesticides. People should protect themselves and the environment from damage.

QUESTIONS

Answer the following questions using complete sentences and correct spelling.

1. What is a plant pest?
2. What are insects? How are they identified?
3. What are the life stages of insects? How do the stages relate to plant damage?
4. What is a beneficial insect? List three ways insects are beneficial.

5. What is a nematode? How do nematodes damage plants?
6. What is a plant disease? What are the signs of plant disease?
7. What are the types of plant disease?
8. What is a weed? What are the kinds of weeds?
9. How do weeds cause problems for plants?
10. What conditions must exist for pests to damage plants? Explain each.
11. How do pests damage plants? How does the damage affect the producer?
12. How are plant pest problems prevented?
13. What methods are used to control pests? Briefly describe each.
14. What is the relationship between the way an insect gets its food and the kind of pesticide that may be used?
15. Distinguish between selective and non-selective herbicides.
16. What is integrated pest management?
17. What are the important safety practices when using pesticides? Select three which you feel are most important.

EXPLORING

1. Make a list of the major plant pests in your community. With each, describe how the pest can be prevented and controlled. (Interview producers, use reference materials, and get the help of agriscientists with this activity.)

2. Visit a farm supplies store or garden center and study the kinds of pesticides available. Note how the materials are packaged. Read the labels for information on approved use and safety.

3. Make a collection of brochures and pamphlets that describe plant pests found in your community. This should include information about the pests and how they are prevented and controlled. Information is available from the Cooperative Extension Service, a university, or supplier of pesticides.

4. Locate a plant that has been damaged by a pest. Determine the nature of the damage and what caused it. Prescribe ways of controlling the damage.

11

APPLYING PRINCIPLES OF ANIMAL SCIENCE

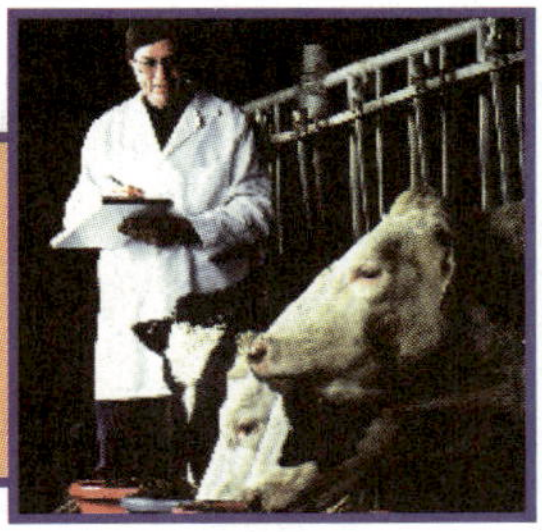

A million different kinds of animals have been identified. Many of these animals provide useful food and fiber products.

Animals come in different forms and live in different places. Some are very large; others very small. Some live in water; others fly in the air and walk on the earth. Some have been domesticated—they grow under the control of humans; others are wild and sometimes dangerous. Humans use the products of both domesticated and wild animals.

Animals have complex bodies, especially those that have been domesticated. Knowing the needs of animals and how their bodies function helps people do a better job of raising them. Some types of animals have been raised for hundreds of years, and the knowledge gathered over the years is most useful. Today, research continues to provide more information about these animals.

11-1. Working cattle on a Texas ranch. (Courtesy, Texas Department of Agriculture)

OBJECTIVES

This chapter presents important principles of animal science. It has the following objectives:

1. Name and describe the major animal groups.
2. Describe the anatomy and physiology of animals.
3. Explain the major organ systems of animals:
   a. skeletal system
   b. muscular system
   c. nervous system
   d. circulatory system
   e. respiratory system
   f. excretory system
   g. digestive system
   h. reproductive system
   i. mammary system

TERMS

alveoli
anatomy
bivalve mollusk
blood
bone
bony fish
brain
breathing
cartilage
crawfish
crustacean
estrus cycle
exoskeleton
external respiration
fetus
gestation
internal respiration
invertebrate
involuntary muscle
lactation
mammal
mammary gland
mollusk
non-ruminant
parturition
physiology
plasma
platelet
pregnant
puberty
receptors
red blood cell
spinal cord
vertebrate
voluntary muscle
white blood cell

GROUPING ANIMALS

All animals are members of the Kingdom Animalia. Most of them share three common traits: They are made of many cells, can move about, and get their food from other sources. (Chapters 5 and 6 provide background information on differences in plants and animals and how they are classified.)

Animals can be grouped in many ways. They share common traits and also have many differences. Animals can be grouped on the basis of body structure and functions, where they naturally live, extent of domestication,

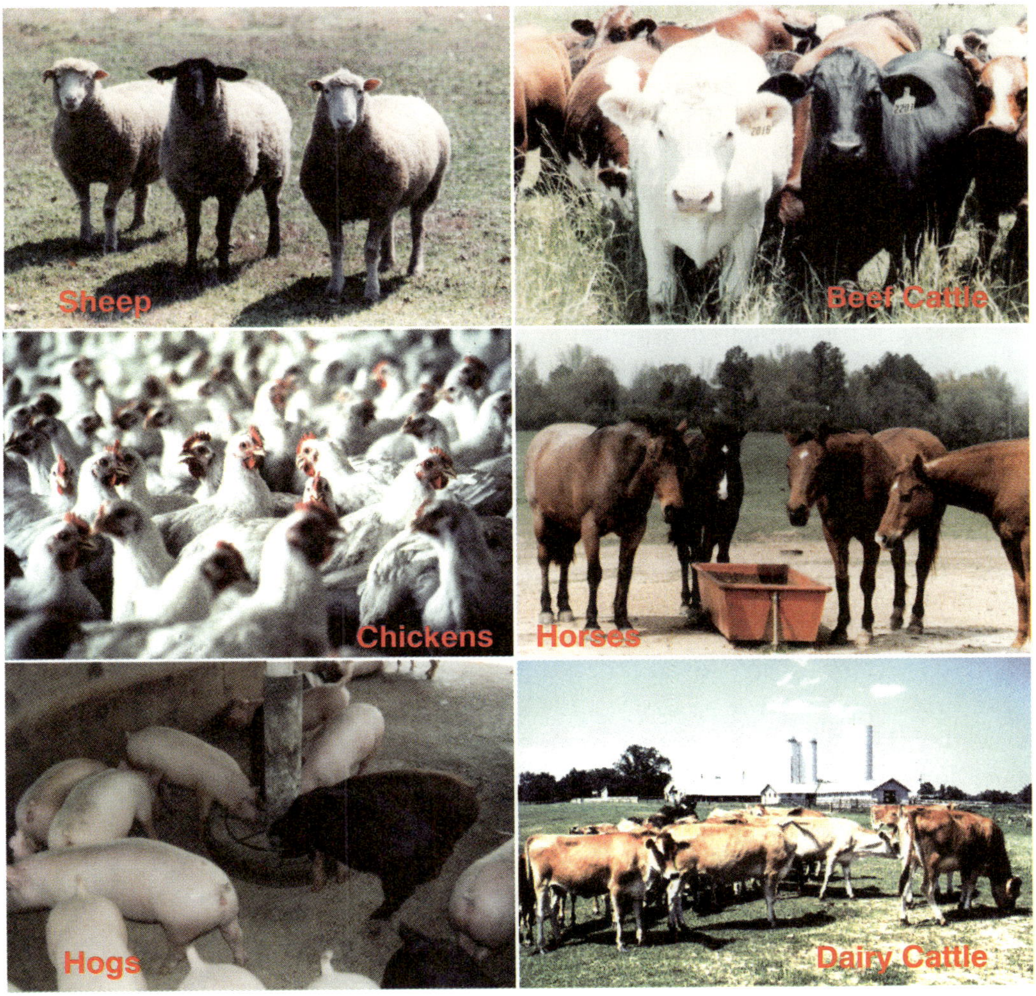

11-2. Important terrestrial agricultural animals. (Courtesy, chickens and dairy cattle, Marco Nicovich, Mississippi State University)

and the products they provide humans. Regardless of how they are grouped, all animals are a part of the earth's ecosystem and contribute to the way people live.

BODY STRUCTURE

Animals have different body structures. The two major groups are ***vertebrates*** and ***invertebrates***. Vertebrates are animals with backbones, while invertebrates do not have backbones.

Invertebrates

The invertebrates are important in agriculture because of what they contribute and also because of the problems they cause. About 97% of all animals are invertebrates. They are a diverse group of animals. Earthworms, spiders, butterflies, snails, and lobsters are examples.

Most of the important agriculture species are in the Phylum Arthropoda. Of course the other phyla also contain important animals, such as the Annelids and Mollusks.

Arthropods. The arthropods have hard outer shells or coverings known as an ***exoskeleton***. These usually fit like a coat of armor. The exoskeleton gives the body protection and support. It keeps the internal organs from being easily injured. The exoskeleton is made of chitin (a tough material made of carbohydrates) and protein. It is similar to the material in the fingernails of humans. Arthropods molt (shed their exoskeleton) as they grow.

The body and appendages (attachments, including legs and antennae) are jointed. These joints allow the animal to move about. Without these jointed segments on the body, the animal would be in a cast similar to that on a broken arm. They couldn't move! Arthropods have simple body systems that allow them to carry out life processes.

The most important arthropods are the crustacea, uniramia, and chelicerata. (These are subphyla in the Arthropod Phylum.)

- **Crustacea**—The ***crustaceans*** are mostly aquatic (water-living) animals with exoskeletons. There are 25,000 species, including shrimp, crawfish, lobsters, crabs, barnacles, and copepods. Crawfish and shrimp are often grown by aquafarmers. Lobster, crab, and others may be harvested from the sea as native or wild animals. Some grow in saltwater; others in freshwater. The freshwater species include crawfish and a kind of shrimp known as prawn.

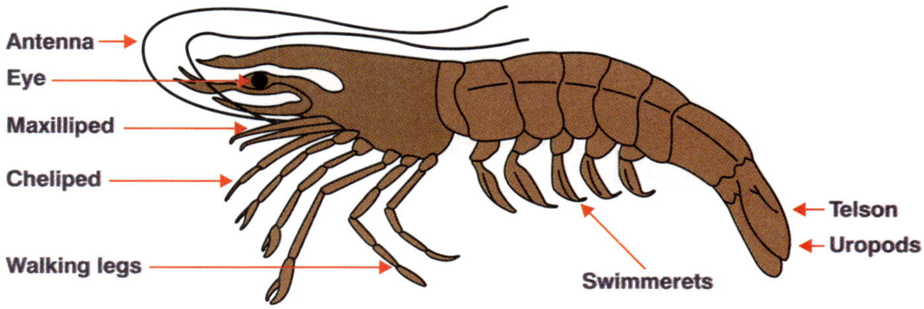

11-3. Major external parts of a shrimp. (Crawfish are similar but have much larger chelipeds, or pinchers.)

A good example of a crustacean is the ***crawfish*** (also known as crayfish). They are found in fresh water all over North America and are grown in Louisiana, Texas, South Carolina, and a few other states. Their bodies, similar to shrimp, are in two sections: cephalothorax and abdomen. The cephalothorax is made of the head and thorax fused together (made as one piece). Antennae, mouth, and other parts are found on the head. The thorax is divided into 13 segments. Five pairs of legs are attached to the thorax. The first pair of legs is large and strong, known as the chelipeds or pincers. Crawfish and lobster have large chelipeds. Shrimp have small chelipeds. The other legs are used for walking. The abdomen is a muscular structure divided into seven segments. Each section has swimmerets except for the last, which has a uropod (tail) that can be used to quickly move backwards through the water. Crawfish are grown for food. The thick muscular part is removed from the shell after boiling the crawfish whole.

- **Uniramia**—This subphyla includes the insects, millipedes, and centipedes. Insects are the most important, since about three-fourths of the animals on earth are insects. Insects are useful as well as harmful. (Previous chapters on plant science gave information about the beneficial insects and those that cause damage to plants.)

  The bodies of insects are divided into three segments: head, thorax, and abdomen. The head has a mouth, antennae, and other structures. Three pair of legs are attached to the thorax as well as two pair of wings. Insects have organ systems to carry out life processes.

  Some insects are farmed. A good example is the honeybee. These insects are kept to help pollinate crops as well as for the honey they produce.

- **Chelicerata**—This subphylum contains ticks, mites, spiders, scorpions, and other species. Many of these are known as pests. They damage plants and animals. Most of these animals are in the Class Arachnida, commonly called arachnids (spiders). There are over 100,000 species of arachnids. They have four pair of legs and some spin webs for homes and to trap their prey of insects.

Annelids. The Phylum Annelida contains earthworms, blood-sucking leeches, and a few others. Some of these do a lot of good, particularly the earthworms in the soil. Leeches are often known as pests, especially when they attack animals such as fish. Some leeches have been used on human wounds to suck out the "bad" blood.

Earthworms are farmed for fish bait. The worms are often grown in enclosed areas where food is provided. Diseases sometimes strike earthworm farms wiping out entire crops.

Mollusks. The Phylum Mollusca includes a number of animals with soft bodies, but most of them have hard shells around them for protection. Slugs and octopuses have soft bodies and no shells. Oysters, snails, clams, mussels, and abalone have hard shells. There are about 100,000 species of ***mollusk***. Most of the mollusks are aquatic animals.

The ***bivalve mollusks*** are a Class of mollusk with two-sided shells that are hinged together. A strong muscular foot opens and closes the shell. Oysters, clams, scallops, and mussels are examples. Most of these are "filter feeders." This means that they do not move around to get food. They get their food by filtering tiny materials from the water that flows through their shells.

Some of these species are farmed; others are captured in the wild. For example, oysters are grown in bay areas of Washington state where the tide gives a good cleaning action to the oyster beds. Oysters must have good water. In feeding, they take in all of the pollution in the water. Humans can get sick eating oysters grown in "bad" water. Oysters harvested from wild stock are more likely to be disease carriers.

Others. Many other invertebrates affect life on earth. They help plants and other animals grow. They cause disease and otherwise damage plants and animals. Some are parasites that live in larger animals. Examples of parasites are the roundworm and tapeworm parasites of hogs, cattle, and other animals. Heavy infestations cause the animal to grow poorly. They may weaken the animal so that it dies.

Vertebrates

The vertebrates are animals with backbones. They are in the Subphylum Vertebrata of the Phylum Chordata.

The vertebrates are named for the vertebrae (bones and cartilage) that surround the nerve cord. They also have a hard skull that protects the brain. They have internal skeletons that provide the framework for their bodies. The part of the skeleton made up of the backbone and skull is known as the axial skeleton. Paired limbs (appendages) are attached to the axial skeleton. Muscles are attached to the skeleton and provide for movement of the animal.

Many important agricultural species are vertebrates. The major vertebrates are presented here.

Fish. Three classes of vertebrates are known as fish. These include Osteichthyes, which are the ***bony fish***, and the Chondrichthyes, which

11-4. Important aquatic animals. (Courtesy, prawns, James Lytle, Mississippi State University)

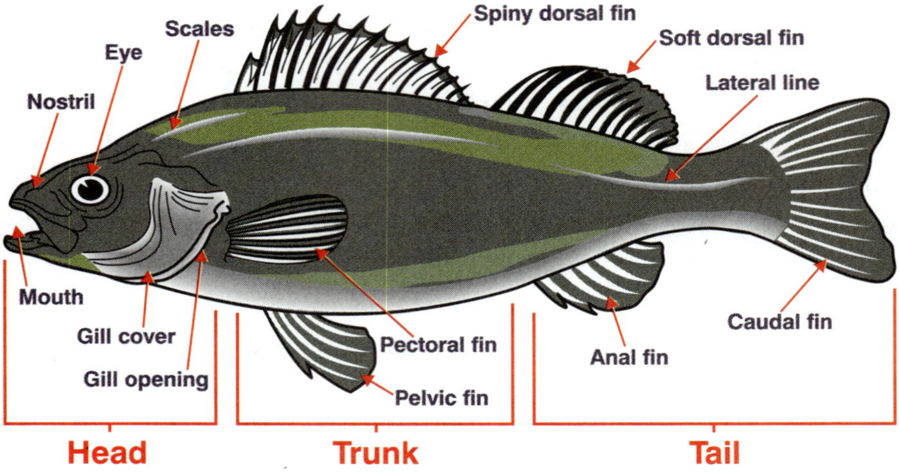

11-5. Major external body parts of a bony fish.

includes sharks, rays, and a few others. The third class is the small Agnatha Class of jawless fish, such as the lampreys.

About 25,000 species of bony fish have been identified. These are important sources of human food. Many are caught as wild fish from rivers, lakes, and oceans. Increasingly, fish are being grown on aquafarms. The most popular cultured fish in North America is the catfish. Over 100,000 acres are devoted to aquafarming this fish. Trout, tilapia, salmon, baitfish, and a few others are aquafarmed. In addition, many species of ornamental fish are kept in homes and businesses as pets.

Fish are covered with scales or skin. Their bodies are divided into three major parts: head, trunk, and tail. The major parts on the head are the mouth, eyes, and gills. The trunk is the part of the body that contains the major internal organs. The tail is a strong muscular structure used to help the fish move about.

Fish have body organ systems similar to other animals. One major difference is how they get oxygen. Fish have gills which "filter" oxygen from the water as it passes over them. Fish reproduce by laying eggs which are fertilized after laying and incubated (tended by the fish before hatching) for several days. Fish are ectothermic animals, which means that their body temperature is regulated by the temperature of the water they live in.

Birds. Birds are in the Class Aves. Birds are adaptable to wider environments than most other animals. Different species live in the air, on the land, and in the water. Scientists have identified about 9,000 species of birds. Many birds are wild, while only a few species are farmed. Chickens, ducks, and turkeys are raised more than the other species. Chicken has grown in popularity as a food in recent years because it is thought to be healthful and free of dangerous fats. However, it can have dangerous organisms that cause food poisoning. Thorough cooking is essential.

The bodies of birds are covered with feathers. The bones are hollow and thin which makes them lightweight so the bird is better able to fly. The wings help lift them in flight. They have two legs with feet and similar internal organ systems to other animals. They reproduce by laying fertile eggs that hatch in a few weeks time.

Mammals. The Class Mammalia includes several important species of animals produced for food and fiber. Most ***mammals*** reproduce by mating of the male and female of the species. The female carries a developing embryo in her uterus for a period of time known as ***gestation***, and gives birth after a time of gestation to a young animal. The females have ***mammary glands*** that produce milk for the newborn young.

Mammals have hair, a heart with four chambers, a lower jawbone with teeth, and a well-developed brain. Mammals are endothermic, which means that their bodies maintain a certain temperature regardless of the temperature of their environment. Examples include cattle, goats, sheep, hogs, horses, and other common farm species.

Several chapters in this book deal with mammals, aves, and bony fish.

AgriScience Connection

PREPARING FOR COMPETITION

Competing in livestock shows is an interesting and educational experience. Many opportunities are available with the different species of animals. Favorites include beef and dairy cattle, swine, and lambs and sheep.

Animals are selected, fed, trained, and groomed so that they look their best. Grooming includes trimming hair and hooves, washing, drying, and other steps.

This shows a lamb being blow-dried prior to showing. The owner is obviously very proud of her lamb. Many owners develop strong bonds with their animals.

11-6. Alligators live in aquatic environments. They are farmed for their hides and meat.

WHERE ANIMALS LIVE

Animals are found in a wide range of environments. Some species require certain kinds of environments if they are to live. Animals can be classified by their habitat and the climate they need for survival.

Habitat

Most animals produced on farms are either terrestrial or aquatic. Terrestrial animals live on land. Aquatic animals live in water or a water environment. A few are amphibious, meaning that they live in water or on land. Some are amphibious only during certain life stages.

Terrestrial animals include the major farm animals: cattle, hogs, horses, goats, sheep, chickens, and turkeys. Ducks live on water and land.

Aquatic animals include fish, alligators, and frogs. These animals are raised for food, leather, or other uses.

Climate

Animals vary in the climates they prefer. Temperature and rainfall are two major factors in climate.

Some animals live well in very cold climates, such as the polar bear. Others prefer temperate climates, such as hogs and cattle. Monkeys, gibbons, and jaguars prefer the warmer climates.

Animals have made adaptations that help them live in wet or dry climates. Those that live in areas with a lot of rain either live in water or have made adaptations to help them escape water. The monkey has a long tail allowing it to go into trees when the water covers the ground. Camels, scorpions, owls, and snakes are adapted to very dry desert climates. Most cattle and other farm animals are adapted to temperate climates.

Cattle with humps are adapted to warmer, wetter climates than those without humps. For example, the Brahman breed of beef cattle is better suited to warm, wet climates than the Angus breed. Ranchers often cross the two breeds to combine the best traits of both in their herds.

PRODUCTS

Animals can be classified on the basis of the products they provide. Some provide food, while others provide fiber and related products.

Food

Several kinds of animal products are used for food. Here are a few examples:

- **Eggs**—Chicken eggs are used in many ways as food. They are often prepared into egg dishes or mixed in with other foods in cooking, such as in making bread. Eggs of some other animals are also used for food, but to a lesser extent. An example is caviar—the eggs of fish.
- **Meat**—Beef cattle, hogs, sheep, chicken, turkey, fish, and others provide meat. The kinds of animals used for meat vary according to the cultural background of the people who consume them. Some people do not eat pork. Horses, dogs, cats, and other animals are eaten in various countries of the world but not widely in North America. Of course, some people use wild animals for food, including deer, rabbits, elk, squirrels, and raccoons.
- **Milk**—In North America, the milk of cows is the major milk food. Some goat milk is used. In other countries, mare's milk as well as the milk of water buffalo and similar domestic animals may be used. People drink milk as a liquid or manufacture it into solid products, such as cheese.

11-7. Uniform, graded eggs are available in modern supermarkets.

11-8. Animals can be trained to do useful work. This shows a Pointer on a quail hunt indicating the location of birds.

11-9. Wool being examined for quality. (Courtesy, American Sheep Industry Association)

Fiber and Other Products

Many non-food products are made from animals. A few examples are listed here.

- **Clothing**—Wool (from sheep) and leather (cattle, alligators, ostrich, and others) are the two major animal products used for clothing. Others include bone for buttons, pearls for jewelry, and cattle horns or tusks for ornaments.
- **Medicine**—Various medicines used by humans are made from animal products. These include bone marrow from cattle for blood disorders, insulin from the pancreas of cattle and hogs, vitamin B_{12} from the liver of cattle, and hormone products.
- **Other Products**—Many products are made from the by-products of animals. Here are a few examples of non-food products made from cattle: perfumes, soap, glue, paint brush bristles, deodorants, pet food, and floor wax.

ANATOMY AND PHYSIOLOGY

Animals are complex living organisms. They are capable of taking in food, transforming and storing it, and releasing its energy.

As with all living organisms, the cell is the basic building block of the animal. Cells divide and divide again. Cells specialize and form masses known as tissue. The tissues form organs. The organs form systems that must work together if the animal is to survive.

General information on the life processes of organisms was presented in Chapter 5. This chapter presents more details about important farm animals.

IMPORTANCE OF ANIMAL KNOWLEDGE

Raising animals requires knowing something about them. More knowledge helps producers raise animals successfully. Knowing what they are

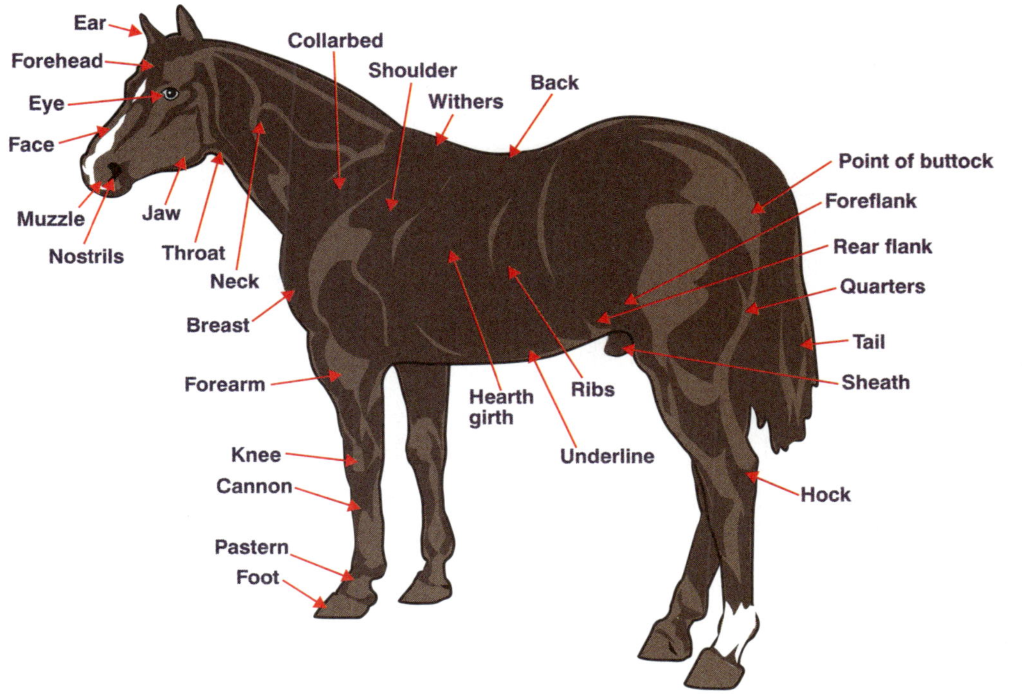

11-10. Major external parts of a horse.

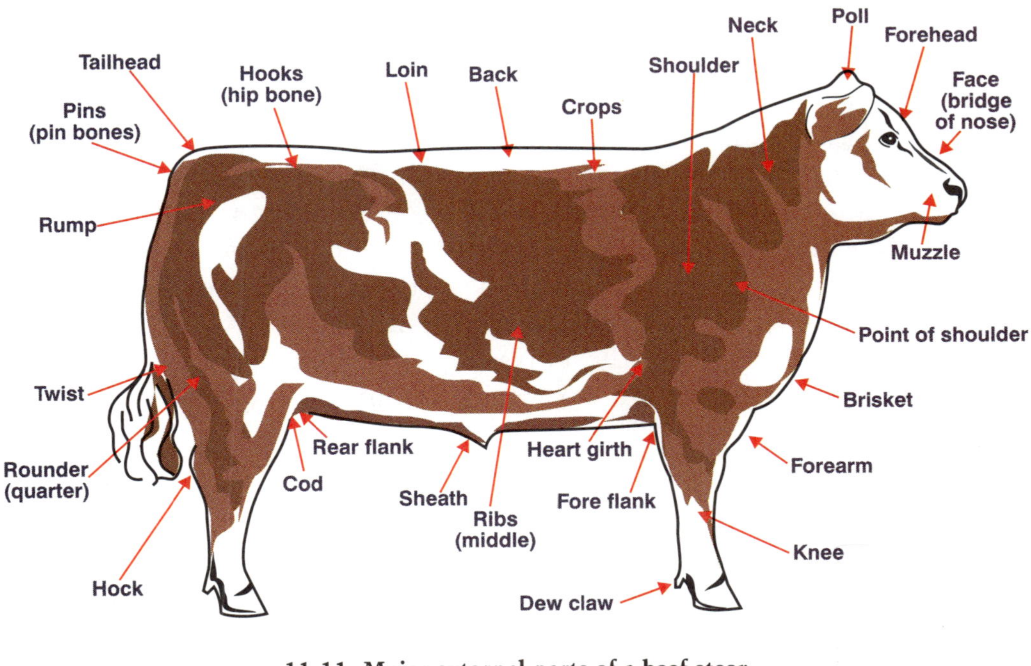

11-11. Major external parts of a beef steer.

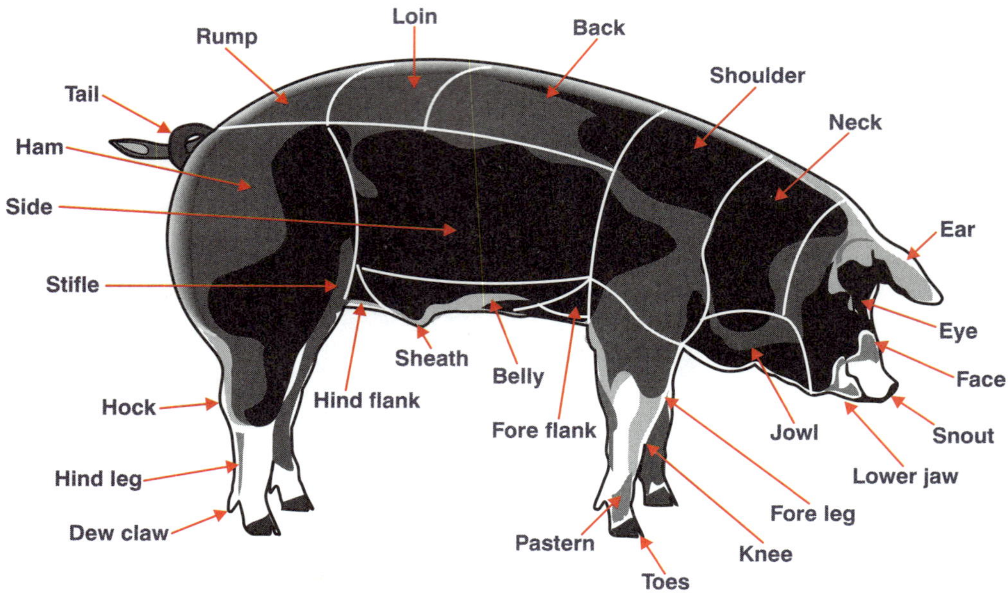

11-12. Major external parts of a hog.

like allows people to help them grow and reproduce. The study of animals often begins with anatomy and physiology.

Anatomy is the study of the form, shape, and appearance of the animal. Gross anatomy deals with structures that can be seen with the human eye. Microscopic anatomy focuses on structures that require the use of magnification to be seen, such as cells and sperm.

Knowledge of external gross anatomy is useful in identifying animals. Horses, cattle, chickens, sheep, and hogs have features that distinguish their species. These parts are also useful in determining the quality of the animal. Note the differences in the parts of the common animals.

Physiology focuses on the functions of the cells, tissues, organs, and systems of the body. It also deals with relationships of the systems to each other. For example, if the lungs of a pig don't take in oxygen, the blood won't be able to absorb oxygen in the lungs. The pig will die!

ORGAN SYSTEMS

The organ systems are designed to carry out specific functions to keep animals alive and growing. The major organ systems of common farm animals give them identifiable characteristics and make it possible for them to carry out life processes. The organ systems are: skeletal system,

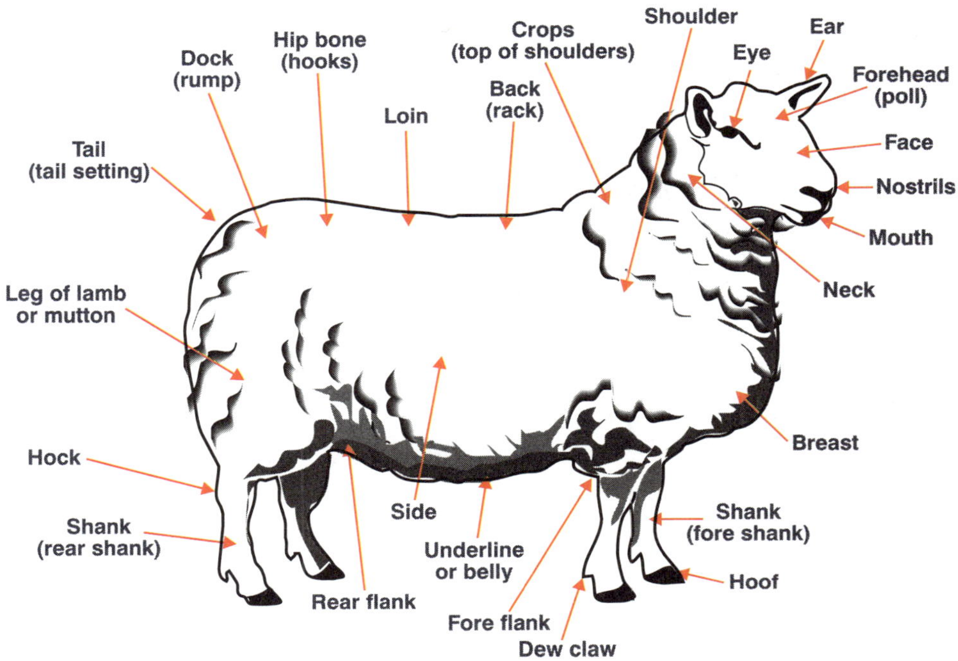

11-13. Major external parts of a sheep.

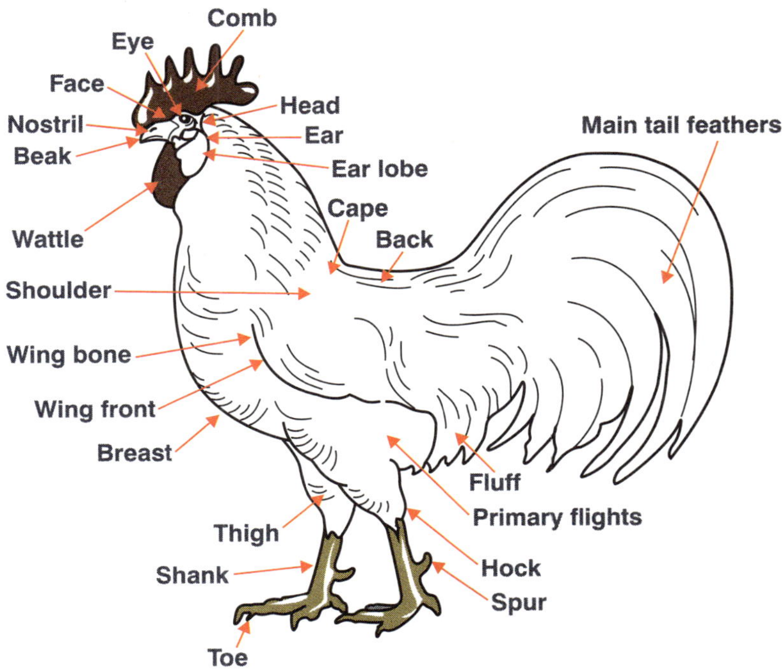

11-14. Major external parts of a chicken.

muscular system, nervous system, circulatory system, respiratory system, excretory system, digestive system, and reproductive system. Female mammals also have a developed mammary system.

SKELETAL SYSTEM

The skeletal system is made of the bones and cartilage that give the body a framework. The skeleton protects the delicate internal organs and helps keep them properly arranged in the body. Bones make it possible for animals to move about and support themselves. Skeletons of different species vary. Chickens and hogs both have skeletons, but they vary a great deal.

Bones are the hard parts of the skeleton. They are made of calcium, phosphorus, and other substances. Bones have several layers: outer membrane (periosteum), which contains small blood vessels; compact bone,

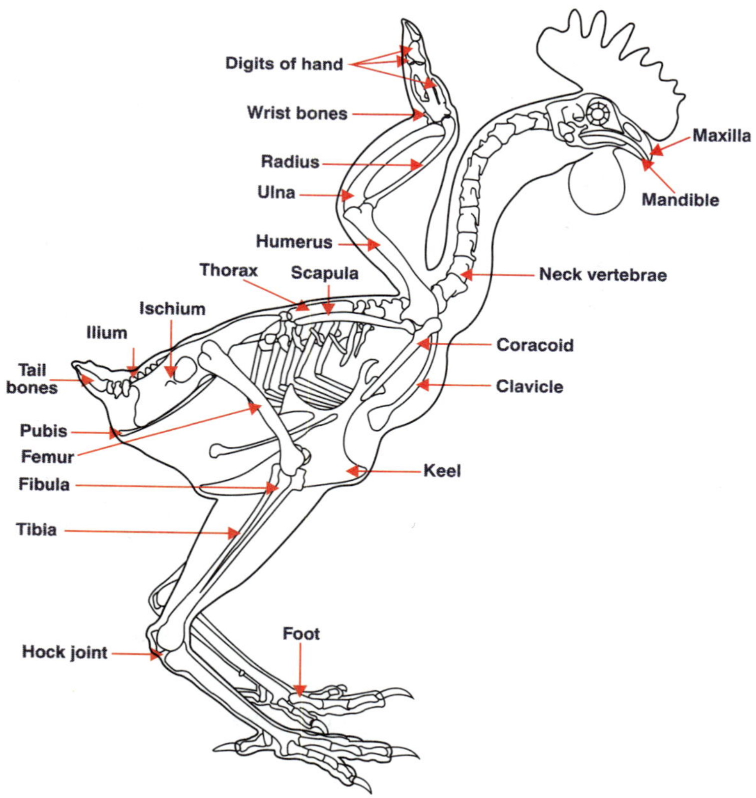

11-15. Skeleton of a chicken.

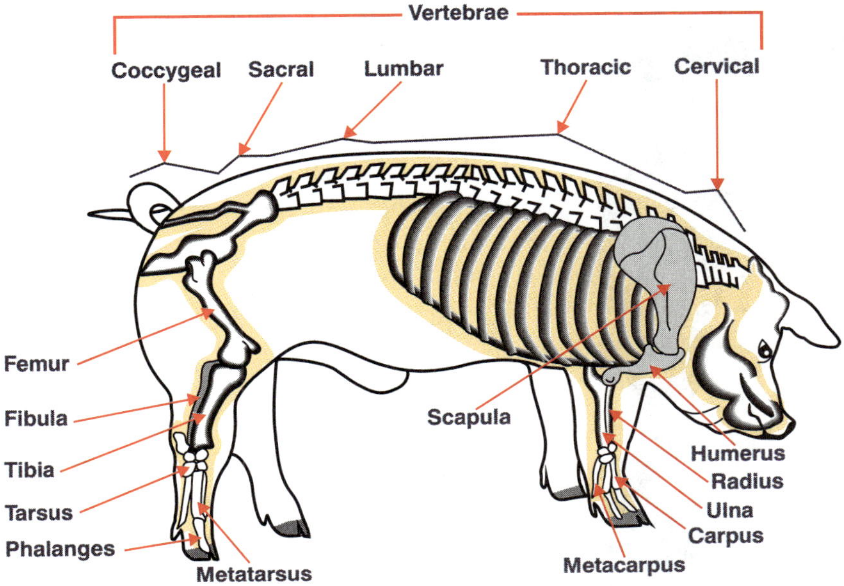

11-16. Skeleton of a hog.

which is made of rings of mineral crystals and protein fibers; and spongy bone, which is the most interior part of the bone. The spongy bone contains red bone marrow, which produces blood cells, and yellow marrow, which serves as an energy reserve for the body.

Cartilage is found at the ends of bones. It is a flexible material that lubricates the joints and cushions shocks. It protects the bones from damage, especially when they move against each other in a joint.

Animals need strong, well-developed skeletons. Proper feed is needed to assure good bone development. Young animals especially need to be fed diets high in calcium and other minerals as well as protein.

MUSCULAR SYSTEM

The muscular system is the largest system in the body. It makes up about 45% of the body weight of hogs, cattle, and chickens. It is important in movement, including locomotion, circulation, digestion, and breathing.

The muscular system is made of muscles and connective tissue. Tendons connect muscles to the bones of the skeleton.

Muscles are of two major types: voluntary and involuntary. ***Voluntary muscles*** are controlled by the "thinking" part of the brain (cerebrum). A message must be sent by the brain for them to move. For example, the

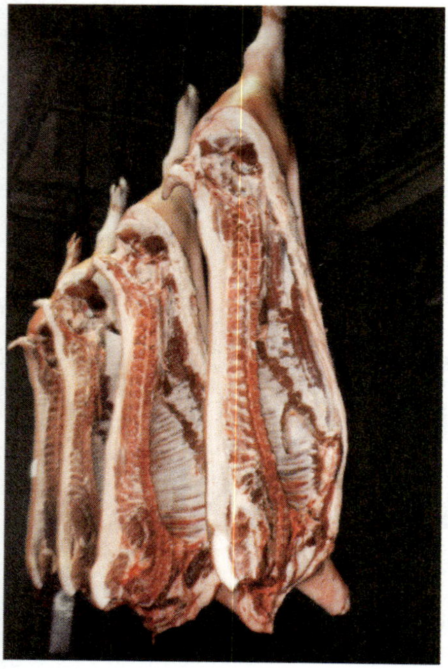

11-17. Hog carcasses showing good muscling with a layer of fat.

muscles used in walking are voluntary muscles. The brain tells them to move and directs the type of movement.

Involuntary muscles are automatically controlled by a lower part of the brain. These muscles operate the heart, intestines, lungs, and other organs. Animals cannot stop these muscles from moving.

Good muscle development is essential in meat animals. The muscle tissues are used for food. People want animals with a lot of muscle tissue, small bones, and less fat.

Muscles that get a lot of exercise are not as tender as those that get less exercise. The muscles in the necks of animals do more work than those in the loin (back); therefore, neck muscles are not as tender.

Tendons are not as tender as muscles. Older animals have more tendons and less muscle tissue. Also, loin muscles and eye muscles of the rib have less tendon and make the best meat, such as beef ribeye steak and filet mignon and pork loin and chops.

Muscles respond to commands from the nervous system.

NERVOUS SYSTEM

The nervous system is made of nerve tissue. This is the tissue that conducts a type of electrical impulse from the ***brain*** to the muscles by way of the spinal cord. The ***spinal cord*** is located in the backbone.

The nerves are made of nerve cells or neurons. These are the strands of cells through which impulses travel. Impulses are short-lived electrical charges. When the brain gives an order, the nerve tissue transmits the order to the muscles.

PARTS OF THE NERVOUS SYSTEM

The nervous system is the most highly developed system in the body. It helps the body to work as a single unit. To do so, it is divided into three subsystems.

- **Central nervous system**—The brain and spinal cord make up the central nervous system. Memory, actions, and reasoning are in the brain. The spinal column is located in the backbone. It is the main tissue through which the brain sends and receives messages.
- **Autonomic nervous system**—This system contains the nerves that are connected to the involuntary muscles and organs. It provides for near automatic operation of the organs. The lower brain retains some control over activities.
- **Peripheral nervous system**—This system deals with the voluntary muscles. It includes all of the nerves outside the brain and spinal cord. The peripheral nervous system carries impulses to and from the brain. Muscle action is sent through these nerves by the brain.

RECEPTORS

Various parts of the nervous system collect information from the environment. These are known as the ***receptors***. The information is processed and acted on by the brain. Neurons must sense and respond to a stimulus as well as carry impulses from one end of the cell to the other. These actions are known as excitability and conductivity. Neurons send impulses to each other through synapses.

The body has five types of sensory receptors (receivers of information): sight, hearing, touch, taste, and smell. These senses are more highly developed in some animals than others. All have good sense of touch. If something attacks an animal, the sense of touch becomes active and, on orders from the brain, responds. All information is quickly processed through the nervous system to the brain, where a response is issued.

CIRCULATORY SYSTEM

The circulatory system is responsible for moving digested food, oxygen, wastes, and other materials around the body. Individual cells need the food and oxygen to live and divide. The circulatory system removes wastes from the cells and carries them to be released by the body. The movement of blood throughout the body is known as systemic circulation.

BLOOD

The "liquid" in the circulatory system is ***blood***. Blood is made of ***plasma*** (the liquid part) and various solid materials. Plasma is 90% water. The

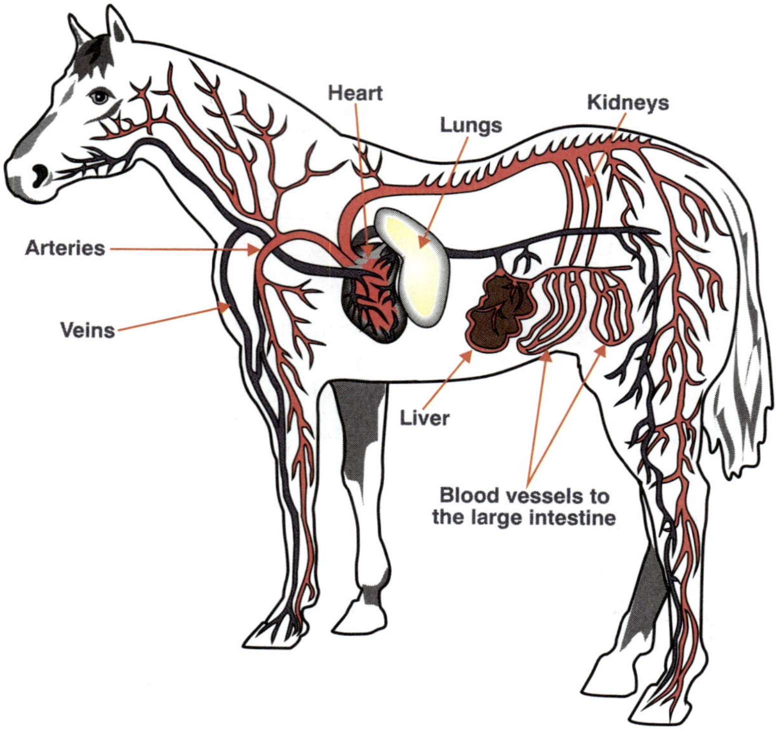

11-18. Circulatory system of a horse. Most other vertebrate mammals have similar circulatory systems.

dissolved substances in the plasma include glucose, vitamins, minerals, and amino acids (proteins). Animals have different types of blood but all blood tends to be similar in overall composition.

The solids in the blood include red blood cells (erythrocytes), white blood cells (leukocytes), and platelets (thrombocytes).

Red blood cells (corpuscles) are made in the red marrow of bones. They contain a protein called hemoglobin. The hemoglobin is the molecule that carries the oxygen in the blood. Red blood cells are much more numerous than white cells. Red cells cannot repair themselves and die after a period of time. More have to be made by the bone marrow. The dead cells are removed by the liver and spleen. The blood in one animal may contain several trillion red cells.

White blood cells help fight off disease. An animal that is diseased will produce a lot more white cells than usual. The pus that forms at wounds is a collection of white blood cells. White blood cells are sometimes called soldiers or scavengers because of their role in keeping down disease.

Platelets are not complete cells but are essential for blood to clot. This is particularly important when an animal suffers a wound, such as being stuck by a thorn or cut by a fence. If the blood doesn't clot, so much blood may be lost that there isn't enough for circulation to continue. The animal will bleed to death.

SYSTEM PARTS

The circulatory system has four major parts.

- **Heart**—The heart is the "pump" that sends blood throughout the system. It is a strong, involuntary muscular organ that contracts (beats) continuously throughout the lifetime of an animal. The beat slows when the animal rests. A heart may beat as much as a billion times during the life of an animal.
- **Arteries**—The arteries are the vessels that carry blood from the heart.
- **Capillaries**—These are the small branches from the arteries that carry blood to the cells. Capillary walls are only one cell thick. This allows easy exchange between the cells and the blood. Blood carries oxygen and energy to the cells and picks up wastes from the cells. After contact with the cells, the blood continues through the capillaries to the veins.
- **Veins**—The veins carry blood back to the heart. This blood has given its oxygen to the cells and picked up waste materials.

The heart sends the blood to the lungs where the carbon dioxide is exchanged for oxygen. Lungs are a part of the respiratory system. After picking up oxygen in the lungs, the blood returns to the heart for circulation throughout the body. Sending blood to the lungs is known as pulmonary circulation.

RESPIRATORY SYSTEM

The respiratory system moves gases to and from the circulatory system. The overall purpose of respiration is to provide the blood with oxygen and to remove carbon dioxide from the blood.

KINDS OF RESPIRATION

Two types of respiration take place in animals: internal and external.

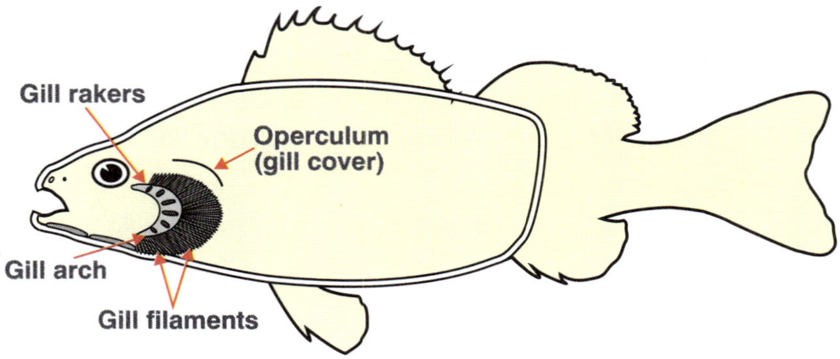

11-19. Gill structure of a fish. (Water flows through the mouth and over the gill filaments where oxygen is removed.)

Internal respiration is the exchange of gases between the cells and the blood within the body. Cells breakdown glucose for energy.

External respiration is the exchange of gases in the lungs between the blood and the atmosphere. Everyone is familiar with the process of breathing.

PARTS OF THE RESPIRATORY SYSTEM

The respiratory system of terrestrial animals has five parts.

- **Nostrils**—These are the openings near the mouth through which gases enter and leave the body. Incoming air is warmed, filtered, and moistened by the hairs and mucous membrane in the nostril area. Of course, some things in the air can't be filtered out, such as harmful gases like carbon monoxide.
- **Pharynx**—The pharynx connects the nose area with the mouth area. It is located at the back of the throat.
- **Larynx**—The larynx is often called the "voice box." Animals produce sounds when air is passed over two ligaments, known as vocal cords. The pitch of the sound varies with the amount of air sent over the cords and the nature of the cords.
- **Trachea**—Also known as the windpipe, the trachea connects the pharynx with the lungs. The trachea branches into bronchi, which further divide into smaller and smaller tubes inside the lungs.
- **Lungs**—The lungs are where gas is exchanged between the atmosphere and the blood. The lungs are located in the chest cavity and supported by the

rib cage and diaphragm. The lungs are surrounded by mucus membranes that secrete liquids to make breathing easier.

Aquatic organisms have different types of respiratory systems. Most fish have a gill system. Oysters, shrimp, and other species vary.

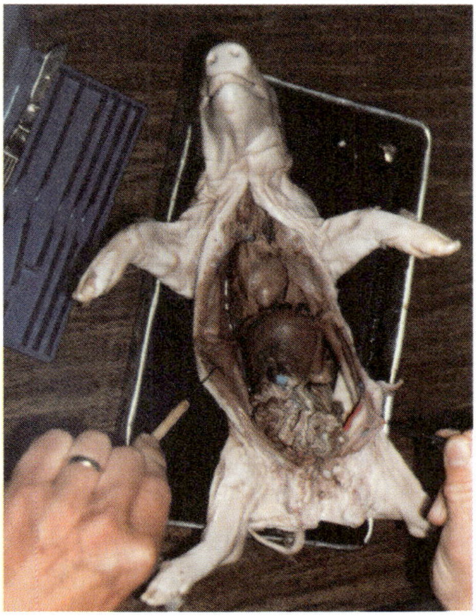

11-20. Several systems can be observed inside this opened fetal pig, including the circulatory, excretory, and digestive systems.

BREATHING PROCESSES

Breathing involves air entering and leaving the lungs. Inspiration is the process of taking air into the lungs. Expiration is the process of air moving out of the lungs. The rib cage, diaphragm, and abdominal muscles are involved in the breathing process. The gases are exchanged in the lungs through the process of osmosis.

The rate of breathing is regulated by the amount of oxygen the animal requires. An active animal breathes faster. As cells burn more oxygen in converting glucose to energy, carbon dioxide is released in the blood. Nerves send a message to the brain that more carbon dioxide needs to be removed and more oxygen is needed. In response, the brain orders more rapid breathing. When the carbon dioxide level has been lowered, the rate of breathing slows upon orders from the brain.

EXCRETORY SYSTEM

The excretory system rids the body of wastes from cell activity. These wastes are known as metabolic wastes. The process of getting rid of them is known as excretion. In addition to excretion, the excretory system helps maintain the proper balances in the animal's body of such things as water and blood substances.

The major products excreted include carbon dioxide, water, and nitrogen compounds from the breakdown of nutrients known as proteins. Carbon dioxide and some water are given off by the lungs which are a part of the respiratory system.

PARTS OF THE EXCRETORY SYSTEM

The major parts of the excretory system and their functions are:

- **Skin**—The skin helps rid the body of wastes through a process known as perspiration. Water and minerals are given off through the skin. In animals that perspire, the skin helps regulate the body temperature. When the body begins to get too warm, the skin perspires and is cooled by the evaporation of the water. The water passes from the skin through pores. Some animals do not perspire, such as chickens and dogs. They pant (breathe rapidly through the mouth) when they need to cool their bodies.
- **Kidneys**—The kidneys filter the blood. They remove wastes created by the cells in the body as well as excess water and minerals. The liquid produced by the kidneys is known as urine.
- **Bladder**—The urine made by the kidneys is stored in the bladder, a muscular sac that is connected to the urethra.
- **Urethra**—The urethra is the tube that carries urine out of the body.

DIGESTIVE SYSTEM

The digestive system prepares food for use by the body. Digestion is the process of breaking food into molecules that the body can absorb. Digestion occurs in the alimentary canal, or digestive tract.

KINDS OF DIGESTIVE SYSTEMS

Animals can be divided into two groups on the basis of their digestive systems: ruminant and non-ruminant. Ruminants are animals that chew cuds, such as cattle, sheep, and goats. ***Non-ruminants***, such as horses, hogs, and chickens, do not chew cuds.

PARTS OF DIGESTIVE SYSTEMS

All animals have digestive systems which are similar in many ways. The major parts are listed here.

- **Mouth**—The mouth is where the food enters the body by ingestion or eating. The mouth typically has jaws with teeth that chew the food. Chewing breaks the food into smaller pieces. Saliva is mixed with food in the mouth

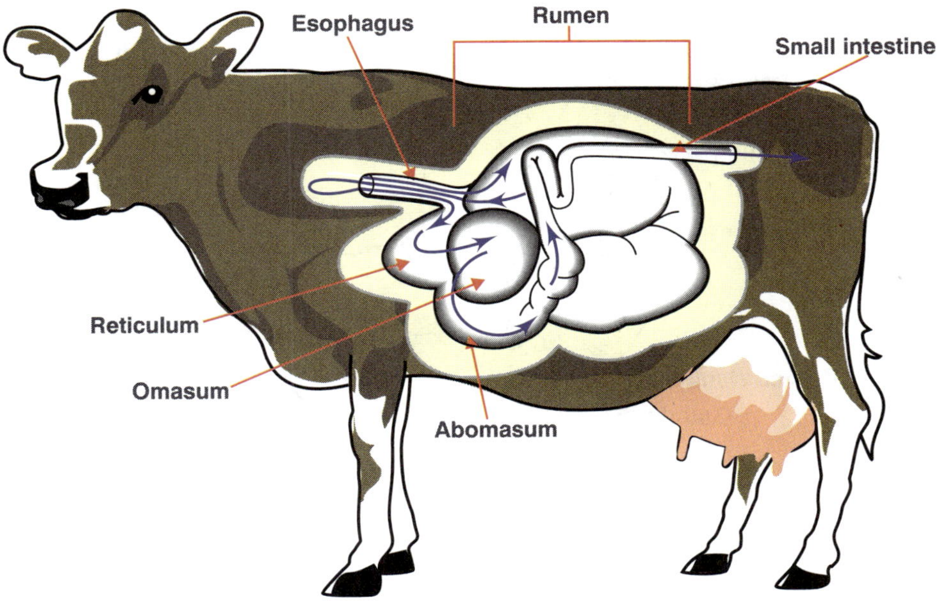

11-21. Digestive system of a cow.

and begins the digestive process and also makes swallowing easier. Saliva is produced by salivary glands in the mouth.

- **Esophagus**—The esophagus connects the mouth and the stomach. Animals move food and water from the mouth by swallowing. Swallowing involves actions of the tongue and pharynx. Muscular contractions in the esophagus move the food along. Food without sufficient moisture (saliva) may be difficult to swallow.

- **Stomach**—Ruminants have compound stomachs. They are divided into four parts, known as compartments. The parts are:

    1. **Rumen**—The rumen is the first compartment. It is also known as the paunch. This is where the food goes when it is swallowed. It is the largest of the four compartments. The rumen has billions of bacteria that break down foods that are high in cellulose into other forms. This is why ruminants can eat large quantities of roughage—feeds that have a lot of volume and relatively low nutritional value.

        When a ruminant has eaten enough food to satisfy its appetite, it goes to a quiet place to ruminate (chew its cud). Food that has been eaten is belched up and rechewed. When ruminants are fed grain, it is not belched up and chewed unless mixed with roughage. Grain must be cracked when first eaten or it will pass through the digestive system whole and undigested.

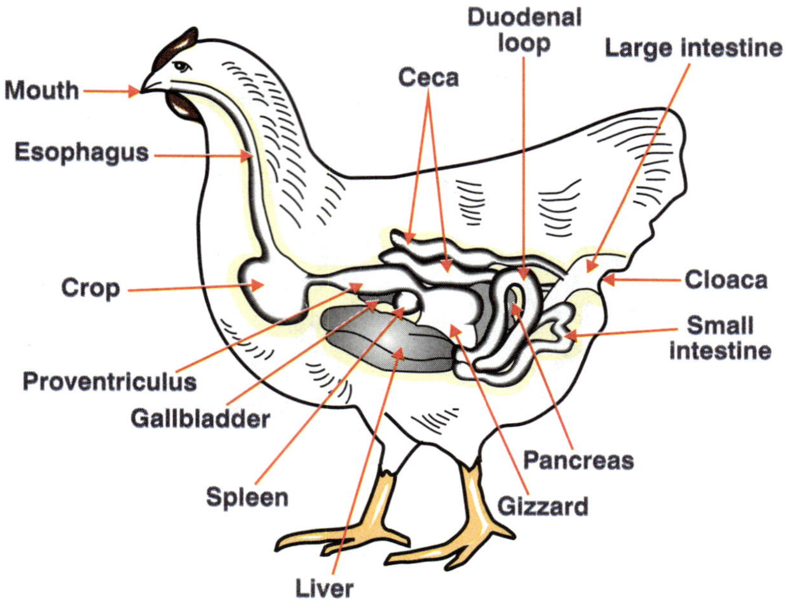

11-22. Digestive system of a chicken.

After rechewing, the food is swallowed and again goes to the rumen. After further digestion, the food goes to the next compartment.

2. **Reticulum**—The reticulum stores food and sorts out foreign materials. Sometimes cattle will eat nails, pieces of wire, or other harmful objects. These are retained in the reticulum.
3. **Omasum**—Also known as the manyplies, the omasum has strong muscular walls that help break the food apart. After this is done, the food passes to the next compartment.
4. **Abomasum**—This is the fourth compartment in the stomach of a ruminant. Here, gastric juices are mixed with the food to help further the digestive process. Next, the food moves to the small intestine.

Non-ruminants are sometimes known as monogastric animals. They have stomachs with one compartment. The digestive process is similar to that in ruminants except that no cud is chewed. Non-ruminants must have food that is of higher nutritive value, particularly the hog. Horses do have a digestive part known as the caecum that allows them to make better use of grass in their diets.

- **Small intestine**—The small intestine follows the stomach. Most of the absorption of nutrients and water takes place in the small intestine. Absorption is the process by which the end products of digestion are transferred into the blood. The liver and pancreas secrete substances into the small intestine that aid digestion. The liver secretes bile which breaks down certain foods, such as fat.

- **Large intestine**—The large intestine follows the small intestine and is the final organ of the digestive system. Slow contractions occur in the large intestine, which move the solid materials along. Mucus is secreted by the intestines and binds the solid materials together. The solid material is called feces.
- **Anus**—The anus is the opening in the body through which the large intestine expels solid wastes.

ADVANTAGES OF RUMINANTS

Ruminants can use large amounts of roughage. Roughage is comprised of grass, clover, and other vegetation. They are able to graze on land that could not be used to produce other crops. This means that they are good at converting forage into higher quality food. Ruminants do not normally need large quantities of grain and other concentrated feeds. They are typically fed grain when in a feedlot to achieve extra growth and tenderness in the meat. Non-ruminants make poor use of roughage. They need grain and other concentrated feed in order to grow rapidly.

SPECIAL ADAPTATIONS IN POULTRY

The digestive systems of chickens, ducks, turkeys, and other poultry vary from those of ruminants and non-ruminants. Fowl have a mouth, crop (where recently eaten food is stored), glandular stomach, muscular stomach (gizzard), and intestines similar to other animals. The food is taken in by the mouth and then is forced down the throat to the crop. The crop is a pouch-type structure that holds and softens food. From the crop, the food goes to the glandular stomach where it is mixed with gastric juices. The food goes from the glandular stomach to the gizzard, a strong muscular organ that crushes and grinds the food. Afterward, it moves to the small intestine where much of the absorption takes place.

REPRODUCTIVE SYSTEM

Reproduction is carried out by the reproductive system. It begins after a male and female animal have mated (copulated). The sperm (male sex cell) unites with the egg (female sex cell) in a process called fertilization. The union of the sperm and egg takes place inside the reproductive tract of the female. In mammals, the new animal is nourished and grows in the female's reproductive tract.

The reproductive system is the only organ system that varies among male and female animals of the same species. These differences determine the sex of the animal. Other systems are similar, such as the digestive and circulatory systems, for both sexes. The male and female systems are designed to make the mating and fertilization processes possible.

The emphasis in this chapter is on the principles of reproduction in livestock.

PUBERTY

Puberty is the age at which an animal is capable of reproduction. Sexual maturity has been reached. Ova (eggs) begin to develop in the female. Sperm begin to be formed in the male. Animals vary in the ages at which they reach puberty. Cattle reach puberty in 8 to 12 months; hogs in 4 to 7 months. Even though puberty is reached, breeding is often delayed. The young animals, particularly the females, may be too small to successfully carry the developing new animal (fetus) and give birth.

FEMALE REPRODUCTIVE SYSTEM

The female reproductive system is designed to make fertilization possible and provide a good environment for the growth of the new animal. The major parts of the female system are listed below.

- **Vulva**—The vulva is the external part of the female reproductive tract.
- **Vagina**—The vagina is the mating organ of the female. It is a muscular, tube-like organ that connects the vulva and cervix. This tube conducts sperm deposited in it by the male to the cervix and is the canal through which the fetus passes at the time of birth.
- **Cervix**—The cervix is the entrance to the uterus. It separates the vagina and uterus.
- **Uterus**—The uterus is a hollow organ with two parts known as uterine horns. This is where the fetus develops after fertilization.

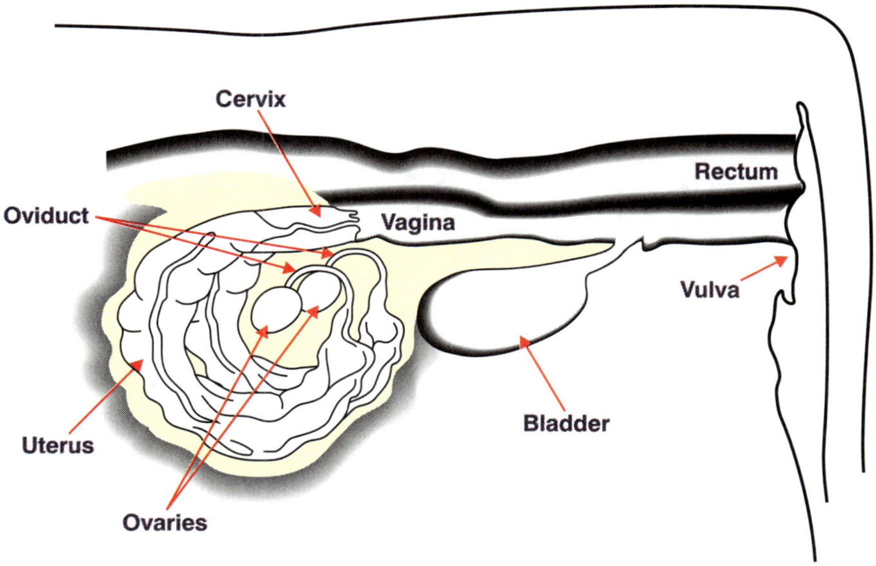

11-23. Parts of the reproductive system of cow.

- **Oviducts**—The oviducts are tubes that run from the uterus to the ovaries. They are also known as fallopian tubes. Fertilization (union of sperm and egg) usually takes place near the upper end of the oviduct.
- **Ovaries**—The ovaries produce the eggs. Females have two ovaries, each attached to an oviduct. Ovaries have many immature eggs which individually mature throughout the life of the female after puberty. The ovaries also produce hormones that regulate the reproductive system. Females usually ovulate (release eggs) on a regular cycle. Mating must occur at the right time in the cycle so that an egg is present to be fertilized.

MALE REPRODUCTIVE SYSTEM

The male reproductive system is designed to produce and store sperm and deposit them in the female reproductive tract. The sperm carries genetic information that joins with that of the egg. A sperm consists of a head filled with DNA and a flagellum (tail). The DNA is a form of nucleic acid that contains genetic traits of the male parent. The flagellum moves the sperm around and gets energy from mitochondria, located just outside the head in the flagellum.

The parts of the male reproductive system are:

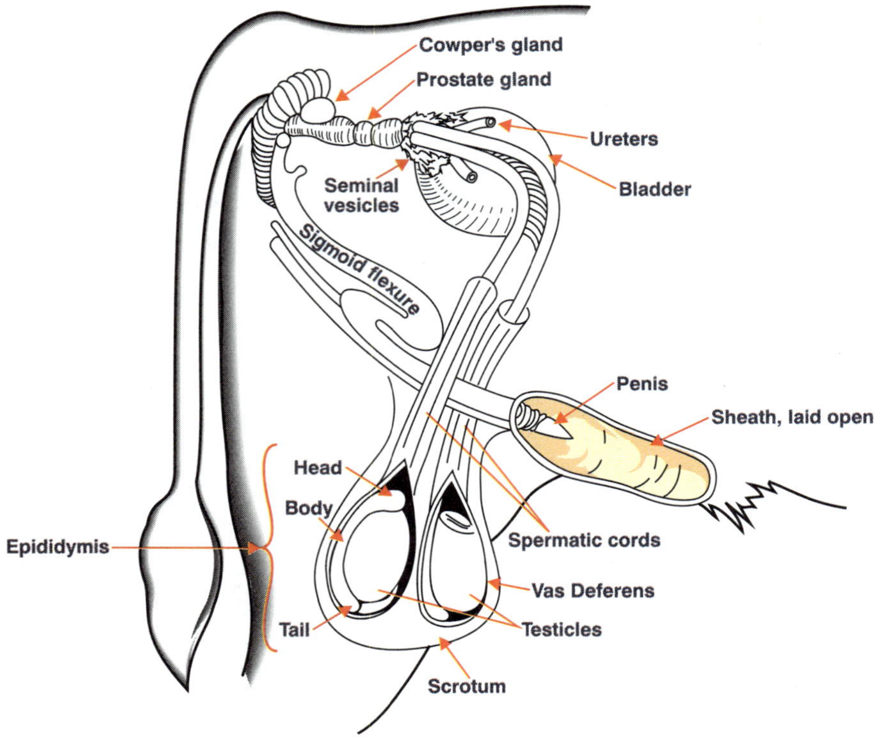

11-24. Parts of the reproductive system of a bull.

- **Penis**—The penis is the part of the male reproductive system that deposits sperm in the vagina of the female. The sperm are in a fluid known as semen. During copulation, the sperm are expelled by ejaculation. Millions of tiny sperm may be released in the semen of one ejaculation.

- **Urethra**—The urethra is the long tube that extends from the urinary bladder to the end of the penis. Semen is ejaculated through the urethra. Urine is also expelled from the bladder through the urethra.

- **Seminal glands**—Two glands produce fluids that help the male produce viable sperm. The cowper's gland is located on the urethra and secretes a fluid that precedes the passage of semen through the urethra. The fluid prepares the urethra for the semen. It neutralizes any urine that could damage the sperm. The prostate gland produces a fluid that carries the sperm. The fluid, when mixed with sperm, is semen. These glands are located on the urethra near the bladder.

- **Seminal vesicles**—The seminal vesicles are located near the bladder and are attached to the urethra. They produce a fluid that nourishes the sperm. The fluid is mixed with that from the prostate gland.

- **Sperm ducts**—The sperm ducts are tubes that connect the urethra with the testicles. The sperm ducts store sperm and move them to the urethra.
- **Testicles**—Testicles, also known as testes, produce sperm. Male animals normally have two testicles. They are located in the scrotum.
- **Scrotum**—The scrotum is a pouch of skin holding the testicles. The scrotum holds the testicles outside of the body so that they have a slightly lower temperature than the body. The lower temperature increases sperm production.

Sperm production begins when the animal reaches puberty and are produced throughout the remainder of the animal's life.

ESTRUS CYCLE

The ***estrus cycle*** is the cycle in the female reproductive system that prepares it for reproduction. During the "heat" part of the cycle, the female is receptive to being bred by the male. The cycle begins with the ripening of an egg or eggs in the ovaries. When mature, the eggs rupture the ovary walls and go into the oviduct. Mating must occur when the ripe egg is present; otherwise, no sperm will be present to fertilize the egg. The eggs are fertilized in the oviducts. The length of the estrus cycle varies among different species of animals. Cattle and hogs have 21-day estrus cycles. The estrus cycle stops when the female becomes pregnant.

The number of eggs released during estrus varies according to the species of animal. Cattle and horses normally release only one egg. Sheep and goats may release two to four eggs. Hogs may release 15 to 20 eggs.

AFTER FERTILIZATION

Fertilization is the union of the egg and sperm. Normally only one sperm enters an egg. After fertilization, the egg travels down the oviduct and attaches itself to the wall of the uterus. Here the fertilized egg gets nourishment and grows by cell division into an animal. While a female has a developing animal in her reproductive tract, she is said to be ***pregnant***. In the first half of the pregnancy the new animal is known as an embryo; it is called a ***fetus*** in the second half of pregnancy.

Gestation is the period between fertilization and full development of the fetus. It ends when the animal is born. In cattle, gestation lasts an average of 281 days. The process of giving birth is known as ***parturition***.

MALE AND FEMALE DIFFERENCES

Males and females vary in the frequency that they can fulfill their roles in reproduction. Males can contribute to reproduction more frequently than females. The roles of the sexes are different. Producers of animals recognize these differences in animal reproduction.

Females vary in the frequency of breeding and number of offspring produced at each breeding. Cows usually have one calf each year. They begin having estrus cycles a few weeks after the calf is born. They can be bred again at this time. Sows can be bred to produce about two litters of pigs a year. Litters are multiple births of pigs, with as many as 20 in a litter. Most hog producers prefer about 12 pigs in a litter. Females are also responsible for providing milk for the newborn animals.

Males can be bred to a number of females each year. Sperm mature in a period of 12 to 24 hours. The male has no role in carrying the developing embryo and fetus to birth. One bull in good condition can naturally mate with about 30 cows a year.

MAMMARY SYSTEM

Only female mammals have developed mammary systems. Male animals have rudimentary or undeveloped mammary tissues.

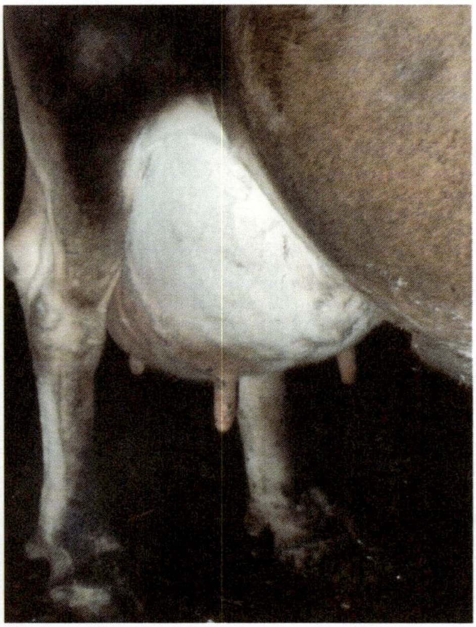

11-25. Udder on a Jersey cow filled with milk.

The mammary system is made of glands that produce milk. All female mammals produce milk as food for their young. In dairying, cows are milked to get milk for human food. Beef cattle, hogs, horses, sheep, and goats all give milk for their young.

The mammary system consists of mammary glands containing ***alveoli***. Millions of alveoli are found in the mammary glands of cows. The alveoli use nutrients from the blood to make milk. A cow has four mammary glands, each has a teat through which the milk passes when feeding the young or during milking. The four mammary glands are bound together to form the udder. Hogs may have 10 to 18 mammary glands along

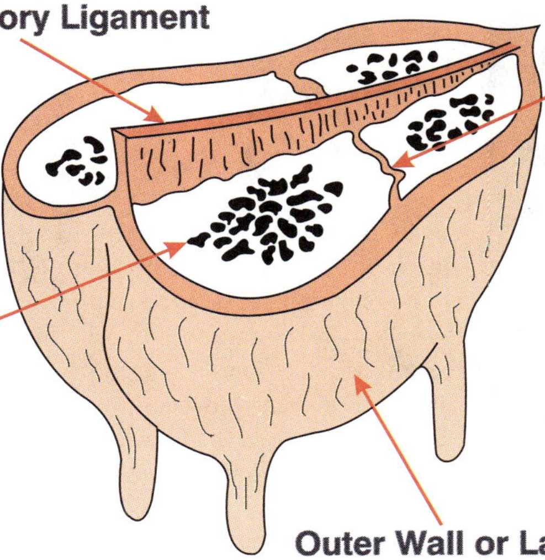

11-26. Structure of the udder.

the underside (belly) of the female. A mare has four mammary glands but only two teats. Females must receive proper feed if they are to produce plenty of milk.

Producing milk is known as ***lactation***. Females lactate only after giving birth to young. Lactation may last for several weeks or months. It stops when the offspring are weaned. Weaning age is reached when the young animal can eat other foods. Modern dairy farms follow practices to assure maximum lactation by the cows.

HOW POULTRY ARE DIFFERENT

Chickens, turkeys, and other poultry do not give birth to live young. They lay eggs which hatch a new animal if fertile and properly incubated. Incubation involves providing the proper temperature and humidity for the egg to develop. Chicken eggs have an incubation time of 21 days; other poultry eggs require a few more days.

In chickens, fertilization involves the sperm of a rooster (male) uniting with the egg as it is being formed in the hen. The male deposits sperm in the hens reproductive tract. Sperm from one mating may live for several days and fertilize several eggs.

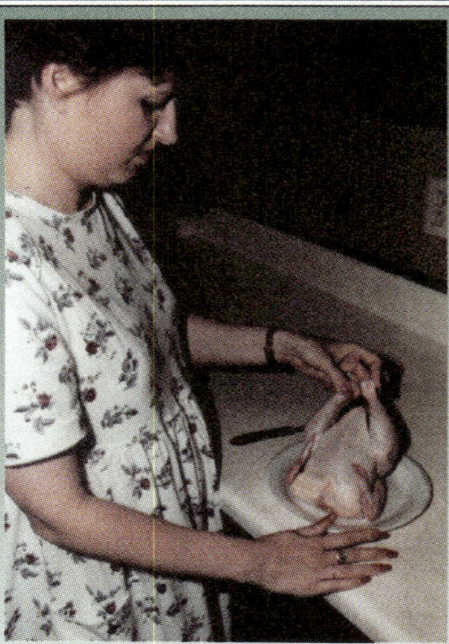

AgriScience Connection

CONSUMERS HAVE A BIG ROLE

Consumers are powerful. They have a big role in the kinds of animals produced and how they are prepared for use. A good example is poultry.

Poultry consumption has increased rapidly in recent years for several reasons. Chickens grow fast and eating them has been promoted as a step toward good human health. Chicken is also a favorite with people because poultry products are made into forms that are easy to prepare. How would you want to get chicken?

This shows a conveniently-processed broiler (young chicken) ready for cooking. The broiler is ready to roast whole or cut into pieces for frying or broiling. The amount of chicken eaten would drop tremendously if consumers had to begin with a live bird and do the killing and plucking!

The sperm fertilizes the female germ cell, which is located on the outer edge of the yolk. Fertilization must take place before the egg shell is fully formed. The female lays the egg and, with incubation, it will hatch into a new chick. Eggs that aren't fertilized will not hatch.

Poultry do not have mammary glands and don't produce milk for their young. The young start eating small particles of solid food shortly after hatching.

11-27. Being small and fragile, newly hatched chicks must be carefully tended for survival.

REVIEWING

MAIN IDEAS

Producers of animals can do a better job if they know the principles of animal science. These help them to understand animals and why they perform as they do.

Animals can be grouped on the basis of several criteria: (1) body structure (vertebrate or invertebrate), (2) where they live, and (3) the products they provide.

The most common farm animals are vertebrates. They have backbones and extensive organ systems that perform important functions.

The important organ systems are: (1) the skeletal system which supports the body and protects delicate organs; (2) the muscular system that provides animals with a means of movement as well as the food products the muscles furnish; (3) the nervous system which sends messages (impulses) throughout the body; (4) the respiratory system that moves gases into and out of the animal; (5) the excretory system that rids the body of wastes; (6) the digestive system which prepares food for use by the body; (7) the reproductive system that creates new animals; and (8) the mammary system which develops only in females and provides milk for young animals.

QUESTIONS

Answer the following questions using complete sentences and correct spelling.

1. What are vertebrates? Invertebrates? Distinguish between the two.
2. What are the distinguishing characteristics of the Phylum Arthropoda? List examples of species in this phylum.
3. What important animals are in the Phyla Annelida and Mollusca?
4. What important animals are in the Phylum Chordata?
5. What important animals are in the following classes: Osteichthyes, Aves, and Mammalia?
6. How can animals be grouped by where they naturally live?
7. How can animals be grouped by the kinds of products they provide?
8. What is anatomy? Physiology?

9. List the organ systems of animals. Briefly describe two important functions of each.
10. Distinguish between voluntary and involuntary muscles.
11. What are the major parts of the nervous system?
12. What is blood? What liquids and solids does it contain?
13. Distinguish between internal and external respiration.
14. What is breathing? What determines the rate of breathing?
15. What wastes are given off by the excretory system?
16. Distinguish between ruminants and non-ruminants. List a major advantage and disadvantage of each.
17. What is puberty?
18. What is the major purpose of the female reproductive system? Male reproductive system?
19. What occurs after an egg is fertilized?
20. How does the frequency of breeding vary between males and females? Why is this important to producers of animals?
21. Why do females have mammary systems? What is lactation?

EXPLORING

1. Dissect a fetal pig, fish, or other small animal to observe its organ systems. (Be sure to follow the appropriate procedures in this activity. Your teacher can demonstrate how it is done.)

2. Make a list of the animals raised in your community. Indicate the kinds of products they provide.

3. Maintain a pair of small animals in the laboratory. Observe their behavior, including breeding, parturition, and how the young are cared for. Be sure to consider the well being of the animals.

4. Tour a dairy farm and observe the milking of the cows. If you have never hand-milked a cow, ask the worker to show you how it is done. Try it for yourself.

12

FEEDING ANIMALS

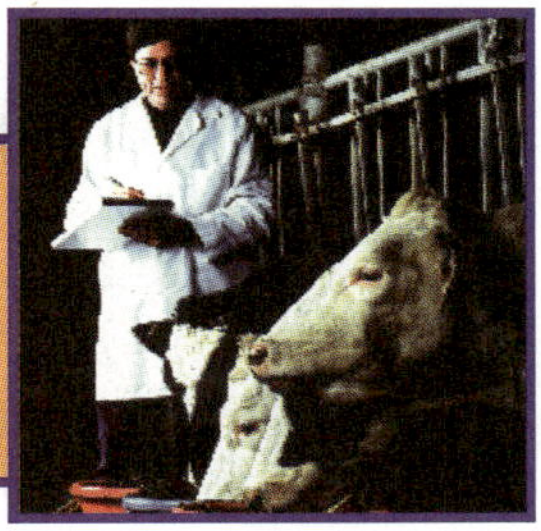

Animals must have the right feed to live and grow quickly. Animals that don't get what they need may be stunted, get sick, or die. Producers can meet the food needs of most animals.

Feed needs vary with the kind of animal. An animal's body structure and organ systems are important. Ruminants can eat large amounts of roughage because they have stomachs with four compartments. They are fed differently than other animals. Non-ruminants usually have stomachs with one compartment and this affects what they are fed. Of course, poultry are different from hogs, horses, sheep, cattle, and fish.

Several things influence what an animal should be fed. Certain feeds can't be used by the bodies of some animals. Also, what an animal needs varies with its stage of life and its activities, such as milk production.

Animals have preferences about feed just as people have their favorite foods. These need to be considered. Cattle will eat about anything if molasses is put on it!

12-1. African hedge hogs are increasingly popular as pets. This one is being fed her favorite food—a worm!

OBJECTIVES

This chapter presents the principles of nutrition and feeding of animals. It has the following objectives:

1. Describe the feed needs of animals.
2. Explain the major nutrients needed by animals.
3. Describe the feedstuffs that provide nutrients.
4. Explain the characteristics of good feed.
5. Describe how animals are fed.

TERMS

additive
balanced ration
carbohydrate
concentrate
cracking and rolling
diet
digestible nutrient
energy
fat
feed
feedstuff
free access
grinding
hay
implant
injection
label
macromineral
maintenance
meal
mineral block
minerals
mixed feed
nutrient
palatability
pasture
pellet
permanent pasture
protein
protein supplement
ration
roughage
scheduled feeding
silage
temporary pasture
vitamins
water

FEED NEEDS OF ANIMALS

Animals should have feed that meets their needs. What an animal needs depends on its age and other functions. Even the weather has something to do with feeding animals!

Animals use feed to maintain their bodies and for growing, reproducing, lactating, working, and, in some cases, for providing other products.

MAINTENANCE

Animals need nutrients to maintain themselves. Nutrients are used to keep the body warm, replace old cells, run the internal organs, and for body movement. Even an animal that is lying still is using energy—its heart is beating! Animals use energy 24 hours a day.

Maintenance is the energy needed to keep an animal from losing weight when not producing milk or other products. Maintenance does not involve any gain or loss in weight.

The amount of feed needed for maintenance varies with the species of animal. Cattle and sheep need nearly 10% more feed when standing than when lying down. This is why animals that are being fed-out (finished) are kept in small pens and encouraged to lie down.

GROWTH

The kind and amount of feed needed varies, since young animals need feed that helps them grow. Of course, animals also need feed to maintain themselves in addition to feed required for growth.

12-2. This sow and pigs need nutrients for maintenance, growth of the pigs, and lactation in the sow.

Growth is an increase in the size of bones, muscles, internal organs, and other parts of the body. In agriscience, one goal is to get animals to grow as rapidly as possible. This makes it very important to provide the proper feed for growth.

Young calves, pigs, and chickens are expected to grow rapidly. They are fed diets that are changed as they grow to better meet their needs. The diet of a very young, growing animal needs careful attention. The older an animal becomes, the more the diet can be like that of an adult. Adult diets provide nutrients for maintenance, reproduction, and lactation.

REPRODUCTION

Producing animals usually involves getting them to reproduce. The goal is to get as many well-developed products of reproduction as possible.

Cattle, hogs, sheep, and horses give birth to live young. This means that the pregnant female must get the feed needed to support her body as well as her developing baby. Feeds that promote growth of the fetus and embryo are needed.

Male cattle, hogs, sheep, and horses also need a proper diet. If they are too fat, they may be incapable of reproducing. Not enough feed may lower the sperm production, resulting in females that are not bred.

In chickens, emphasis is on egg production. The eggs may be fertile and used to hatch chicks. Most eggs, however, are used as human food.

LACTATION

A female mammal must receive a proper diet in order to produce sufficient amounts of milk. Dairy cows often have special diets designed to maximize milk output. A cow can't produce milk unless she gets plenty of the right feed.

Females nursing young animals need to have feed that helps them provide enough milk for their young. Sows may have 10 or more offspring to feed. Horses and cows usually have one foal or calf, while sheep may have two or three lambs.

WORKING

Working animals are used for power, pleasure, and other purposes, such as the herd dog that watches sheep. Animals must get feed for maintenance as well as to provide energy for physical activity.

Animals used for power include horses, oxen, and water buffalo. These animals need feed that provides enough energy. In North America, animal power has largely been replaced by mechanical power, such as tractors.

Pleasure animals include horses used for riding and racing. They must have the feed needed for activity as well as to maintain the body.

12-3. This horse needs energy to carry its rider and do what the rider commands it to do. (Courtesy, American Quarter Horse Association)

OTHER PRODUCTS AND USES

Animals that are producing wool need to have diets that include important nutrients associated with growth. If the animal doesn't receive the right feed, not only is the rate of growth reduced, but the quality of the wool may be low.

Animals that are used as pets or for showing must have diets that meet their individual needs. The calves, lambs, and pigs at a livestock show have often received specialized feeds designed to help them grow and put on fat.

12-4. Dairy cows often learn a feeding routine as part of each day. These Jerseys are patiently waiting to be fed.

NUTRIENT NEEDS OF ANIMALS

A ***feed*** is a product containing nutrients. The ingredients used to make the feed have various nutrients in them. Different ingredients have different kinds and amounts of nutrients.

Nutrients are the chemical substances in feed that support the life processes of the animal. Nutrients nourish and support the growth of animals. Nutrients make it possible for animals to maintain themselves, grow, reproduce, and perform other functions.

A ***ration*** is the total amount of feed an animal gets in a 24-hour period. The feed can all be given at one time or at several smaller feedings

AgriScience Connection

ATTRACTING WILD BIRDS

Setting up a bird feeder near your home will help attract wild birds. The feeder shown here is designed to keep the feed dry, provide a perch for the birds to rest on while feeding, and keep rodents from getting into the feed. Using a prepared wild bird feed mix is usually satisfactory. One drawback: Any uneaten seed may fall to the ground and germinate, causing weeds in the lawn or flower bed.

Wild animals have similar nutrition needs to domesticated animals. These animals often need feed in addition to what nature can provide. Keeping a bird feeder will help the wild birds get the needed nutrition.

throughout the day. A ***balanced ration*** is one which provides all of the nutrients needed by the animal in the right amount and proportion. Providing too much of some nutrients results in waste. A deficiency of one nutrient can be a limiting factor in animal growth and production.

A ***diet*** is the type and amount of feed and water an animal eats. Diets vary according to the nutrient content of the feed ingredients. Diets should be based on the needs of the animal.

The major nutrient needs in the diets of animals are energy nutrients, protein, minerals, vitamins, and water.

ENERGY NUTRIENTS

Energy supports all life processes of animals. A deficiency (not having enough) stunts growth, lowers production, and causes loss of body tissue. Animals without enough energy certainly can't do as much work as they could if fed a diet high in energy nutrients.

Energy is provided by two nutrients: carbohydrates and fats. In addition, some energy comes from protein.

Carbohydrates

Carbohydrates are sugar, starch, and related substances. They are formed in plants by photosynthesis. Animals that eat leaves, stems, roots, and other parts of plants get the carbohydrates. Animal bodies use up most of the carbohydrates in the diet. They are converted to glucose which is found in the blood and provides much of the energy animals need. The remaining carbohydrates are stored as body fat.

Carbohydrates are made of carbon, hydrogen, and oxygen. They are often consumed as glucose, sucrose, and other forms of sugar, including starch.

The major sources of carbohydrates are corn, oats, hay, soybean oil meal, and grain sorghum. Most adult cattle can get adequate carbohydrates from the grass and other roughages in a good pasture.

Fats

Fats are similar to carbohydrates, but are more concentrated sources of energy. Fats have 2.25 times as much energy as carbohydrates.

Fats provide energy for the animal. Excess fat in the diet of animals is stored as body fat on the animal. Animals being fattened for slaughter often receive diets that are high in fats and carbohydrates.

The major sources of fats are meat scraps, tankage (processed meat and bones), cottonseed, and fish meal.

PROTEIN

Protein is needed for maintenance, growth, reproduction, and other functions of animals. Protein is used to build new tissue and repair old tissue. Young, growing animals especially need protein. When not needed for building cells and tissue, it can be used for energy.

Diets must contain enough protein for animals to grow and reproduce. Animals need protein in order to form muscle tissue (meat). Bones, hair, hooves, skin, organs, and other parts of the body require protein for growth and maintenance. Production of milk, eggs, and wool also requires protein.

The major sources of protein are tankage, soybean oil meal, legume hay, blood meal, feather meal, fish meal, and skim milk.

Protein quality varies according to its makeup. Protein is made of amino acids. Twenty-three amino acids have been found in protein. The amino acids are known as the building blocks of protein. Ten amino acids are essential; the others are known as nonessential. Some feeds have small amounts of certain amino acids; others have higher amounts. Proteins that provide the right balance of amino acids must be included in the diet.

Table 12-1
Essential Amino Acids

Arginine	Isoleucine	Lysine	Phenylalanine	Tryptophan
Histidine	Leucine	Methionine	Threonine	Valine

Different animals make use of different protein sources. Ruminants can form some of the amino acids in their digestive processes. Roughages low in protein are converted to higher quality protein in the rumen. Animals with simple stomachs cannot make protein, so it must be obtained in their diets.

When animals get more protein than they need, the liver converts the excess to energy. The nitrogen in the protein is excreted by the kidneys (this makes the urine yellow). Since protein costs more than fats and carbohydrates, balanced diets should be fed to keep costs down.

MINERALS

Minerals are important in the maintenance, growth, reproduction, and other functions of animals. They are used for teeth and bones, in body fluids and secretions, and in soft tissues of the body.

Without enough minerals, the body will not grow properly. Deficiencies result in weak and crooked bones, less milk production, fewer eggs, breeding problems, and disease. The disease problems may be indirect or direct. Indirect disease problems occur when animals are weak and causes of disease (bacteria, etc.) can invade the body. Diseases caused directly by a lack of minerals are known as nutritional diseases.

Eighteen minerals are needed by animals. Some of them are needed in larger quantities than others. Those needed in larger amounts are known as major or ***macrominerals***. Six minerals are macrominerals and are needed in amounts of one or more grams in a ration (daily amount). Calcium and phosphorus are often needed in the largest amounts. Analysis of animal bodies shows that they contain slightly over 1% calcium and slightly less than 1% phosphorus. Calcium and phosphorus are primarily found in teeth and bones.

Minerals needed in smaller amounts are known as trace or ***microminerals***. Very small quantities of these minerals are needed in a ration. Most feedstuffs grown in soil that contains these minerals will furnish enough to meet the needs of animals.

Table 12-2
Essential Minerals Needed by Animals

Macrominerals (Major)	Symbol	Microminerals (Trace)	Symbol
calcium	Ca	chromium	Cr
salt (sodium chloride)	NaCl	cobalt	Co
phosphorus	P	copper	Cu
magnesium	Mg	fluorine	F
potassium	K	iodine	I
sulfur	S	manganese	Mn
		molybdenum	Mo
		selenium	Se
		silicon	Si
		zinc	Zn

Feed ingredients high in minerals include alfalfa and soybean hay, cereal grains, bone meal, molasses, and salt. Mineral supplements may be given, such as the block of mineral salt that is in the pasture for cattle to lick. Minerals are often added to manufactured feeds. Mineral supplements include finely ground limestone, oyster shell, and ground rock phosphate.

VITAMINS

Vitamins are important in regulating body functions, keeping the body healthy, and developing resistance to disease. Most rations contain enough vitamins; however, shortages may occur under certain conditions. Additional vitamins are needed when animals are expected to grow quickly. Fast growth usually requires vitamin supplements. The absence of one or more vitamins can result in problems, such as disease or inability to breed.

Plants naturally make some vitamins, such as vitamin C. Animals that eat the plants ingest the vitamins.

12-5. To be healthy, animals must receive proper nutrients. This orangutan shows that it has a good diet.

**Table 12-3.
Vitamins Classified by Solubility**

Fat-Soluble Vitamins	Water-Soluble Vitamins
Vitamin A	Biotin
Vitamin D	Choline
Vitamin E	Folic acid (folacin)
Vitamin K	Inositol
	Niacin
	Vitamin B-1 (thiamin)
	Vitamin B-2 (riboflavin)
	Vitamin B-3 (pantotheni acid)
	Vitamin B-6
	Vitamin B-12
	Vitamin C (ascorbic acid and dehydroascorbic acid)

Agriscience has developed pure forms of vitamins known as crystalline vitamins. These synthetic vitamins are often sprayed on or mixed with feeds at the feed mill.

Some vitamins are fat soluble. This means that animals can store them in the fat in their bodies. Other vitamins are water soluble and aren't stored in the bodies of animals. Vitamins A, D, and the B vitamins are probably most important in animal health.

Vitamin A

Vitamin A is required by all animals. Since plants don't make vitamin A, it is made by animals in their bodies from carotene. Carotene is sometimes called provitamin A. All plant leaves and yellow foods are high in carotene. Yellow corn is a common source of carotene. Carrots, squash, yellow sweet potatoes, and pumpkins are all high in carotene.

Cattle grazing on green pastures are able to make all of the vitamin A they need. High quality grass or legume hay and silage also provide provitamin A.

Vitamin D

Vitamin D is required by all farm animals. Most feed contains little vitamin D. However, this lack in the feed doesn't pose a problem. The skin of animals contains provitamins, which act in the presence of ultraviolet (sun) light to form vitamin D. The most important sources of vitamin D are hay and roughages that have been cured in the sun.

B Vitamins

Eleven different substances are known as B vitamins. Numbers are used after the "B" to distinguish between the forms of vitamin B. Some of the B vitamins are known by their chemical names, such as thiamin.

Animals vary in their need for the B vitamins. Ruminants can make it in their compound stomachs. Animals with simple stomachs can't make the B vitamins; therefore, they must get the vitamins in their feed. Since the B vitamins are water soluble, they aren't stored in the body fat of animals. This means that they must get some of the B vitamins every day in their rations.

WATER

Water is probably the most important of all nutrients. Most animals could live longer without feed than water. Animal bodies contain high amounts of water; a market-ready hog is 40% water and water makes up 70% of a newborn calf.

Water is an important part of all body processes. It is found in all cells and all body fluids. It serves as a carrier to move materials around the body and eliminate wastes. It helps regulate body temperature and is a part of all chemical processes. Water softens food and aids in digestion. Excess water leaves the body as urine, in perspiration, in feces (solid body wastes), and as water vapor from the lungs.

12-6. Most cattle are provided free access to water. These cattle have access to a tank that is automatically kept full of fresh, clean water.

Animals should be given high-quality water. It should be free of dangerous chemicals and excess minerals. The source of the water should be dependable because animals need a steady supply.

Most animals are given all the water they want. It is made available in tanks, ponds, or through automatic watering systems. The watering system must be suited to the animal.

SOURCES OF NUTRIENTS: FEEDSTUFFS

Feed is what animals eat to get energy and for other body functions. Some animals may be left on their own to graze the pasture for their feed. Other animals may be confined (kept in pens) and fed carefully prepared rations. What animals eat is important to their growth and health.

Table 12-4
Nutritional Value of Selected Feedstuffs[1]

Feedstuff	Digestible Protein (%)	Total Digestible Nutrients (%)	Fiber (%)
Animal Protein Sources:			
blood meal	58.4	60.4	0.9
fish meal	53.6	70.8	0.9
tankage	50.5	65.8	1.9
Plant Protein Sources:			
cottonseed meal	35.9	72.6	11.0
soybean meal	42.0	78.1	5.9
corn gluten meal	36.7	79.7	3.8
Grains:			
corn	6.9	82.5	2.4
oats	9.0	68.5	12.1
wheat	11.1	80.0	2.6
Roughages:			
alfalfa hay (mature)	15.2[2]	55.0	29.3
Bermuda grass hay	9.2[2]	43.0	28.4
soybean hay	14.1[2]	49.0	31.0
wheat straw	3.2[2]	40.0	41.7

[1]Some rounding of information has been included.
[2]Reported as crude protein.
Sources: M.E. Ensminger. (1987). Beef Cattle Science. Danville, Illinois: Interstate Publishers, Inc. and L.P. Jacks. (1976). Basic Principles of Animal Science. Mississippi State University.

Animals need balanced rations. This is a ration that meets the animal's requirements for energy, vitamins, and minerals. Balanced rations vary according to the kind of animal, its age, and what it is doing. A balanced ration provides the right amounts of the needed nutrients for a 24-hour period.

Various ingredients are used in feeds. Some of these are high in nutrition value; others are low.

The feed ingredients, also known as ***feedstuffs***, in a ration are classified by their fiber content. Fiber is what remains of feed after the sugar, protein, and other substances have been removed. Fiber is made of the plant cell products, such as cellulose and lignin. The two general classes of feedstuffs are roughage and concentrate. Table 12-4 shows the nutritional value of selected feedstuffs.

ROUGHAGES

Roughages are feedstuffs that are high in fiber and low in energy. They have low digestibility. A ***digestible nutrient*** is the part of a feedstuff that can be digested, or broken down. The fiber in roughages is not easily digested. Ruminants and horses are efficient users of roughage. This makes roughage a more appropriate feed for these animals than a ration with a high amount of grain. Of course, these animals are sometimes fed grain to gain body weight more quickly.

Examples of roughages include hay, green pasture grasses and legumes, and silage. Some by-products are used as roughage, such as dried sugar beet pulp and the left-overs from vegetable processing.

Pasture

Land where grasses and other plants grow for animals to graze is ***pasture***. It is the most popular way of providing feed for cattle, sheep, and horses. Nearly 30% of the land area in the United States is covered with pasture. Pasture is of little value to chickens and turkeys. Adult breeding hogs are sometimes allowed some pasture.

The nutritional value of pasture depends on the kinds of plants and their vegetative stages. Pastures in areas with high rainfall are often planted with grasses and legumes that grow rapidly. The tender plants are high in water (sap). Dry areas (less than 20 inches of rainfall a year) may have native plants that grow more slowly and contain less water. Pasture plants must be adapted to the environment.

12-7. These sheep have an excellent pasture. Note that a water trough is provided and that the pasture has a good fence.

Kinds of Pasture. Two kinds of pastures are typically used: permanent and temporary. Livestock producers may use both types in their system of production.

- **Permanent pastures—*Permanent pastures*** are planted with grasses and legumes that will live and grow for years. Many pastures grow native plants. Pastures in areas with more rainfall may be seeded, fertilized, and otherwise improved. Permanent pastures are often seeded with warm season grass and legumes. Bermuda grass, fescue, Bahia grass, Kentucky bluegrass, dallis grass, and other native grasses may be used. Common legumes include red clover, alsike clover, and sercea lespedeza.
- **Temporary pastures—*Temporary pastures*** may be planted for winter, summer, or semipermanent grazing.

  Winter temporary pastures are planted with crops that grow in the winter. Rye grass, oats, wheat, and other cold-tolerant plants may be used. Winter pastures are often used in the southern part of the United States.

  Summer temporary pastures are planted in the spring for grazing in the summer and early fall. Millet and grain sorghum are most widely grown as summer pastures.

  Semipermanent pastures are planted with grasses and legumes and grazed for several years before being plowed. This type of pasture is not as widely used as permanent and temporary pastures.

Pasture Management. Pastures must be established and maintained. Establishing a pasture involves preparing the land and selecting the plant varieties to be sown. Of course, good fences must be in place to keep the livestock in the pasture.

12-8. This beef animal is starving! The pasture is short and dead leaving nothing to eat. The cow will soon likely die from starvation.

Pastures must be maintained. Weeds must be kept down. Fertilizer may be needed. In some cases, diseases may need to be controlled. Permanent pastures are usually clipped once or twice a year to cut weeds and bushes and remove the old growth from pasture plants. Livestock may need to be rotated to keep them from over-grazing (eating the plants too close to the ground).

Hay

Hay is made of the leaves and stems of plants that have been cut and dried for feed. Most hay is cut with a mowing machine or conditioner and dried in the sun. Special hay crops may be planted or hay can be cut from pastures that grow more vegetation than the cattle eat.

The best hay is made from the leaves and tender stems of legume plants. Examples include alfalfa and soybeans. Other hay of lower nutri-

12-9. Quality alfalfa hay. Note that tender leaves are still attached to the stems.

12-10. Cut grass hay curing in the sun.

tional value is made from rye grass, oats (plants as well as immature grain heads), Bermuda grass, and Johnson grass. Older plants with a lot of stems aren't as good as younger plants with few stems.

The best quality hay is free of weeds. Some weeds don't have much nutritional value; others are poisonous and can kill livestock. If possible, hay should be protected from rain after it is cut and until it is stored. Hay should be dry before baling and storing. Hay that isn't dry enough may catch fire by spontaneous combustion. Harvested hay should be stored to protect it from rain, rotting, and the loss of nutrients. Hay may be baled in rectangular or large round bales. It is usually cut in the spring and summer and stored for feeding in the winter. Some farms feed hay all year, such as those raising dairy cattle.

Hay is fed to cattle, horses, and sheep. Legume hay is higher in minerals and has more protein than non-legume hay.

Silage

Silage is made from cut green plants by chopping them into small pieces and placing the "chop" in a silo. Water is usually added and the chopped plants are fermented to preserve the nutrients. Silage is cut in the spring and summer when the plants are green and fairly tender. Fermentation produces an odor that some people find unpleasant.

Corn is used most frequently to make silage, followed by grain sorghum. Most corn is cut for silage after the ears have formed but before they begin to dry. Silage is high in water content.

Silage is often used to feed dairy cattle and in feedlots where animals are being fed for slaughter.

12-11. Corn silage. Note that the silage is chopped stalk, leaves, and ears, with corn kernals throughout

Other Roughages

A variety of other roughages are used to feed certain livestock. Four examples are listed here.

- **Crop residues**—Crop residues are the parts of plants left in a field where a crop has been harvested. For example, only the seed of soybeans are harvested. The residue consists of leaves, stems, and pods where the beans grew. Cattle may be allowed to graze on the fields or the residue may be gathered and brought to the cattle.
- **Haylage**—Haylage is a form of silage with low moisture. The grass or legumes are wilted before they are put in the silo. Haylage use is increasing for dairy cattle.
- **Soilage**—Soilage involves cutting and chopping plants in the field and hauling them to feed animals that are confined, such as in feedlots. This is also known as green chop. Soilage is fed without storage or fermentation.

12-12. Bunker silo made with concrete walls and floor. The silage is moved with a front-end loader to a wagon for hauling to the cattle.

12-13. Six common feedstuffs.

- **Poultry litter**—Chickens are grown in close confinement in houses where the floor is covered with wood shavings and sawdust. After the chickens have been grown and sold, the house is cleaned out. Molasses may be added to the remaining wood shavings, chicken manure, and uneaten feed and the mixture is then fed to cattle.

CONCENTRATES

Concentrates are feeds that are low in fiber (less than 18%) and high in energy. Concentrates have high digestibility, known as total digestible nutrients (TDN). Concentrates are fed to hogs, fish, and chickens and used to supplement roughages for cattle, sheep, and horses. The feedstuffs used for concentrates are grains and protein supplements.

Grain

Grains are high in TDN but do not provide a balanced ration. Grains provide energy and should not be fed without other feedstuffs. Most grains are low in protein and must be supplemented. Since grains are high in energy, they are used to add body fat to animals.

Cattle, hogs, and sheep as well as poultry that will be slaughtered are usually fed diets high in grain to help fatten them. Fat deposits help make the meat tender and better tasting. Of course, too much fat is not desirable.

Examples of grain concentrates include corn, oats, wheat, and grain sorghum. Corn is the most widely used grain, especially for beef cattle and hogs. Corn is low in fiber and high in TDN. Grain sorghum has increased as a feed in some locations, particularly for poultry. Dairy cattle, horses, and other animals may receive corn. Fish feed used on catfish, tilapia, and trout farms is often high in corn and soybeans.

Protein Supplements

Protein supplements are high in protein and TDN. They may be mixed with grain or fed separately. Young animals that are growing rapidly are usually fed rations high in protein supplements. Three types of protein supplements are used: animal, plant, and synthetic.

Animal sources of protein supplements are fed to hogs, chickens, and fish. The protein content ranges from 40 to 60%. These sources quickly correct any deficiencies in a ration high in grain. Examples of animal sources of protein are: meat scraps and tankage (from meat packing plants), blood meal (dried blood collected from packing plants), fish meal, and skim milk.

Research has shown that certain species of fish fed some fish meal grow better than those that do not receive fish meal. The amount of fish meal is kept low. Fish meal is a by-product from the processing of fish or it is made from trash fish that are harvested wild from oceans and lakes.

Plant sources of protein supplement include soybean oil meal, cotton seed oil meal, and various grain by-products from manufacturing, such as corn gluten meal. Digestible protein content ranges from about 20 to 45% for most plant protein supplements.

Beef cattle, dairy cattle, horses, and sheep do well on plant protein supplements. Soybean meal can be satisfactorily used for hogs.

Synthetic supplements may be used to feed ruminants. Examples include urea and molasses, rice hulls, and citrus pulp that have been treated with ammonia. Urea is a by-product of nitrogen fertilizer plants and has been widely used in supplements for ruminants. Exposing hay to a gas form of ammonia enhances nutrient content and digestibility.

SUPPLEMENTS

Mineral and vitamin supplements are often used. They help ensure that the animals are getting the essential nutrients for their body functions. The method used depends on how the animals are fed.

When cattle are pastured, mineral blocks, or supplement feeders may be used. Cattle have free access to the supplements and can get them anytime they want to.

In mixed feeds, the supplements are blended with the concentrates as the feed is being manufactured.

AgriScience Connection

FEEDING A LOT OF FISH

Feeding organisms that live in water is different from those that live on land. Feed has to get into the water where the fish or other species are located. Feeding large numbers by hand is impractical.

Fish producers have developed special methods of handling feeding work. The feed is stored in bulk bins near where fish are grown, usually large ponds (as is the case here). Trucks with specially-designed hoppers that will blow feed out into the water are used.

Fish feed is usually higher in protein than feed for many other animals and is more expensive. Plus, feeding more than the fish eat will damage the water. Computers are used to control the feeding process so that little is wasted and the fish get what they need.

FEED ADDITIVES AND IMPLANTS

12-14. Feedstuffs are stored outside at this large feed mill in the dry climate of Fresno, California.

Animal growth and production are important to producers. Various products are available to help animals grow better or produce more. These help producers make a profit from their farming businesses. In many cases, these are added to feeds or are given to the animals.

Additives

Additives are placed in feed while it is being manufactured to preserve it and enhance the growth of the animals. Many kinds of additives are used. All must be approved by the Food and Drug Administration (FDA). They should only be used for the approved purposes.

Additives are used for various reasons. Here are a few examples.

- **Medications**—Antibiotics and other medications may be added to feed to keep disease under control among the animals.
- **Wormers**—Most animals are susceptible to internal parasites. Wormers are added to some feeds during manufacture. In most cases, wormers are administered separately from the feed.
- **Marketing enhancement**—Consumers often want products with certain characteristics. Using chicken feeds that are high in carotene and xanthophyll makes the leg shanks and egg yolks have a darker yellow color.
- **Antioxidants**—Antioxidants are put in feed to keep the feedstuffs from spoiling. These are widely used in fish meal, tankage, and other animal concentrates.

Implants and Injections

Various products may be placed directly into the body of animals. These are intended to make the animal more productive or protect it from dangers in its environment.

Injections involve using hypodermic needles and syringes to get a substance into the body system of an animal. ***Implants*** involve placing solid materials under the skin. Most implants release substances over a period of time.

As with feed additives, the Food and Drug Administration rules on the use of implants and injections related to animal nutrition. Questions about the use of these processes have not been totally answered.

Animals may be injected with hormones to increase growth or productivity. The use of bovine somatotropin (bST) increases milk production in dairy cows. BST naturally occurs in cows but when they are given more, their bodies undergo change. The cows eat more feed and give more milk. A similar product has been used for hogs to increase milk production in nursing sows and growth in pigs. Also, female fish may be injected with human gonadotropin to encourage spawning (egg laying).

GOOD FEEDSTUFFS

Some feedstuffs are better than others. Animals require different kinds of feeds. Select quality feedstuffs for the specific animals that are to be fed. The feed must be one the animal can efficiently use to grow and perform other functions.

FEED REQUIREMENTS

All nutrients are important. If one is missing, the other nutrients won't be used efficiently. Selecting feed involves knowing the needs of the animal as well as the feed ingredients.

Nutrient Content

Feed must provide the nutrients needed by an animal. What is needed is influenced by the functions the animal needs to perform, such as maintenance, growth, and reproduction.

Feed must be nutritionally complete. Protein must be provided in the form the animal can use in its digestive system. Sufficient energy must be provided. Energy is measured in calories. Laboratory procedures can be used to assess the calories in feed. (A calorie is the amount of heat required to raise the temperature of 1 gram of water 1°C.) Minerals, vitamins, and water must be provided.

Some feedstuffs have a lot of nutrients; others don't. Grain and other concentrates are highly nutritious. Wheat straw, corn stalks (fodder), and similar plant materials don't provide much nutrition.

Animals should get balanced rations. Agriscientists have chemically analyzed various feedstuffs to determine the nutrients in them. This information is used in selecting and mixing feeds. It should result in a balanced ration for the animals being fed. (Advanced books on animal science and feeding often give detailed information on how to balance a ration.)

Palatability

Palatability is how well an animal likes a feed. How well does it feel against the palate in the mouth? Taste, odor, temperature, and texture are important. A good ration is of no benefit if the animal won't eat it!

In general, animals tend to like feeds they are used to eating. If feed is to be changed, do so gradually. Introduce some of the new feed along with the old.

Some of the feeds that cattle like best are green pasture grasses and legumes, corn, oats, good silage, and molasses. Feeds that are palatable will be chosen by the animal over those that aren't.

Free of Hazardous Materials

Poisons, spoiled feed, and contamination with foreign objects can make a feed dangerous to animals. Feed ingredients should be free of dangerous

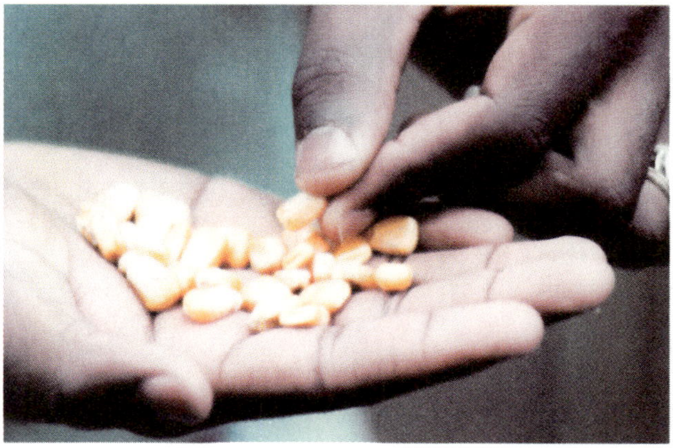

12-15. Examining the quality of corn to determine suitability as livestock feed.

12-16. This water trough is growing algae and should be cleaned.

poison residues. Feed should never be stored near pesticides. Only ingredients that are free of mold and disease sources should be used.

Feed should not be contaminated with metal fragments or other materials. For example, a small nail in feed can injure a cow's stomach and cause a condition known as "hardware disease."

Water should be clear and not muddy. It should be free of pesticides and other hazards.

Variety

Some variety of feedstuffs is needed. Rather than a ration being all one feedstuff, it should contain several ingredients. Variety helps to ensure that all of the essential amino acids are present. Though not as important for ruminants, animals with simple stomachs need variety in their feeds. Hogs and chickens are two important examples. Variety makes feed more

12-17. This dairy farm uses on-farm bulk storage for concentrates. The feed is being moved from the bin to a wagon for hauling to the cattle.

palatable, which means that animals will eat more. When they eat more, they grow faster! Variety also helps provide needed vitamins and minerals.

Bulkiness

Bulkiness has to do with the proportions of concentrates and roughages. Roughages are bulky, while concentrates aren't. Ruminants especially need feeds that are bulky. Chickens and hogs need less bulk.

Cost

Feed cost is important. It determines if the producer will make a profit. Of course, the cheapest ration may not always be best because the quality of the feed may be low. The best way to assess cost is to look at the net returns (profit) when the animals are sold.

Feed Storage

Feed must often be stored on the farm. It should be protected from anything that would lower its quality. Feed should not get wet, be damaged by rodents and insects or be contaminated with fertilizer and pesticides. Some feed shouldn't be stored long because the vitamins and other ingredients lose their quality fairly rapidly. Feed is often hauled from mills to feed troughs. This saves loading and unloading the feed as well as retains the nutrients until it is eaten. Feed may be stored in bags or in specially-designed bins in unbagged form.

12-18. Feed may be put into large paper bags during manufacturing.

FEED MANUFACTURING

Feed may be manufactured on the farm or ranch or by a commercial feed mill. Commercial feed mills are in the business of making feed that is sold to animal producers.

12-19. Large amounts of feed are delivered to farms in bulk and unloaded directly into a feed storage bin.

Manufacturing Processes

Manufacturing involves preparing the feed in the form that is needed. It includes several activities: (1) getting the feedstuffs (ingredients), (2) grinding or otherwise preparing the feedstuffs, (3) mixing the ingredients in the proper amounts based on nutrients needed in the feed product, and (4) preparing the feed in the form that it will be fed.

Feed should be properly formulated to provide the needed nutrients. Chemical laboratory analysis may be needed. Feed is bagged or stored in bulk. Proper labeling of bags is essential.

On-Farm Manufacturing

Larger farms and ranches may prepare their own feed. This involves buying the necessary equipment and hiring people to operate it. The ingredients may be grown on the farm or purchased. In most cases, at least some of the ingredients must be purchased, such as mineral and vitamin mixes. The feeds must be formulated to provide the needed nutrients.

It is usually impractical for small farms to have feed processing equipment. In some cases, farmers form cooperatives that make feed for a number of different people but do not sell to the public.

FORMS OF FEED

Feeds are manufactured into various forms. The feeds may be mixtures of various feedstuffs or one feedstuff that is prepared to improve its quality as a feed.

Mixed feeds are made of a variety of ingredients. The ingredients are designed to provide the nutrients needed by animals. They may be prepared in different forms or particle sizes. The appropriate form should be selected to meet the needs of animals as well as provide the necessary nutrients. Convenience of feeding is also a consideration.

Roughages

Roughages are often fed to cattle, sheep, and horses without harvesting. The animals graze in the pasture and do the harvesting themselves! The most common harvested roughages are hay and silage.

Bales. Hay is usually made into bales. The bales can be moved for storage and feeding. Hay is cut, dried by the sun, made into bales, and hauled for storing out of the weather. Increasingly, hay is stored in the field and covered with plastic or a tarpaulin. In dry climates, hay may not be covered.

Hay stored outside in the weather can deteriorate in quality in climates with rain. The hay gets wet and begins to decay. Its color and consistency change. Animals don't find it palatable or nutritious.

Loose Chop. Silage is usually cut and chopped green into small one- to two-inch pieces. It is not bound together in any way. The chopped silage is stored in a silo where water is added and it ferments. This process is known as ensiling. Silos are upright, round structures or trenches in the earth, or they are built above the earth out of concrete.

Pellets and Wafers. Roughages are sometimes made into pellets or wafers. ***Pellets*** are feed pieces made by forcing ground forage through holes in a heavy steel plate to give them shape. A binder (material to hold the pellet together) is added to the feed. Pellets for cattle are about one inch in diameter and one to two inches long. Several feedstuffs may be mixed together in pellets to assure a more complete diet. For example, some corn

12-20. Large round bales of hay stored in a barn to protect them from rain.

12-21. Three forms of feed are shown here (from the left): meal, small pellets, and range cubes. The meal is a range supplement of cotton seed meal and salt for grown cattle. The pellets are high-protein feed for young heifers. The range cubes are large pellets intended for feeding to cows on pasture to supplement winter grazing.

and soybean meal may be added during the grinding process. Pelleting is fairly expensive because it requires fine grinding of the ingredients.

Wafers are similar to pellets, except slightly larger and they don't require the ingredients to be ground as finely. This method is increasing in use with good quality hay crops.

Concentrates

Concentrates are made into forms of feed that maintain the nutritional quality, are easy to use, and provide animals with the needed nutrients.

Cracking and Rolling. ***Cracking and rolling*** are used to break the hard outer coatings on grain. The kernels of grain may be broken into one or more parts by the process. Animals can more easily digest grain kernels when the outer covering is broken. Grains that aren't cracked can pass through the digestive tract of cattle whole and not be digested. Corn and oats are examples of grains that should be cracked.

Grinding. ***Grinding*** involves making the particle size of feed smaller. Small animals require small feed sizes. For example, day-old chicks must have feed in particles sizes that they can eat. They can't swallow big pieces of feed!

Grinding also cracks grains open so that they are more digestible. It permits feed ingredients to be mixed together and made into other forms, such as pellets. Ground feeds are sometimes known as ***meal***.

Extruding. Extruding produces feed in particles similar to pellets. The ingredients are ground, mixed, heated with steam to soften the particles, and forced through a steel plate to form pieces of feed. Extruded pellets may range in diameter from $^1/_8$ to $^1/_2$ inch. The length may range from less than $^1/_2$ inch to an inch or more.

Extruding allows feed ingredients to be mixed together to provide a proper diet. These are known as "all-in-one rations." They are often made for various animal species at different stages of life. All of the nutrients needed are in the pellets. The animal merely needs to eat enough pellets to meet its nutrient needs.

A good example is fish feed. The extruded pellets contain all of the nutrients a fish will need. The amount of protein and other nutrients is varied depending on the age and size of the fish. Most fish need feed that has 28–40% protein; small, young fish need higher protein.

Supplements

Supplements are prepared in forms that are easy for the producer to use and that help meet the nutritional needs of animals. Most supplements are used for cattle that are on pasture to provide nutrients not sufficiently available in the plant materials eaten by the animal. Supplements may be provided in several forms.

- **Blocks**—Minerals and other materials may be compressed into large blocks. Some blocks weigh 50 pounds and are about 8 inches wide, 8 inches thick, and 12 inches long; others may weigh as much as 500 pounds. Most cattle producers are familiar with 50 pound salt and ***mineral blocks*** that are put in the pasture. Cattle lick the blocks to get the needed nutrients.
- **Liquids**—These supplements are in liquid form. They often contain water, molasses, and other ingredients that carry the minerals and other nutrients. Molasses makes the liquid more palatable to cattle. Liquid supplements are 50 to 75% molasses. Minerals and vitamins are mixed with the liquid material.
- **Mixes**—Various mixes of loose, ground ingredients may be fed in troughs or other containers. The cattle have free access to them. Salt, protein sources (such as cotton seed meal), and other materials may be in the mixes.

FEED LABELING

All feed should contain the nutrients that animals need. Commercial feeds must meet certain standards. On bagged feeds, ***labels*** are used to provide information about the feed.

A label tells the ingredients, nutritional content, and the weight of the feed product. Other information about the feed may be included.

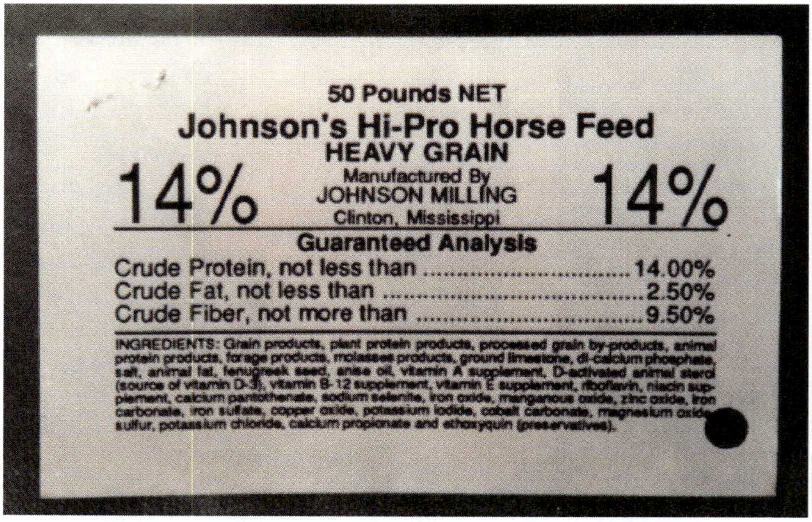

12-22. Sample feed label showing nutrients and ingredients.

HOW ANIMALS ARE FED

Feed is of little value if animals don't get it. Also, they must get the right feed for the functions they are performing, such as maintenance, growth, reproduction, and lactation. Many factors are involved in how animals are fed. Producers follow different production practices.

FREE ACCESS

Free access means that animals have access to the feed when they want it. The feed is provided in feeders or troughs and always available. Hay may be provided in racks where animals can get to it. Of course, cattle, sheep, and horses on pasture can graze when they want to!

Livestock on pasture may be provided free access to mineral blocks and other supplements. When using free access, the feed must be protected from the weather and damage. Special feeders may be used.

12-23. Hogs in confinement must have all the feed they want in order to grow rapidly. These hogs have free access to feeders that are automatically filled.

Automated feeding systems may be used to regularly add feed to the containers. Some are controlled by computers; others require hand feeding.

Animals in confinement (pens) must be provided the feed they need. They can't graze as they want to. Hogs being fed for market in pens, cattle in feedlots, and chickens being grown in houses for meat are usually provided access to all the feed they will eat. Rapid growth occurs when they get all the feed they want with the right nutrients.

SCHEDULED FEEDING

Some animals are fed only at certain times of the day. Management of the animals doesn't allow free access to feed. Feeding times may be based on the needs of the animals as well as controlling animal behavior.

12-24. These caged laying hens are provided free access to feed. Note the eggs that have been laid. An automated feed trough is just above the eggs.

12-25. Timed feeder for scheduled feeding of fish in a tank.

12-26. With a demand feeder fish can have feed at any time. They must wiggle a "trigger" suspended from the bottom of the feeder into the water.

Scheduled feeding is often used with cattle and fish. Dairy cows are often fed at the time of milking. This allows the cow to eat and "let-down" the milk. The way they are fed varies with the management system being used. Dairy cows that are on pasture may only be given a concentrate supplement at milking. Those in confinement have hay or silage throughout the day. It also makes it possible to measure the amount a cow is fed. Beef cattle producers may provide supplemental feeds at certain times of the day, such as each afternoon in the winter.

Some fish are fed at certain times of the day depending on their needs. Others are provided feed on a demand basis. Very young, newly hatched fish, known as fry, are fed once each hour throughout the day. The feed they get is in very small particles. Their digestive systems are small and won't hold much. Little energy is stored in their bodies; therefore, they must be fed frequently. As they grow, the time between feedings can be lengthened. Fish several inches long in growing ponds may be fed once or twice a day. They are fed no more than they will eat in a few minutes. Feed they don't eat is dissolved in the water and wasted. Also, uneaten feed can cause problems in the water, such as oxygen depletion.

REVIEWING

MAIN IDEAS

Animals need feed for several purposes: maintaining their bodies, growth, reproduction, lactation, working (race horses, draft horses, and herd dogs), and to produce other products, such as wool.

Feed is used to provide the nutrients animals need. Nutrients are chemical substances that support the life processes of the animal. These make up the rations, which are the amount of feed an animal gets in a 24-hour period. A diet is the amount and type of feed the animal eats. The major nutrients needed by animals are: energy, which is primarily provided by fats and carbohydrates; protein, which is used for growth; minerals; vitamins; and water.

Nutrients come from the feeds animals ingest. The major feed ingredients are roughages and concentrates. Roughages are high in fiber and make good feed for ruminants. Concentrates are low in fiber and are needed by hogs, chickens, and fish. Supplements and feed additives may be used to help animals grow.

Animals need to receive a balanced ration. The ingredients must be palatable to the animal. They should be free of pesticide residues and weeds, provide a variety of feedstuffs, have bulkiness appropriate for the animal, and be economical in cost.

Feed may be manufactured on the farm or by commercial feed mills. The forms of feed used should be appropriate for the animal. Animals may be provided feed on a free access basis or feeding may be scheduled. Producers must select the best systems for their situations.

QUESTIONS

Answer the following questions using complete sentences and correct spelling.

1. How do animals use feed? How do these uses vary with the age and size of the animal?
2. What are the nutrient needs of animals? Briefly describe each.
3. What feedstuffs provide the nutrients needed by animals? Distinguish between the feedstuffs and give examples.

4. What kinds of pasture may be used?
5. What determines the quality of hay?
6. What are feed additives and implants? Distinguish between how the two get into the bodies of animals.
7. What are the basic requirements of feed?
8. How is feed obtained? What is involved with the sources?
9. What is a mixed feed?
10. How are concentrates made into feed?
11. What is a feed label? What information does a label have on it?
12. How are animals fed? Distinguish between the methods.

EXPLORING

1. Visit a local feed source. Determine the kinds of feed that are sold and the animals they are for. Look at how the feed is packaged and the ingredients in the feed. Determine the protein, energy, mineral, and vitamin content of a feed. Prepare a written report of your findings. (Remember, most supermarkets have large sections of pet feed.)

2. Tour a feed mill. Determine the feedstuffs that are used, how the feed is made, and the forms of the final product. Write a report on your findings.

3. Visit a farm or ranch that uses pasture for animal feed. Interview the farmer or rancher about the kind of pasture used and the practices that are followed in growing it. Prepare a report on your findings.

4. Make a list of the plants that can be grown in your area for animal feed. List the kinds of animals that eat the plants. Briefly describe how the parts of the plants are prepared for use as feed.

13

BREEDING ANIMALS

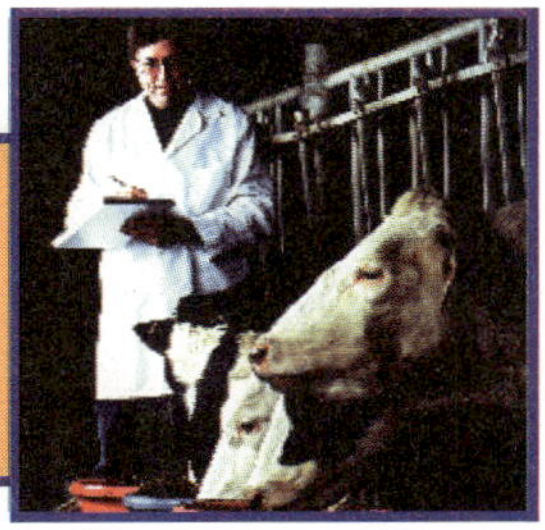

Each year in the United States the average person eats 112.4 pounds of prepared red meat (beef, pork and lamb and mutton). If this amount is multiplied by 260 million (the number of people), the result is that over 292 billion pounds of meat are eaten. A lot of animals are required to produce that much meat!

Producing new animals is an important part of animal science. New animals must be regularly and efficiently raised. Without them, the supply of meat, wool, milk, eggs, and other animal products would soon be used up.

Reproducing animals is supporting the creation of new life—getting more animals. Animals need to be properly fed and otherwise cared for to reproduce. Producers can help assure that animals are able to reproduce.

Most of the animals on farms have similar reproductive processes. Horses, cattle, sheep, and hogs are mammals and give birth to live young. However, chickens and fish aren't mammals, and reproduce by hatching fertile eggs.

13-1. Texas Longhorns have been around a long time and are receiving more attention from producers. (Courtesy, Texas Department of Agriculture)

OBJECTIVES

This chapter focuses on reproducing mammals. It has the following objectives:

1. Explain breed and bloodlines.
2. List examples of breeds.
3. Describe different breeding systems.
4. Distinguish between methods of insemination.
5. Describe management practices in breeding animals.

TERMS

artificial insemination
beef-type
bloodline
breeding
broiler
castration
colostrum
cow and calf production
crossbreeding
dairy-type
debeaking
draft horse
estrus synchronization
feeder pig
feedlot
inbreeding
hybrid
lamb
light horse
market hog
mutant
mutton
natural insemination
needle teeth
outcrossing
pedigree
placenta
polled
pony
pregnancy testing
presentation
production system
purebred
purebred breeding
spay
steer
stocker calves
straightbreeding
upgrading

BREEDS AND BLOODLINES

Different animals are produced for different purposes. First, most farm animals are produced on the basis of what their species will provide for the benefit of humans. Secondly, breeds and bloodlines have been developed based on animals that produce more of what people want. In recent years, various crosses and hybrids have been developed to further improve productivity.

Consumers are important. They want more and better products at the lowest possible prices. They want products that are healthful and easy to prepare. Recent trends are for foods that have less fat and can be grown with less stress on the environment.

BREEDS

A breed is made up of animals of the same species that share common traits. They usually have a common origin and similar characteristics. The similarities, passed from generation to generation, are known as inherited traits.

Bloodlines

Bloodlines are groups within breeds. All members of a bloodline tend to have one common ancestor. In some cases, it might be an animal with outstanding qualities, such as a horse that wins the Kentucky Derby.

AgriScience Connection

ANIMAL IDENTIFICATION

Owners of animals need to be able to identify each animal. This is especially true with breeding animals when pedigree or production information is important.

Several methods are used, such as the ear tag shown here. Each animal in a herd has a tag with a different number. Other methods of identification include tattoos, making nose prints, and using DNA tests. With hogs, ears may be notched to help with identification. Brands are mostly used to identify the owner of an animal rather than the specific animal within a herd or flock. (Courtesy, James Lytle, Mississippi State University)

Inherited Traits and Mutations

Offspring inherit traits from both parents. These traits make them members of their breeds and useful for specific purposes. Several inherited traits are:

- **Color**—Animals of the same breed commonly have the same color. Red parents usually produce red offspring.
- **Milk capacity**—Animals that produce a lot of milk will generally have offspring that produce a lot of milk.
- **Horns**—Cattle that have horns will produce offspring with horns in most cases. Likewise, cattle that are naturally without horns will produce offspring that won't grow horns. (Of course, if cattle have had their horns removed [known as de-horning], they are considered to be horned cattle.)
- **Size**—Large parents will frequently produce offspring that will grow to be large animals. The reverse is true for small animals.
- **Type**—Animals are of certain types. This means that they have characteristics that make them useful. Dairy cattle are useful because of the large amount of milk they give. Beef cattle are good for meat because they have a high proportion of muscle tissue in their bodies.

Exceptions occur to inherited traits. Offspring will sometimes have traits that aren't obvious in their parents. These traits were likely present two or more generations ago in the ancestors and have occurred again. Sometimes a trait might show up for the first time. Offspring that have traits genetically different from their parents are known as ***mutants***. Mutations can occur naturally or be created by radiation or chemicals.

13-2. Donkeys have traits that make them easy to identify: big ears, small frame, brownish to grayish color, and a sometimes stubborn disposition. This animal is also known as a burro.

Purebreds

Animals registered in a breed or eligible for registry are known as ***purebreds***. The breed must be recognized and criteria established for registration. Purebred offspring are eligible for registration if both of its parents were registered. Registration is not automatic. Papers must be filled out on the offspring and submitted to the respective breed association. Most associations charge a fee to register animals.

Beef Cattle

Beef cattle are raised for their ability to produce beef. Consumers want a lot of lean muscle and less fat. ***Beef-type*** is indicated by good muscling in the loin and hind quarters.

Most breeds of beef cattle originated in Europe. A few breeds have been developed in North America using the European breeds and crossing them with cattle imported from India.

More than 25 breeds of beef cattle are found in North America. A few of the breeds are more widely raised. These are known as the major breeds. The major breeds of beef cattle are:

- **Angus**—Originating in Scotland, the Angus has been a popular breed in North America. The cattle are black, ***polled*** (naturally without horns) and have a smooth coat of hair. They grow well and produce high quality beef. Since angus have a dominant polled gene, nearly all crossbred (bred to other breeds) offspring will be polled. One variation in the Angus is the Red Angus. This breed is similar to Angus but is red in color.

13-3. Angus bull. (Courtesy, American Angus Association)

13-4. Brahman cattle have humps and loose skin.

- **Brahman**—Brahman cattle are known for their loose skin and large humps over the shoulders. Commonly, their color varies from light gray to nearly black, but some have red coloring. They are tolerant of heat and insects. Brahman cattle are often crossed with other cattle. The breed was developed in the United States from cattle imported from India. Mature bulls weigh up to 2,200 pounds and cows up to 1,400 pounds. They are sometimes known for a mean disposition—ready to fight!
- **Brangus**—The Brangus breed was developed by crossing Brahman and Angus cattle. Brangus are solid black and polled. They tend to have skin and hump traits of the Brahman, but not to the same extent.
- **Charolais**—Charolais cattle originated in France. The are white to a light straw color with pink skin. They are among the larger beef cattle, with bull weights reaching 2,500 pounds and cows weighing up to 1,800 pounds. Most are naturally horned. Charolais make good use of feed and have a lot of muscle.

13-5. Grade Charolais cow.

- **Chianina**—Originating in Italy, Chianina were first brought to North America in 1971. The cattle are white except for the switch (end) of the tail, which is black. Their skin has a black pigment. They are said to be gentle and tolerate both hot and cold weather. Chianina cattle are the largest beef breed, with bulls reaching as much as 4,000 pounds and cows reaching 2,400 pounds. Chianina are often used in crossbreeding because they make good use of pasture.
- **Hereford**—The Hereford breed originated in England and was brought to the United States about 1817 by Henry Clay. Herefords have white faces and red bodies. They also have white on the legs, switch and belly. A horned breed, Herefords have been very popular in North America. They are hardy and vigorous. Mature males weigh about 1,800 pounds, while females weigh about 1,200 pounds. They are sometimes called "white faces."
- **Polled Hereford**—The Polled Hereford has the same traits as the Hereford except without horns. Polled Herefords were developed in the United States by selective breeding of Herefords that didn't have horns. Polled Herefords have been very popular.

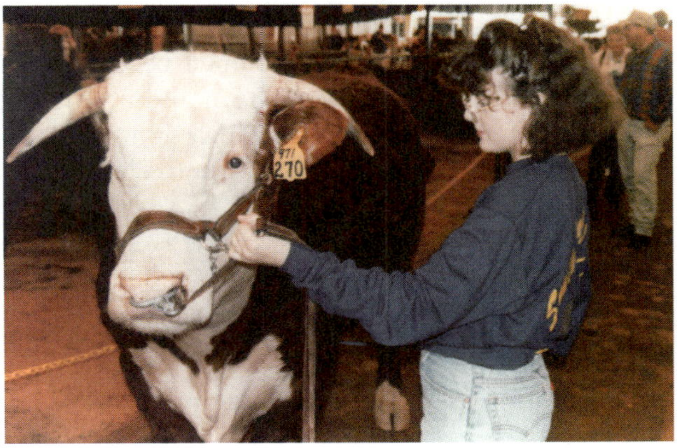

13-6. Horned Hereford bull being prepared for showing.

13-7. Polled Hereford cows. (Courtesy, American Polled Hereford Association, Kansas City, Missouri)

13-8. Limousin cow with calf. (Courtesy, North American Limousin Foundation, Englewood, Colorado)

- **Limousin**—Limousin originated in France hundreds of years ago. The first breed association was established in 1886. The Limousin was first brought to the North America in 1969. Colors vary. Most Limousin in France are red but may be light yellow or black in North America. The variations are due to selection of the breeding stock. They usually have horns. The cattle are lean and have a large loin area (valuable muscle along the back). Weights range up to 2,400 pounds for bulls and 1,350 pounds for cows.
- **Santa Gertrudis**—Developed by the King Ranch in Texas, Santa Gertrudis are crosses of the Shorthorn and Brahman breeds. Santa Gertrudis are cherry red, usually horned and have loose hide, similar to the Brahman. The bulls have small humps over their shoulders. They are similar in size to the Brahman.
- **Shorthorn**—Shorthorn cattle originated in England and were brought to Virginia in 1783. The cattle are red and white, with a red-white mix known as roan. Shorthorn cattle have horns, except for the Polled Shorthorn breed. The cattle are adapted to a wide range of climates and the cows make good mothers. Mature weights range up to 2,400 pounds for bulls and 1,500 pounds for cows.

13-9. Grade Santa Gertrudis cattle.

- **Simmental**—Originating in Switzerland, Simmental date back hundreds of years. They were brought to Canada in 1967 and then to the United States in 1971. Simmental semen was first introduced in the United States in 1968. Their faces are white or light straw in color and their bodies are red to dark red and may appear spotted. Bulls weigh up to 2,600 pounds and cows weigh as much as 1,800 pounds. Simmentals are known for rapid growth on roughage. They develop good muscle with little fat.

- Other Beef Breeds—Some other beef breeds include: Beefmaster, Braford, Red Brangus, Maine-Anjou, Gelbviah, Galloway, Charbray, Devon, Pinzgauer and Texas Longhorn.

(More details on the breeds of beef cattle are available in books on beef cattle production. A good example is *Beef Cattle Science* by M. E. Ensminger, Interstate Publishers, Inc., Danville, Illinois.)

Dairy Cattle

Dairy cattle are raised for the milk they produce. Dairy cows must have the capacity to produce a large amount of milk, which is known as ***dairy-type***. In addition, some of the cattle are slaughtered for beef. Cows that are no longer milked are sold for meat. Five major breeds of dairy cattle are raised in North America. A brief description of the breeds is presented here.

- **Ayrshire**—The Ayrshire breed was introduced into North America in 1822 from Scotland. Colors of ayrshire vary. They may be shades of cherry red, mahogany, brown, or white. Black and brindle (blackish streaks) are not acceptable. They have horns of medium length that turn up and out. Ayrshires rank third among the dairy breeds in the amount of milk given, at 11,700 pounds a year with 4.0% milk fat.

- **Brown Swiss**—The Brown Swiss originated in the Alp Mountains in Switzerland and were brought to North America in 1869. They are solid brown; the shade of brown ranges from light to dark. The noses and tongues are black. The medium horns are turned forward and upward. Brown Swiss rank second in average milk production per cow at 12,100 pounds with 4.1% milk fat.

- **Guernsey**—Guernsey cattle came from the Isle of Guernsey off the coast of France. They were brought to the United States in 1831. Guernseys may be any shade of fawn with white markings. Their skin is yellow. The horns turn outward and toward the front. Guernseys rank fourth in volume of milk produced at 10,600 pounds a year with 5% milk fat. Guernsey milk

13-10. Holstein cows are the largest dairy breed and produce the most milk.

has a slight yellow or golden color which lends the name "Golden Guernsey" to the milk.

- **Holstein-Friesian**—An old breed of cattle from the Netherlands, the breed was brought to North America in the mid-1600s. Distinctive black and white color patterns make them a breed that is easy to identify. They are good grazers and can eat a lot of feed. Holsteins are the predominant breed of dairy cattle in the United States, with an annual milk production of 14,500 pounds per cow with 3.5% milk fat.
- **Jersey**—The Jersey breed ranks fifth among the breeds in pounds of milk produced but is first in the percent of milk fat. Jerseys average 10,000 pounds of milk with 5.4% fat a year. The breed originated on the Isle of Jersey in the English Channel and was first brought to North America about 1815. Jersey colors range from cream to almost black. The cows are the smallest of any of the dairy breeds. They have horns which curve in toward the head and forward.

13-11. Jersey cows are among the smallest in size but produce milk with the highest fat.

Hogs

Ten breeds of hogs, also known as swine, are found in North America. There are close to 400 breeds throughout the world. Except for producers of registered stock, emphasis on growing specific breeds of hogs has declined. Various crosses are being used because of better growth and type (more muscle and less fat). The preferred hog is one with a lot of lean (known as meat type) and little surplus fat. Those with large amounts of fat are known as lard-type hogs. A third type is the bacon-type hog—a long, lean hog. The major breeds of hogs are listed here.

- **American Landrace**—This breed of hogs is white with black skin spots sometimes appearing. The breed has long sides and a level top. The American Landrace produces large litters of pigs and makes efficient use of feed. The ears tend to droop over the eyes.
- **Berkshire**—The Berkshire's ears stand erect over its eyes. The color is black with six white points: each foot, some white on the face and a white tail switch. A hardy hog, Berkshires don't grow as large as other breeds. They have long, deep sides and are moderately wide across the back. Berkshires sometimes have small litters but this has largely been overcome through selective breeding.
- **Chester White**—This breed is popular in the northern parts of the United States. Chester Whites are white but bluish freckles on the skin are acceptable. The females are exceptionally good mothers and produce large litters.
- **Duroc**—The Duroc is the most popular breed of hogs in North America. Its color may be various shades of red. The ears of the Duroc droop over

13-12. Duroc barrow.

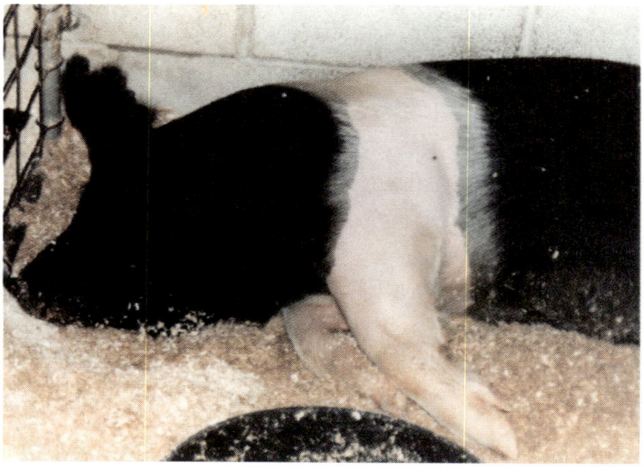

13-13. Hampshires are easy to identify—black with a white belt around the shoulders.

the eyes. Durocs have a number of traits that make them popular: large size, good appetite, and hardy and large pig litters. Though body type varies within the breed, Durocs are good meat-type hogs.

- **Hampshire**—With distinctive color marking—black with a white belt around its shoulders—the Hampshire is a meat-type hog. The breed is trim, with little extra fat. The ears stand erect above the eyes. Hampshire sows raise a high proportion of their pigs and are well suited to confinement. They are known as active hogs.
- **Poland China**—The Poland China is black with six white points: the feet, tip of tail and nose. Years ago the Poland China was a short, fat-type hog. Through selective breeding the breed is now a good meat-type hog. Poland Chinas are often crossed with other breeds of hogs.

13-14. Yorkshire boar confined with a strong fence to prevent unwanted breeding of sows and gilts.

13-15. Yorkshire gilt.

- **Yorkshire**—Yorkshires are white hogs and those without freckles on the skin are preferred. Erect ears are important in the identification of Yorkshires. Yorkshire sows farrow (give birth to) large litters and raise them well because they are good milkers (give a lot of milk). The body is long and has long, deep sides. Slaughtered Yorkshires yield a quality carcass (yield good meat).
- **Other Breeds**—Other breeds include: Spotted Poland China, Tamworth, and Hereford.

Hybrid Hogs. Improved hogs known as hybrid or crossbred hogs have been developed through breeding programs. Several companies have developed hogs that have exceptional ability to produce high quality meat hogs. The hogs have genetic material from two or more breeds. The boars are sold to pig producers for mating with sows. These hybrid hogs often go by the name of the company that produced them, such as a "DeKalb" for a boar from DeKalb Swine Breeders.

(More information on breeds of hogs is available in various books. A good reference is *Swine Science* by M. E. Ensminger and R. O. Parker, published by Interstate Publishers, Inc., Danville, Illinois.)

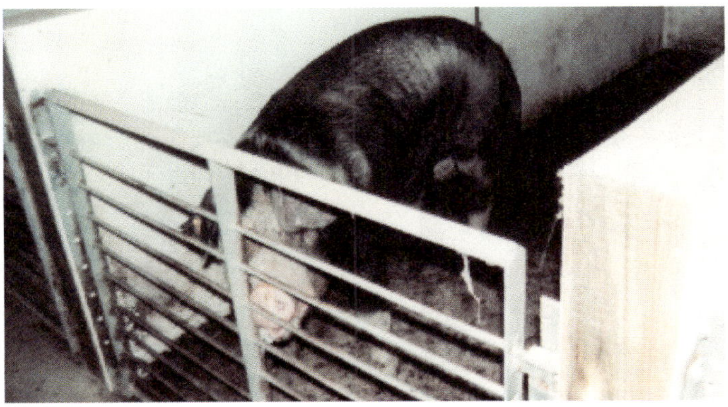

13-16. Boar used in a special genetics program to produce high-performance meat hogs.

Sheep

Sheep are raised for wool and meat. The meat of a young sheep is known as ***lamb***, while that of older sheep is ***mutton***.

Sheep are classified on the basis of the quality of their wool: fine wool, medium wool, long wool, crossbred wool, carpet wool, and fur sheep. Examples of breeds in each type are listed below:

- **Fine wool sheep**—The breeds of fine wool sheep include Rambouillet, American Merino, Delaine Merino, and Debouillet. These sheep produce wool with a heavy yolk (oil content) and are also used for meat.
- **Medium wool sheep**—Breeds of medium wool sheep include Cheviot, Dorset, Finnish Landrace, Hampshire, Shropshire, Southdown, and Suffolk. This group of sheep is used more for meat than for wool.
- **Long wool sheep**—Examples of long wool sheep are the Cotswold, Leicester, Lincoln, and Romney. These breeds produce a long wool with coarse fiber.
- **Crossbred wool sheep**—Breeds of crossbred wool sheep include Columbia, Panama, and Southdale. These sheep have excellent banding (flocking together) instincts and are adapted to the range in the western United States.
- **Carpet wool sheep**—Only one breed of carpet wool sheep is found in the United States: Black-faced Highland. The fleece (wool) is coarse, wiry and tough.
- **Fur sheep**—The Karakul breed is the only sheep used for fur in the United States. Young lambs are killed and the fur pelts are used to make coats.

13-17. Dorset sheep are medium wool sheep.

13-18. Suffolk sheep have distinct black faces and legs.

13-19. Katahdin ewe with lamb.

- Hair sheep—One breed of sheep with hair is important in the United States—Katahdin. It is not necessary to dock (remove tail) of Katahdin lambs because hair does not get dirty as does wool.

(More information on sheep breeds is available from in-depth books on the subject. One example is *Animal Science* by M. E. Ensminger, published by Interstate Publishers, Inc., Danville, Illinois.)

Goats

Goats are raised for three uses in North America: milk, meat, and mohair. Like sheep, goats are ruminants. They are often raised on land where brush and weeds grow. Goats eat shoots, leaves, twigs, briars, and other things that cattle won't eat. Some have even been unfairly accused of eating cans!

Goat breeds are placed in three classes or groups, as follows:

13-20. Spanish goat.

- **Mohair and cashmere goats**—Mohair is the long hair of Angora goats. Cashmere is a fine quality hair produced by the Cashmere goat, mostly in China and Iran.

13-21. The Alpine is used as a milk goat.

Mohair is used for clothing, furniture upholstery, and other products. The only breed that produces mohair in North America is the Angora. Most of these are grown in Texas and other southwestern states.

- **Dairy goats**—Dairy goats are raised for milk. They are usually raised on small farms with little space. The breeds of dairy goats include LaMancha, Alpine, Nubian, Saanen, and Toggenburg. Colors vary a great deal, but are often white with gray, cream, tan, and black coloring.
- **Spanish goats**—Spanish goats are kept for meat and to control unwanted vegetation. They survive with little care and are prolific (produce many offspring). Spanish goats have many colors ranging from solid white to black and with spots.

13-22. Large flocks of Spanish goats are found in some areas of the United States. (Courtesy, Texas Department of Agriculture)

Horses

13-23. Paint horses have distinctive color markings. (Courtesy, Barlink Ranch, Madras, Oregon)

Many breeds of horses are found in North America. Formerly kept for doing work (providing power), they are now often kept as pets, to race or work rodeos and to herd cattle. Crosses of horses and donkeys have been used to produce the mule—a steady, durable work animal.

Horses are often classified into three groups: light horses, ponies, and draft horses. A brief description of each group is presented below.

- **Light horses—*Light horses*** weigh 900–1,400 pounds and are used for riding, driving, racing, and other purposes. They are typically longer-legged and able to move fast. The Quarter Horse has emerged as a popular breed for pleasure purposes and working cattle. They are receptive to training and following commands of the rider. Other common breeds are: Arabian, Paint Horse, Appaloosa, Tennessee Walking Horse, and Palomino.
- **Ponies—*Ponies*** are small horses weighing under 900 pounds. Small miniature horses are included here. The Shetland Pony is a popular small horse for children to ride and to keep as a pet.

13-24. Quarter Horse mare with colt. (Courtesy, American Quarter Horse Association)

- **Draft horses—*Draft horses*** were developed to pull heavy loads, such as wagons, plows or logs through the woods when timber was cut. The use of animals for power has declined since the increased use of tractors and other vehicles with engines. Draft horses weigh upwards of 1,400 pounds, with some much heavier. Draft horses are kept as pets and for recreation, such as competition in pulling contests. A popular breed is the Clydesdale, but the Belgian, Percheron and Shire breeds are also still around.

(More information is available in detailed books on horse production, also known as equine science. One example is *Horses and Horsemanship* by M. E. Ensminger, published by Interstate Publishers, Inc., Danville, Illinois.)

13-25. The White Pekin duck is the most widely grown breed of ducks in North America. They weigh about 8 pounds when mature.

13-26. Ostrich production is expanding in the United States. An ostrich egg is larger than a human hand.

Poultry

Poultry is a broad group of domestic birds (fowl) including the chicken, duck, and turkey. Of these, the chicken is by far the most important bird. Turkeys and ducks are raised for meat; chickens are raised for meat and eggs. Other poultry animals include quail, guinea, ostrich, and emu.

Breeds of poultry are often further divided into varieties. Over 200 varieties of chickens are found in the United States. Of these, only five breeds are of major importance: White Leghorn,

13-27. Chickens have been bred for production, with purebred breeds declining in importance. The White Rock is the most important bloodline of meat chickens. (Courtesy, Marco Nicovich, Mississippi State University)

White Rock, Rhode Island Red, Barred Rock, and New Hampshire. Chickens have been highly selected and bred for one of two uses: eggs and meat. A chicken used to produce eggs is not considered a meat-type chicken, though they may be used for meat.

Cross and hybrid versions of the White Leghorn are used for egg production. Likewise, the White Rock has served as the major breed for developing meat chickens. The breeding of chickens has received more attention than the breeding of any other animal.

13-28. More channel catfish are produced than any other species of fish in North America. This shows fish being dipped from a pond to a truck for hauling to a processing plant. (Courtesy, *Progressive Farmer Magazine*, Birmingham, Alabama)

Fish and Other Aquacrops

The farming of fish is known as aquaculture. Though practiced in other nations for centuries, aquaculture has emerged in North America as an important area of agriculture since 1970. Research in the last few years has improved the quality of fish stocks considerably. More attention has been given to breeding trout than any other fish. In recent years, catfish and tilapia have been studied. Hybrid striped bass have been developed by crossing white bass and striped bass. These hybrids are thought to have much potential.

13-29. Tilapia are being more widely grown in the United States. These tilapia are in an aquarium.

Other aquacrops include shrimp, oyster, crawfish, red fish, snails, crabs, alligators, and frogs. All of these have specific environment and cultural requirements. People considering aquafarming should study the species they are considering very carefully.

BREEDING SYSTEMS

Breeding is the process of helping animals reproduce. In addition to providing valuable offspring, systems of breeding are used to maintain and improve the quality of the animals being raised. Breeding is sometimes used to create new breeds, such as Santa Gertrudis cattle.

PEDIGREE

The genetic traits of an animal come from its ancestors. The names of the ancestors of an animal form its ***pedigree***. Most pedigrees go back no more than five generations of animals. The pedigree of an animal is important in some breeding systems. Some producers keep careful pedigree records of their animals.

KINDS OF BREEDING SYSTEMS

A breeding system is the way animals are selected for mating to get certain results. Breeding systems help producers control the heredity of their animals. One system is not necessarily better than the other; producers should use the system that helps them meet their goals.

Two breeding systems are used: straightbreeding and crossbreeding. Each has variations.

Straightbreeding

Straightbreeding is the mating of animals of the same breed. Several different approaches are used in straightbreeding: purebred breeding, outcrossing, and inbreeding.

Purebred Breeding. ***Purebred breeding*** involves mating purebred animals. These are the registered animals of certain breeds, such as the Polled Hereford cattle or Hampshire hogs. Purebred animals are raised to provide breeding stock of known quality and ancestry. (The major breeds of livestock were presented earlier in the chapter.)

The offspring of a purebred male and female of the same breed should be eligible for registration as a member of the breed. The exact parents of offspring must be known. Controlling the access of males to females is essential. Only one male should be with a female in order to be certain of the parents of an offspring. If two bulls are in the pasture with a herd of cows, the exact sire (father) of the calves will not be known. This means that the offspring can't be registered.

Occasionally some offspring aren't eligible because of traits known as disqualifications. These are often color markings, ear shapes, or other traits that are undesirable. Disqualifications occur because of a genetic factor known as recessive genes. Registration standards for purebred animals are set by a breed association.

Purebred breeding is intended to emphasize the desired traits of animals. It is used to produce animals of high quality that provide the best products for people. Just as the desired traits are emphasized in purebred animals, the undesired traits may also be made more prominent. The animal is said to be homozygous (has identical genes for a certain trait).

If the goal is to furnish the products that people want, using purebred animals may not be the best choice. Crossing purebreds with other breeds or with grade (not purebred) animals may result in the best animals for meat. This has been widely used with hogs and chickens.

Outcrossing. ***Outcrossing*** is mating animals of the same breed but of different families in the breed. Families are animals that share the same bloodline. The animals that are mated do not have a close relationship in their pedigrees.

Outcrossing is a good way to expand the genetic background of purebred animals. This approach is used to reduce or limit a weakness or undesirable trait. Outcrossing is used to conceal undesirable genes.

A variation of outcrossing is linecrossing. It involves mating animals of the same breed that are as different as possible. The pedigrees of animals must be studied to determine common ancestry. When linecrossing, as much difference as possible is wanted in the ancestry.

Inbreeding. ***Inbreeding*** involves mating animals of the same breed with closely related animals. For example, sires may be mated with their daughters or sons may be mated with their mothers. This breeding system is not widely used except by producers of foundation (original) purebred stock. Inbreeding emphasizes both the desired and undesired qualities of animals.

Crossbreeding

Crossbreeding involves mating animals of different breeds (but definitely of the same species). It may involve mating two purebred animals or a purebred male with a grade (not purebred or mixed breed) female. It is used to improve the quality of the products yielded by the offspring. For example, crossbreeding may be used to improve rate of growth, use of feed, or resistance to pests.

Crossbreeding often involves ***upgrading*** or grading up. This is when a purebred male is mated with a grade female. It is intended to improve the quality of the offspring. For example, grade cows mated with a purebred bull will produce calves with half the inheritance of the bull. This approach

13-30. Crossbreeding often produces hogs with more desirable traits, such as this Yorkshire–Hampshire cross gilt.

is used to produce calves that have more meat, no horns or to accomplish other specific genetic purposes. Cattle producers are often reminded to have quality bulls—they're half the herd!

SEXUAL CLASSIFICATION OF ANIMALS

Most farm animals are classified on the basis of their age and sexual condition. In breeding, sexually mature animals are used to produce new animals. Only those capable of regular breeding should be kept. It's expensive to feed animals that don't reproduce!

In some cases, animals are sexually altered. This usually involves removing the testicles from the males, known as ***castration***. Sometimes females are neutered (made incapable of reproducing) through a process known as spaying. The ovaries of the females are removed. This practice is often used with pet animals, such as dogs and cats. The animal is known as a ***spay***. Male animals are more likely to be neutered than females. Castration is fairly easily done, while spaying is more involved.

Animals are castrated or spayed to keep them from reproducing. In meat animals, castrated males produce better meat and don't try to breed the females.

The sexual classifications of selected animals are shown in Table 13-1.

Table 13-1
Sexual Classification of Selected Animals

		Mature		Castrated Male	
Species	Young Animal[1]	Female	Male	Young	Mature[2]
Cattle	calf	cow	bull	steer	stag
Hog	pig	sow	boar	barrow	stag
Sheep	lamb	ewe	ram	wether	
Goat	kid	doe	buck	wether	
Chicken	chick	hen	rooster	capon	
Horse	foal	mare	stallion	gelding	stag
male	colt				
female	filly				

[1]Young animal of either sex, except as indicated for horses.
[2]Castrated after secondary sexual characteristics have developed.

PRODUCTION SYSTEMS

Breeding systems and production systems aren't the same. Breeding systems focus on the ancestry of animals to produce the best possible quality. A ***production system*** focuses on the outcome of production—the kind of product produced.

Production systems are used to describe beef cattle, hog, and sheep production. Though there are similarities, each is different due to qualities unique to the species of animal. All have the production of meat as a common goal. How it is done differs by species.

The production systems for beef cattle are purebred, cow and calf, and finishing or fattening. Hogs and sheep have similar systems. For each species, there are producers of (a) purebred animals, (b) young animals for growing into meat animals, and (c) meat animals ready for slaughter.

PUREBRED PRODUCTION SYSTEMS

Purebred production systems are used to produce purebred animals. The purebred animals are used by producers of animals that will be used for meat, milk, or other purposes. Purebred producers must study the breed being produced and keep careful records. Many purebred producers enjoy working with a definite breed. They may compete with other purebred producers at shows to see who has the "best of show."

AgriScience Connection

WHY MUZZLED SHEEP?

If you go to a livestock exhibit you may see sheep with muzzles, much like the lambs in this photograph. The muzzles serve useful purposes and do not harm the sheep. In fact, they protect the sheep from dangers.

The muzzles keep sheep from eating shavings and other materials that might make them sick. Shavings are needed to provide comfortable, clean bedding. Fresh shavings sometimes have a taste that lambs will eat. Muzzles may also help reduce the amount of dust that is inhaled in a dusty barn. The muzzles are regularly removed for eating feed and drinking water.

Purebred producers raise both male and female animals. Careful screening of the animals is done to select the best ones. Those that aren't the best may be grown for meat.

Male purebred animals are usually in greater demand because of their use in breeding. Males usually bring more money when sold.

Purebred systems are used for beef and dairy cattle, hogs, and sheep.

MEAT-ANIMAL PRODUCTION SYSTEMS

Animal production systems are designed to produce offspring. These animals are grown to produce meat, milk, or other products. Cows are kept to produce calves, sows to produce pigs and ewes to produce lambs. Calf and pig production is the most important among the meat animals.

Cow and Calf Production

Cow and calf production involves keeping cows to produce calves that will primarily be used for meat. Calves may be born weighing 75 pounds or more. Calves are weaned at about 500 pounds. After weaning, they may be slaughtered as baby beef or used as stockers. ***Stockers calves*** are calves that are put in a feedlot or on pasture for additional growth and fattening. The cattle are said to be commercial because they are not purebred. Often, a purebred bull is used with the grade cows.

The system usually involves a number of management practices that help the cows reproduce and the calves grow well. In general, the cows graze permanent or temporary pasture much of the year. In some climates,

13-31. A brindle steer. (Brindle means that the animal is a crossbreed with irregular brown or gray streaks.)

they are fed hay and supplements in the winter. They are bred so that the calves are born in the winter or early spring. This time of calving allows the cow to use the growing vegetation for milk production so the calf can grow as rapidly as possible. The calves should weigh about 500 pounds by late fall of the year. The cow is bred in the late spring or early summer to calve again in winter or early spring.

Some producers carefully select their best female calves (heifers) for herd replacements. A cow will produce about 10 calves over her lifetime. Producers must continually replace cows that don't calve or that get old. Producers who don't raise their own herd replacements must buy them from other producers.

Male calves (bulls) are usually castrated when a few days old. The testicles are removed so that the animal doesn't develop the characteristics of a bull. A calf that is castrated before sexual maturity is a ***steer***. Steers cannot breed.

Calves that are heavy (400 to 700 pounds at weaning) may be slaughtered. These calves are known as baby beef. The meat is lean and tender. Other calves are sent to feedlots for more growth and fattening or are grown on improved pasture.

Feeder Pig Production

Feeder pig production is raising pigs that are sold to other producers to grow to meat-size hogs (220–260 pounds). The feeder pigs are sold at weights of 30 to 60 pounds. Much as a cow and calf system, feeder pig production involves keeping females (sows) to farrow (have pigs). The sows are selected to produce large litters of quality pigs. Producers like to have about 12 pigs in a litter. Boars of the desired type are mated with the sows. Some feeder pig producers use carefully-selected, genetically-improved boars. Most sows will produce two litters of pigs in a year.

Replacement females may be selected from the female pigs. Selection involves looking at type, strength of the body and number of nipples. A female should have at least 12 well-developed nipples (six on each side) so that large litters of pigs can be fed.

Male pigs are usually castrated when a few days old. They are fed as barrows (male pigs castrated before sexual maturity). Barrows can be fed in the same pens as the females.

A number of management practices must be followed to successfully raise feeder pigs. Baby pigs are fragile. They must be protected from harsh weather and from the sow. Heavy sows will accidentally mash baby pigs when they lie down. Mashed pigs may be injured or killed. Newborn pigs have eight sharp teeth, known as ***needle teeth***. These teeth are usually clipped off shortly after the pig's birth. If not clipped, the teats of the sow will be punctured and cut when the pigs nurse. Baby pigs are also given iron and other injections to keep them healthy.

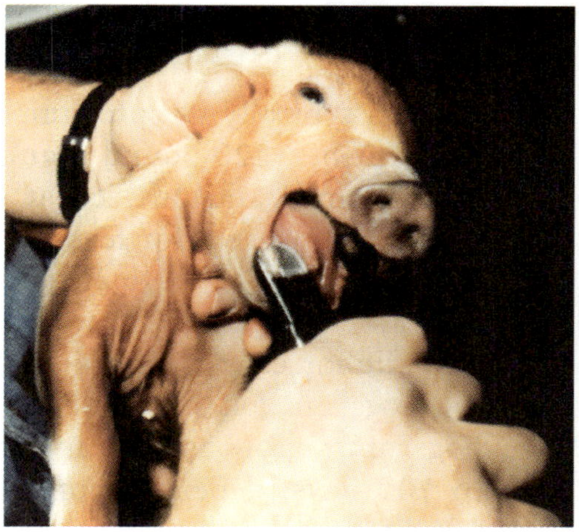

13-32. Clipping a baby pig's needle teeth. (Courtesy, Jim Floyd, Extension Veterinarian, Auburn University)

FINISHING SYSTEMS

Finishing systems feed young animals to the right size and weight for marketing. In most cases, finishing involves feeding diets high in concentrates so that animals will put on body fat. In both beef and pork, fat improves the flavor and palatability of the meat. Of course, too much fat isn't good. People want meat that has just enough fat to provide some marbling (small fat streaks in the lean) but not a lot of excess. Some people feel that too much fat in the human diet is not healthy.

Beef Cattle

Beef cattle are finished in ***feedlots*** and on pasture. The aim is to get the animals to grow and put on body fat. High energy rations are fed. Feedlots are confinement systems. Pastures don't confine the animals to small spaces. Most feedlots can handle over 1,000 head at a time.

Most cattle finishing is concentrated in a few states: Texas, Nebraska, Iowa, Kansas, California, and Colorado. The grain (corn and grain sorghum) used as feed is grown in these states. Many states provide the stocker calves that are fed-out in the feedlots. Many of them are trucked from the cow and calf systems of the South and West to the feedlots.

Stockers begin finishing at weights of 400–700 pounds. They are fed high-energy rations for 90 to 120 days or more until they weigh 1,000 pounds or more. Feedlots expect to get 2.5 to 4 pounds of gain per day. Large amounts of feed and water are needed by feedlots. Disposal of wastes requires lagoons and other measures. Feedlot wastes cannot be run directly into streams. Finished cattle are sold directly to packing plants.

Market Hogs

Pigs typically go into finishing weighing 30 to 60 pounds and leave as well-finished 220-pound ***market hogs***. Before going to finishing, they have been weaned and on feed at the feeder pig farm. The pigs are placed in confinement areas for feeding out. The confinement building may be built on concrete slabs and partially covered with a roof or completely enclosed. Free-access feeding and watering systems are provided. Also, they should be built for easy cleaning and moving of the hogs.

Hogs are fed high energy rations of corn and related feedstuffs. The hogs must grow as well as put on body fat. Most hogs are on feed for 70 to 100 days, depending on their size and the daily rate of gain. Market weight is 220 pounds or a little above. Packing plants penalize growers by reducing the price of hogs that are too big.

Disease control is essential. An outbreak of disease can quickly kill a large number of hogs grown in confinement. Hogs are usually grown in pens of 20 to 50 animals.

METHODS OF INSEMINATING LIVESTOCK

Insemination is placing the male's semen in the reproductive tract of the female. Two methods are used: natural and artificial.

NATURAL INSEMINATION

Natural insemination involves the male depositing semen in the vagina of the female during copulation. The semen must be present during the fertile time of the female. Males may be in the same pasture or pen or they can be kept separately and the females brought for breeding when they are in heat.

ARTIFICIAL INSEMINATION

Artificial insemination involves collecting the semen from a male and depositing it in the reproductive tract of the female. Beef and dairy cattle, horses, hogs, and other animals can be inseminated artificially. The first major use was in dairy cattle. Now it is used with other species. Even chickens and turkeys are artificially inseminated.

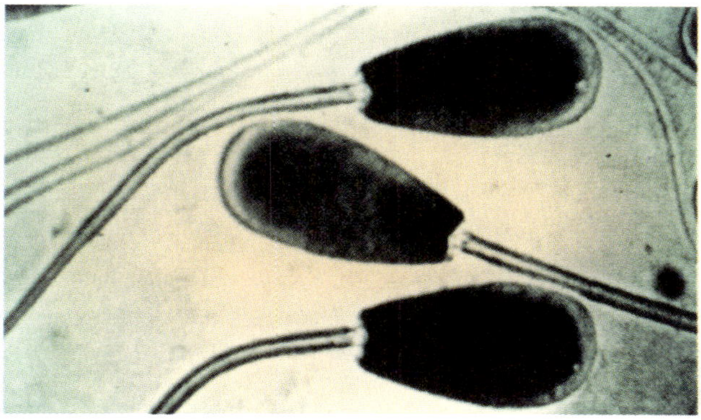

13-33. Greatly magnified bull sperm. (Courtesy, Jim Floyd, Extension Veterinarian, Auburn University)

The use of artificial insemination has increased in recent years. Artificial insemination allows the use of semen from superior males that are owned by another party. It also makes it possible for a male to breed many more females than could be done naturally. Semen contains millions of tiny sperm. It can be diluted so that more females can be bred.

Semen Quality

The semen used in artificial insemination must be of good quality. It should be from a male of proven ability to produce the kind of animal that is desired. Semen is collected and carefully stored to keep the sperm alive. Semen can be stored for a week at 41°F or for several months frozen at –320°F (in liquid nitrogen).

Technicians who collect semen must have proper training in the procedures to assure good sperm. The general procedure is to use an artificial vagina heated with warm water. With cattle, the bull mounts a steer and

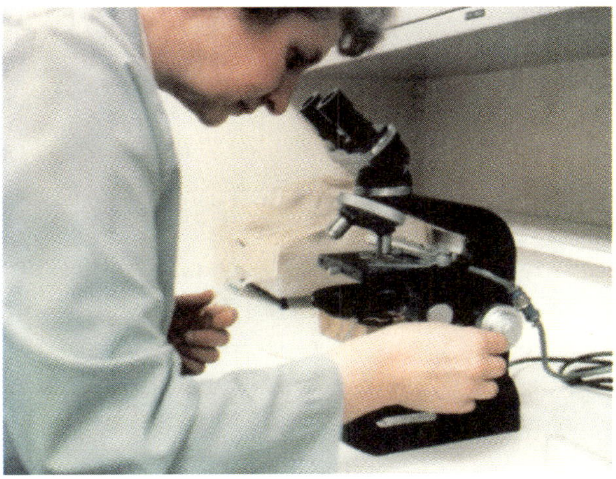

13-34. Sperm motility (movement) is examined with a microscope before processing and use in artificial insemination. (Courtesy, American Breeders Service, DeForest, Wisconsin)

13-35. Cross-section view of a semen tank. Semen is stored in liquid nitrogen at –320°F. (Courtesy, Jim Floyd, Extension Veterinarian, Auburn University)

the penis is placed in an artificial vagina. Electrical stimulation may be used to encourage ejaculation. After collection, the semen is checked with a microscope for live, motile (active) sperm.

Portable semen tanks and other equipment are used to maintain quality. The amount of semen needed for one breeding is placed in a "straw" at a laboratory that has proper facilities. Most semen is provided through cooperatives or associations, such as the American Breeders Service of DeForest, Wisconsin.

Estrus Manipulation

The estrus cycle is the cycle in females that prepares the reproductive system for fertilization. Estrus (also known as heat) is the time in the cycle when the female is receptive to breeding. For natural breeding, the female will allow the male to mount her. In artificial insemination, breeding is done at the time eggs are released by the ovaries for fertilization. For best conception rates the cow should be inseminated about 12 hours after heat is first detected.

Females can be injected with certain hormones to change their estrus cycle. ***Estrus synchronization*** involves using hormones to get several females to come in heat at the same time. It is important when using artificial insemination, advanced breeding procedures such as superovulation (getting the female to produce a number of eggs at one time), and embryo transfer.

Insemination Procedures

Artificially inseminating cattle involves placing the semen in the cervix using the rectovaginal method. The procedure involves the inseminator placing a plastic or rubber glove over one hand. It is then inserted in the rectum of the cow. This hand guides the insemination tube through the vagina into the cervix. The semen is released from the tube between the middle of the cervix and the uterus.

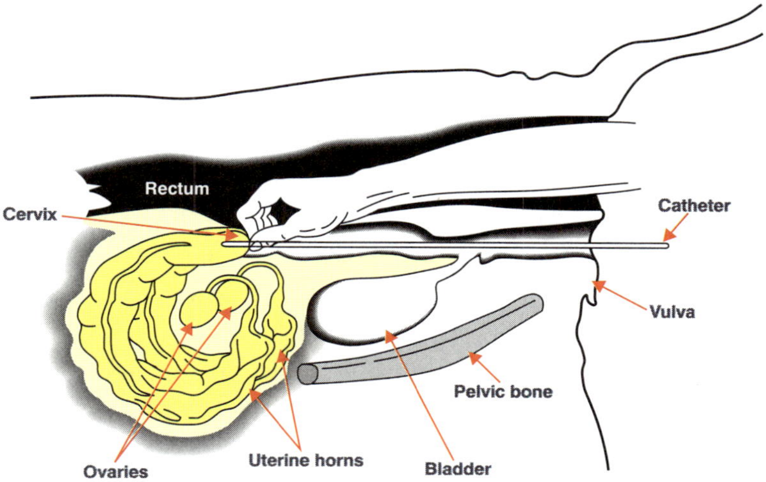

13-36. Artificial insemination in a cow.

All processes in artificial insemination require training. Safety practices should be followed, because animals can be dangerous. Training programs on artificial insemination are regularly held by private breeding companies and agriculture colleges.

MANAGING BREEDING ANIMALS

Animals must be carefully managed if they are to efficiently produce new animals. This includes knowing when to breed, how to manage the pregnant female, the procedures at the time of birth, and how to care for

Table 13-2
Important Information in Breeding Animals

Species	Recommended Age to Breed	Estrus Cycle (days)	Gestation (days)
cow/heifer	14 months	21	283
sow/gilt	12 months	21	114
ewe	17 months	17	148
doe	18 months	20	151
mare	2–3 years	21	336

the newborn animals and mother after the birth. Of course, animal needs vary by species and, sometimes, by breed. Table 13-2 summarizes breeding information on selected animals.

The information presented here primarily applies to beef cattle.

WHEN TO BREED

Females should be bred at the time when they are most capable of carrying the embryo and fetus to birth.

Heifers should be bred when they will be old enough and sufficiently mature to carry the calf and give birth. Heifers should weigh 600 to 750 pounds when bred. They should calve when they are about two years old. A heifer that isn't mature and is small will have difficulty calving. The bone structure of a small heifer may not be large enough to allow the calf to be born. Some producers use bulls that sire smaller calves on first-calf heifers.

Cows are bred at the next heat after they have calved or at such a time so that calves are born when wanted. Most cattle producers want calves born in the winter or early spring. This means that cows should be bred about 283 days ($9^1/_2$ months) before the calf is to be born. (Hogs have a gestation period of 144 days—3 months, 3 weeks, and 3 days.)

PREGNANCY TESTING

Pregnancy testing is used to determine if females are pregnant. Several kinds of tests are available. Some require laboratory testing using samples of blood or urine from the female. Milk can be tested to see if a cow is pregnant. Progesterone (a hormone) levels in milk increase when a cow is pregnant.

For cows, the most common and reliable way of pregnancy testing is using the rectal method. The hand is covered with a rubber glove and inserted in the rectum several inches. The hand feels the uterus through the rectum wall. Changes in the size and location of the uterus indicate pregnancy. Training is needed in order to use this test accurateluy.

Several other methods are used, such as "bumping"—pushing inward with the hand on the lower right flank (experienced cattle producers can detect a fetus)—and checking for a fetal heartbeat using a stethoscope.

Pregnancy testing is used to determine the status of the female. She might need to be bred again. Some producers sell non-pregnant females to a packing plant. There is no need to feed an animal that isn't producing!

13-37. Normal presentation of calf in uterus.

MANAGING THE PREGNANT FEMALE

A pregnant female must receive proper nutrition to develop the new animal(s) in their bodies. Good pasture normally provides adequate nutrition for cows and sheep. Other animals may require concentrates. The female usually doesn't need additional special care.

As the time of parturition (giving birth) gets close, the female should be regularly observed. In some cases, the female may need to be moved to an appropriate place. Most cows can remain in the pasture. Hogs may need to be placed in farrowing crates. Regardless, the overall goal is to help the female give birth to healthy new animals that will live and grow rapidly.

Signs of approaching parturition are: enlarged udder, swelling of the vulva, and hollowness in front of the pin bones. Nervousness, voluntarily going away from the herd, and muscular exertion and distress are signs that the birth is near.

GIVING BIRTH

Most animals give birth without assistance. However, some assistance may be needed, particularly for pigs and lambs. With cows, the calf should be born in about an hour after abdominal straining begins. After an hour, the cow may need help.

Calves are normally delivered with the head between the two front legs. If other parts are visible, help may be needed. Cows that have calves in abnormal fetal positions (fetal position is known as ***presentation***) may need help. If no help is given, the calf and/or cow could be lost during parturition.

Many cattle producers are skilled in assisting with the birth process. If not, a veterinarian can do what is needed. Pulling on the new animal may assist the female in giving birth. This should be carefully done and only in an approved manner. Sanitation is very important. Also, people can contract diseases from animal fluids. They should be careful to protect themselves.

AFTER THE BIRTH

Most animals can usually take care of themselves fairly well after they are born. Pigs may need special attention. All newborn animals should be protected from harsh weather. In some cases, they may need help to start nursing their mother.

The first milk given is known as ***colostrum***. It is high in antibodies and other substances that help the new animal survive. Early nursing is essential to be sure that the new animal gets the colostrum.

In the uterus, a fetus develops outside of the ***placenta*** (bag structure) which nourishes and protects the developing animal. The placenta is expelled by the female during or shortly after giving birth—usually within 3 to 6 hours. Sometimes the placenta (commonly called afterbirth) is retained by the female. Only people who know how to remove retained afterbirth, such as animal technicians and veterinarians, should attempt to do so.

Some animals require management practices shortly after their birth. This includes clipping the needle teeth of pigs, identifying the animals, and castrating the males.

Cows, sows, and other mammal mothers need plenty of feed to help them produce milk for their young. Properly feeding a female during lactation is more critical than during gestation. Cows on pasture usually get enough nutrients. Dairy cows definitely need feed supplements high in energy, protein, minerals and vitamins. Sows need feed with grain, protein, minerals, and vitamins. All nursing females need free access to water.

MANAGING THE BREEDING MALE

More attention is given to managing breeding females than males. The male, however, needs to be properly fed and managed to assure reproductive capacity.

Bulls may be kept in separate pens from cows or pastured together with them. If separate, the bulls should be kept in well-fenced pastures or pens. They should get plenty of exercise. When pastured with cows, the bull is with them only during the breeding season. The remainder of the year the bull is kept in a separate pen.

Bulls need feed with energy, protein, minerals, and vitamins. They should be kept healthy and in vigorous condition. Without adequate nutrition, bulls may fail to breed the cows.

Boars are usually kept separate from sows except at breeding time. They can have mean dispositions and be dangerous to people. They should get plenty of exercise and a diet with energy, protein, minerals, and vitamins. Some boars are allowed access to green pasture and provided a concentrate feed supplement. Boars should not get too fat or be neglected.

POULTRY PRODUCTION FOR MEAT

Producing poultry is very different from producing beef cattle, hogs, and other animals. There are also differences between producing chickens for meat

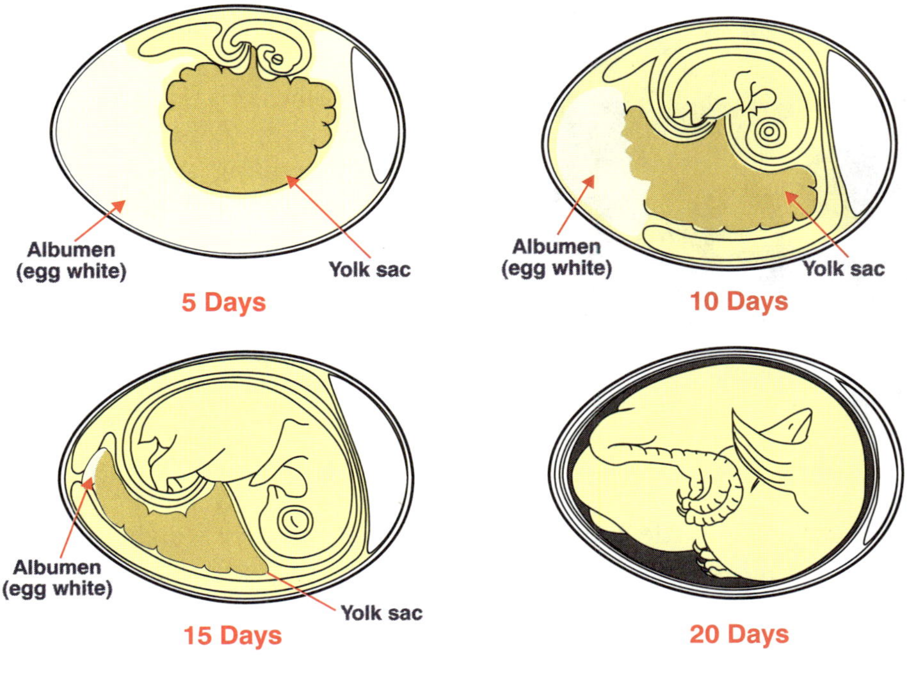

13-38. Development of chicken embryo in egg. Chicks hatch 21 days after incubation begns.

13-39. Small incubator as used in laboratories.

and eggs. The emphasis here is on chickens raised for meat. Most meat is from ***broilers***—chickens 6 to 12 weeks old weighing $2^1/_2$ pounds or more.

13-40. Day-old chick. (Courtesy, McCarty Foods, Jackson, Mississippi)

The production begins when fertile eggs are produced on breeder farms. The eggs are taken to a hatchery for incubation. Eggs hatch into chicks in 21 days. The chicks may be subject to several management practices, such as ***debeaking*** and injections. (Debeaking is done to remove the tip of the beak. This keeps them from attacking and pulling the feathers out of other chickens.)

A day or two after hatching, the chicks are taken to broiler farms. The farms may have one or more broiler houses (buildings where broilers are grown). Several thousand chicks are placed in one house. Feed and water are provided free access. Heat or cooling may be needed, depending on the weather. In 5 to 6 weeks or a little longer the birds reach broiler weight (frying size) of 3 to 5 pounds. They are ready for hauling to a processing plant.

REVIEWING

MAIN IDEAS

The goal of breeding animals is to provide food and other products for people. Knowing what is involved helps producers do a better job raising animals.

Several breeds of animals exist among all species of commonly produced animals. Some producers raise purebred animals; others raise grade animals. Various breeding systems are used to improve animal quality. These include variations of straightbreeding and crossbreeding.

Production systems are used to produce young animals and grow them to the right size. Beef cattle typically begin with cow and calf systems. After weaning, the calves go to a feedlot for finishing. In finishing, the calves are fed high energy feeds to get rapid growth and the development of body fat. Hogs go through similar procedures, with feeder pigs being produced for finishing in confinement. Poultry is different from beef cattle and hogs, though it also involves producing the young and growing them out.

QUESTIONS

Answer the following questions using complete sentences and correct spelling.

1. What is a breed? Distinguish between bloodline and breed.
2. What are the common breeds of beef cattle? Name one major trait that would help you identify each breed.
3. What are the common breeds of dairy cattle? What are the advantages of each breed?
4. What are the common breeds of hogs? What major traits help identify them?
5. What are the classes of sheep on the basis of wool quality? Give an example of one breed in each class.
6. What are the three classes of goats? How do the classes differ?
7. What are the three groups of horses? Briefly distinguish between each group. Which is most widely found today?
8. What is a breeding system? Name and distinguish between the common breeding systems.

9. What is a production system? How do production systems differ from breeding systems?
10. What production systems are used to produce beef? Briefly describe each.
11. What production systems are used to produce pork? Briefly describe each.
12. What methods are used to inseminate animals? Distinguish between the two methods.
13. What are the major management practices in breeding animals? Briefly describe each.
14. What is the general production system for broilers?

EXPLORING

1. Survey your community and identify the animals that are produced. List the breeds and/or crosses that are involved and the production systems that are used.

2. Visit a local farm or ranch and observe the artificial insemination process. Of course, many farms do not use artificial insemination. Determine why artificial insemination is used on the farm you observe.

3. Observe the management practices used on a farm or ranch in breeding animals. This involves clipping needle teeth, castrating, and giving injections.

4. Collect labels from feed for newborn, young, mature, and lactating animals. Study the amounts and kinds of nutrients in the feed. Compare the feedstuffs that are used.

14

KEEPING ANIMALS HEALTHY

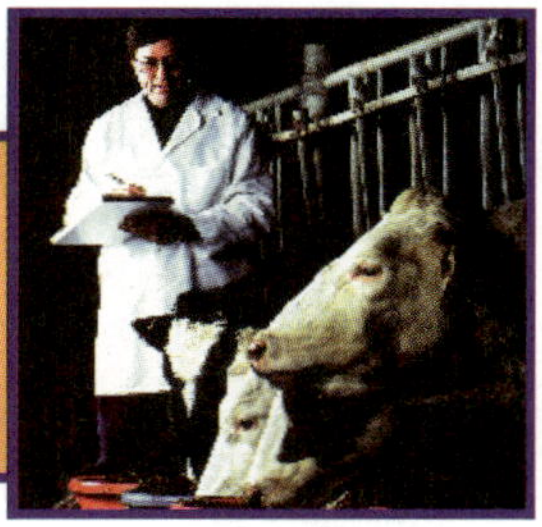

Healthy animals grow fast and are productive. Slow growth and lack of production, such as not giving much milk, are signs that the animal may not be healthy. Producers can use practices to keep animals healthy.

The role of the producer is to provide a good environment for the animal. Various things can be done to assure a good environment for animals.

Not keeping animals healthy costs about $10 billion a year in the United States. It also results in lower quality meat and other products. Sometimes the products are condemned because they aren't good and might pose danger to people if they eat them. Who wants food that isn't safe?

Many of the practices to keep animals healthy can be carried out by the producer. In some cases, the assistance of animal technicians or veterinarians may be needed.

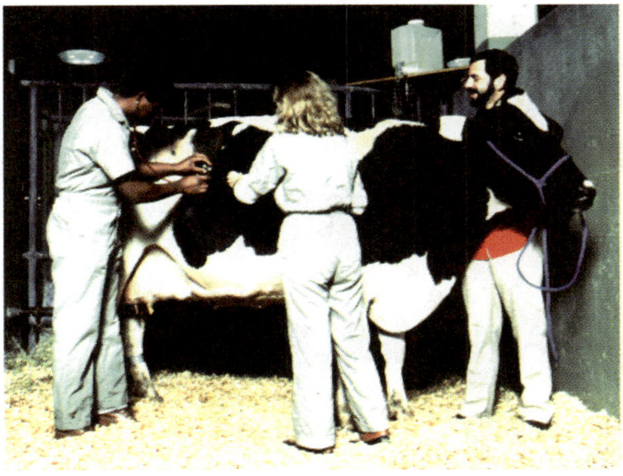

14-1. Good animal health requires helping animals resist disease. This cow is being examined by a veterinarian. (Courtesy, Veterinary Medical Center, Texas A&M University)

OBJECTIVES

This chapter presents general information on helping animals grow and produce as they should. It has the following objectives:

1. Explain health and the signs of good and ill health.
2. Describe environmental influences on animal health.
3. List and explain the losses caused by poor animal health.
4. Explain how good health is maintained.
5. Describe the kinds of diseases and give examples of each.
6. Explain general methods of disease control.

TERMS

acquired immunity
acute disease
antibody
body system disease
chronic disease
contagious disease
disease
endotherm
ectotherm
external parasite
health
immunity
infectious disease
injection
internal parasite
isolation
natural immunity
non-contagious disease
parasite
pathogen
phagocyte
preconditioning
primary defenders
pulse rate
respiration rate
sanitation
secondary defenders
symptom
systemic medicine
topical medicine
vital signs

ANIMAL HEALTH

Health is the condition of the body and a measure of how the functions of life are being performed. Good health means that all life processes are being performed normally.

The health condition of animals can vary widely. Some animals are healthy—free of disease and pain. Others may be unhealthy and, to some extent, suffer disease and pain. Some ill health problems can be prevented and corrected fairly easily; others are more difficult.

A ***disease*** is something that causes pain, injury, or inability to do normal things. The diseased animal deviates from normal health.

GOOD HEALTH

Good health is the absence of disease. The animal has normal body conditions. Producers of animals need to know the signs of good health.

Here are several signs of good health:

- **Eating**—A healthy animal will have a good appetite. The animal will eat when feed is offered. In ruminants, chewing the cud (known as cudding) is a sign of good health. When they don't cud, a disease should be suspected. In fish, feeding is used as an important management practice. When fish come to the top of the water for feed, they are in good health. A general practice is to use floating fish feed. Healthy fish will quickly eat the feed until they have all they want. If the fish don't feed, a problem likely exists.
- **Alert and content**—Healthy animals are alert to what is going on around them. They are also content and appear satisfied. Healthy hogs have curled tails. Healthy cattle will stretch when they get up. Nervous animals may have health problems.
- **Bright eyes and a shiny coat**—Eyes give a lot of information about health. They should be clear and alert. The membranes around them are pink and moist. There are no discharges from healthy eyes. The hair is bright and has good color. Old hair sheds away when the new coat is grown.

14-2. This two-day old colt shows signs of good health and is ready to grow into a sound horse.

- **Normal feces and urine**—The presence of blood, pus, or mucus in the feces is a sure sign of problems. The consistency of the feces should be normal for the animal and the feed it is getting. Producers usually know the normal characteristics of feces and urine. Any changes are obvious.
- **Vital signs**—These are the signs of life. Important ***vital signs*** are breathing, pulse rate, and body temperature. Increases or decreases in the vital signs indicate a problem. Of course, vital signs must be judged according to the species and activity of the animal. Several normal vital signs of animals are given in Table 14-1.

Table 14-1
Normal Signs of Selected Animals

Species	Average Normal Temperature	Normal Pulse Rate	Normal Respiration Rate
	(rectal °F)	*(rate/min.)*	*(rate/min.)*
Cattle	101.5	60–70	10–30
Hog	102.6	60–80	8–15
Sheep	102.3	70–80	12–20
Goat	103.8	70–80	12–20
Horse	100.5	32–44	8–16
Chicken	106.0	200–400	15–30

Pulse rate is movement in the arteries due to heart beats. All animals have ranges of normal heart beats. Rates that are too slow or fast are signs of possible problems. Pulse rate is measured as the number of heart beats in a minute. Pulse rates are normally taken on the soft part of the leg just above the foot. The artery should be prominent and easily felt with the fingers to count the heart beats.

Breathing is known as ***respiration rate***. It is measured as the number of breaths taken in a minute. Increased breathing indicates that the animal is stressed in some way or has been actively exercising. The respiration rate is determined by observing the rise and fall of the animal's sides (flanks).

When taking pulse and breathing rates, a clock or stopwatch should be used to allow an exact minute for the count.

All animals have normal body temperature ranges. These are usually taken using a thermometer placed in the rectum, known as a rectal thermometer. Temperatures are usually reported in degrees F. Increases in temperature indicate a health problem, such as infectious disease.

14-3. These crossbred heifers show signs of good health.

- **Reproduces**—Normal sexually-mature animals reproduce if allowed to mate. An animal can have diseases in the reproductive system that cause them not to reproduce. Females may not cycle and get pregnant after mating. Males may not produce sufficient sperm or may have an injury that prevents them from mating. Reproductive failures may be the result of several different health problems.

ILL HEALTH

If the signs of good health are not present, the animal is said to be ill. Producers should observe their animals each day for health signs.

The signs of ill health include: (1) not eating normally or drinking water, (2) sunken eyes and possible discharge from the eyes, (3) discharge from the nostrils or mouth, (4) inactive—stands in one place drawn-up, (5) rapid breathing, (6) rapid pulse rate, (7) higher than normal body

14-4. The spots on the fish are signs of disease. (Laboratory analysis indicated that this was a disease known as ESC—enteric septicemia.) (Courtesy, Fish Farming Experimental Station, Stuttgart, Arkansas)

temperature (fever), (8) dull hair, (9) lumps or protrusions on the body, (10) open sores, (11) going away from the rest of the animals, (12) unusual feces or urine—may contain blood, and (13) loss of production, including drop in milk production and not gaining weight.

ENVIRONMENTAL INFLUENCES ON HEALTH

Animals prefer certain environments. When environments change or the animal is in one that is not suited to it, the animal is stressed. Stress occurs when an animal cannot adjust to its environment. While stressed, the animal is susceptible to disease.

Environmental health factors include temperature, light, moisture, being moved, and pollution.

TEMPERATURE

Animals live best at certain temperature ranges. Temperatures that are too high or low may cause health problems.

Most farm animals are ***endotherms***. This means that they maintain a certain body temperature. Fish and most other aquatic animals are ***ectotherms***, meaning that their body temperature adjusts to their environment.

Endothermic animals adjust as best they can to hot and cold weather. In hot weather, they may sweat, pant, or wade into water to cool their bodies. In cold weather, they burn more energy to keep their bodies warm. They must be fed better and allowed to exercise. When an animal shivers,

14-5. Careful attention is given to maintaining a constant temperature for these growing broilers.

14-6. These chickens are too warm. (The sign to look for is obvious panting through open beaks.) (Courtesy, Marco Nicovich, Mississippi State University)

the muscles are contracting involuntarily to burn energy and keep the body warm.

Young animals cannot tolerate cold weather as well as older animals. In cold weather, young animals often need to be protected from the low temperatures. Young pigs are often kept in barns with heat. Calves can withstand more cold than pigs, but should be sheltered from extremely low temperatures.

During warm weather, confined animals need care to keep their body temperatures down. Chicken houses may be ventilated with fans. Hogs may have a fine mist of water sprayed on them. Cattle in pastures will take advantage of shade trees during hot weather.

Extreme temperatures stress animals, but those that are properly fed are better able to withstand the extremes.

14-7. Cattle get under trees in hot weather to help stay cool. As ruminants, cattle chew their cuds in the shade during the middle of hot days.

LIGHT

Sunlight and length of day influence animals. Some animals, particularly those with light skin pigments, may develop skin problems from exposure to the sun.

Lengthening the day can increase growth and production of some animals. Chickens are particularly sensitive to light. Both broiler and layer houses may be lighted for a few hours after dark. In laying hens, the light causes more hormones to be dumped into the bloodstream by the pituitary gland. This results in the chickens growing faster and laying more eggs. Of course, chickens have limits in their production.

14-8. Electronic controls are used to regulate temperature and light in a broiler house.

MOISTURE

Moisture may be in the form of humidity or precipitation (rainfall, snowfall, etc.). Air with high humidity does not cool an animal as well as air with low humidity. The perspiration doesn't evaporate at the same rate.

Confined animals may be more affected by humidity. Some chicken houses and barns can build up a lot of moisture. One cow breathes nearly $1\frac{1}{2}$ gallons of water into the air in a day. Ventilation is needed to move the excess moisture out of barns and enclosures and bring drier air in.

Disease agents live better in moist environments. Most causes of disease do not survive as well in low humidity.

Animals may need to be protected from rain and rising water. Cattle can be drowned when water overflows from a creek into their pasture.

MOVING

Animals may be stressed when they are corralled, hauled or otherwise moved. Chasing animals to get them into pens or barns raises their body temperature. Loading and hauling them in trucks can expose them to wind, heat, cold, rain, and other conditions.

Animals going through sale barns or feedlots are exposed to health hazards. The concentration of animals from many different places promotes disease. While stressed, an animal has lower resistance and is more likely to contract a disease. Some producers of stocker cattle precondition them so that the stress of moving won't make them sick. Animals in good condition are better able to respond to stress.

14-9. Herding animals through an auction sale creates stress. (Courtesy, Texas Department of Agriculture)

POLLUTION

Pollution and other hazards in the environment may create disease. Animals are sometimes exposed to poisons in their water and feed. Pastures sprayed with pesticide can cause poisoning. Industrial wastes in the air and water can damage animals. Storing feed in places where poisons have been kept can contaminate the feed. Hauling feed or feedstuffs in box cars or trucks with residues of harmful materials can contaminate feed.

AgriScience Connection

DISPOSING OF DEAD ANIMALS

Even with good care, an animal in a flock or herd will occasionally die. Properly disposing of dead animals is important. Not only does proper disposal minimize odor, it also prevents the spread of disease to healthy animals. Fish and poultry producers must often dispose of dead animals.

Poultry producers commonly use one of two methods to dispose of dead birds: incineration and composting. Incineration involves burning the remains of an animal to reduce it to ash and destroy all disease-causing organisms. A fuel-powered incinerator is needed for this, as shown in the side photograph.

Composting involves using natural decay processes with dead animals. The dead animals are placed in a compost pit or bin along with wood shavings and soil high in bacteria. The remains of dead animals will decay in a few days. A compost bin is shown in the top photograph.

LOSSES CAUSED BY POOR ANIMAL HEALTH

Poor animal health results in losses for the producer. Some of the loss is easy to see; other loss may go unnoticed. The greatest losses may be those that aren't easily seen.

DEATH

The most obvious loss is the death of animals. Dead animals are of no value. They can't be sold and their disposal creates additional costs. Chicken producers frequently have deaths of their birds. Nearly all chicken farms have installed compost pits, incinerators, or other means of disposing of dead birds.

Dead animals should be properly disposed of by burying or burning. Producers should get information on the approved ways of disposing of dead animals in their local area. If dead animals aren't properly disposed of, they can create more disease. Additionally, decaying animals create bad odors and attract other unwanted animals, such as buzzards and wild dogs.

LOWER PRODUCTION

Poor health causes lower production in many ways. Some of them are fairly easy to see; others aren't. Here are a few examples: (1) animals fail to breed and bear young; (2) animals grow slowly; (3) production is reduced, such as milk; (4) products are of lower quality, such as meat may be condemned if diseased; and (5) increased costs of production, such as drugs and labor to treat animals.

Healthy animals provide their owners with greater returns—more money!

14-10. Dead fish floating on the water are signs of disease. (Courtesy, Fish Farming Experimental Station, Stuttgart, Arkansas)

HUMAN DISEASE

Some animal diseases can be transmitted to humans. When this occurs, humans are diseased and suffer. Some of the diseases cause severe problems in humans.

A good example is brucellosis in cattle, also known as Bang's disease. Humans can contract undulant fever by drinking milk from cows with the disease or by letting fluids from the animal get into open wounds on the human body. Dairy cattle are regularly checked for Bang's disease. Milk from cows with the disease cannot be used. Pasteurization (heating milk to 145–150°F for 30 minutes) will help make the milk safe.

14-11. Signs of eye disease include a swollen eye lid and discharge from the eye.

Another example is trichinosis. This is caused by a small worm in the flesh of some hogs. It can be transferred to humans if the pork meat is not cooked thoroughly. Very few hogs going to market have trichinosis. A decline occurred when feeding raw garbage to hogs was made illegal. (Garbage might contain raw pork with trichina. If so, hogs would be infected when they ate it.)

MAINTAINING GOOD HEALTH

Producers can help animals have good health. Regular practices to keep down disease should be a part of the production system. The major goal is to help animals have good health.

14-12. Fenced lagoon used to dispose of the wastes from a pig parlor.

SANITATION

Sanitation is keeping areas where animals are being raised clean. Filth harbors disease and carriers of disease. Sanitation is important because it reduces the number of sources of disease. Animals that don't come into contact with the causes of disease aren't likely to get the disease!

Dead animals should be removed and properly disposed of. Barns should be cleaned out. Manure and other wastes should be properly managed.

14-13. Sanitation may include washing the tires of a vehicle with a disinfectant solution.

Many farms use fenced lagoons for the disposal of wastes from pig parlors, feeding pens, and dairy barns.

Rodents should be controlled. Rats, mice, and other animals are carriers of some diseases. Keeping their numbers down will keep the diseases away.

Some producers use disinfectants as a part of their sanitation programs. A disinfectant is a substance or process that destroys the causes of disease. Animal supplies stores have disinfectants for sale.

PROPER NUTRITION

Animals that receive proper feed and water have the nutrients for good health. Animals with good health are not as likely to get a disease.

Good nutrition helps an animal resist disease. Poorly fed, weak animals don't have the resistance they need. (Chapter 12 presented important principles of nutrition.)

ISOLATION

Isolation involves separating diseased and non-diseased animals. Sick animals are removed from the herd Diseases are often spread by animal contact. Isolation protects healthy animals from contact with diseased ones.

New animals brought to a farm or ranch should be isolated for at least three weeks. In isolation, the animal will likely show evidence of disease if it has one. If no disease is evident after the isolation period, the animal can be introduced to the herd or flock.

RESTRICT TRUCK AND EQUIPMENT TRAFFIC

Trucks, farm equipment, and other vehicles can bring disease to a farm or ranch. Trucks should be disinfected between trips to different farms, feedlots, or other animal producers. Mud on the truck as well as manure and sawdust in the back of the truck can transport disease.

RESTRICT HUMAN ACCESS

People who go from farm to farm may carry diseases on their shoes or boots. Every place they go they pick up and leave small particles that

14-14. Pan with disinfectant solution at the door to a poultry house. Every person entering must dip their boots or shoes in the solution.

stick to their boots. Some farms have boot tubs at entrances to animal facilities to disinfect the boots. A solution of disinfectant is in the tub for dipping the feet. A few farms use disposable plastic boots. Some farms don't allow visitors.

PRECONDITIONING

Preconditioning is used with stocker calves and other animals that are going to be placed under stress. It prepares the animal for the stress. Performing routine management well ahead of the time of stress will reduce the amount of stress. For example, castration, dehorning, and other practices should be done by two months of age. Calves should be weaned and started on feed 30 days before shipping. The amount of time required for hauling and moving the animals should be kept to a minimum. Other health care practices to prevent disease can be used.

IMMUNIZATION

Immunity can be developed in some animals for certain diseases. Immunity means that the animal is resistant to the disease. The animal can withstand exposure to the disease if it comes into contact with the cause.

Animals are immunized with vaccines or naturally develop immunity. Natural immunity often develops if an animal has a disease and recovers from it. In most cases, animals must be artificially immunized against certain diseases.

Immunizing involves helping animals develop immunity to a disease. Animals may be injected with vaccines, consume the vaccine in their feed or be given the vaccine in some other way. Most vaccines contain dead or live organisms, or their products, that cause the disease. The animal's body uses these in developing immunity.

KINDS OF DISEASE

Animals are subject to different diseases. These diseases show different signs or symptoms. A ***symptom*** is the way a disease shows itself. Some diseases may have several symptoms.

CLASSIFICATION OF DISEASES

Diseases may be classified as contagious or non-contagious. Knowing how a disease is classified helps in preventing and controlling it.

Contagious Diseases

A ***contagious disease*** is one that can be spread by direct or indirect contact of the animals. It is caused by small, microscopic organisms such as bacteria and fungi or by larger organisms such as worms. Controlling the organisms helps prevent contagious diseases from attacking animals.

Examples of contagious diseases in cattle are blackleg, Bang's, and anthrax.

The major causes of contagious diseases are:

- **Bacteria**—Bacteria are one-celled organisms, sometimes called germs. Some bacteria are beneficial and don't cause disease. The concern here is with those that cause disease, also known as pathogens. Tuberculosis and tetanus are two examples of bacterial diseases.

  Pathogenic bacteria produce poisons within the body of the animal. Some of the poisons may be isolated to one place on the body, such as a sore. Other poisons may be moved throughout the body by the blood in the circulatory system.

- **Protozoa**—Protozoa are one-celled organisms that cause diseases different from bacteria. Anaplasmosis in cattle and coccidiosis in poultry are two examples that are discussed later in the chapter.

- **Viruses**—Viruses are very small, incomplete cells. They cannot be seen with ordinary microscopes. Viruses require a host (another cell or organism) in order to survive. Animals infected with bacterial diseases are particularly susceptible to viral diseases. Examples of viral diseases include rabies, hog cholera, foot and mouth disease, and distemper.

- **Fungi**—Fungi are one-celled organisms that primarily cause diseases on the external parts of the body. Fungal diseases are seldom fatal. Examples include ring worm and certain ear and eye infections.

- **Parasites—*Parasites*** are usually multi-celled animal organisms that live in or on other animals. A few are single-celled, such as protozoa. Parasites get their nutrition from the animal they live on, known as a host.

  Internal parasites live inside an animal. The internal parasites may be in the digestive system, muscle, or other tissues of an animal. Some internal parasites are taken in through the mouth, such as when an animal eats contaminated feed or water with the organisms in it. Most of the internal parasite contamination is due to feces in the feed. An example is when cattle come into contact with manure while grazing. Pastures that are short due to over-grazing force the animal to eat closer to the ground. This increases the likelihood of the animal getting some manure into its mouth. Examples of parasites that are ingested include cattle grubs (larva of the warble fly), tapeworms, trichina, round worms, and hook worms.

  External parasites live on the external parts of an animal. They get their food from the blood and tissue of the animal. Examples include ticks, fleas, lice, and mites. External parasites can transfer contagious diseases or parasites from one animal to another, such as mosquitoes and other biting insects spreading anaplasmosis among cattle. Leeches are examples of external parasites on fish.

14-15. Greatly enlarged drawing of a dog flea—a common external parasite of pets.

Non-Contagious Diseases

Diseases caused by conditions or substances that aren't transferred from one animal to another are known as ***non-contagious diseases***. Animals don't give these diseases to each other.

Non-contagious diseases typically involve: (1) injuries, such as broken bones, bruises, and cuts; (2) poisons, such as pesticides, fertilizers, and substances in plants (an example is prussic acid which forms in sorghum and Johnson grass grown under certain conditions); (3) faulty body processes, such as milk fever, heart failure, and acetonemia; (4) congenital diseases or those present at birth, such as defects in parts of the body; and (5) inadequate nutrition, such as mineral and vitamin deficiencies.

One form of injury results from feed that contains metal pieces, such as nails, wire, or parts of machines. These objects can lodge in the digestive system or puncture holes through the stomach or intestine walls. This condition in cattle is known as hardware disease.

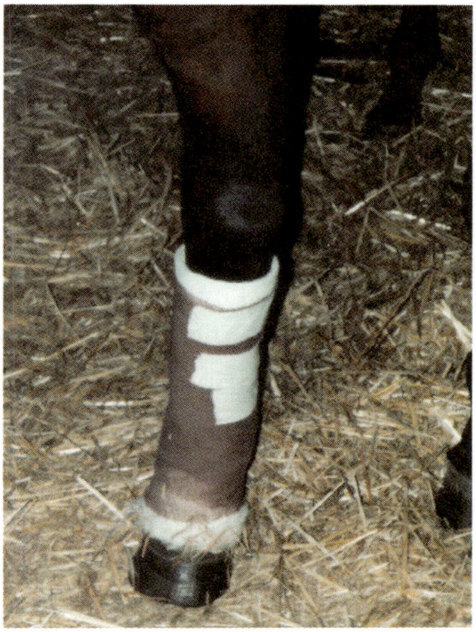

14-16. The non-contagious disease—a leg injury—of this horse has been appropriately treated by a veterinarian.

INFECTIOUS DISEASES

Diseases caused by organisms that get inside an animal's body are known as ***infectious diseases***. These diseases enter, grow, and are active in the animal's body. Infectious diseases have symptoms such as fever, abnormal breathing, and sores.

The microorganisms that get into the body are known as ***pathogens***. They cause the infectious diseases.

Infectious diseases aren't always contagious. Sometimes they begin without being caught from another animal.

TIME OF AFFLICTION

Some diseases last for a short time; others last a long time. ***Chronic diseases*** afflict an animal for a long time. ***Acute diseases*** afflict an animal for a short period of time and may be quite severe.

Chronic diseases may gradually begin and cause few or no symptoms for days, weeks, or months. For example, a few worms in a hog may cause no real problem. Chronic diseases can weaken animals so that they are susceptible to acute diseases.

Acute diseases begin quickly and show symptoms almost immediately. They are often the diseases that are likely to cause loss of production and death of an animal in a short amount of time.

BODY SYSTEM DISEASES

Diseases known as ***body system diseases*** may attack only certain systems of the body. The disease may be confined to the organs in these systems. All body systems are subject to disease.

Diseases of the digestive system occur when the system fails to function properly. The problems may be associated with the mouth, stomach, intestines, and the glands or organs attached to the digestive system.

Genitourinary diseases are in the reproductive or urinary system of animals. Reproductive diseases result in the animal not reproducing properly, if at all. Urinary problems may be infections or blocked urine flow.

Respiratory diseases interfere with breathing and the oxygenation of blood.

Circulatory system diseases interfere with blood circulation or qualities of the blood. Restrictions in the flow of blood damage all parts of the body. A common circulatory system disease in cattle is anaplasmosis.

Nervous system diseases attack the brain and spinal cord or the nerves which spread throughout the body. Movements of the body may be impaired.

SELECTED DISEASES OF ANIMALS

Some diseases attack many different animals; other diseases attack only certain species of animals. For example, roundworms are found in hogs, cattle, sheep, horses, poultry, and fish. Blackleg is primarily a disease of cattle, but sheep and goats are sometimes affected.

Anaplasmosis

Anaplasmosis is a parasitic disease caused by protozoa that attack red blood corpuscles. Cattle are primarily affected. In chronic anaplasmosis, animals become anemic and may overcome the disease on their own. In acute forms, animals have rapid heartbeat, muscular tremors, and loss of appetite. Sick animals may become aggressive and want to fight. Death may occur in a few days. The disease can be prevented by immunizing cattle. Some treatments are available if the disease is caught early.

Anthrax

Anthrax is an acute infectious disease that affects most endothermic animals. It most frequently affects cattle during the summer when they are on pasture. Affected animals have fever, rapid respiration, and swellings on the neck. Animals usually die suddenly. Prevention involves vaccinating animals, controlling flies, and sanitation. Treatments are effective if large doses of penicillin are administered early.

Blackleg

Blackleg is an acute, highly infectious disease that usually results in death. It primarily affects cattle, though sheep and goats can get it. Symptoms include high fever; swelling in the neck and shoulder area; muscles in the neck, shoulder, and thighs crackle when mashed; loss of appetite; and death. Few animals recover from blackleg. Blackleg is prevented by vaccinating calves.

Bang's Disease (Brucellosis)

Bang's disease affects cattle, sheep, goats, and hogs. The disease causes female animals to abort. The reproductive tract of the female is infected. Cattle may have to be bred several times to conceive (get pregnant) and the pregnancy may end in a few weeks due to the disease. Prevent Bang's by vaccinating young heifers. Sanitation, testing cattle, and bringing only Bang's-free cattle into a herd help control the disease. Government programs have helped eradicate Bang's disease.

Hog Cholera

Hog cholera is a viral disease that is highly contagious. Hogs get high fevers, lose their appetite, get weak, and drink a lot of water. Considerable effort has been spent eradicating hog cholera in the United States through federal government programs. There is no known treatment. Vaccines formerly used are no longer allowed in the United States. Any hogs with the disease are destroyed.

Coccidiosis

Coccidiosis is a parasitic disease of poultry, affecting chickens, turkeys, ducks, geese, and game birds. It is caused by a protozoan parasite. Symptoms include bloody droppings (feces), ruffled feathers, unthrifty (sick) appearance, and pale coloring. Medications known as anticoccidials are put in feed and water to prevent and treat coccidiosis. Since it is transmitted by infected birds, wild birds should be kept away from poultry flocks and infected areas sanitized.

Equine Sleeping Sickness

Horses are affected by sleeping sickness. It is caused by a virus transmitted by insects that bite horses, mules, and wild rodents. Affected animals walk aimlessly crashing into things; later they appear sleepy, can't swallow, grind their teeth, and possibly go blind. Some horses recover, but those that don't usually die within two to four days. No effective treatment has been developed. Horses can be vaccinated to develop immunity to the disease.

Foot and Mouth Disease

Animals that have cloven (divided) feet get foot and mouth disease. It is a highly contagious disease caused by a virus. There is no known treatment or vaccine. Quarantine is the best control at the present time. No animals or meat can be imported to the United States from countries with the disease. Animals with the disease get watery blisters in the mouth and on the skin around the feet. Teats of females may have blisters. Animals will have high fever. Foot and mouth disease is not currently a problem in North America but is significant in other countries.

Grubs (Warbles)

Cattle grubs are internal parasites caused by heel flies. They primarily affect cattle. The fly lays its eggs in the summer around the heels of cattle. When they hatch as larva, they go into the skin and move through the body until they reach the back in a few months. In the back, the grubs (larva) cause bumps to swell in late winter and spring. The larva (about 3/4 inch long) will come out of the bumps and grow into an adult fly and

14-17. The cattle grub is caused by a fly that lays eggs on the heels of cattle. Several months after hatching and entering the body, the larva shows up as a grub under the skin of the animal. The larva damages the hide and the meat of the animal.

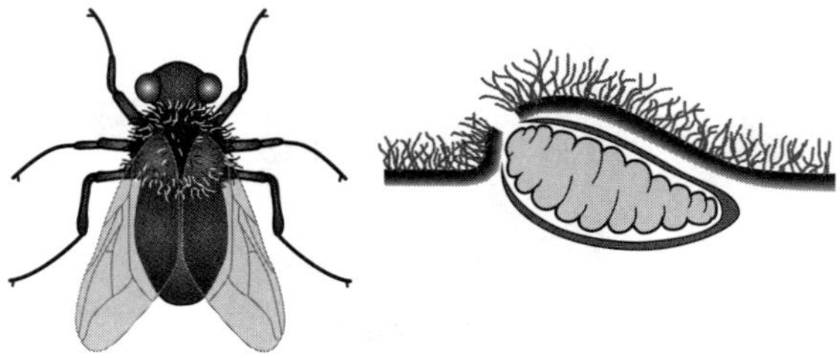

repeat the cycle. A good fly control program will prevent grubs. Several treatments are available and should be used only as approved. Grubs damage the hide by making holes in it as well as the flesh around it. Cattle grubs cause more injury to cattle than any other insect.

Leptospirosis

Leptospirosis is a bacterial disease that affects cattle, sheep, and most other farm animals, including dogs. Symptoms include high fever, poor appetite, bloody urine, and pregnant females that abort their fetuses. Antibiotics are sometimes used to treat leptospirosis. Animals can be vaccinated against the disease.

Lice

Lice are external parasites that attack cattle, hogs, and other species. Lice are small insects that suck the blood of their host. Animals may become anemic. Lice cause animals to itch. Animals with lice are often scratching or rubbing against trees or posts. Lice usually appear in the winter. Back rubbers or other means can be used to administer a range of insecticides on cattle to control lice.

Mastitis

Mastitis is a bacterial disease that affects female cattle, sheep, goats, and swine. It is an infection in the udder which interferes with milk production. The milk may be thick or lumpy in cattle with chronic mastitis. Acute mastitis is indicated by fever and a hard, very warm udder that is sore. Acute mastitis that is untreated can result in death. Mastitis is treated by injecting medicines into the udder through the openings in the teats. Milk from infected cows cannot be used.

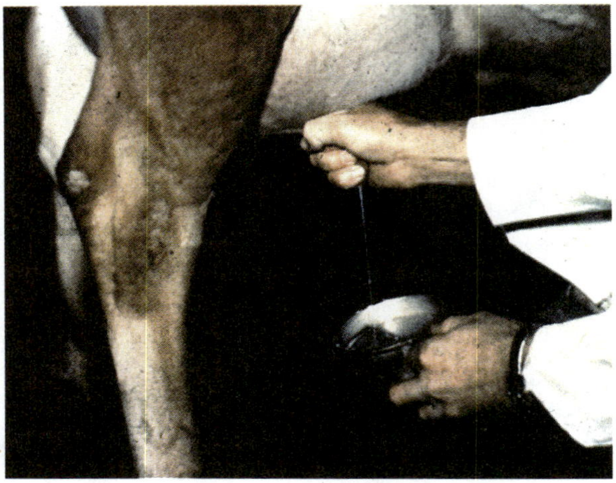

14-18. Strip cup tests may be used to identify cows developing mastitis. (Courtesy, Marco Nicovich, Mississippi State University)

Roundworms and Tapeworms

Roundworms, tapeworms, and other worms are parasites that can infect all animals. They are found in the stomach and intestines. Tapeworms can grow several feet long. A worm this large uses a lot of feed that the animal has eaten. Animals with low infestations of small-size worms may not show symptoms. As the number and size of worms in the digestive tract increases, the animal loses weight, becomes anemic, and may have diarrhea. Sanitation is a good control procedure. Treatments are available but only approved products should be used.

Shipping Fever

Cattle and sheep of all ages can get shipping fever, though it is a particular problem for younger animals. It is an environmental disease caused by conditions animals encounter when hauled or sold through a sale barn. Animals have a high temperature, discharge from the eyes and nostrils, and may cough. Most have difficulty breathing. It is more likely to afflict thin, under-fed animals that are hauled long distances by truck. Cattle should be vaccinated three to four weeks before being moved. Some cow and calf producers use a preconditioning program, which includes steps to keep animals healthy while being shipped.

Others

Many diseases affect animals. All reduce their productivity. Individuals wishing more information should refer to comprehensive books, such as

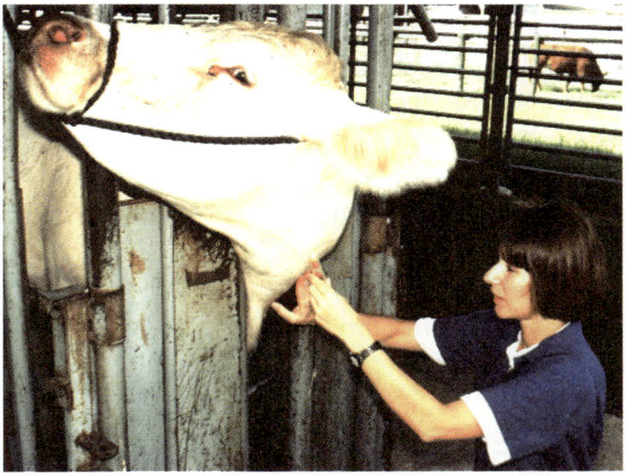

14-19. A blood sample is being taken for laboratory testing by a veterinarian to accurately determine the health problem of an animal. (Courtesy, Veterinary Medical Center, Texas A&M University)

The Stockman's Handbook by M. E. Ensminger, published by Interstate Publishers, Inc., Danville, Illinois.

HOW ANIMALS RESPOND TO DISEASE

The bodies of animals defend themselves against infectious diseases. Strong animals with immunity are usually able to avoid most diseases or at least to overcome them.

PRIMARY DEFENDERS

The ***primary defenders*** against disease are the skin and mucous membranes of the body. Skin forms a barrier against disease. If it is cut, disease can enter the wound.

Mucous membranes that line the soft tissues at openings to the body secrete fluids that defend against disease. The stomach secretes gastric juices, the eyes have tears, and the urinary tract gets washed out by the urine.

SECONDARY DEFENDERS

If a pathogen gets past the primary defenders, the body has ***secondary defenders*** against disease. These secondary defenders try to keep the animal healthy. They are natural responses to many diseases. Also, animals may be immunized against some diseases.

One form of secondary defender is the ***antibody***. Antibodies are produced by organisms in response to an invasion of pathogens. The body may produce up to six different kinds of antibodies, depending on the kind of disease.

Another kind of secondary defender is the ***phagocyte***. These are cells that surround and destroy disease-causing microorganisms. Some phagocytes move around the body in the blood, known as white blood cells. Wounds on the body often collect pus, which are white blood cells. Other phagocytes don't move about in the body. They attack pathogens that come past them in certain organs, such as the liver, spleen, and bone marrow.

DISEASE CONTROL

Preventing a disease is far better than trying to treat it. Producers need to focus on prevention, though they may need to treat animals when they get infected. Everything possible should be done to keep animals disease free.

IMMUNITY

Immunity is the ability of an animal to resist a disease. It is important in keeping animals healthy.

Natural immunity is when an animal is immune to a disease. Some animals won't get certain diseases; other animals get the disease and overcome it. Overcoming a disease often causes the animal to develop

14-20. These newly hatched chicks are being vaccinated at the hatchery before delivery to a broiler grower.

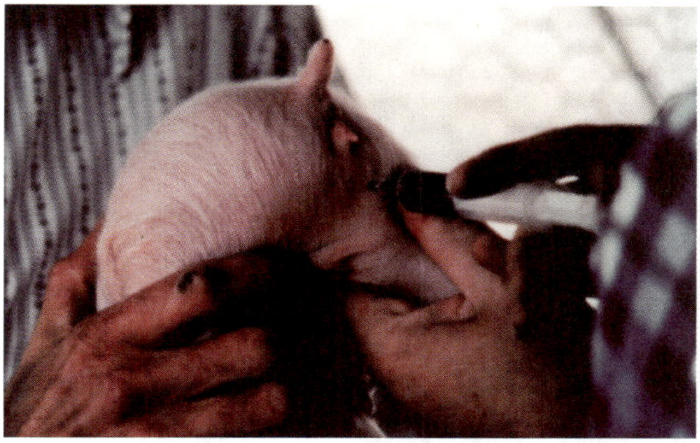

14-21. This young pig is being vaccinated to help it develop immunity to disease.

immunity to getting the disease again in the future. This is known as acquired immunity.

Acquired immunity may be permanent or temporary. Permanent immunity means that the animal will never get the disease again. Temporary immunity lasts for a short period; afterward the animal may again be susceptible to the disease.

Producers can help animals develop immunity by using immunizing agents, also known as biologicals. The common kinds of immunizing agents are:

- **Vaccines**—Vaccines contain some form of the organism that causes the disease. The organism in the vaccine may be live, killed or modified. The animal's body uses the vaccine to develop immunity. Animals develop immunity 7 to 21 days after vaccination. Some vaccines create permanent immunity; others protect for only a few weeks or months. Animals must be vaccinated again if the immunity lasts only a short time.

- **Serums**—Serums are made from the fluid part of the blood which contains antibodies. Serums do not contain pathogenic organisms, as do vaccines. Serums help the body's secondary defenders in their work. Serums are injected into the body.

- **Bacterins**—Bacterins are fluids that contain dead pathogenic organisms. They are sometimes known as suspensions. Animals are injected with the fluid and may gradually develop resistance.

- **Toxoids**—Toxoids are poisons that are no longer poisonous. They are produced by pathogens and altered to eliminate the poison effect.

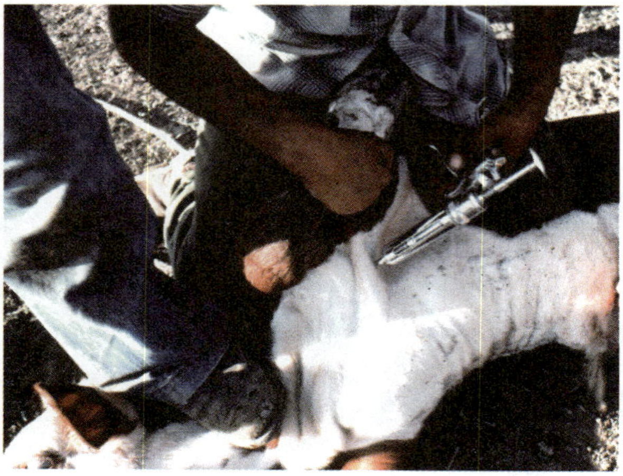

14-22. Calf being vaccinated. (Courtesy, Texas Department of Agriculture)

DESTROYING ORGANISMS

Once the cause of a disease is present, various procedures and substances can be used to try to get rid of it. Sanitation, isolation, and other practices are effective in keeping diseases away.

After an animal gets a disease, it needs help in overcoming it. The location in the body and cause of the disease must be considered. For example, most internal parasites aren't controlled with materials for use on external parasites.

A few examples of ways organisms are destroyed and animals are helped to overcome disease are given here.

Antibiotics

Antibiotics are germ-killing substances made from bacteria and fungi. The best known antibiotic is penicillin, made from a mold known as *Penicillium notatum*. Other well-known antibiotics are terramycin and streptomycin. These are effective on some diseases that don't respond to penicillin.

Most antibiotics are given to animals by injection. Some forms are added to feeds, especially when the animals are small and hard to inject. For example, fish and poultry may have antibiotics mixed with their feed.

Pesticides

Pesticides include insecticides, which kill insects; miticides, which kill mites; and other poisons. Pesticides are used to kill external and internal parasites.

Dietary Supplements

Animals may be given vitamins, minerals, and sugar solutions, such as dextrose. The supplements help animals overcome anemia and provide energy. They are used with weak animals and to treat diseases such as acetonemia in cattle.

Others

The immunizing agents may also be used to treat disease. These include vaccines, serums, bacterins, and toxoids. They were explained in the previous section of the chapter.

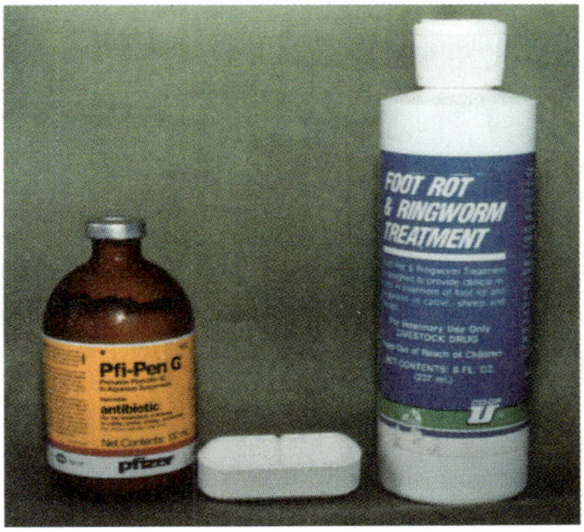

14-23. Three forms of animal medicine are shown here. On the left is an antibiotic, which is injected into the bloodstream of the animal with a hypodermic needle and syringe. In the center is a bolus (large pill) that contains wormer and is placed in the digestive tract of the animal. On the right is a topical liquid that is applied to diseased areas on animals.

HOW MEDICATIONS GET WHERE THEY ARE NEEDED

Medications must get where they can help the animal overcome the disease. The nature of the disease is a major factor in determining the medicine to use. (Note: Only medications approved by the Food and Drug Administration should be used and only as recommended.)

- **Systemics—*Systemic medicines*** are absorbed into the bloodstream of the animal. They may be poured on the animal, given through its mouth or rectum, by injection through the skin, or applied in other ways.

  With pesticides, the poison kills insects or other parasites that live within the host or that suck its blood. The cattle grub is commonly controlled with a systemic poison. Of course, the poison used must not damage the animal in any way. Also, poisons should only be used as approved by the Food and Drug Administration.

- **Topical applications**—This involves applying therapeutants to the skin or outer surface of animals. Absorption into the body is very slow. In most cases, topicals are applied to control diseases on the skin.

  Topical medicines may be sprayed, dusted or poured on the animal. Animals are sometimes dipped into big tubs or vats of pesticide solution.

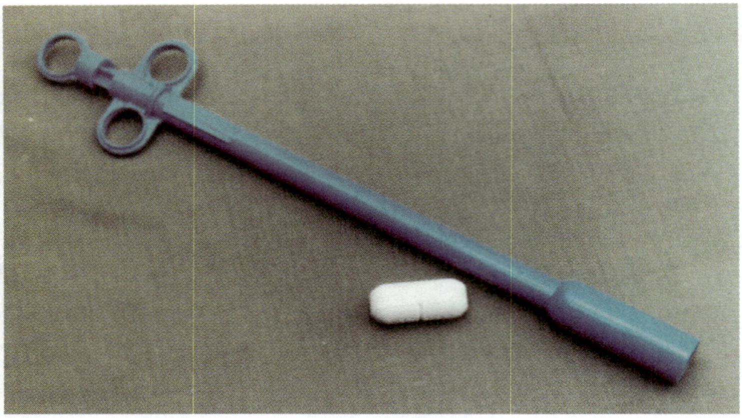

14-24. Balling gun with bolus. A balling gun is used to get the bolus into the digestive tract of cattle. The bolus is placed in the end of the gun and released when the gun is in the animal's throat. Without the balling gun, it is very difficult to get bolus medicine into the animal's digestive tract.

Sometimes backrubbers are put in pastures and near places where cattle gather. The fabric in the backrubber may be saturated with the pesticide. Cattle and other animals rub against the material to get the pesticide on their bodies. Cattle rub and scratch when they have lice, and the backrubber may be effective. Dipping and spraying are thought to be the most effective ways of applying topical medicines. Some topical medicines come in aerosol cans for easy application to small wounds.

- **Internal medicines**—These are substances that are placed inside the animal's body. They may be added to feed or injected into various parts of the body. In some cases, they are systemic medicines. Sometimes they are given through the mouth (orally) as liquids or solids, such as a big tablet or bolus. Dose syringes or balling guns may be needed to get the medicine inside the animal. These procedures may be used to treat parasitic or pathogenic diseases. For example, dose syringes are often used to administer liquids to control worms.

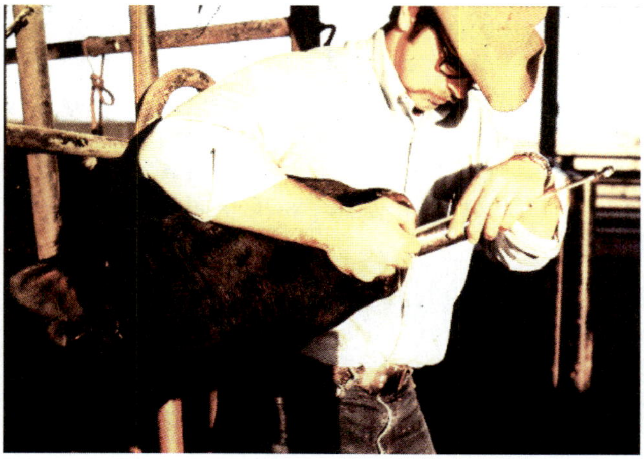

14-25. Dose syringe being used to administer medicine.

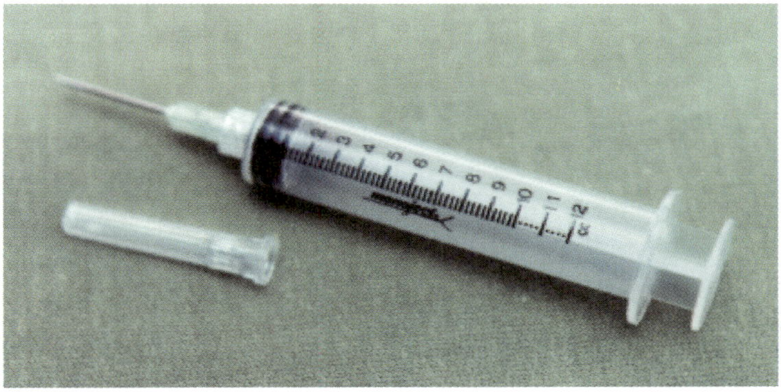

14-26. Syringe with hypodermic needle used to give injections.

Kinds of Injections

Giving medicines directly into the bloodstream is known as ***injection***. It provides a fast way of getting antibiotics and other substances into the body systems of the animal. A hypodermic needle and syringe or other device is used to puncture the skin and dispense the medication. How injections are made depends on the animal and the kind of medication.

- **Subcutaneous injections**—These injections are made just beneath the skin. They are given where the skin is somewhat loose and in easy-to-get-to places, such as the neck or behind the shoulder of beef cattle.
- **Intramuscular injections**—These injections are made through the skin into the muscle. Penicillin is injected into the muscle of the hindquarters and shoulders of most animals.

14-27. Animal health care may require a veterinarian to come to the farm or ranch. This photograph shows a well-equipped truck for veterinary medicine work.

- **Intravenous injections**—These injections are made into the veins of an animal. Intravenous injections are used when a quick response is needed to save the life of an animal.
- **Intraperitoneal injections**—These injections are made through the belly wall of cattle and a few other animals. Intraperitoneal injections are given in the hollow of the flank. They are used when a quick response is needed.
- **Intranasal injections**—This involves inoculating cattle through the nose.
- **Intramammary injections**—This method is used to treat diseases in the udder, such as mastitis in cattle.

14-28. Good habitat helps to keep wild animals healthy.

REVIEWING

MAIN IDEAS

Good health is a must if animals are to grow and produce as they should. Producers that spend time around animals soon learn what animals in good health are like. They also learn how to quickly see when an animal is sick.

A healthy animal has the following: (1) a good appetite (eats when fed); (2) bright, alert eyes; (3) shiny coat; (4) normal feces and urine; and (5) good vital signs—pulse rate, respiration rate, and temperature appropriate for the species.

Animals are more likely to maintain good health in the right environment. Sanitation, nutrition, isolation, restricting truck and human traffic, preconditioning, and immunization are important in keeping animals healthy.

Diseases may be contagious (spread by direct or indirect contact) or non-contagious (not spread from one animal to another). Some diseases affect more than one species; others just affect one species.

Contagious diseases are caused by organisms that either (1) infect the animal—known as infectious disease—and usually produce poisons; or (2) live in (internal) or on (external) the animal.

Some diseases attack a certain body system. Of course, when one system doesn't work well, the entire animal is sick. Diseases may be chronic or acute. Chronic diseases often last a long time and are less severe. Acute diseases develop quickly and are more severe.

Animals respond to disease. Their primary defense is their skin and mucous membranes which keep out pathogens. Their secondary defense is the ability to produce antibodies and phagocytes, especially white blood cells.

Prevention is far better than trying to cure the disease after animals get infected. Animals develop natural immunity and can be immunized. Immunization involves injecting the animal with a substance that helps it develop resistance to the disease.

Medications are given to help animals fight-off disease. Medicines need to get where they can help the animal. Some medicines are systemics and get in the blood system and are carried throughout the body. Others kill causes of disease on contact and do not get carried around the body. Various methods of giving medicines are used including oral, rectal, topical, and injection.

QUESTIONS

Answer the following questions using complete sentences and correct spelling.

1. What is health? Disease?
2. What are the general signs of a healthy animal?
3. How does the environment influence animal health?
4. What losses occur due to poor animal health?
5. What practices can be followed to help assure good animal health?
6. Distinguish between contagious and non-contagious disease.
7. What is an infectious disease? Pathogen?
8. How are diseases related to body systems?

9. What are some common diseases of animals? What animals can have each of these diseases? What are the symptoms? How are the diseases prevented and treated?
10. How do animals defend themselves against disease?
11. What is immunity? What are the common kinds of immunizing agents?
12. What may be used to help an animal overcome disease?
13. What kinds of medications may be used?
14. What are the kinds of injections? Describe how they are used.

EXPLORING

1. Tour a veterinary medical clinic. Observe the equipment used and procedures followed. Observe the treatment of an animal. Write a report on what you saw.

2. Interview a local animal producer. Determine the health problems that have existed and how they were controlled.

3. Check the vital signs of an animal. Record its pulse and respiration rates and rectal temperature. Compare what you found with what is normal for the species.

4. Observe an animal that is in good health. Make a list of the signs of good health. Compare your list with those in a book or the opinion of a person experienced with the species.

15

USING BIOTECHNOLOGY TO IMPROVE LIFE

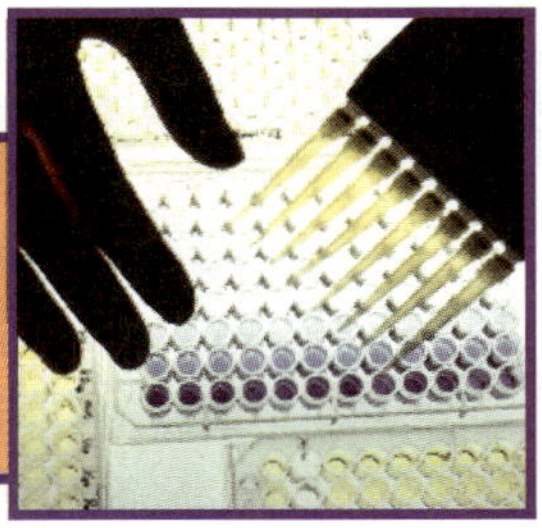

Plants and animals can be made better! For years, agricultural research has studied how to improve them and help them grow. Many improvements have been made.

In recent years, people have gotten interested in new approaches to making improvements—biotechnology. Agriscience is reaching the point that being able to dream it means being able to do it! A little time may be required but it can be done.

How about tomatoes that don't rot or corn that doesn't need nitrogen fertilizer? Making small changes in plants and animals can produce big results.

All of this is of little benefit if it doesn't make life better. This is a big issue with some people. They are afraid of the new life forms that may be created. But, weren't some people afraid of the first automobiles? Just because a few people don't like something, doesn't mean it should not be done. Think about it: Making life better by making better plants and animals—it has great potential!

15-1. Students often do biotechnology skills. The test tube contains DNA extracted from *E. coli* bacteria. The cloudy, fibrous substance in the liquid is the DNA. You can get the same results in your school's lab.

OBJECTIVES

This chapter provides background and applications of biotechnology. It has the following objectives:

1. Define biotechnology and describe how it is being used.
2. Identify issues associated with biotechnology.
3. Distinguish between two major areas of biotechnology.
4. List and explain examples of organismic biotechnology.
5. Describe the role of genetics, cells, and genomes in molecular biotechnology.
6. Describe the process of genetic engineering and the use of recombinant DNA.
7. Identify areas of agriscience being developed through genetic engineering.

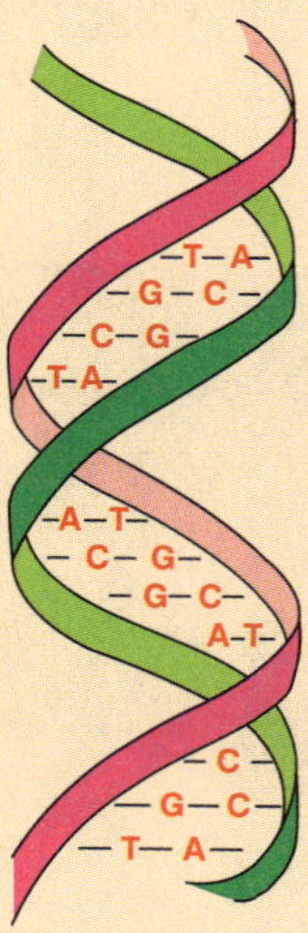

TERMS

allele
cleavage
cloning
computer simulation
cropping models
diploid cell
dominant
donor
donor cells
double helix
embryo splitting
embryo transfer
flushing
gene expression
genome
genotype
gonadotropin
growth chamber
haploid cell
heterozygous
homozygous
immunochemistry
interferon
in vitro
ligation
molecular biotechnology
mutation
organismic biotechnology
phenotype
plant tissue culture
recessive
recipient
recombinant DNA
superovulation
synthetic biology
transformation
transformed cells
transgenic animals
ultrasonic transducer
variability
vector
vesicle

BIOTECHNOLOGY: NEW AND BETTER LIFE AND THE ISSUES

Biotechnology is using biological processes to get new or better plants and animals. Sometimes the nature of the organism is changed. Other times, new substances are produced.

Biotechnology is viewed differently by scientists. Some have a broad view, meaning that biotechnology is concerned with all of the ways plants and animals are improved and used to make products. Other scientists view biotechnology as creating new forms of life.

Agriscientists have a broad view: Biotechnology is all of the applied operations in producing food, fiber, and shelter. New plants and animals, new cultural practices for plants and animals, and new ways of preparing food are included. All plants and animals and their products are part of the study of biotechnology.

SYNTHETIC BIOLOGY

An area involving skills similar to biotechnology is synthetic biology. ***Synthetic biology*** is creating lifelike characteristics with chemicals. The characteristics of birth, growth, and death have been created with collections of chemicals and processes.

A major part of synthetic biology is the creation of vesicles. ***Vesicles*** are tiny sphere-shaped structures with cell-like characteristics. Cell membranes are imitated using chemical processes somewhat like creating soap bubbles.

Vesicles are very small—often smaller than what can be seen with an ordinary microscope. Powerful microscopes and attached computers and video systems enlarge the images for study. With these procedures, physical organic chemists use nonliving materials to study living processes.

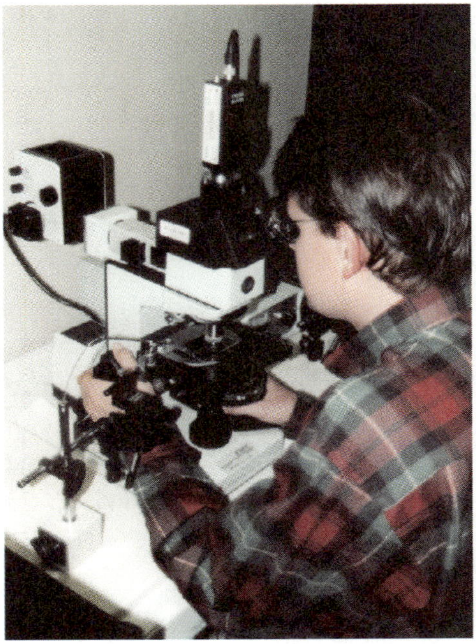

15-2. Synthetic biology involves using powerful microscopes with associated computer and video systems.

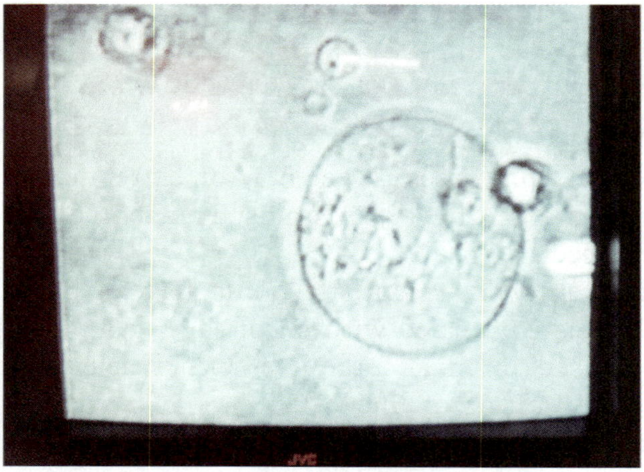

15-3. A greatly enlarged vesicle as viewed on a computer screen.

Using chemicals to create living material in a laboratory still has a way to go and may never be possible. Some people doubt if it can be done. Why not? Living material is nothing more than a collection of chemical processes!

BIOTECHNOLOGY BENEFITS

Biotechnology is being used in agriscience and in other areas, such as human medicine and chemical engineering. Biotechnology in agriscience involves: developing and using new forms of plants and animals, modifying plants and animals to improve their life processes, making substances from natural biological processes, and improving the quality and kind of foods.

All of these are used to provide more of what people want at the lowest possible cost. They help overcome things that limit production and quality.

Some forms of biotechnology have been used a long time. As mentioned in Chapter 2, biotechnological processes are used in baking bread, making cheese, and using microorganisms to dispose of wastes. Modern laboratory equipment and procedures have made new biotechnological methods possible. The overall goal is to have more and better food, fiber, and shelter.

BIOTECHNOLOGY ISSUES

Just because something can be done, doesn't mean it should be done! Or, should it be done?

Using biotechnology is acceptable in many areas of life. Most people like foods made with biotechnological processes. They like having plenty of food at a reasonable cost. Also, practices usually gain acceptance over a period of time.

15-4. The carnations this young woman is holding were propagated using tissue culture.

- **Uncertainty**—Some people are frightened by new practices. They don't like biotechnological processes that they feel may harm the environment. They don't want "strange" things released into their world. The effects of changed plants and animals on the environment are unknown.
- **Dangers in New Life Forms**—Some people fear that new life forms may have dangers that aren't seen at first. "Monsters" could be released into the world! On the other hand, agriscientists conduct careful research on the organisms. They feel that no such monsters exist in the new forms they produce.
- **Keep Organisms "Natural"**—Some people feel that plants and animals should be "natural." They don't think that people should change them so that they are no longer what they would have been in nature. Other people feel that as long as plants and animals are made better, every thing is fine.
- **Unhealthy Food**—Some people feel that plants and animals produced with artificial means are inferior and unhealthy. They feel that unhealthy food has been produced. They don't want foods grown with the help of fertilizer, pesticides, and animal medicines. Other people feel that foods produced with these substances are just as good as the others. The use of these materials makes it possible to produce enough food for the world's rapidly increasing population.
- **Harm to Handlers**—Some people feel that preparing foods that are produced with biotechnological methods may be harmful to the cook. In one major city, the chefs of the restaurants agreed that they would not prepare foods that had been grown using certain processes. Other people feel that the products will cause no such harm.
- **Labeling**—Should food products grown with the help of biotechnology have labels informing people of this? Some people want "natural" foods that haven't been changed. Labeling would allow them to select those that haven't been changed using biotechnology. The problem: Nearly all foods

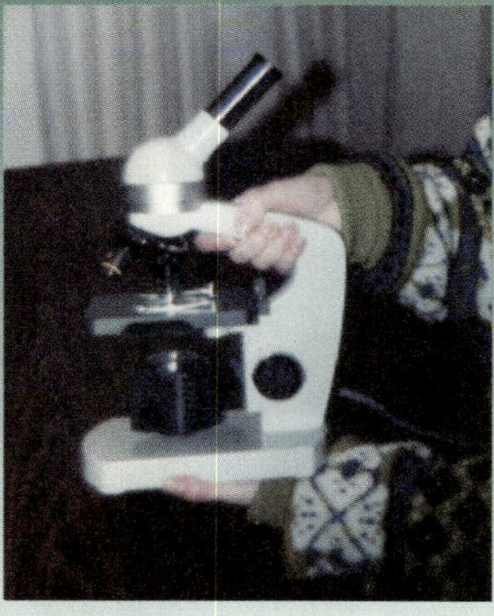

AgriScience Connection

HOW TO CARRY A MICROSCOPE

Microscopes are important and delicate pieces of equipment. They must be accurate and easy to use. It is frustrating to use a simple microscope that has been damaged. Have you ever tried to see something and could not find it because the microscope was not working properly?

Keep a microscope in good condition by handling it properly. When moving a microscope from one place to another, put one hand under the base and grasp the arm of the microscope with the other hand. Only carry the microscope and no other items such as slides or books.

Keep the lens and stage area clean. Avoid touching a lens area. If it does get dusty, use lens paper to clean the area. Keep fluids off the stage as these can stain and make the microscope dirty. Some substances can also cause the microscope to corrode.

are produced using practices that alter how the plant or animal would have naturally grown. Any regulations would be issued by the U.S. Food and Drug Administration.

- **Ethics of Joining Plants and Animals**—Some biotechnology involves moving animal "parts" into plants. Many people have questions about such processes. They feel that food will be changed dramatically. In most cases, only one tiny trait is moved. The trait becomes nearly lost among all the thousands of other traits of the organism. (More information is presented later in the chapter on the parts that are involved.)
- **Lack of Information**—People need to get good information before they develop opinions about biotechnology. Approval of new products takes a long time. Research is expensive work. Decisions need to be made based on good information; otherwise, investments in developing new things may be wasted.

TWO AREAS OF BIOTECHNOLOGY

As agriscientists gain more understanding of life, they are better able to use biotechnology. Using advanced forms of biotechnology requires advanced education. Agriscientists must be well trained and have modern equipment.

Biotechnology is a combination of biology and technology. Technology is the application of science processes. Biology deals with living organisms—both plants and animals. Bringing the two together offers exciting opportunities in agriscience.

Work in biotechnology is carried out in many ways. In this book, two major levels are discussed: organismic and molecular.

ORGANISMIC LEVEL

Organisms are complete living things. ***Organismic biotechnology*** deals with intact or complete organisms. The work involves helping them live and grow better as well as perform useful processes. Working to improve plant or animal growth using nutrients or by controlling hazards is important.

More organismic than molecular biotechnology is used in agriscience. Examples are presented later in the chapter.

MOLECULAR LEVEL

Molecular biotechnology involves changing the structure and parts of cells. It is often studied by molecular biologists.

All life can be reduced to molecular levels. This means that cells—the building blocks of organisms—and their parts can be divided into smaller pieces, known as molecules. A molecule is the smallest particle of a substance. When a molecule is divided, it is broken down into two or more atoms.

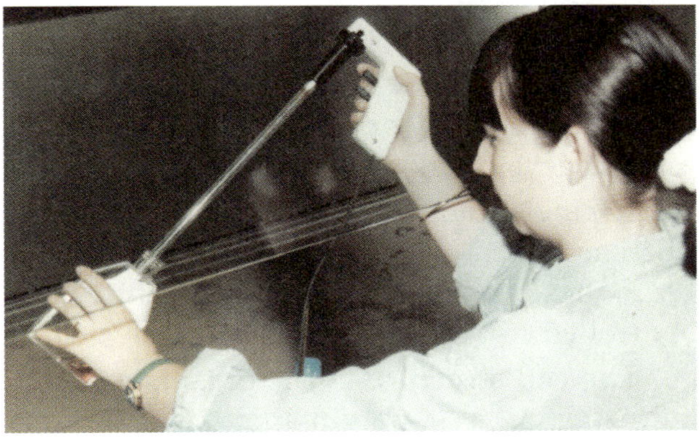

15-5. Biotechnology research requires careful attention to the methods used. This work is being carried out in an aseptic environment (free of organisms).

An atom is the smallest unit of an element. Elements cannot be broken into other substances. Elements are substances that are studied by chemists.

All things (living and nonliving) on earth are made of elements. Only 109 elements have been identified, and some of them are very rare. Most of the things in our daily lives are made from no more than 20 elements. More things are made of carbon, hydrogen, and oxygen than any other elements. (A later chapter in the book will include principles of chemistry in agriscience.)

The processes are complex. Molecular biotechnology begins with cells and the materials that are in them. This includes the information in cells and their organization. A background in genetics and molecular biology is important.

ORGANISMIC BIOTECHNOLOGY

Organismic biotechnology deals with helping organisms live and grow better. It takes the organism as it is and makes improvements. Many of the practices carried out in producing plants and animals could be included in this area of biotechnology.

Several examples of organismic biotechnology are presented here.

Cloning

Cloning is a process of asexually reproducing organisms. No union of male and female germ cells occurs. The new organisms are identical to their parents. The genetic information in the organism has not changed. Three examples are tissue culture, embryo splitting, and apomixis.

Plant Tissue Culture. Plants can be reproduced using tissues or cells. ***Plant tissue culture*** involves propagating plants using single cells or small groups of cells. Laboratory procedures are needed. Sanitation and nutrition for the developing organism are needed.

A single cell or small number of cells is removed and placed in a nutrient solution. The solution usually contains nutrients, vitamins, plant hormones, and sugar. The temperature and photoperiod must be properly controlled for the plant being propagated.

The cell structure of the organism is not changed or modified by tissue culturing. It is simply a method of reproducing identical plants, known

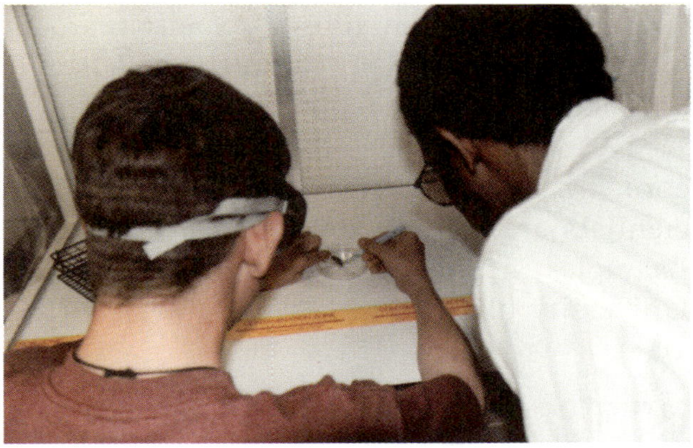

15-6. Student learning to prepare a tissue culture using a Lamnar flow hood.

as clones. Tissue culture methods are used with several plants, including carnations, ferns, and orchids. Proper equipment and some training are needed.

Embryo Splitting. An embryo of a few cells can be split into two parts which develop into a complete animal. With cattle, ***embryo splitting*** involves removing the embryo from a cow seven days after conception. The embryo is cut or split in half. Each half is placed in the uterus of another cow. These cows are known as ***recipients*** because they receive the split embryos. The embryos grow to full-term and produce two identical calves. The genetics of the recipient cows are not transferred to the calves. The cow providing the embryo (donor cow) and the bull providing the sperm in breeding give the genetic information to the new animals.

Embryo splitting is expensive and requires considerable skill. It is used to increase the number of offspring of valuable animals. Only veterinarians with high levels of training can perform embryo splitting.

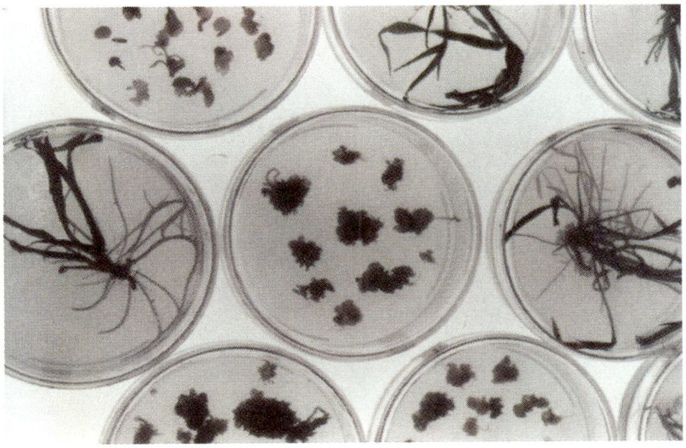

15-7. Cloned corn being cultured in petri dishes. (Courtesy, DEKALB Plant Genetics, DeKalb, Illinois)

Apomixis. Apomixis is asexually reproducing plants by seed. The ovule is not fertilized with sperm in the pollen. All of the plants that are produced are identical to the female plant. They are clones of the female.

Nearly 300 species of plants can be reproduced by apomixis, including corn, wheat, strawberries, and beets. Though a natural phenomena, agriscientists are using apomixis to develop improved varieties of some crops. Examples include buffelgrass at Texas A&M University and millet at the University of Georgia.

Greater Fertility

Various methods are used to increase the reproductive capacity of superior animals. Valuable cattle, hogs, and horses can become more valuable when they are able to produce more than the normal number of offspring.

Getting more offspring involves sexual reproduction. Male and female sex cells unite to form a new organism. Two examples are given here. Both examples require advanced training in reproductive processes.

Superovulation. Most females naturally produce a certain number of eggs during their estrus cycles. The reproductive potential of a female can be increased through procedures known as ***superovulation***.

A good example is in cattle. Cows normally produce only one egg (ovum) at a time. By injecting the cow with a hormone, such as ***gonadotropin***, the cow may release 8 to 20 eggs (ova) during estrus. While in heat, the cow is inseminated and the eggs are fertilized.

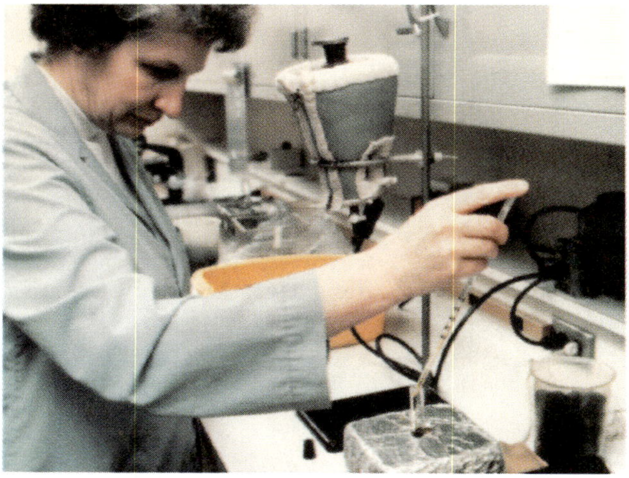

15-8. Adding antibiotics to freshly collected bovine semen protects the fragile sperm. (Courtesy, American Breeders Service, De Forest, Wisconsin)

Seven days after fertilization, the developing embryos are flushed from the cow (known as a ***donor***). ***Flushing*** involves using a saline solution (water with some salt) in the uterus of the cow. The embryos are "washed out" and removed from the solution. They are very small—no bigger than the tip of a small match. The embryos are immediately put into another cow (known as a recipient cow) to develop into full-term calves. In some cases, the embryos are frozen and later placed in recipient cows. Embryos can be frozen in straws to make future implantation easier. Surgically removing the embryos has also been done, but it is more expensive and likely to injure the cow.

Embryo Transfer. This involves removing an embryo from its mother and placing it in another female to develop. ***Embryo transfer*** is a part of the process of superovulation but can be used without superovulation.

With cattle, an embryo could be flushed from a donor cow and put into a recipient cow to allow the donor cow to begin estrus. Most cows will begin estrus 21 to 60 days after the embryo is removed. This allows a valuable cow to produce more offspring. Recipient cows are less valuable but are good mothers.

15-9. Genie, a mature sow with pigs, is a transgenic animal. Her milk contains a human blood protein needed in human medicine. Genie was developed at Virginia Tech University. (Courtesy, Virginia Tech and Rick Rudd, University of Florida)

More Production

Various processes can be used to increase production from plants and animals. A simple example is using fertilizer around crops. Fertilizer gives the plant more nutrients which helps cells grow better. This changes the normal growth of the plant—growth that would occur without the fertilizer. Three biotechnological examples using animals are given here.

Milk Hormones. A product named bovine somatotropin (bST) can be given to cows to increase the amount of milk they produce. This hormone naturally occurs in cows and is present in all milk. Injecting a cow with additional bST causes her system to become more productive. Of course, more feed must be given if more milk is to be produced. Research has shown that cows will produce 25% more milk when given bST.

Meat Hormones. The pituitary glands of hogs produce porcine somatotropin (pST). This hormone regulates the growth of pigs. Injections of pST cause hogs to produce more muscle cells. This results in the valuable cuts of pork being larger than they would be otherwise. pST also decreases the amount of body fat a hog has. Larger pork chops and hams give the producer a higher price for the animal when it is sold. Methods of mixing pST with feed haven't been developed. Daily injections are required.

Growth Implants. Implants are small pellets that are placed under the skin of animals. They are used primarily to provide a substance to the animal that will promote growth. Most implants go under the skin behind one of the ears. The U.S. Food and Drug Administration has approved several items for this use.

The materials used in implants are intended to make more efficient use of feed. Some of the implants are hormones; others are derived from molds that are grown on corn products. Most implants provide slow release of the materials over a period of several weeks.

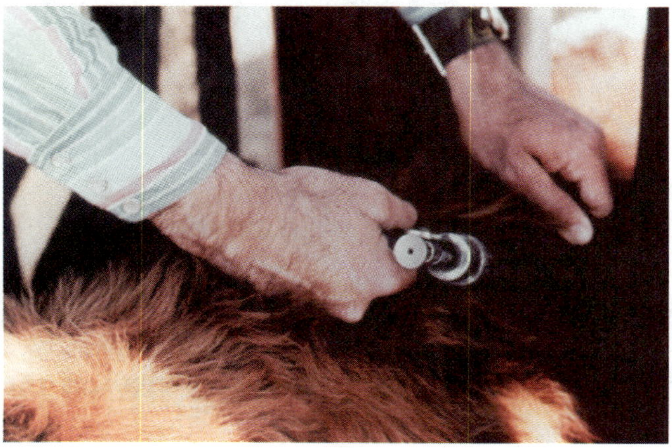

15-10. Implanting a beef calf with a growth enhancer. (Courtesy, Marco Nicovich, Mississippi Cooperative Extension Service)

Aquaculture

Freshwater aquaculture has grown rapidly in North America. Many new approaches for growing fish have been developed.

Spawning. Spawning is the process of female fish producing eggs that are fertilized with sperm from male fish. Aquafarmers are increasingly controlling the process of spawning. This involves a lot of "fish science," known as ichthyology.

Spawning naturally occurs at certain times of the year, usually in the spring. Aquafarmers are able to induce spawning at other times by injecting the female fish with gonadotropin. Spawns (egg masses) may contain thousands of eggs. The size of the eggs depends on the size of the fish. These eggs must be fertilized naturally by a male or artificially with sperm taken from male fish. Fish usually need to have a spawning place provided for them.

15-11. Two large catfish spawns (egg masses) are ready to be moved to a hatchery. One female broodfish produced the eggs in each spawn. One spawn may have 5,000 to 10,000 eggs or more. (Courtesy, *Progressive Farmer Magazine*, Birmingham, Alabama)

Hatching. Hatching involves getting baby fish from fertilized eggs. Eggs may be naturally hatched or moved to a hatchery where they are artificially hatched. In a hatchery, conditions must be used that duplicate natural hatching. If not, the eggs won't hatch. This includes regulating water movement, temperature, and oxygen level. Fish eggs usually hatch in a week or less.

Nursery Management. Caring for fry (tiny newly-hatched fish) requires a lot of science knowledge. The fry must be fed. Diseases must be prevented. Oxygen must be maintained in the water. The water temperature must be right and cleansed of waste materials. This is done in the hatchery and later in nursery ponds or tanks.

Sex Changes. Some species of fish don't grow well when the males and females are left together. In order to prevent problems, researchers

15-12. Careful management of a hatchery and nursery includes regular cleaning of the hatching and growing trougs. (Courtesy, *Progressive Farmer Magazine*, Birmingham, Alabama)

have developed ways of changing the sex of newly-hatched fry. This is particularly important with fish that mature sexually when still very small, such as tilapia where all-male populations grow best.

Ultrasonics

Ultrasonics has many potential uses in agriscience. It is a technology that is now used for assessing product quality and fetal development in pregnant females. Many additional uses are likely in the future.

Ultrasonics involves using high-frequency sound waves that people can't hear. Most ultrasound is produced by ***ultrasonic transducers***. These devices convert electric energy into ultrasonic waves. The waves are sent out and received back. Instruments show the presence of objects and may be used to indicate movement.

Product Quality. Ultrasound can be used to assess the amount of fat on cattle and other animals. Though not widely used on market animals,

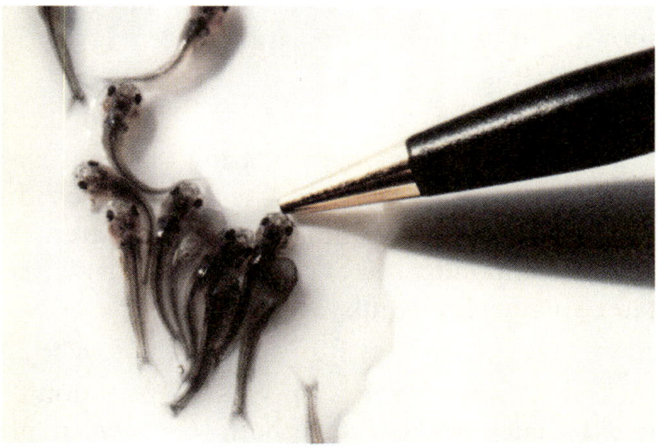

15-13. Tiny fry (recently hatched fish) must receive proper care to grow. Note that these fry are barely larger than the tip of a writing pen. The developing internal organs and skeletons are visible through the skin of the fry. (Courtesy, *Progressive Farmer Magazine*, Birmingham, Alabama)

it provides the opportunity to assess the thickness of fat and muscle. It is more widely used in selecting breeding stock because they will transmit traits of fat and lean to their offspring. It is also used on animals in livestock shows to assess their quality.

Pregnancy Assessment. Ultrasound can be used to determine the development of an embryo and fetus in a pregnant female. It has been used on cattle and other more-valuable animals, such as registered breeding stock. The ultrasound can be used to develop an image of the unborn animal on a screen similar to a television.

15-14. Ultrasound procedures are being used on this beef animal. (Courtesy, American Polled Hereford Association, Kansas City)

Locate Fish. An adaptation of ultrasonics is used to locate objects under water, such as wild fish in the ocean. The equipment is known as a sonar device. It sends out sound waves through the water. The presence of fish will result in a sound wave being sent back, similar to an echo. Sonar will indicate the depth of the fish in the water.

Other Uses. Other uses of ultrasonics include: testing pipe, concrete, and other materials for defects; cleaning delicate instruments; and detecting and treating diseases.

Marketing Strategies

Food products are produced in different shapes and forms to encourage people to buy them. These forms don't necessarily have more nutrients; they are just different in shape or other traits.

Miniature vegetables have been developed to get people's attention. Examples include "baby" corn and carrots.

The shapes of some vegetables have been changed. One example is the round carrot. These carrots can be sold for a higher price than the traditional long carrots. However, the nutritional value of the round and long carrots is about the same.

15-15. Animal nutritionists study the nutrient needs of growing animals. (Courtesy, Agricultural Research Service, USDA)

Animal Nutrition and Feeds

Many approaches have been used to improve the growth of animals, especially ruminants. These have included treating feedstuffs before they are fed to cattle as well as intervening in the digestive process.

Ammoniating Hay. Low quality grass hay may be treated to make it more nutritious. The hay is usually treated with anhydrous ammonia in specially-designed ammoniation chambers. Research at Texas A&M University and Mississippi State University using ammonia on Bermuda grass hay has shown that the protein content in the hay is increased up to 3.5%. Further, the digestibility of protein in the hay increases up to 14%.

This biotechnology process increases the feed value of low quality feedstuffs.

15-16. Chamber for ammoniating large round bales of hay.

15-17. The steer fitted with a cannula is being prepared for digestibility tests. The steer usually grazes in the pasture with other beef animals. It is in the squeeze only for a few moments for research.

Digestibility Testing. Digestibility studies may involve direct intervention in the digestive systems of cattle.

One procedure involves creating an opening from the outside into a part of the digestive system. Surgical procedures are used to make a fistula, which is an opening in the hollow stomach to the outside. A cannula (tube) is placed in the fistula. The cannula is made of rubber or glass and has a cover that can be used to close the opening. The cannula usually has a diameter of three to four inches.

Feed can be taken from or put into the digestive system through the cannula. Feed is sometimes placed in the system using a mesh cloth bag attached to a string that is fastened to the cover of the cannula. This allows easy retrieval of the partially-digested feed sample.

The fistula is made by a veterinarian using careful surgical procedures. The animal experiences no discomfort from the cannula when it is properly placed.

15-18. The cattle in this feedlot have been fitted with sensors. The sensors are on soft plastic chains around the animals' necks. They allow the animal to eat at only one place so that the amount of feed can be studied in relation to rate of growth.

Controlled Feeding. Improving nutrition and feeding involves careful study of animal feed consumption and growth. Cattle are sometimes fitted with sensors that allow the animal to eat only feed prepared for it. The kind and amount of feed can be varied and compared to the rate of growth of the animal. The animal can get feed only at the feeding station the sensor will allow to open.

Predicting the Future

Biotechnology methods can be used to predict life in the future. Scientists indicate that global warming is underway and that the amount of carbon dioxide is increasing. How will a warmer earth with more carbon dioxide affect life?

Plant Growth Chambers. ***Growth chambers*** can be used to study the effects of changes in the environment on plants. These are structures that control and monitor the environment. Computers and other instruments are used to assure that the environment is constantly maintained. Regular measurements are made of the plants. With large plants, such as trees, only a portion of the plant may be in the chamber. This allows comparison of growth in the regulated environment with growth in the outside environment.

15-19. A simple growth chamber can be made out of a large plastic bucket. This "grow bucket" has a light as well as containers for growing media.

Computer Simulations. ***Computer simulation*** is being used with various crops. The simulations gather information to answer questions about growing the crop. In some cases, ***cropping models*** have been developed. These allow producers to get information on the weather, soil moisture, and stage of growth of the crop and make decisions about what to do with the crop. Should an insecticide be used? Should irrigation be used? These models help producers choose the right practices at the right times.

A computer simulation on erosion can be used by agriscientists in preventing soil loss. Known as SLOSS (a short form of "Soil LOSS"), the model helps estimate average soil loss under varying conditions. The conditions include the kind of crop grown on the land, climate, and practices used to grow the crop.

15-20. A growth chamber that completely controls the environment of the growing plants is show here. All processes are controlled by computers. This chamber can regulate temperature, carbon dioxide, and other environmental factors.

Other Examples

Almost every practice used to produce plants and animals has some organismic biotechnological implication. Of course, it must be science-based and not modify the genetics of the organism.

Additional examples include: (1) using growth regulators to get plants to grow or to inhibit growth, such as using gibberellic acid on plant cuttings to get them to root; (2) forcing plants so that they grow at times of the year when they wouldn't normally grow, such as tomatoes in a greenhouse; and (3) using wastes as feed to reduce disposal problems and provide economical feed. (An example is feeding chicken litter to cattle.)

15-21. This growth chamber includes only part of a tree. In the research underway here, the effects of global warming are being measured. The growth chamber is fitted with a computer-controlled heater that keeps the air in the chamber 2°F warmer than the environment.

MOLECULAR BIOTECHNOLOGY: IMPORTANCE OF GENETICS

Genetics is concerned with the traits parents pass to their offspring. How these traits are transferred is known as heredity. The outcomes of genetics and heredity are obvious in daily life.

Plants or animals have offspring that have many of the same characteristics as their parents. Hogs always have pigs and cattle always have calves. The two species are different, yet each species is much alike. Within each species, each animal is different. This difference is known as ***variability***.

CELLS

Plants and animals are made of living cells. The cells are often specialized into tissues and organs. They grow and divide as the organism grows and reproduces.

Cells are made of various materials that direct life processes. Changing plants and animals involves changing the processes that occur. These changes can be simple or complex. A simple example is giving a plant fertilizer so that it has more nutrients to carry out life processes. A more complex example is changing the structure and arrangement of the substances in the cells.

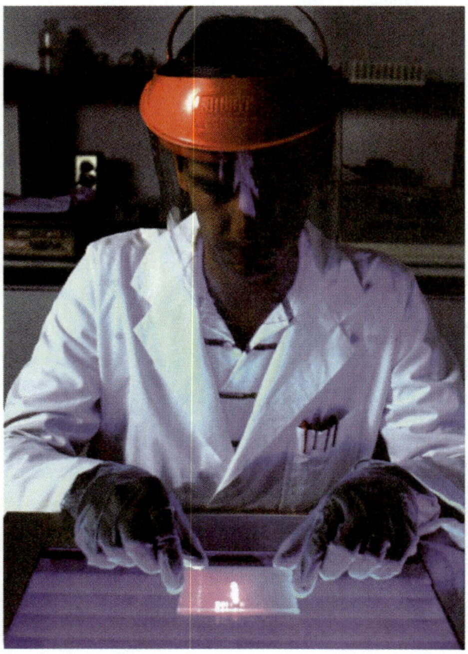

15-22. A plant technician using stain and ultraviolet light to study soybean dwarf viruses. (Courtesy, Agricultural Research Service, USDA)

GENOMES

The heredity material in a cell is a ***genome***. The genome is made of chromosomes. All of the chromosomes in a cell form a chromosomal set. Chromosomes serve as the genetic link between parents and their offspring.

New body cells are formed by mitosis (cell division). Every new body cell formed in an organism has an identical genome. All of the genomes are copies of each other.

When organisms sexually reproduce, the genome of the offspring is a combination of that of its parents. All of the cells formed in the offspring have identical genomes.

Chromosomes

Chromosomes are thread-like parts inside a cell nucleus that contain genetic material and protein. Chromosomes are found in pairs. The genetic material is made of DNA (deoxyribonucleic acid). Advanced biotechnology often involves studying and changing DNA.

Another nucleic acid in the cell is RNA (ribonucleic acid). Similar to DNA, RNA takes information from the genes for producing proteins. Several kinds of RNA are found, based on their roles in the cell.

The cells of organisms usually have several chromosome pairs. Corn cells have 10 pair, mice cells have 20 pair, and humans have 23 pair of chromosomes in each cell. Cattle and horses have 30 pair of chromosomes, swine have 20 pair, and sheep have 27 pair. The number of chromosomes determines what organism is produced.

A cell with all of the chromosomes characteristic of a species is a ***diploid cell***. The chromosomes are arranged in pairs in the nucleus. These identical pairs are known as ***homologous chromosomes***. The pairs are identical in nearly all species except in some species the sex chromosome pair is different between males and females.

Genes

Genes are the parts or segments of chromosomes that contain the heredity traits of organisms. Each chromosome has many genes. Since chromosomes are in pairs, genes are also in pairs. The different forms of genes are known as ***alleles***.

The alleles may be different or the same. When they are the same, TT or tt, the organism is said to be ***homozygous***. When they are different, the organism is said to be ***heterozygous*** (Tt).

Genotype. The genetic makeup of an organism is known as ***genotype***. Its appearance is known as ***phenotype***. Offspring, as products of their parents, have both genotype and phenotype heredity. If the parents have similar genes, genotype and phenotype of the offspring are like the parents. There is no variation; therefore, the offspring are homozygous. If the parents' genes are different, the offspring could have genotypes and phenotypes different from the parents, and be heterozygous (different gene types). Systems of breeding, as presented in Chapter 13, make use of similar and different gene types. Outcrossing is used to get heterozygous offspring. Homozygous offspring have like genes. Inbreeding is used to develop homozygosity.

Dominant and Recessive Factors. Some traits are dominant; others are recessive. ***Dominant*** traits are caused by alleles that cover up or mask the ***recessive*** traits of other alleles. Stated another way, recessive traits are covered up by dominant alleles. Recessive traits can be expressed if both parents have the recessive alleles. This happens when solid black animals that have alleles of white faces are mated. Their offspring could have a white face, but not always.

A good example illustrating dominant and recessive genes is mating horned and polled cattle. When a purebred polled animal is mated with a horned cow, the offspring will be polled. Some of the offspring may have scurs or small nubs of horns. Not all polled cattle are genetically purebred polled. Some of them carry a recessive gene for horns. When mated with an animal with horns, the offspring could develop horns.

Another example of recessive traits is the Red Angus. Angus cattle have been black for hundreds of years. Yet, a red-colored Angus is occasionally born. Records show that about one out of every 400 calves is born red. Mating two red animals will always produce red offspring because it is a recessive trait. This has helped the Red Angus breed develop.

Dwarfism is sometimes a problem in cattle. Producers of grade cattle often control for it by outcrossing. Purebred producers should not sell cattle with recessive dwarfism genes for breeding stock. They should be used for beef.

Double Helix

DNA forms a structure known as a ***double helix***. Two strands of DNA are twisted, forming a spiral structure. The spiral strands are made of subunits known as nucleotides. They are joined together by bonds between pairs of nitrogen base materials. Some people refer to the double helix as a twisted ladder. The side supports are made of sugar and phosphates.

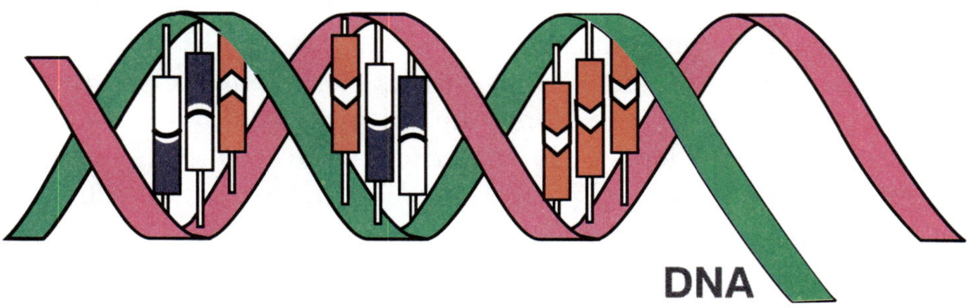

15-23. Double helix showing DNA.

The rungs are made of nitrogen base pairs. Because of the "rung" structure, segments of DNA can be cut out and new ones inserted. (This is described later in the chapter.)

Genetic Code

The information in an organism that causes it to function is its genetic code. The code makes amino acids that, in turn, make proteins. An organism is what it is because of its genetic code. Cattle have one genetic code, horses another, and corn plants still another. People are what they are because of their genetic code. If the code was different, people would be something else!

Mutation

A ***mutation*** is a change that naturally occurs in the genetic material of an organism. The genes and chromosomes are changed. They may appear suddenly and without any advance signs. A mutation often has no cause other than the fact that things just happened differently in that organism.

The mutations may be genetic and physical differences (genotype and phenotype).

Mutations can also be induced through radiation and other means. The things that cause mutations are known as mutagens. Birth defects due to exposure to radiation are examples. Of course, not all mutations are viewed as defects.

Mutations sometimes produce new and better plants and animals. Organisms can be more desirable because of a mutation. It is likely that the first polled Hereford resulted from a mutation. Larger fruits and nuts may also be mutants. The pecan is a common example. Just one tree with extra-large pecans can occur by chance or by mutation. Using vegetative propagation, this one tree can become many trees. The many trees produce valuable extra-large pecans.

GROWTH AND REPRODUCTION CELL PROCESSES

Biotechnology involves getting organisms to grow and reproduce. Growth is when organisms either increase in size or yield products that can be used. Reproduction is when organisms give rise to new organisms.

Growth and reproduction involve cell division. Two different processes are involved: mitosis and meiosis. Cell division for growth and repair is mitosis. Meiosis is cell division for reproduction.

MITOSIS

Mitosis is a sequential process of cell division. It is the phase of the cell cycle when one cell becomes two cells. The process is repeated over and over during the life of an organism.

Mitosis occurs in four steps:

- **Prophase**—Prophase involves the development of chromosomes. Replication of DNA occurs at the end of Interphase. This step follows a time of rest for the cell, although some scientists indicate that the cell never rests.
- **Metaphase**—During the metaphase, the chromosomes move to the middle of the cell along an invisible line called the equator.

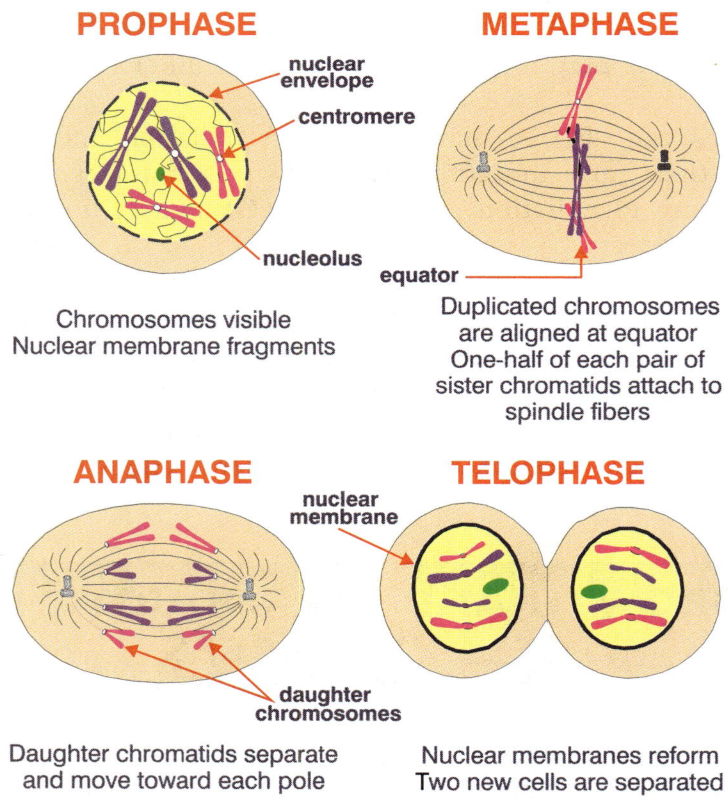

15-24. Mitosis involves one cell becoming two cells.

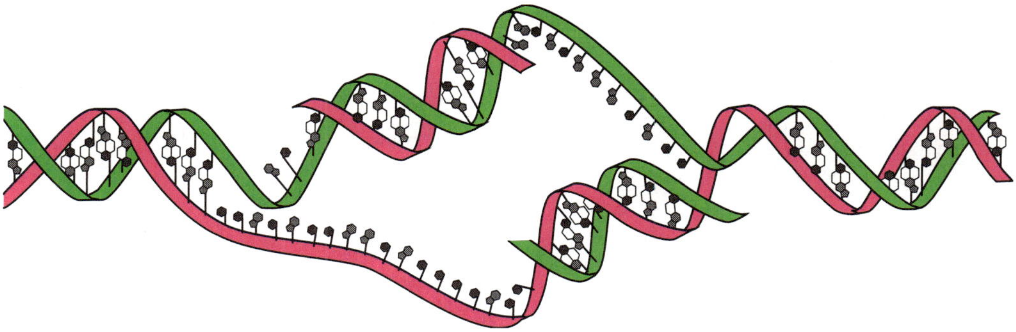

15-25. Making another DNA strand involves a DNA strand splitting apart and one old strand producing a new strand. This process is known as replication.

- **Anaphase**—In this step, the chromosomes separate and move to opposite sides of the cell pulled by shortening spindle fibers.
- **Telophase**—In telophase, the chromosomes undergo additional maturity, the nucleus reforms, and a membrane appears between each mass of chromosomes and divides the cytoplasm. The two separate cells can repeat the process.

MEIOSIS

Meiosis is the division of cells in the reproductive process. The cells produced are known as gametes. Male gametes are known as spermatozoa or microgametes. Female gametes are known as ova or macrogametes. Spermatozoa (sperm) are produced in the testicles of males and ova are produced in the ovaries of females.

The cells produced through meiosis are known as ***haploid cells***. This is because they have single chromosomes rather than homologous pairs. When a sperm unites with an ovum, a new individual is formed having diploid cells.

Meiosis I:

- Prophase I—DNA is replicated at the end of interphase before prophase begins. The formation of thread-like spindle fibers that will attach to the chromosomes begins. The homologous chromosome pairs (the ones that are alike) are located together. Synapse or crossing-over of the chromosomes can occur in this phase which can give a different combination of genes. The nuclear membrane and nucleolus will be dismantled by the end of the phase.

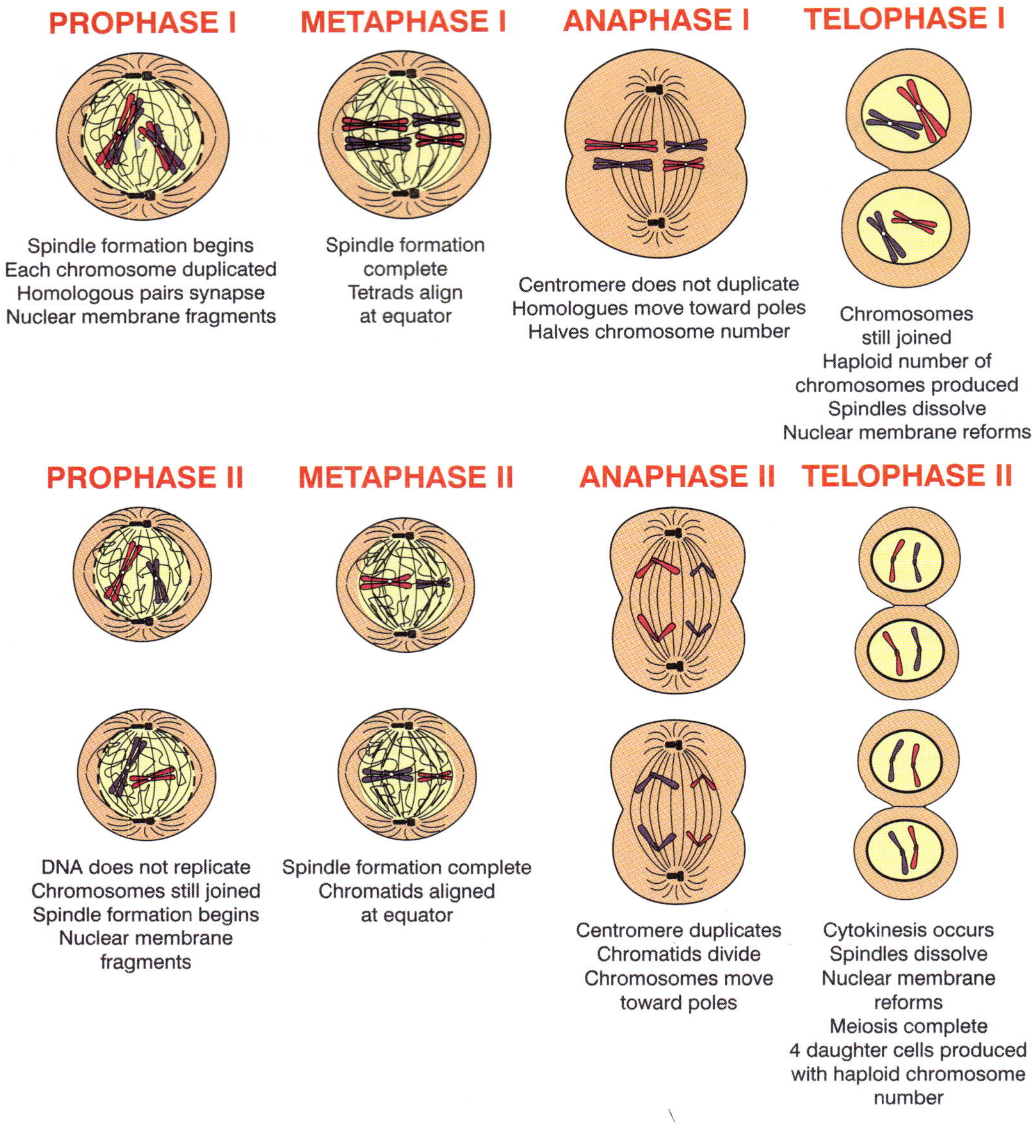

15-26. Divisions and phases of meiosis.

- Metaphase I—The tetrads (group of four chromosomes that are alike) align at the equator with spindle fibers attached from the centromere to the poles. The centromere is the point of attachment between the two chromatids.
- Anaphase I—The centromere does not duplicate, the paired chromosomes are pulled to the poles by the shortening spindle fibers. This halves the chromosome number.

- Telophase I—Chromosomes are still joined with half the original chromosome number (haploid number). The nucleolus and nuclear membrane reform. Two daughter cells are produced with one-half the original chromosome number.

Meiosis II:

- Prophase II—DNA does not replicate before the phase begins. Spindle formation begins.
- Metaphase II—The chromatids align at the equator of the cell with spindle fibers attached from the centromere to the poles.
- Anaphase II—Centromere duplicates, which divides the chromosomes with each half pulled by the spindle fibers to the poles.
- Telophase II—A cell membrane forms across between the two developing nuclei. The nucleolus and nuclear membrane reform. Cytokinesis occurs. Four daughter cells are produced containing the haploid number of chromosomes.
- The spermatozoa and ova are the cells of sexual reproduction. When an ova (egg) with the haploid number of chromosomes and a sperm with the haploid number join, a new individual with the diploid number of chromosomes for that species is produced.

MOLECULAR BIOTECHNOLOGY: GENETIC ENGINEERING

Genetic engineering is a molecular form of biotechnology. The genetic information in the cell is changed or used to make a product. Sections of DNA molecules may be cut out and new sections inserted.

Chromosomes contain thousands of genes. Finding the right gene to remove or add requires considerable searching. A method known as ***gene expression*** is used to isolate a gene that does certain things. A similar process involves removing a gene to see what the organism does afterward. If it has a certain quality afterward, the correct gene has been found.

Genetic engineering is also known as gene manipulation. Moving a gene from one organism to another is known as gene transfer.

RECOMBINANT DNA TECHNOLOGY

Recombinant DNA is taking a tiny amount of DNA from one chromosome and moving it to another. The DNA of two different organisms is combined. This involves removing part of the DNA and adding DNA from another cell. Scientists find moving the DNA a challenge because it is so small.

All cells have membranes. In addition, plant cells have walls which give them rigidity. The membrane and wall are intended to protect the cell. Invading the nucleus of the cell with new DNA is not easy. DNA is a rather large part of a cell. Several approaches have been used, including mixing the DNA in a solution and microinjection (injecting with a very small instrument). With both methods, many cells die. Many of those that survive may not have the new DNA in their nucleus. Success is often slow.

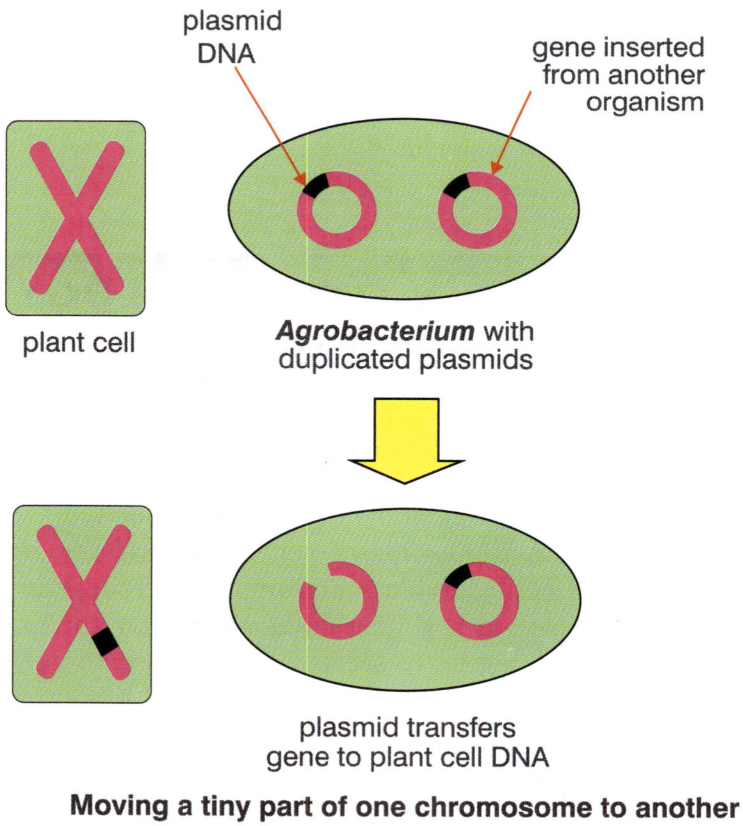

15-27. Recombinant DNA

Vectors

Vectors carry new DNA into a cell. Yeast cells and the bacterium *Escherichia coli* are often used as vectors. The *E. coli* bacteria are most often used as vectors. They are commonly found in the intestinal tract of humans and other vertebrates.

Vectors are cells that are easy to work with. They lend themselves to being carriers.

Breaking and Joining

Ligation involves uniting or attaching the two DNA fragments. It includes cutting the DNA in the cell that is to be transformed and joining it with the new DNA. Ligation is often done ***in vitro***, which means in glass (usually a test tube).

Cells that receive new material are known as ***transformed cells***. The process of changing the cell is known as ***transformation***. Cells that provide the new material are known as ***donor cells***.

Getting the new DNA to join with the DNA that has been cut is known as ***cleavage***. Several laboratory procedures have been used in this process. Refinements of these procedures are needed. Various substances are being used to try to more efficiently unite strands of DNA.

15-28. Microprojectile units are used to send tiny particles coated with DNA into cells during genetic engineering. These units are also known as particle guns. (Courtesy, Ann Stevens, BIO-RAD, Hercules, California)

USING GENETIC ENGINEERING IN AGRISCIENCE

Genetic engineering is a new technology that is thought to have a great deal of potential for plants and animals. Several efforts are well under way that could change how food and fiber are produced. A few examples are given here.

Herbicide-Resistant Plants

Various herbicides are used to kill weeds, especially in fields of crops and greenhouses. In some cases, the weeds and

15-29. Cotton produced by an experimental herbicide-resistant cotton plant. (Courtesy, Monsanto Company, St. Louis, Missouri)

the crop plants are much alike. Even when selective herbicides are used, both the weed and crop plants are killed.

Recombinant DNA can be used to move the gene from a plant that is resistant to the herbicide into the crop plant that isn't. The resulting crop plant would not be killed by the herbicide. Research by the Monsanto Company has produced plants that are herbicide-resistant. These are now being field tested for possible future use.

Insect-Resistant Plants

Many different kinds of poisons are used to control the insects that attack plants. Genetic engineering is being used to develop plants with a natural resistance to insects. This will result in plants that won't have to be sprayed with insecticide.

Considerable effort is being made with cotton, since about 40% of all pesticides used each year in the United States are used on cotton. Similar procedures are being used on food crops, including potatoes and tomatoes. No pesticides would need to be used to grow them!

Bollgard® is a patented gene bred into cotton to help it resist damage by the bollworm. The gene is from *Bacillus thuringinesis* (bacteria) and is commonly called the Bt gene. Transgenic Bt cotton varieties will completely control the worm. This saves having to apply costly pesticides and protects the environment from the residues. Scientists are unsure as to how long the Bt cotton will control the worms. Pests often develop resistance to methods used to control them.

Strong interest in insect resistant plants is due to insects that are resistant to insecticides. Over 500 insects have developed resistance to some kind of insecticide, including roaches, mosquitoes, and flies. Biotechnology will help overcome these insect problems.

Disease-Resistant Plants

Plants that can resist disease should be more productive and less costly to grow. Further, the quality of the produce should be improved. No chemicals to control diseases will need to be sprayed on the crops. Major efforts have been made to control viral diseases in tomatoes and tobacco.

Transgenic Animals

Research is under way using a process known as embryo microinjection, to introduce growth hormone genes into livestock. The animals (known as ***transgenic animals*** because their genes are changed) are injected when they are embryos. This process can also be used to enhance the quality of lean meat to body fat. Animals with more lean meat bring higher prices.

Frost Protection

Plant crops lost to frost cost producers millions of dollars each year. Plants with resistance to frost would prevent these losses.

Dew naturally forms on plants and other objects at night. It happens as moisture from the atmosphere collects on plants when the temperature cools at the end of the day. If the weather is cold, the dew freezes to form frost. Oranges, tomatoes, strawberries, lettuce, and many other crops can be killed or damaged by frost.

15-30. Bags of genetically improved potatoes at an Idaho Farm. (Courtesy, Ashton Hi-Tech Seed Company, Ashton, Idaho)

Some bacteria have been found useful in protecting crops from frost damage. The bacteria have a protein that prevents the formation of ice crystals. Agriscientists are experimenting with the bacteria by putting it on crops when temperatures below freezing are expected.

Development is well under way on a tomato that will resist frost. The work has involved moving a gene from an arctic flounder (kind of fish) to a tomato. The gene produces a substance that acts like antifreeze.

Longer Storage Life

Many foods in fresh form can be stored only a few days before they begin to go bad (rot). Efforts in molecular biotechnology have focused on improving the shelf-life of products. A good example is the tomato.

Research by the Calgene Company of Davis, California, has produced a tomato that has a longer storage life. The tomato keeps its flavor longer and doesn't begin to spoil as soon. Known as MacGregor's tomatoes, it is the first widely available food developed through molecular biotechnology.

New Animal Products

Several new ways of making medical products for humans using animals have been developed. Work in this area is sometimes called ***immunochemistry***. The major goal is to develop immunizing agents for human health. Some of the substances being developed are proteins known as ***interferons***. Some interferons may be used as anticancer drugs. The use of these processes could result in new kinds of farms—immunochemistry animal farms! What the animals produce might be sold much as milk is today.

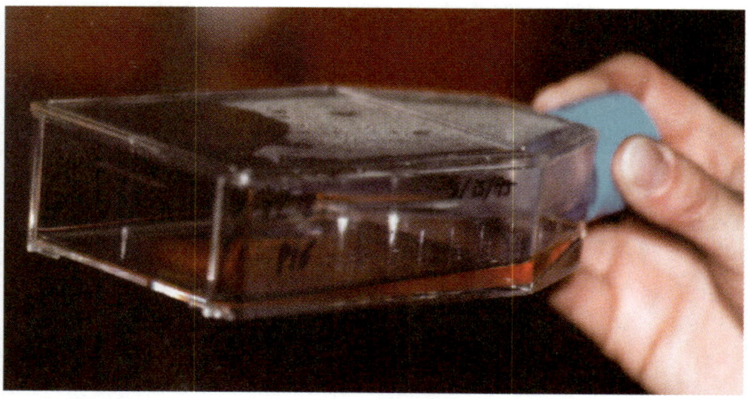

15-31. Enzymes extracted from bovine pituitary glands are being grown in a laboratory. The research is to develop a product to help reduce human pain from disease.

A LONG WAY TO GO

Even though genetic engineering has received widespread attention, practical uses are still limited. More research and field testing are needed to make usable forms available. There is a lot of work to do!

REVIEWING

MAIN IDEAS

Biotechnology is using biological processes to change plants and animals or the products they produce. At the organismic level, it deals with helping organisms grow better and produce more. At the molecular level, it deals with changing the genetics of plants or animals.

Many organismic biotechnological practices are used in producing plants and animals. These include cloning, greater fertility, more production, and predicting the future.

Molecular biotechnology changes genetic information in cells. A thorough knowledge of genetics, genomes, chromosomes, DNA, and genes is needed.

Genetic engineering is molecular biotechnology. It uses recombinant DNA procedures. With gene splicing, a gene can be removed and another inserted in the DNA of an organism.

Most forms of genetic engineering are still in developmental stages. Widespread use of a few products is near.

People need good information about biotechnology. They need to be able to weigh the advantages and disadvantages of the processes involved.

QUESTIONS

Answer the following questions using complete sentences and correct spelling.

1. What is biotechnology?
2. What issues are related to biotechnology?
3. Name and distinguish between the two major areas of biotechnology.
4. What are four areas where organismic biotechnology is making contributions to agriscience?

5. What is genetics? Why is it important in biotechnology?
6. What is a genome? What are the contents of a genome?
7. Distinguish between genotype and phenotype.
8. What are dominant and recessive factors?
9. Distinguish between homozygosity and heterozygosity.
10. What are mutations? When are they useful?
11. What is genetic engineering?
12. What is recombinant DNA? Relate the process of gene splicing to vectors and breaking and joining.
13. What areas of agriscience have developments under way using genetic engineering?

EXPLORING

1. Clone a plant using tissue culture. Have your agriscience or biology teacher help you. Possible plants include carnations, carrots, and African violets. Use references on tissue culturing. Several weeks will be required.

2. Write a short paper on the use of embryo splitting in beef cattle. Use magazines, books and authorities on the subject to get information.

3. Make a survey of students in your school or class to get their opinions about the use of biotechnology. Your class may wish to work as a group on this project. Carefully develop the questions you want to ask. Collect and compile the answers.

4. Develop a family tree for yourself. Note traits that have been inherited that may be dominant or recessive. How do your traits differ from other members of your family? (Things to look for are eye color, height, weight, hair color, and overall bone structure.)

16

APPLYING PRINCIPLES OF EARTH SCIENCE

The earth has many living and nonliving things. All of these must share the earth's space. They must exist together.

Plants and animals depend on the earth for their life-supporting nutrients. They use the resources that are available to support life. First, are plants. They take nutrients and through photosynthesis make sugar and other substances. Secondly, animals eat the plants for their food. Upon completing their life cycles, both return to the earth as elements. New plants and animals grow and complete the cycle again.

Space on earth is limited, and so are its resources. Yes, the earth is big—25,000 miles (40,074 km) around at the equator! But, there are many people who need food, fiber, and shelter. This requires making wise use of the earth's resources.

16-1. Information about the earth includes the weather. Here, equipment to record weather information is being monitored. (Courtesy, Agricultural Research Service, USDA)

OBJECTIVES

This chapter presents important information about earth science as related to producing plants and animals. It has the following objectives:

1. Describe major features and resources of the earth.
2. Explain changes that occur in the earth.
3. Describe atmosphere and its importance in agriscience.
4. Explain the importance of climate in agriscience.
5. Identify and describe major factors in weather.
6. Explain succession and its relationship to the earth.

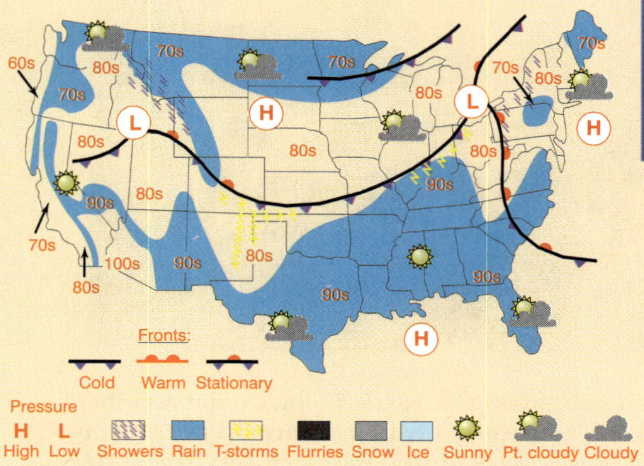

TERMS

air pressure
atmosphere
barometer
biotic community
chemical weathering
climate
climate zone
cloud
continental crust
core
crust
dew
galaxy
humidity
mantle
mechanical weathering
polar climate zone
precipitation
primary succession
relative humidity
revolution
rotation
secondary succession
solar system
stratosphere
succession
temperate climate zone
thermometer
tropical climate zone
troposphere
universe
weather
weathering
wind

THE EARTH'S RESOURCES

The earth is a part of the universe. The ***universe*** is best described as all that exists—both known and unknown. It is made of galaxies, which are clusters of dust, gases, and stars.

The earth is found in the ***galaxy*** known as the Milky Way. This galaxy has many objects—too many to count or see. A galaxy is further divided into solar systems. A ***solar system*** is the group of bodies that are held together or served by a sun. One sun, nine planets, over 40 moons and more than a 100 billion stars make up the solar system that the earth is in.

The earth is unique among all planets in its solar system. It is the only known planet that supports life. Life as we know it requires water and oxygen. The earth has liquid water and an abundance of oxygen. All of the earth's energy is from the sun. The energy can be stored, as in oil, or is used daily for life processes, such as photosynthesis.

ITS STRUCTURE

The earth can be viewed as a large, almost round object having three major parts: a surface, an interior, and an atmosphere. Life on earth uses resources from all three.

The earth's structure can be compared to an orange fruit. The orange has: a peel—which can be compared to the earth's surface; pulp, juice, and seed—which compares to the earth's interior; and surroundings—which compare to the earth's atmosphere.

The Earth's Surface

The surface of the earth is made of mountains, valleys, plains, streams, lakes and oceans. Combined, these are known as the earth's ***crust***.

The crust is made of minerals, known as inorganic compounds, and other materials. Inorganic materials are formed of many different elements and make compounds that don't always contain carbon. Rocks are common examples of inorganic materials.

Kinds of Crust. Some of the crust is above water; other parts of the crust are covered by oceans. The crust that is land is known as the

16-2. The earth's crust supports a wide range of plant and animal life used to meet the needs of humans. (Courtesy, U.S. Department of Agriculture)

continental crust. This is the area where crops are grown, animals are raised, and humans live.

The crust beneath the ocean is known as the oceanic crust. Of course, the waters of the oceans support plant and animal life, such as fish and seaweed. Some grow attached to the oceanic crust.

Continental Crust. The ***continental crust*** is made of inorganic materials as well as materials from plants and animals. Those from plants and animals are known as organic materials. Carbon is one of the elements present in all organic materials, such as coal. Many organic materials are found in the soil. (The soil is discussed in more detail in Chapter 17.)

More than 2,000 different kinds of minerals are found in the crust. Fewer than 20 minerals make up most of the crust. Minerals are typically thought of as rocks or rock-like substances. The continental crust varies from 12.4 to 43.5 miles (20 to 70 km) thick. (The crust is thinner under the deep oceans.)

The surface of the earth is important because it supports life. Soil is the part of the surface made of both inorganic and organic materials. Plants put roots into the soil and get nutrients from it.

The surface also has water which may be in liquid, solid (frozen water or ice), or gas (vapor or steam) form. Liquid water is found in lakes and streams on the surface, as molecules in the soil or as aquifers deep in the earth. Frozen water is found in higher elevations and in colder climates. Gaseous water is found in the atmosphere. Liquid water is needed by plants and animals for their life processes. In addition, water is the living environment for some plants and animals, known as aquatic life.

The earth's crust is continually changing. One of the processes of change is weathering. The action of water and other forces on minerals changes them into forms that can support life. Rocks become soil particles through weathering.

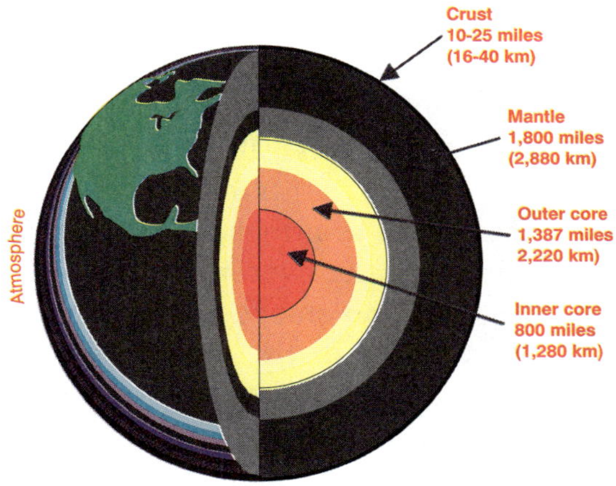

16-3. Structure of the earth's interior.

The Earth's Interior

As a large nearly-round ball, the earth averages 7,908 miles (12,735 km) in diameter. The center has a hot ***core***—7,772°F (4,300°C) and is made of dense materials, such as iron. The materials become less dense near the crust.

Between the core and crust is a layer known as the ***mantle***. This is a hot area, with some molten (melted) rock. The temperatures are as high as 2,030°F (1,110°C).

The Earth's Atmosphere

The air that surrounds the earth is the ***atmosphere***. It is a collection of invisible gases. Water vapor and solid particles (particulate) are in the air. Oxygen and nitrogen are most abundant in the air near the earth. Nitrogen makes up slightly over 78% of the air, with oxygen making up

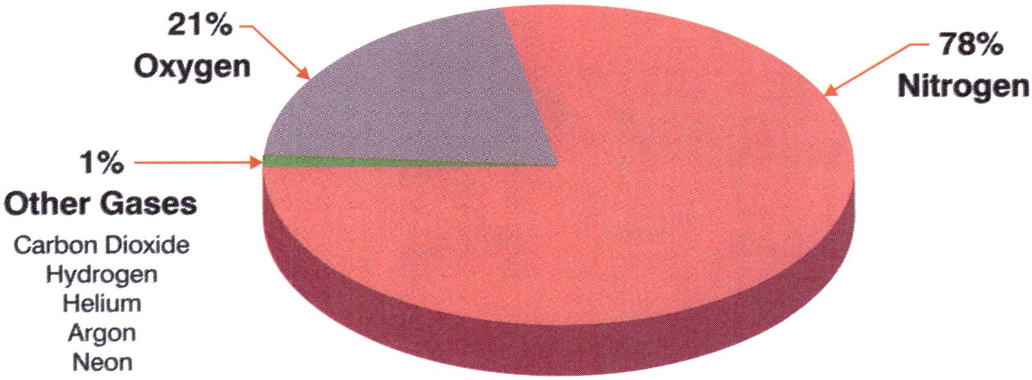

16-4. Gas composition of the air.

nearly 21%. The other gases are present in very small amounts, such as carbon dioxide which makes up approximately .03% of the atmosphere.

The gases are held close by the earth's gravity. They are not found in the atmosphere above 30 miles from the earth.

The amounts of water and solids vary with the surface composition beneath the atmosphere. For example, the amount of water vapor in the air is greater over a tropical forest than over a desert. Solid particles, such as dust and smoke, are known as pollution or air pollution. There is more particulate in the atmosphere above cities and dry cropland.

The atmosphere and the things that happen in it are important in the production of plants and animals. Temperature, wind, and precipitation are results of activity in the atmosphere. Understanding these activities is important in the successful production of plants and animals.

CHANGES IN THE EARTH

The earth moves and changes. It rotates and revolves. These movements cause day and night and the seasons of the year—both important to plants and animals. The structure of the earth also changes as new products are formed.

MOVEMENTS OF THE PLANET EARTH IN THE SOLAR SYSTEM

The earth moves in several ways. It changes its position in the solar system as well as undergoes changes in its composition.

16-5. Movement of the earth can be studied using a globe and map. (The equator is being pointed out on the globe.)

Day and Night—Rotation of the Earth

Rotation is the turning of the earth. One complete rotation (turn) is made every 24 hours—1 day.

Almost everyone is familiar with a globe, or model of the earth. Locations on the globe include the North and South Poles. An invisible axis extends from each of these poles. The earth rotates on this axis. A good way to see how this works is to stick two pencils into opposite sides of an orange and turn the pencils between your fingers. The orange rotates!

Seasons—Revolution of the Earth

Revolution is the movement of the earth in space around the sun. One year (365.24 days) is required for the earth to make one revolution around the sun. Seasons of the year are based on the position of the earth in its revolution.

The earth is divided into two parts by the equator (the middle point around the earth): Northern Hemisphere and Southern Hemisphere. Summer occurs in the hemisphere of the earth that is tilted toward the sun. Seasons are reversed in the hemispheres, meaning that summer in the Northern Hemisphere is winter in the Southern Hemisphere and vice versa. When the sun is directly over the equator, day and night are of equal length, known as equinox.

The spring season begins in the Northern Hemisphere on the day when the North Pole begins to slant toward the sun. This is known as the vernal equinox, usually March 21.

Fall begins in the Northern Hemisphere when it tilts away from the sun and the Southern Hemisphere tilts toward the sun. This is known as the autumnal equinox. It is the beginning of the fall season in the Northern Hemisphere and is usually September 21.

Why Revolutions and Rotations Are Important

Life processes are regulated by the movement of the earth. Light from the sun affects both plants and animals because they are able to carry out maintenance and growth processes using the sun's energy.

Day Length. Plants must have light for photosynthesis. Longer days provide more light for the process to occur in plants. This allows them to grow and store food. As the days get shorter, plants carry out less photosynthesis.

Animals also respond to light. Life processes occur more rapidly in light. Darkness is associated with resting. Producers of poultry are well aware of the importance of providing artificial light to help birds grow faster and lay more eggs.

Seasons. Plant and animal life cycles are related to seasons of the year. Some plants prefer warm weather; others cool weather.

The major growth of warm season plants, such as corn, is in the spring and summer. They stop growing in late fall and winter.

Cool season crops, such as winter wheat, prefer the shorter days of winter. Cool season plants may stop growing and die when the days get longer and the weather becomes warmer. In the case of winter wheat, it produces a seed head (grain), dies and is harvested.

Animal production is based on seasons of the year. Cows are bred to give birth in the late winter and spring. The reproductive cycles of some animals are related to seasons and hours of light. In addition, baby animals aren't stressed by cold weather when born at warmer times of the year. Some young animals, such as pigs, must be kept in warm places if born during cold weather.

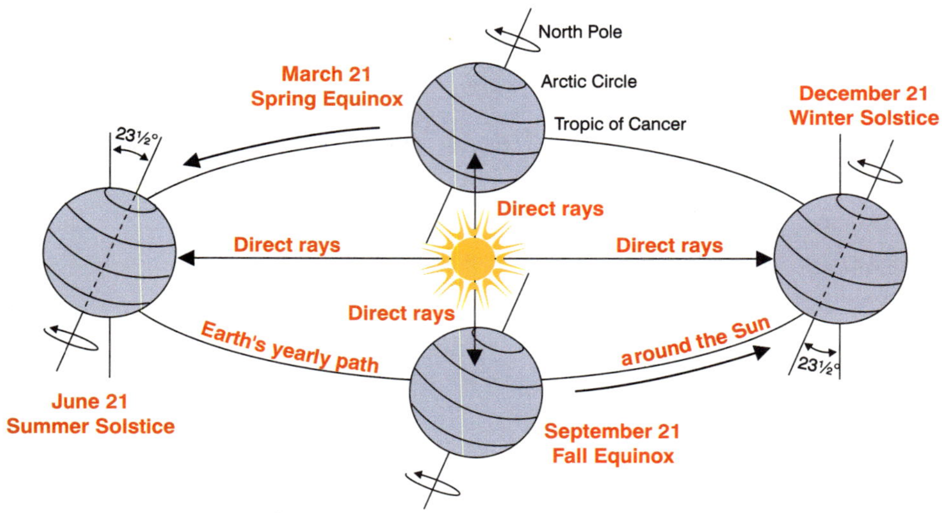

16-6. How revolutions of the earth cause seasons.

MOVEMENTS OF THE EARTH'S INTERIOR

The features of the earth's surface are due to changes in the earth's interior and processes that occur on the surface. Mountains, valleys, and plains were caused by movement.

The internal layers of rock in the earth heat and cool. This causes them to move and change forms. As they warm, they shift positions and sometimes become molten. Molten rock pushing upward through cracks and rock layers may form volcanoes. These forces cause the surface to change. Mountains may be formed as rocks push upward and bend.

Rocks sometimes break under great pressure. This creates fractures (breaks) and faults. A fault is when surfaces continue to move after being broken. The movement may take place only occasionally and create earthquakes. Some movements may not be noticed. Major movements result in earthquakes that shake the earth's surface. Crops can be destroyed and animals killed by major earthquakes.

WEATHERING

Weathering is how the rocks and minerals on the crust of the earth are changed. Rocks break into smaller pieces when exposed to water, air, and changes in temperature. Weathering is a major factor in the formation of soil.

Chemical weathering is when new substances are formed by chemical reactions that occur when rocks and minerals interact with carbon dioxide and oxygen. The presence of water causes reactions of carbon dioxide and oxygen to take place. Oxygen may join with minerals to form oxides, like iron oxide (rust). Water also acts as a solvent to dissolve minerals from the rocks. Acid can be formed by carbon dioxide dissolving in water. The carbonic acid formed causes weathering to occur more rapidly, and other substances in the water may speed decomposition as well.

Mechanical weathering occurs when larger rocks break into smaller rocks. The rocks form smaller and smaller pieces until they are the size of soil particles. This is sometimes called disintegration. Rocks break apart when the water on them freezes. Water freezes in the small cracks and, as it expands, the pressure forces the rocks to crack apart. Rocks also break apart after falling or bumping around in streams of flowing water.

Weathering is a slow process. Thousands of years may be required for rocks to become particles of soil. However, acid rain, sudden temperature changes, and other atmospheric conditions may cause faster weathering.

For instance, heavy rainfall increases the amount of water flowing over rocks and speeds weathering.

Rocks made of different materials weather at different rates. Limestone is a relatively soft rock and weathers more quickly. Granite is a hard rock that weathers slowly. The differences in rocks make them useful for different purposes. Buildings are often made of granite because they will last a long time. Limestone is often crushed and used in soil to improve fertility because it reduces the level of acidity.

HOW THE ATMOSPHERE IS IMPORTANT IN AGRISCIENCE

The atmosphere is the air that surrounds the earth. Many things take place in the atmosphere that effect agriscience.

The atmosphere is made of five layers. The layer closest to the earth is the ***troposphere***. It extends 4.4 to 10 miles (7 to 16 km) out from the earth (generally thought of as up). Above this layer is the ***stratosphere***, which extends out about 30 miles (50 km) from the earth. As the distance from the earth increases, the temperature cools. For example, the temperature in the stratosphere averages about –67°F (–55°C). The troposphere, stratosphere and other layers beyond serve important functions for life on earth. These layers protect and sustain the earth. For example, ozone in the outer part of the stratosphere protects the earth from ultraviolet rays of the sun. Life on earth could not exist without ozone to absorb the ultraviolet rays. Some scientists feel that the ozone layer is being destroyed by gas and chemical pollution.

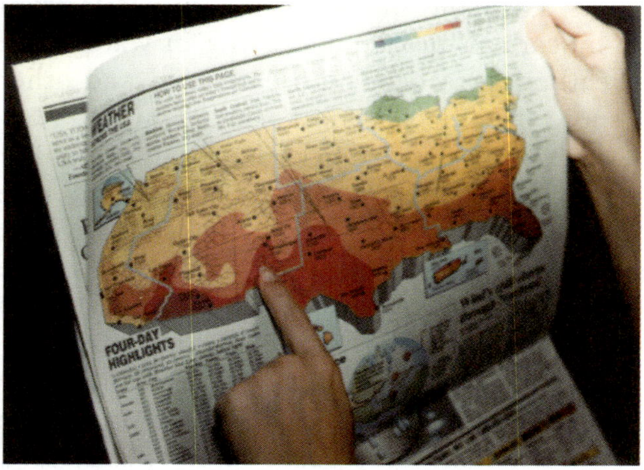

16-7. Daily newspapers carry detailed information about the weather. This map shows weather conditions at different locations. Accompanying articles describe weather trends and conditions.

WEATHER

Weather is the conditions that exists in the atmosphere including temperature, air pressure, wind, moisture, clouds, and precipitation. These conditions affect the growth of plants and animals on the earth.

Temperature

Temperature is a measure of the heat energy of molecules as they move about. The molecules of gas that form the air are continually moving. The faster they move, the higher the temperature. The rays of the sun cause them to move faster by transferring energy to them.

Temperature is measured with a ***thermometer***. Different kinds of thermometers are used. Most have a small sealed bulb at the bottom of a tube that contains mercury or alcohol. As discussed earlier in the book, temperature is measured in degrees Celsius (°C) or Fahrenheit (°F).

The earth and its atmosphere are heated by radiation. The radiation is the energy of the sun that travels as waves. Energy can also be absorbed by the earth and transferred to other objects, including the air. Three methods are used to transfer heat: radiation, conduction and convection. Conduction is the transfer of heat from one object to another when the objects are touching. Convection is the uneven transfer of heat by a carrier, such as when air moves about the earth creating wind currents.

The temperature varies in different places on the earth. The portion of the earth titled toward the sun is warmer. Also, different materials absorb heat differently. Dark colors absorb heat, while light colors reflect heat. In warm climates, the roofs of buildings are often painted white or silver to reflect heat.

16-8. Thermometers. The thermometer on the left is a sealed bulb mercury thermometer. The one on the right has two knds of metal which expand differently with heat. Thermometers should be located in the shade and not attached to buildings or other structures that may cause them to give an inaccurate reading.

16-9. A soil thermometer is used to measure the temperature of soil. A soil thermometer has a 4-inch (10 cm) long shaft that is pushed into the soil. Crops need certain soil temperatures in order to germinate and grow. Most seed germinate best when the soil temperature is 68 to 86°F (20 to 30°C).

Plants and animals are adapted to different temperatures. Some can tolerate low temperatures; others can't. Selecting which plants and animals to produce requires a knowledge of their temperature requirements.

Air Pressure

Air pressure is the force caused by the weight of the air on an object due to the earth's gravitational attraction. The denser the air above a place, the higher the air pressure. The top of a mountain has less air above it and, therefore, a lower air pressure.

Air pressure is measured with a ***barometer***. A barometer has a glass tube that is sealed at one end; the other end rests in mercury. As the air pressure goes up, the mercury rises higher in the tube because the air pressure outside the tube is greater than inside. Because mercury is poisonous, barometers are also made using other materials. Air pressure is measured in pascals (Pa), which are units of the metric system, or pounds per square inch (psi), which are units of the English system. Air pressure in tires is measured in pounds per square inch, such as 30 psi. Average air pressure at sea level is 101,300 Pa. Equivalent units of air are: 760 mm Hg (mercury) = 101,300 Pa = 1 atm (atmosphere).

Barometric pressure is used to determine elevations (distances above sea level) and changes in the weather. As elevation increases—when climbing a mountain or flying an airplane for example—barometric pressure goes down.

Variations in air pressure also help in forecasting the weather. Lower pressure means that the weather is humid, that clouds are forming, and that rain is likely. Increasing barometric pressure indicates clearing weather. Low pressure usually comes before a storm that could damage crops and livestock.

Areas of high and low air pressure are used to forecast the weather. Maps on television and in newspapers show the locations of highs and lows. As these fronts move across North America, weather forecasts are made for areas ahead and behind the pressure systems.

Producers can use information about barometric pressure to help make decisions about plants and animals. For example, hay should not be cut if the barometric pressure is going down. Rain will be likely, and this will damage the hay before it has cured.

Air pressure influences the boiling point of water. At sea level, water boils at 212°F (100°C). It boils at lower temperatures as the elevation (altitude) increases. Water on the Rocky Mountains may boil at 200°F! Cooking and keeping food safe may require longer boiling points to kill any germs in it.

Wind

Wind is moving air. Differences in air pressure cause wind. Heat also causes air to move by convection. Wind is measured in two ways: direction and speed. Wind direction is measured with a wind vane. Speed is measured with a device called an anemometer.

Winds are classified in three types: local, global, and jet stream.

Local wind is caused by differences in temperatures between land and water and different elevations in the land. Since land warms faster than water, breezes often blow from bodies of water toward the land. This is more likely to happen during the day. At night, since land cools faster than water, the wind may blow from the land to the ocean on the coastline.

Global wind has effects over large areas of the earth. These winds are the result of rotations of the earth and differences in temperature on the earth. These winds can cause conditions that lead to major storms, such as hurricanes.

16-10. Damage caused by straight-line wind in Minnesota during July 1995.

16-11. Two devices for measuring wind: anemometer (left), which measures speed, and wind vane (right), which measures direction.

Jet streams are important factors in weather. They are located at the top of the troposphere and travel at high speeds. They are often shown on weather maps to give weather forecasts for several days. In the northern hemisphere, when the jet stream moves down toward the equator colder weather can be expected.

Humidity

Humidity is the amount of moisture in the air. The moisture is in vapor form, known as water vapor. Measuring humidity also involves knowing the temperature. Warmer air will hold more water vapor.

The amount of water in the air is compared with the air's ability to hold water vapor. This is known as ***relative humidity***. Relative humidity is measured as a percentage. Higher humidity means that the water vapor in the air is nearing to 100%. Rain or other precipitation occurs at 100% humidity. As air cools, its capacity to hold water goes down and rain is likely. This is the reason rain or snow comes before cooler weather.

Clouds are condensed water vapor. They are made of water droplets and pieces of ice floating in the air. Some clouds have more moisture than others. Stratus clouds are widespread clouds that are used as a basis for

16-12. A sling psychrometer can be used to determine relative humidity.

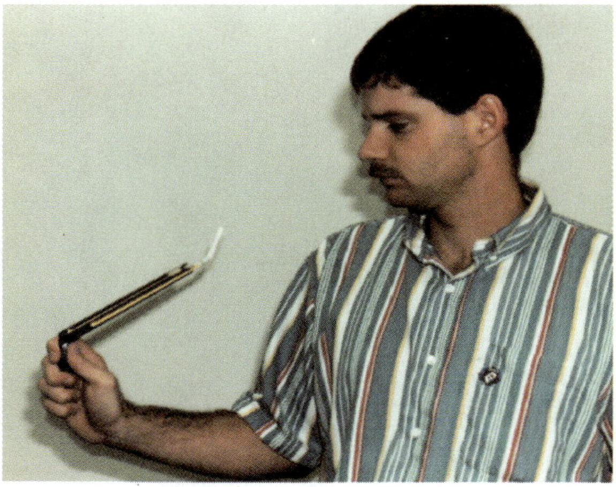

16-13. Cumulus cloud.

predicting rain. Stratus clouds create overcast weather that block out the sun's rays and bring rain. Cirrus and cumulus clouds generally indicate fair weather. Cirrus clouds occur at high altitudes and appear wispy. Cumulus clouds are fluffy clouds that are found at all altitudes.

16-14. Stratus clouds indicate that rain is likely.

Precipitation

Precipitation is any form of water that falls to the earth. Rain is the most common, but snow, sleet, and hail also bring water to the earth.

Liquid Precipitation. Rain is liquid precipitation. Raindrops form when clouds are cooled and the tiny cloud droplets merge to form larger droplets or ice. As they merge, they get heavier. When they get too heavy to stay afloat in the air, they form a teardrop shape and fall toward the earth.

16-15. Rain gauge mounted to a post. Rain gauges such as the one shown here should be emptied before the weather is cold enough to freeze. If water stays in it, the freezing of the water will crack the gauge.

Some will evaporate before reaching the earth; others fall in the form of rain.

Precipitation levels range from a few inches in desert areas of North America to several hundred inches on mountain tops in Hawaii. The eastern half of North America receives an average of 30 to 60 inches of precipitation a year. The amount of rain decreases west of the Mississippi River to 10 inches or less in the drier areas of the west.

Precipitation is measured with rain gauges. These give the depth of the rainfall in inches or centimeters. Rain gauges should be located where they can accurately measure the rainfall. Buildings, trees, and other structures may block rain or cause runoff that gets into the gauge. When this happens, the rainfall isn't measured accurately.

Simple rain gauges can be put up at home. Weather stations use more sophisticated approaches.

Frozen Precipitation. Snow, sleet, and hail are frozen or solid forms of water. Snow and sleet fall in cold weather. Sleet is usually the size of raindrops. It forms when rain leaves the cloud and freezes before it reaches the ground.

16-16. Snow accumulations can quickly make travel difficult.

Hail falls in warm weather during severe storms. Hail pieces may be an inch or more in diameter and can cause considerable damage to crops and buildings. Strong upward wind lifts the falling particles back into the cloud where they get larger. When they become too large for the wind to support, they fall to the earth.

Snow is a very important source of water in western areas that rely on irrigation. For example, snowfall in the mountains of eastern and northern California is essential to supply water for crop irrigation in the valleys the next summer. Some of the mountains in California may have snow accumulations of 10 feet or more during winter.

16-17. A rain gauge positioned on the ground by pushing the spike into the earth. Note that this gauge can also be used to measure the amount of water added by sprinkler irrigation.

Precipitation Contents. Precipitation brings with it contents of the air. Rain sometimes brings nitrogen; other times it brings smoke particles, acids and other solids. In very stormy weather, rain may appear to bring large objects. For example, a kind of storm known as a tornado can lift water from a pond or stream. Fish and other objects are lifted with it. When it falls, people think it has "rained fish." Actually, storm debris has fallen to the earth.

Some people have become concerned about acid rain. This rain contains pollution that forms acid in water. The acid rain damages crops, the soil, and the structures that it falls on.

16-18. Weather station rain gauges like this are used by meteorologists. Water collects in the top and is funneled into a gauge.

Dew point. ***Dew*** is a form of liquid precipitation. It is water that forms on the leaves of plants and other materials when the temperature of the air cools so that it cannot hold as much moisture. Plant leaves become wet with dew drops. Dew forms on clear nights when there is no wind. It is dried the next day by the sun.

AgriScience Connection

EVIDENCE OF POOR INSULATION

Buildings need to be properly insulated. Good insulation saves valuable energy and cuts the cost of heating during the winter.

Energy loss in cold weather shows where frost and snow first melt. Heat loss through a roof melts frozen water. With good insulation, less heat would be lost and the roof would not show patterns of melting.

This building has poor insulation. Look at the photo and explain why. What could be done to prevent this heat loss?

Dew point is the temperature at which dew will begin to form. It is lower than the temperature of the air or the same as the air if relative humidity is 100%.

Frost forms when the weather is cold and the dew freezes. Frost is made of tiny ice crystals that can damage tender crops.

In humid climates, as much as 5 inches of precipitation may be formed annually by dew. Dew is an important source of moisture in dessert areas.

Many agricultural operations are scheduled by dew and the amount of dew that has collected. Cut hay left in the field may be covered with dew. It cannot be baled until the sun dries the dew. Harvesting of grain and cotton may have to be delayed until the dew has dried. Also, harvesting into the night may have to be stopped when dew has formed.

Weather Changes

The weather changes because one or more of the conditions in the earth's atmosphere have changed. When changes in temperature and air pressure occur, other things are likely to happen.

Large air masses can develop; some have warm air and others have cold air. When the two masses touch, fronts result. A front is the boundary between the two masses. Fronts may be warm or cold. Warm fronts follow cold fronts, that is, warm air replaces cold air. Cold fronts result as cold air replaces warm air.

Changes in the air go on all the time. The nature of the changes vary with the seasons and other conditions in the atmosphere. Thunderstorms can develop when cold air bumps into warm air. The cold air pushes the warm air upward, where it forms heavy clouds and creates wind. Lightning may occur when electrical charges develop in clouds. Tornadoes are violent, funnel-shaped storms. They have high winds that can damage trees, crops, buildings and other structures. Hurricanes are large tropical storms that can do extensive damage. Hurricanes are known as typhoons in the Pacific Ocean area.

Scientists are continually studying the weather. There is still a lot to be learned, such as why a hurricane forms. People are also interested in ways to create rain. In dry weather farmers often need a rain. Lacing clouds with certain chemicals, called salting, has been tried as a way to make rain. Silver iodide and sodium chloride have been used with some positive results.

CLIMATE

Climate is an average of all the weather conditions for an area over a period of time, usually a year or more. Climates vary in temperature and amount of rainfall. Various regions of the earth are known for certain climates because of (1) their latitudes (distance from the equator); (2) altitudes (height above sea level); and (3) surrounding geography (mountains, valleys, and nearby lakes or oceans).

Climate Zones

The earth is divided into three ***climate zones***: tropical, temperate and polar. These zones are based on temperature.

In ***tropical climate zones*** the average temperature is above 65°F (18°C). The tropical zone is a wide band around the earth on both sides of the equator.

Temperate climate zones are bands around the earth between the tropical and polar zones. The average temperature in temperate zones is based on winter and summer average temperatures. Temperate zone temperatures average above 65°F (18°C) in the summer and below 50°F (10°C) in the winter.

16-19. Crops must be adapted to the climates where they are grown. This abundant soybean crop is the result of the right climate and good weather throughout the growing season. (Courtesy, American Soybean Association, St. Louis, Missouri)

Polar climate zones are those where the average temperature for a year is below 50°F (10°C). The polar zones include the North and South Poles.

Climate Factors

The weather conditions in an area are described by measures of the temperature and precipitation. These are the result of several factors.

- **Sun rays**—Rays that strike the earth directly, as they do at the equator, are warmer. Sun rays that strike the earth at an angle are cooler. The round structure of the earth and how it is tilted in relation to the sun cause differences in the temperature and amount of light produced by the rays. The angle of the sun's rays is related to latitude. Latitude is the distance of a place north or south of the equator. It is measured in degrees.
- **Altitude**—Altitude is the elevation or distance above sea level. The higher the elevation, the cooler the weather. The tops of tall mountains may stay so cool that snow stays on them year round. For example, mountains in the Rocky Mountain National Park in Colorado have snow year round. The peaks are also above the tree line, meaning that only tundra plants grow there.
- **Water areas**—Large areas of water influence climate. Oceans and large lakes tend to make the climate near them less extreme. Water is slower to cool than the land. On warm days, the water cools the air as it moves to the land. On cold days the water warms the air as it moves toward land. Large areas of water tend to keep the nearby land area more nearly the same temperature year round.

16-20. Melting snow feeds a small stream in the mountains of northeast Georgia.

LIFE SUCCESSION: USING THE EARTH'S RESOURCES

Sustaining life requires using the resources of the earth. Plants and animals live together using these resources. The physical features of the

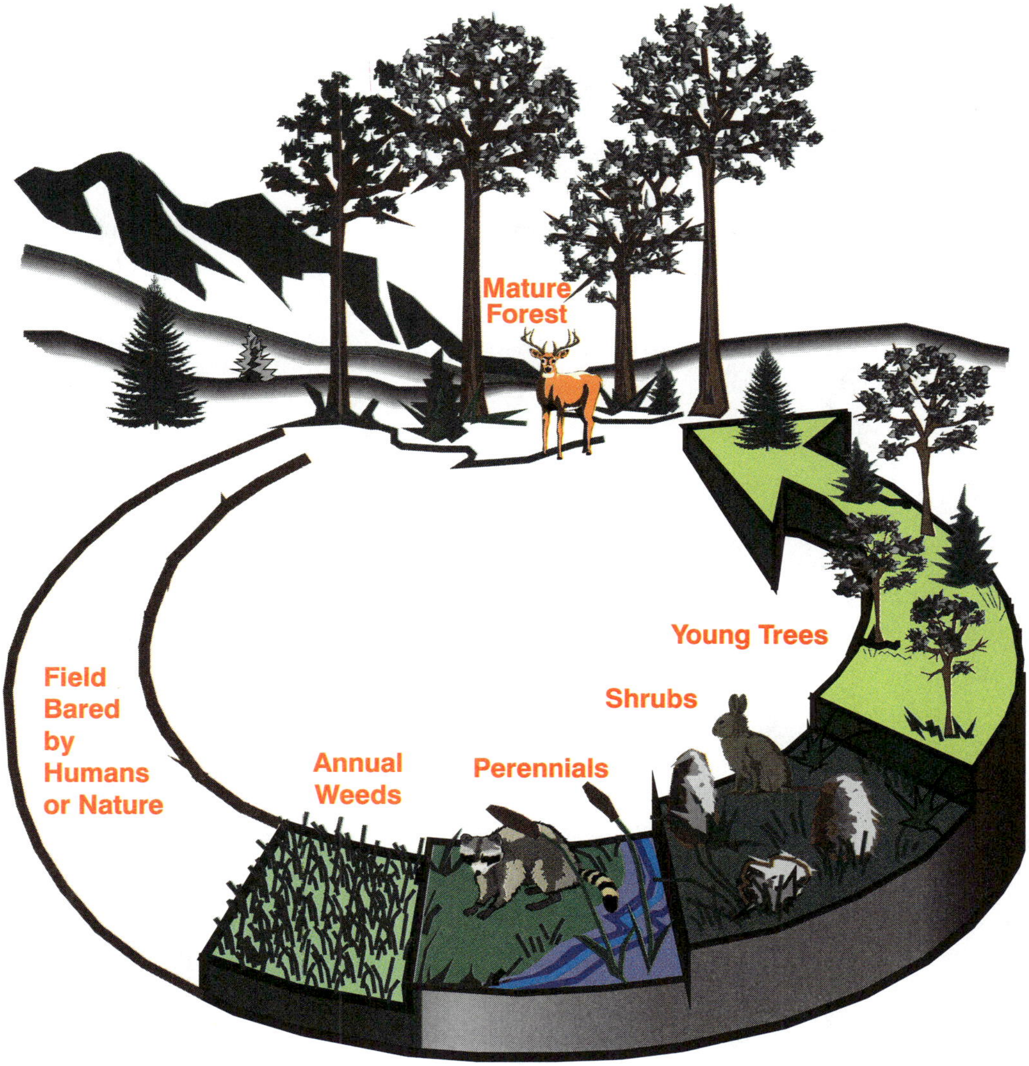

16-21. Succession of plant and animal life.

earth promote growth and, at the same time, resist change. The earth has ways of dealing with changes. New ecosystems (communities of organisms) grow when old ones are destroyed. (See Chapter 3 for information on ecosystems.)

SUCCESSION

Plants and animals follow certain patterns of growth. The process of growing and changing is known as ***succession***. It involves the natural continual replacement of the organisms that live in a place.

Disasters can destroy an ecosystem. Fires burn range land. Hurricanes destroy forests. Volcanic eruptions destroy everything in the path of the lava. After a disaster, the resources of the earth begin to build a new ecosystem.

Replacement is always going on in an ecosystem. New plants and animals are produced and grow. Old ones mature and die. The decomposition of the old plants and animals provides organic matter. This material reacts with the earth's minerals to cause change in the composition of the crust of the earth.

Primary Succession

Primary succession is the development of a biotic community where none existed. ***Biotic communities*** are the groups of plants and animals that live together. In most places on the earth, established communities have been around for hundreds or even thousands of years.

Here is how it might have happened in North America: The glaciers that covered much of the region likely began to disappear about 15,000 years ago. This left large areas of exposed land and rock. Low forms of life began to grow first, such as lichens. These were gradually replaced by mosses, which were followed by ferns, grasses, shrubs, and trees. As the trees grew, the grasses and other small plants were shaded out. First insects, and then other animals began to grow as the plants provided food. Mature plants and animals died and new ones grew in their places. Throughout this process, rocks, and minerals were weathering. Soil was being formed that could support additional life. The decomposition of the living organisms also added humus to the soil.

Secondary Succession

Secondary succession occurs when an area is destroyed by a natural disaster or by people. The plant and animal life sequentially develop much as during primary succession. These occur all of the time, although it is easy to overlook what is happening. Here are three examples.

- **Abandoned field**—Cropland that was plowed and planted for years could be abandoned. The land is no longer farmed. Over the years, new plants grow and develop. First, are grasses and other small plants. These are followed by small trees and larger trees. New animal life moves in to live in the plants. Some people make use of the succession by planting trees on old fields. They are able to "tree farm" and gain good production from the land.
- **Clear-cut forest land**—Some methods of harvesting timber involve cutting all of the trees on the land. No trees are left. Tree farmers will further

16-22. Clear-cut forest areas will be carefully re-planted in Oregon. (Courtesy, John D. Berry, Clackamas Ranger District, Forest Service, USDA)

clear the land and plant young trees, known as seedlings. These will grow in a few years to dominate the land. Until the trees are large enough, grasses and weeds will grow and compete with them for space. Undesirable trees may also grow, and should be cut out to help the new forest grow properly. If the land is not planted with trees, natural succession will take place. Grasses and small plants will grow. Different species of trees will also grow and overtake the grasses and small plants. Some of the tree species will be dominant; others will be suppressed.

16-23. Dangerous lightining often accompanies thunderstorms. It may set fire to a forest or range area.

- **Burned land**—Lightning may set fire to a forest or range area. The fire may be intense and destroy all of the trees and other plants. The land may have nothing left growing on it. In one growing season, grasses and small plants will begin to sprout. In another season, trees will begin to grow. Over a period of time the forest or range land will re-grow. Those species that are dominant will survive until they die and new ones take their place. (Note: Fires are particular problems in dry areas of western North America. Not only do fires destroy plants and animals, they also cost money. People who own the timber lose what it was worth.)

REVIEWING

MAIN IDEAS

The earth is one planet in the Milky Way galaxy. It is in a group of other planets, moons and stars with one sun that form a solar system.

The earth has three major areas: crust, interior, and atmosphere. The crust is important because it supports plant and animal life. Through weathering, rocks and other minerals have formed soil. The interior of the earth has a very hot core of heavy materials in the center. Between the core and the earth's crust is a layer known as the mantle. The atmosphere is made up of the air that surrounds the earth.

Many kinds of changes occur in and on the earth. First, the earth is constantly moving. The turning of the earth is known as rotation. One rotation is made every 24 hours. The movement of the earth in the solar system is known as revolution. One revolution is made every 365.24 days. Rotations and revolutions are measured in days and years. These movements are important in producing plants and animals.

Weathering is the process that changes the crust of the earth. Rocks are broken and changed into other substances. Rocks become soil over many, many years of weathering.

The atmosphere affects the conditions in the environment on the earth. Weather is the condition that exists in the atmosphere. Temperature, air pressure, wind, humidity and precipitation make up the weather.

Climate is the average of all the weather conditions over a year or several years. The earth has three climate zones: tropical, temperate, and polar. The tropical zone is a band on both sides of the equator that extends around the earth. Temperate bands are found on both sides of the tropical zone. The polar zones are located between the temperate zones and the poles.

The earth's resources are used to support a succession of plant and animal life. Succession is the growth and change of patterns in plant and animal life. The plant and animal life is known as a biotic community. Primary succession occurs on land where biotic communities have not existed before. Secondary succession follows primary succession and takes place after an area is disturbed by a disaster or people.

Understanding earth science helps agriscientists in using the earth's resources to produce plants and animals. It also provides a background for learning about soil—the part of the earth's crust that supports terrestrial plant and animal life.

QUESTIONS

Answer the following questions using complete sentences and correct spelling.

1. What is the universe? How is the earth a part of the universe?
2. What are the parts of the earth's solar system?
3. What are the three major parts of the earth? How do these parts relate to each other?
4. Distinguish between continental and oceanic crust.

5. How does the planet earth move in the solar system? Name the kinds of movements and explain each. Why are these movements important in producing plants and animals?
6. What is weathering? What are the two major kinds of weathering? Why is weathering important?
7. What are the major elements of the weather? Describe how each is measured.
8. What is climate? List and briefly explain three climate zones.
9. What is succession?
10. What is primary succession? When does it happen?
11. What is secondary succession? When does it happen?
12. How can agriscientists get involved in succession?

EXPLORING

1. Make a collection of weather maps from the local newspaper for a week. Review the maps and note trends in the weather.

2. Develop a log of the weather at your home for a week or longer. Use a thermometer to measure and record the temperature at the same time each day. If possible, use a rain gauge and measure any precipitation. Other measurements can be used if the equipment is available to make them. (Alternative: Rather than make and record the readings, compile the information from the newspaper. For each day, chart the high and low temperatures, as well as precipitation, barometric pressure and relative humidity.)

3. Using reference materials, determine if your home is in the tropical, temperate, or polar zone. Assess the features of the climate where you live when making the determination. Write a report on your findings.

4. Arrange to visit a weather station. Have the meteorologist explain the equipment and how it is used in studying the weather.

17

APPLYING PRINCIPLES OF SOIL SCIENCE

Most life depends on soil. Directly or indirectly the food of both plants and animals comes from the soil.

Having plenty of food, fiber, and shelter requires productive soil. Nations where people are hungry don't have good productivity from their soil. Various conditions can limit soil productivity. Managing the resources in the soil is essential.

Using the soil makes it easy for it to be lost. Plowing loosens the surface so that it can be washed away by a heavy rain. Growing crop plants takes nutrients out of the soil. Producers need to follow practices that return the nutrients. If nutrients aren't returned, the soil loses its fertility.

Soil varies from one place to another. Different practices are needed for different soil types. The irrigated soil of western North America must be managed differently from that in eastern North America where producers do not rely as much on irrigation.

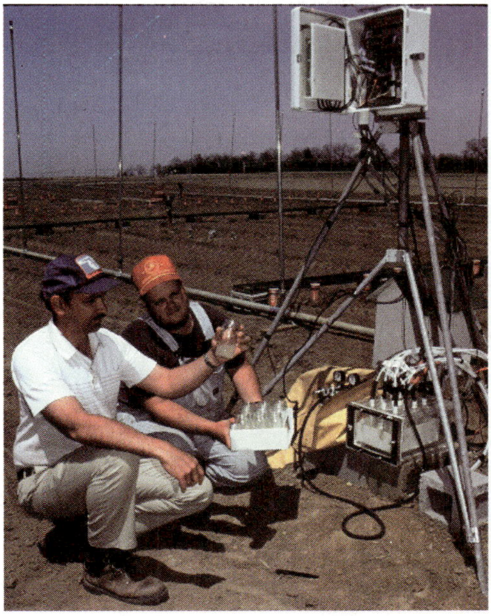

17-1. A vacuum system is being used by these soil physicists to study the water content. (Courtesy, Agricultural Research Service, USDA)

OBJECTIVES

Knowing how soil is formed and learning methods of management are essential for producing plants and most animals. This chapter has the following objectives:

1. Describe the materials that make soil.
2. Explain the different ways that soil can be described.
3. Explain how soil is formed.
4. Describe soil profile.
5. Explain the kinds of soil water.
6. Describe soil pH and its relation to plant growth.
7. Explain the relation between soil and land.
8. Describe important soil management practices.

TERMS

agricultural drainage
artificial erosion
capillary water
contour plowing
crop rotation
gravitational water
horizon
humus
hydroscopic water
internal drainage
land capability classes
land forming
loam
mineral soil
minimum tillage
mulching
organic soil
parent material
soil aeration
soil pH
soil structure
soil texture
soil tilth
soil profile
strip cropping
terracing
water table

Soil Contents

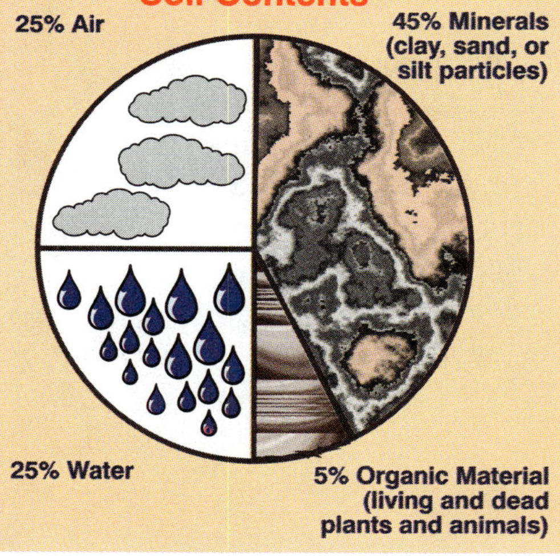

SOIL: WHAT IT IS AND WHAT IT CONTAINS

Soil is the top few inches of the earth's crust that support plant growth. It is the medium in which plants grow. Soil varies widely throughout North America, yet has similar ingredients everywhere.

Soil is made of four ingredients: minerals, organic matter, water, and air. The minerals and organic matter are the solid particles in the soil. The water and air fill the spaces between the particles.

In addition to the four ingredients, the soil also contains living organisms. Some are very tiny; others are larger, such as earthworms.

MINERAL MATERIALS

Minerals are naturally occurring inorganic elements or compounds that sometimes form crystals. Crystals are made of chemical elements that have joined together to form solid materials. Examples include quartz and feldspar.

Crystals usually have a definite structural arrangement and different types mix to form rocks. Sandstone rock, for example, is a mixture of quartz and other materials.

The mineral materials are from rocks and other minerals that have weathered. These materials form particles known as sand, silt, and clay. In addition, many other minerals are found in the soil.

Most of the soil in North America is known as ***mineral soil***. This is because it is higher in minerals than other materials. Soil that is higher in organic matter is known as ***organic soil*** (organic soil must have over 20% organic matter). Mineral soils usually have no more than 6 to 12% organic matter.

Sand

Sand is the largest-sized mineral compound in soil. The particles are 0.05 to 2.0 millimeters (mm) in diameter. Most sand is made of quartz but it can be of many different materials. Its chemical name is silicon dioxide (SiO_2).

Sand is an important ingredient in soil. The amount varies; some soil is mostly sand. Soil with over 85% sand is known as sand.

Soils that are high in sand dry out quickly. The sand particles do not hold water well. Soils high in sand aren't very fertile because water can move through them quickly and wash the nutrients away.

Silt

Silt particles are smaller than sand but larger than the smallest particle, known as clay. Silt particles range from 0.002 to 0.05 mm in diameter. They are small enough to fill spaces between sand particles.

Silt particles are sometimes deposited by water when land is flooded or washed into the oceans where streams empty. The large, fertile Mississippi River delta is a product of silt settling from flood water of the Mississippi River. Hundreds of years may be required for large land areas to develop by silting.

Clay

Clay is the smallest particle in soil. It is less than 0.002 mm in diameter. Various minerals are found in clay; kaolin and vermiculite are two examples.

Clay fills spaces between silt and sand particles. It gives soil the ability to hold water. Soils that are high in clay are slow to dry out after a rain. Clay also holds the minerals that plants need for growth.

Clay soil particles are small and have more total surface space and space between them than the larger sand particles. This additional space helps the clay hold water.

17-2. Three soil components (from left to right): organic matter, clay, and sand.

Other Minerals

Many other minerals are found in soil. These are very important for supporting the growth of plants. In fact, these are the minerals that become plant nutrients. They are often added in fertilizer to give plants more nutrients. The major minerals, in element form, include calcium, phosphorus, potassium, and nitrogen. Some of these are easily lost from the soil, such as nitrogen; others tend to remain in the soil longer, such as calcium. (Chapter 9, Producing Plants, has information on plant nutrients.)

ORGANIC MATTER

Organic matter is made of plant and animal remains in various stages of decay. Decaying organic matter releases nutrients into the soil. Well-decomposed organic matter is known as ***humus***. Soils with a high humus content are usually high in the nutrients that plants need. Examples of soils high in humus are peat and muck. These soils are dark brown or black and hold a lot of water.

17-3. Decaying leaves are found in forests under trees. The leaves on top are just beginning to decay. More decay has occurred deeper into the leaves. Leaves next to the soil are decayed organic matter. Leaf covers may be four or more inches thick. This is a continuing process. New leaves are added each year to the material, sometimes known as leaf mold.

Soils that have organic matter are more productive. The soil is more easily tilled and made into a loose seedbed for planting. High organic matter also increases the chemical capacity of the soil known as ionic exchange. This increases the availability of nutrients and helps plants grow better.

The organic matter in soil naturally develops when plants and animals die. Plowing and growing crops on the soil changes the natural state. This means that producers need to use practices that encourage the development of organic matter.

Producers often leave crop residue and plow it into the soil. This residue is organic matter that will decay and form humus.

17-4. Fertile soil in the Salinas Valley of California produces excellent vegetable crops, such as the lettuce shown here.

Manure may be applied to fields. Cattle manure from feedlots is often used in solid or liquid forms. The litter may be cleaned out of chicken houses and spread over the land. Even home owners buy bags of dried manure to add to the soil in their gardens and flower beds.

Forests naturally provide organic matter. The leaves shed by trees fall to the ground and begin decay. Depending on the kind of tree, a few years may be required for the leaves to become humus.

WATER

Water in the soil is an important nutrient for plant growth. The water is held between the soil particles. Water adheres (sticks) to the surfaces of the particles. Soils that have a high clay content will hold more water. Soil that is too wet or too dry will not grow most crops and ornamental plants. Of course, the amount of moisture needed varies with the kind of plant. Some plants don't need much water; others need a lot. Some plants can grow in an aquatic environment, such as water cress. Some terrestrial plants are grown using hydroponics, or without soil in a water solution. An example is lettuce.

Water enters the soil in two ways:

- **Soaking downward**—Water soaks downward (percolates) into the soil after precipitation or irrigation. When soil gets all the water it will hold, the water runs off into streams and lakes. Applying too much irrigation water is a waste. If irrigation water runs off a field, too much has been put on the land.

- **Soaking upward**—Water can move upward in the soil from higher concentrations deeper in the earth. However, this is not the major way most crop and pasture plants get their water.

AIR

Just as water fills spaces between soil particles, air also fills these spaces. When water is added to soil, the air is driven out. As soil dries out, it contains more air. Plant roots need air to live and grow. Most plants grow better when moisture and air are about equal in the soil.

Some plants adapt to different environments. For example, cypress trees grow in wet places. In order for the roots to have air, primarily oxygen, the roots grow structures known as "knees." These rise above the water level and may be several feet long.

Most crop plants must have a balance of air and moisture in the soil to grow well. Crops in soil that is too wet for them may develop a yellow color rather than a healthy green. A good example is corn. If water stands around corn plants for a few days, the leaves turn yellow. Once the water is gone and air returns to the soil, the corn plants will regain their green color. In the meantime, however, the yellowing of the leaves slows growth of the plant and reduces yield.

AgriScience Connection

GROWING PLANTS WITHOUT SOIL: HYDROPONICS

Terrestrial plants can be grown without soil! Hydroponics is growing plants in soilless media—water solutions that provide the essential nutrients. (Growing terrestrial plants with roots in soil is know as geoponics.)

This shows lettuce being grown with hydroponics as well as artificial light. To do so, the natural cultural needs of the plants must be met. A method of anchoring the plants must be used so that the roots are in the solution and stems are upright.

SOIL MAKE-UP

Soil forms naturally on the earth's crust, but it varies widely in composition. Soil can be described by its physical, chemical, biological, and drainage characteristics.

PHYSICAL NATURE OF SOIL

Soil has two physical qualities that are important: texture and structure.

Soil Texture

Soil texture is the proportion of sand, silt, and clay in the soil. These materials give soil its physical appearance. In nature, all soils are mixtures of sand, silt, and clay. Of course, some soils may be almost exclusively one kind of material, such as sand or clay.

Soils are named on the basis of content: sand, loam, and clay. ***Loam*** is made of about equal parts of sand, silt, and clay particles. Different names are used to indicate the percentages of the various materials. Here are a few examples:

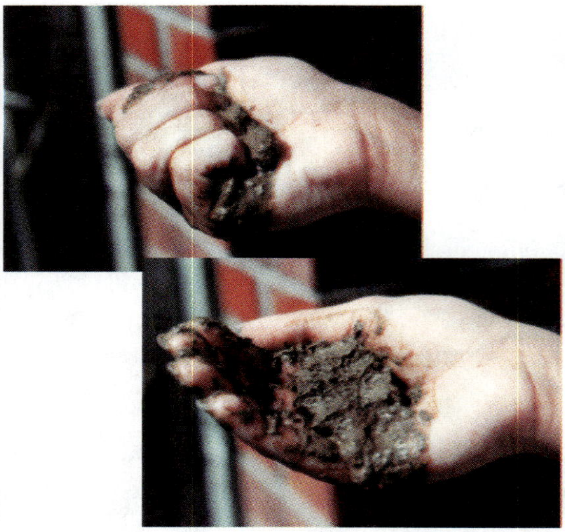

17-5. Soil high in clay. Soil texture may be assessed by feeling a sample of the soil. Rubbing moist soil between the fingers is sometimes known as the ribbon test. Soil high in clay will feel slick, while that high in sand will feel gritty. Soil high in clay will hold more moisture and have a tendency to be "soupy."

17-6. Soil high in sand. The sample holds little water and has a gritty feel.

- **Sandy loam**—Sandy loam is made mostly of sand with a little silt and clay. When dry, this soil forms clods that are easily broken.
- **Silty clay loam**—Silty clay loam is made mostly of silt with some clay and a little sand.
- **Silt loam**—Silt loam is made mostly of silt but has some clay and sand. Dry clods of silt loam are difficult to break. When wet, the soil is "buttery" or "velvety."
- **Sandy clay loam**—Sandy clay loam is made mostly of sand with some clay and a little silt.
- **Clay loam**—Clay loam is made mostly of clay with some silt and sand. Dry clods are very difficult to break.
- **Sand**—Sand is soil that is 85% or more sand. It may contain less than 15% silt or less than 10% clay.
- **Loam**—Soil materials are present in approximately equal amounts.
- **Clay**—Clay soil is made of 40 to 100% clay. Some sand and silt may be present. Clay soils with sand feel gritty when touched. Clay soils with silt feel smooth when wet.

The soil triangle is used to explain the make-up of soil. The triangle illustrates how soil textures are related to the different materials in the soil.

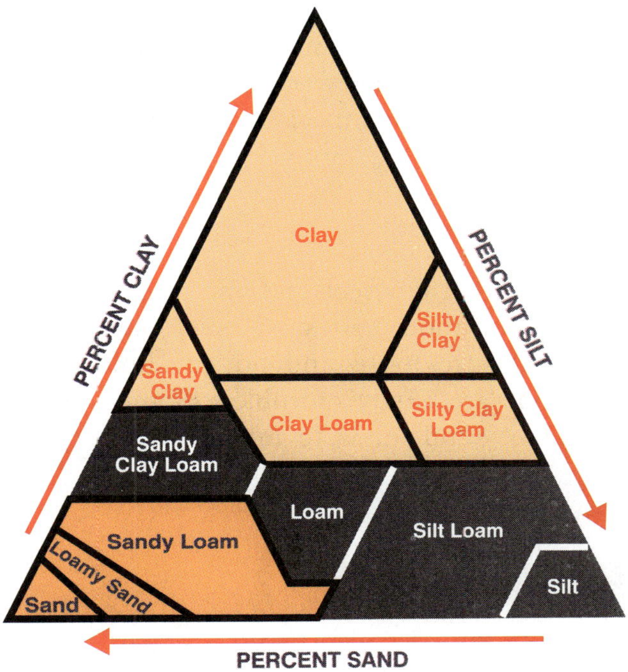

17-7. The soil triangle is used to describe the make-up of different soils.

17-8. This cut along a roadway shows structures forming in the soil. Note that big cracks have developed as this high-clay soil has lost its moisture.

Soil Structure

Soil is more than its parts of sand, silt, and clay. It also has structure, which is the arrangement of the soil particles into shapes or pieces. ***Soil structure*** is more complex than texture.

Soil structure may be in layers, blocks, grains, and other shapes. Detailed study of soil involves learning about these structures.

Soil Tilth

Soil tilth is the physical condition of the soil. It is an artificial condition made by plowing the soil. Various disks, harrows, and other implements may be used to provide good tilth. Good tilth is important in preparing a seedbed. Cultivation equipment may be used to keep weeds down and prepare a soil mulch for crops.

Soil with good tilth is loose and not too wet or dry. There are no large clods. It is easily planted and otherwise used in production.

Soil Consistence

Soils vary according to their feel and the amount of effort needed to break a small clod. The amount of sand, silt, clay, and water influence consistency. Some soils are loose and do not hold together in a clod. Others form clods that are easily crumbled. Some soils have clods that are difficult to crumble. Sand tends to make soil loose, while clay makes it hard to crumble.

CHEMICAL NATURE OF SOIL

Soil contains various chemical elements. Some of these are essential for plant growth, such as nitrogen, potassium, and phosphorus. Others

17-9. The soil in this flower bed is being tilled to develop good tilth. It was first dug to a depth of six inches. The soil was pulverized into small pieces using a hand hoe and rake.

are merely present in the soil, but some elements may actually damage crops. Too much of any element can pose a problem. (The nutrients needed by plants were covered in Chapter 9.)

In addition to the elements used for plant growth, soil also has an overall chemical quality known as pH. The pH of soil refers to its acidity or basicity. Most plants have a preferred range of pH in which they grow best. (Soil pH is covered in more detail later in this chapter.)

BIOLOGICAL NATURE OF SOIL

Most soil has a lot of plant and animal life in it. Organisms vary in size from very small to larger sizes, such as earthworms, ants, and rodents. The plants and animals do valuable things in the soil.

Break Down Organic Matter

Bacteria, fungi, and algae are the micro-organisms that act on leaves, stems, and plant and animal remains. They cause the decay. Organic matter in the soil is a result of the decay process. Years may be required for materials to decay and become soil.

Without the action of microorganisms, plant and animal remains would accumulate and not decay.

Other animals help break down organic matter. Beetles and termites break wood apart. Birds and squirrels peck and dig into wood and the earth. The activities of these animals increase the rate of change.

17-10. Earthworms help develop good soil.

17-11. Crawfish make deep burrows and bring soil to the top forming mounds which may be 6 to 10 inches high.

Aeration of the Soil

Good soil contains air. Tiny cracks and other breaks in the soil allow air to enter. Tillage is an artificial method of ***soil aeration***. The process breaks the soil apart and allows air to enter.

Small animals live and burrow in the soil. This creates holes that let air and water into the soil. Earthworms, ants, crawfish, and other animals move the soil around and make air holes. Their activity also improves the tilth of the soil. Particles from several inches or feet below the surface may be brought up.

Add Fertility

Some microorganisms add fertility to the soil. A good example is found where legume plants grow. These plants, in the presence of rhizobium bacteria, fix nitrogen from the air in the soil. Examples of such plants include clover, alfalfa, soybeans, and all beans and peas.

SOIL FORMATION

Soil has gradually developed over thousands of years. It is a process that goes on all of the time. Several activities go on that make soil:

17-12. Sprinkling a mild formaldehyde solution on soil helps in studying the biological nature of soil. (Courtesy, Agricultural Research Service, USDA)

changing parent material, weathering, climate, plant and animal life, and slope and drainage.

PARENT MATERIAL

Parent material is the mass of material from which soil develops. These materials are the rocks, other minerals and, in some places, peat. (Peat is a dark material made of decaying plants that grew in wet places, such as bogs.) Most parent material is a few feet below the top of the ground. In some places, outcroppings of parent material show up at the surface.

Common parent material includes shale, limestone, and sandstone. Shale and slate form soils that are high in clay. Limestone forms shallow soil that may not be very good crop land. Sandstone forms soils that are sandy.

Parent materials may be moved about before soil is formed. Areas of North America which were once under water may have marine deposits as parent material. Shells and other fossils may be found in the material. Some parent material may be washed from one place to another, such as the delta areas along rivers. Other parent material may be blown by huge, long-time winds from one place to another. Wind-blown material is known as loess. Large areas of loess (a deep, fertile soil) are found just east of the Mississippi River. Glaciers, gravity, and lake deposits also help make parent material.

Some parent materials must change a lot, such as rocks, to become soil. Others need little change, such as loess. Both chemical and mechanical

17-13. This eroded hillside shows a very thin topsoil over parent material of shale. Marine fossils are quite prominent in the shale.

processes break the parent material into smaller and smaller pieces. (These processes and weathering were explained in the previous chapter.)

WEATHERING

Weathering is the process of changing materials into soil. Soil doesn't just suddenly appear; it goes through stages of gradual development. Rocks break apart and change their forms. Water movement through the soil leaches (carries) materials with it.

Older soils have often lost nutrients to leaching and plant growth. The rate of plant growth slows down. If new nutrients aren't added, the soil becomes less productive. It isn't as fertile as it once was.

CLIMATE

The weather conditions in an area influence how fast soil is formed and lost. Heavy rains, freezing temperatures, and other features cause processes to go fast or slow.

Heavy rains can leach nutrients out of the soil. This is especially true of sandy soils. Producers often use fertilizer to add nutrients back into the soil.

PLANTS AND ANIMALS

The organisms that grow on and in soil influence its development. Plants make leaves, stems, roots, and other structures that decay to form organic matter. Dead animals also decay to help form soil.

Some animals create conditions that have a big influence on soil formation. For example, when beavers dam a creek the surrounding land area may be flooded. Decay processes of the organic matter are changed. Some plants are killed because they can't grow in the wet land. The land may cease to be productive.

SLOPE AND DRAINAGE

Lay of the land has a lot to do with how soil is formed. Land that slopes will have faster water runoff and dry out more quickly. Land in creek and river bottoms will stay wet longer and form a different kind of soil.

Hilly land often loses good soil to erosion. Water running down the hill picks up soil particles and carries them off. Level to gently rolling land is usually more fertile.

Soils that hold water (known as poor internal drainage) are slow to form and are low in fertility. Air can't get into the soil because of the water. A lack of air slows the decay of the organic matter. When organic matter doesn't decay, nutrients aren't released into the soil.

SOIL PROFILE

A ***soil profile*** is a vertical section of the soil at a particular place. The profile shows layers of materials from the surface down several feet. Profiles are divided into horizons.

HORIZONS

Soils have four ***horizons*** (layers) of materials, as follows:

- **A Horizon**—The A horizon is the topsoil. It is the upper few inches of the soil that is rich (high in nutrients). The roots of most grasses and other small plants grow in the topsoil. The thickness of topsoil varies. In areas where the soil has eroded, there may be no topsoil. The soil has lost its fertility. In other places, the topsoil may be several inches thick. Topsoil is usually gray to brown or black in color. The A horizon joins the B horizon. The exact location is sometimes difficult to determine because the two horizons blend together.
- **B Horizon**—The B horizon is known as subsoil. It is below the topsoil and contains more clay and less organic matter. The subsoil usually has a

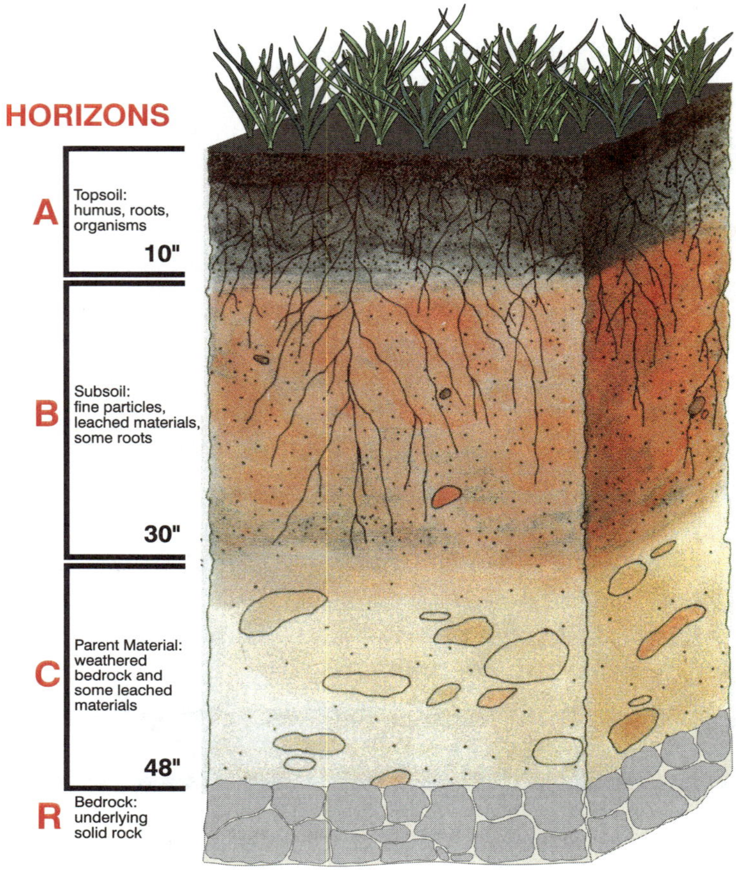

17-14. Locations of soil horizons.

brighter color because it is higher in minerals and lower in organic matter. Colors may be red, yellow, brown, or various combinations. The B horizon influences the percolation (movement) of water in the soil. Sometimes this horizon becomes hard and compacted, forming a layer known as a hardpan. The hardpan restricts water percolation and plant root growth. Subsoiling with deep chisel plows to a depth of about 18 inches is used to break the hardpan. Some hardpans are caused by heavy machinery compacting the ground.

- **C Horizon**—The C horizon is the parent material. It lies between the B horizon and the earth bedrock. The C horizon is important because it will later become soil. It also influences the fertility of the subsoil and, later, the soil. The C horizon is sometimes known as unconsolidated material.
- **R Horizon**—The R horizon is the bedrock that is below the C horizon.

17-15. A deep, productive soil with thick A and B horizons. (Courtesy, David Pettry, Soil Scientist, Mississippi Agricultural and Forestry Experiment Station)

SOIL PRODUCTIVITY

Productivity and soil profiles are related. A soil with a thick A horizon has more topsoil and is more fertile. The B horizon also influences soil fertility because it later becomes topsoil.

A soil's profile is closely related to how it can be used. Soils with a thin A horizon don't have much topsoil and won't produce large crops. Producers need to use practices that keep the topsoil from being lost.

Internal drainage of the soil has a lot to do with productivity. (Internal drainage is discussed later in the chapter.)

SOIL WATER

Plants need water to grow. They get their water from the soil. Soils vary in the amount of water they have and can provide for plants.

WATER TABLE

Water table is the depth of the natural level of free water below the surface of the earth. In humid climates that have heavy rainfall, the water table may be only a few feet down. In dry areas, such as deserts, the water table might be hundreds of feet below the surface.

Land near large bodies of water, such as lakes, oceans, and rivers, is likely to have a higher water table than land at larger distances from natural bodies of water. Sandy soils near bodies of water may have a water table that is only three feet deep.

17-16. These students are using a "pit" to learn about soil formations. (Courtesy, Jeannie May, U.S. Soil Conservation Service)

The water table is determined by digging a hole in the ground. The distance down to the point where water will collect and stand in the hole is the water table.

FORMS OF WATER IN THE SOIL

Soil with good tilth usually contains about 25% water. Plants need plenty of water to grow and produce large crops.

Three forms of water are found in the soil: capillary, hydroscopic, and gravitational.

Capillary Water

Soil particles are coated with thin layers of water, known as ***capillary water***. The water is free to move from one soil particle to another by capillary action because the water sticks to the surface of soil particles. Smaller particles have more surface and have greater capillary pull. This means that soils high in clay (the smallest soil particles) hold more water.

Hydroscopic Water

Hydroscopic water is a thin layer of water that sticks to soil particles. It does not move about, as does capillary water. Even the driest soils have some hydroscopic water. Plants are able to use some of this water but it depends on the amount in the soil. Clay and silt soils are better in dry weather because they hold more hydroscopic water.

Gravitational Water

Gravitational water is the water that fills cracks and air spaces between soil particles. In low places after rains, water may stand on the land. This water causes problems when trying to grow crops. Crop plants may become yellow and die from lack of air in the soil.

HOW WATER IS LOST

Water may be lost from the soil and not available for plant growth. Even in locations with an abundance of rain, water is lost so that the soil may be dry at times of the year. Water is lost in the following ways:

- **Runoff**—Runoff is excess water that runs from the land and doesn't soak into the earth. Heavy rains or melting snow create runoff. Streams can sometimes flood their banks if the runoff is too great. Various earthen structures can be built to slow runoff rates and allow more time for water to soak into the soil. Examples include terraces and farm ponds.
- **Evaporation**—Evaporation is water that turns to vapor (gas form) and goes into the atmosphere. Large amounts of water can be lost from the soil on hot days through evaporation. Mulch helps prevent evaporation.
- **Transpiration**—Transpiration is when plants release water into the air through their leaves. The tiny stomata on the leaves open and close to help the plant adjust to changing temperatures.
- **Percolation**—Percolation is the downward pull of gravity on water. Water soaks into the soil at the surface and goes down into the lower soil profiles by percolation.
- **Harvested crops**—Many plants contain a high percentage of moisture. When harvested, the moisture is carried with the crop. Tomatoes and watermelons are much higher in moisture content than wheat and soybeans.

INTERNAL DRAINAGE

Internal drainage is how readily water moves through the soil. Some soils don't allow the water to move very fast. Hardpans and shallow bedrock can hold water in the soil. Internal drainage is also determined by the texture of the soil, such as sand, silt, and clay content.

Land is often classed by how it drains internally. Some land has too much drainage, which means that the soil doesn't hold enough water for crops to grow. Soils that have a lot of sand tend to have too much internal drainage. Other soils have very poor internal drainage. These tend to have a lot of clay and hold water on the surface for a long time. Layers of material under the surface may restrict the percolation of water. Most crops prefer land that is well drained with neither too much nor too little internal drainage.

Ditches and drain tiles are sometimes used with land that has poor drainage to make it better.

SOIL pH

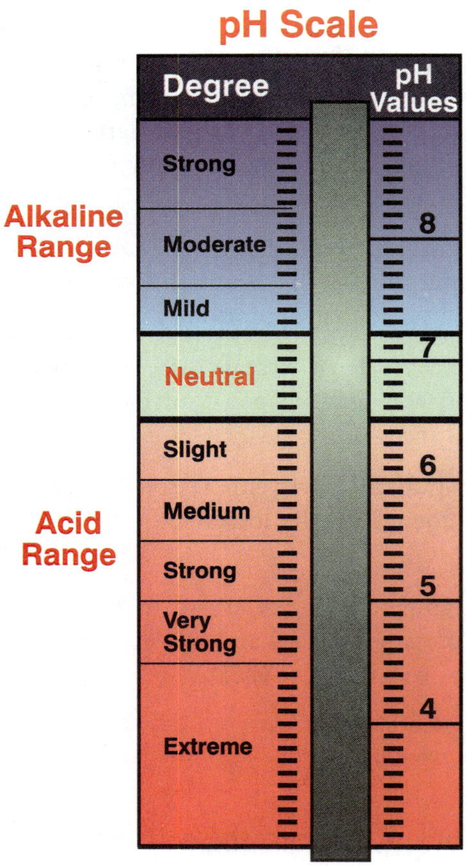

17-17. Soil pH as related to alkalinity and acidity. Few crop plants would tolerate a pH below 4.0 or above 10.0.

Soil pH is a chemical way of expressing the acidity and alkalinity of soil. pH refers to the hydrogen ion concentration in the soil solution. A 14-point scale is used to express pH. Zero is extremely acid, 7.0 is neutral and 14.0 is extremely basic. All soil has a pH between 0 and 14. Most soil is in the pH range of 4.0 to 8.0 (strongly acid to slightly alkaline).

Important chemical reactions take place between the soil and plants. The pH of the soil affects this reaction. Some nutrients aren't available at the "wrong" pH. The best pH range for a good supply of nitrogen (N) is 6.0 to 8.0. Phosphorus is tied up when the pH is below 5.5 and above 7.5. Even when the soil is in the best pH range, the nutrients may not be available if they aren't in the soil. More plant nutrients are available when soil is at a pH level of 6.5 than any other pH level.

HOW pH IS DETERMINED

The pH of soil is determined by testing a sample. This may involve using a pH testing kit in the field or sending the

17-18. Moss growing on this area is a sign of low pH and little or no topsoil.

sample to a soils laboratory for analysis. Careful interpretation of test results is essential. Trained specialists should be consulted to assess what to do to solve soil pH problems.

HOW pH IS MODIFIED

Soil pH can be changed by using certain materials on the land. What is used is determined by the pH of the soil and the direction of change that is desired. How much is used is determined by the amount of change needed in the pH of the soil. Major change requires more soil additive.

Soils that are too acid are made less acid by adding materials that are alkaline. For example, ground limestone ($CaCO_3$) (often called agricultural lime) can be added to acid soil to raise the pH (raising the pH makes it less acidic). Lime neutralizes acidic soil. Lime is high in calcium.

The amount of lime to use depends on the kind of soil and the amount of change to be made in pH. Sandy soils need less lime than clay soils. The size of the lime particles also affects how well they work. Finer lime particles act faster but are used up more quickly. Examples of the amount of lime to use are shown in Table 17-1.

Soils that are too alkaline are made more acidic by adding sulfur. The sulfur forms sulfuric acid in the soil. Sulfur may be added in granular or powder forms. Aluminum sulfate [$Al_2(SO_4)_3$] and iron III sulfate [$Fe_2(SO_4)_3$] are sometimes used. The amount to apply depends on the results of soil testing and how much the pH is to be lowered.

Table 17-1
Amount of Lime Required to Raise pH

Kind of Soil	Amount to Raise pH from 4.5 to 5.5 for 1,000 sq. ft.	Amount to Raise pH from 4.5 to 5.5 for one acre
	(pounds to use)	*(tons to use)*
Sandy loam	45	1.00
Silt loam	80	1.75
Clay loam	100	2.20

Note: The information represents approximate amounts of agricultural lime to use to raise the top 7 inches of soil from 4.5 to 5.5 pH.

CROP pH PREFERENCES

Crops vary in their pH preferences. Some grow better in somewhat acid soils; others prefer slightly alkaline soils. The producer must match crop and pH in order to get the best growth.

Some plants will grow reasonably well in either acid or alkaline soil, but their flowers or fruit may vary. An example is the hydrangea, an ornamental shrub. The flowers will be pink if grown in basic soil and blue in acid soil. (This means that if you want pink hydrangea flowers, add sulfur to the soil because sulfur makes the soil more acid. If you want blue flowers, add lime because it makes the soil more basic.)

Table 17-2 gives pH preferences for several plants.

EFFECTS OF FERTILIZER

Some kinds of fertilizers cause changes in soil pH. Fertilizers high in nitrogen tend to cause acids to form in the soil and lower the pH. Fertilizers with sulfur also lower the pH. Continuous use of acid-forming fertilizer will require the addition of lime to neutralize the acid.

Soil testing is used to determine the amount and kind of fertilizer to use. The same testing can also provide pH information. Both pH and fertilizer information can be studied by a soil specialist. Then recommendations can be made on the products to use. The specialist will also need

Table 17-2
Suitable pH Ranges for Selected Plants

Plants	Strongly Acid (pH 5.0–5.5)	Medium Acid (pH 5.5–6.0)	Moderately to Slightly Acid (pH 6.0–6.5)	Slightly Acid to Slightly Alkaline (pH 6.5–7.5)
Azaleas	X			
Irish potatoes	X			
Watermelons	X			
Bermuda grass		X		
Oats		X		
Peaches		X		
Pecans		X		
Rye		X		
Strawberries		X		
Apples			X	
Cantaloupes			X	
Corn			X	
Cotton			X	
Peanuts			X	
Soybeans			X	
Tomatoes			X	
Wheat			X	
Alfalfa				X
Cabbage				X
Red clover				X
Spinach				X

to know the crop that will be grown on the land. Recommendations are based on what is in the soil as well as what the crop will require.

The processes that occur in the soil are very complex. Always use the best available information when applying materials to change the nature of the soil.

SOIL SALINITY

Soil can have too much salt (salinity) to grow crops. The salt may be naturally present in the soil or added through contamination. A lot of soil has been damaged by using irrigation water containing various forms of salt.

HOW SALT AFFECTS PLANTS

Plant roots absorb water by osmosis. This process depends on the osmotic pressure in the roots. Water enters roots when the osmotic pressure is greater than in the soil. Salt in the soil raises the osmotic pressure of the soil. Even a little salt means that less water can flow into the roots. Plants aren't able to carry out their functions as well as they could if they got more water. Excessive salt can kill plants. Of course, some plants are more tolerant of salt in the soil than other plants. Also, the stage of growth of the crop is important. Some plants will grow in salty soil but their seed do not germinate and grow well. (Osmosis and plant water were covered in Chapter 7.)

SOURCES OF SOIL SALT

Salt can get into the soil in several ways. Some soil near areas of natural salt outcroppings has salt in it. Industrial accidents can pollute the soil with salt. The major source on irrigated land is the water used for irrigation.

The selection of irrigation water is very important. Even small amounts of salt in water can eventually add up to large amounts of salt on an acre of land.

Irrigation water is measured in inches or acre-inches. An acre-inch is one acre of land covered with one inch of water. Acre-inches are often converted to acre-feet. One acre-foot is an acre covered with 12 inches (1 foot) of water. One acre-foot of irrigation water containing 750 ppm salt will add a ton of salt on an acre of land. If more than an acre-foot is used on land, the amount of salt deposited in the soil by the irrigation water goes up.

17-19. Very little vegetation grows in the saline sand beach on the Gulf Coast of Florida.

LAND: MORE THAN SOIL

Land is all of the characteristics of a site that make it suitable for use. Soil is a major factor, but there are many others: climate, slope, water supply, and location in relation to other features.

All of the factors that make up the land contribute to its productivity. Agricultural land is often classed on the basis of its productive capability. These are known as ***land capability classes***. The classes consider the soil as well as all of the other features.

LAND CLASSIFICATION FACTORS

The agencies of the U.S. Department of Agriculture have established classes of land based on capability. Major factors in capability are:

- **Surface texture**—This is the texture of the soil on the surface. It is the percentage of sand, silt, and clay. The soil triangle is used in texture classification.
- **Internal drainage**—This is the permeability of the soil. It is the ability of the soil to take up water and air and make it available for use by growing plants. Soil internal drainage is classed as very slow, slow, moderate, and rapid. Of course, clay soils are slow and sandy soils are rapid. Parent material also influences internal drainage. Layers of rock can block the internal movement of water.
- **Depth of topsoil and subsoil**—Depth refers to the part of the soil that plant roots can readily use. Deep soils usually have topsoil and subsoil that is over 30 inches deep. Moderately deep soils are 20 to 30 inches, shallow

17-20. Though a fertile soil is present, the use of this land is impaired by flooding from the Yazoo River.

soils are 10 to 20 inches and very shallow soils are less than 10 inches deep. Hardpans can restrict plant growth and limit soil depth.

- **Erosion**—Erosion refers to the amount of soil that has been lost. Thin topsoil results when erosion carries it away. Land that has lost topsoil has eroded. It is not as productive as land that has plenty of topsoil. Plants grow best in topsoil that has the essential nutrients. None to slight erosion exists when less than 25% of the topsoil has been lost and there are no gullies. Moderate erosion is when 25 to 75% of the surface soil has been lost and a few small gullies are present. Severe erosion exists when 75% of the surface soil has been lost and small gullies are present. Very severe erosion exists when over 75% of the topsoil has been lost and large gullies are present.

- **Slope**—Slope refers to unlevel land. The land may be gradually sloping, hilly, or have severe slopes, as on the sides of mountains. In classing land, slope is based on the number of feet land falls over a distance of 100 feet. How land is classed varies a little from one state or region to another. In general, land that falls less than 2 feet in 100 feet is nearly level. Land that falls over 12 feet in 100 feet is steep or very steep. Other slopes are gently sloping, which is 2 to 5 feet fall over 100 feet of distance, and moderately sloping, which is 5 to 8 feet fall over 100 feet of distance.

- **Surface runoff**—This is the rate at which water soaks into and runs off of the surface of the soil. It deals mostly with excess water—that which doesn't soak in. Surface runoff can be good (no water problems), fair (water is removed slowly), poor (water is removed so slowly that soil is wet a lot of the time), and excessive (water is removed too rapidly causing erosion). Level or near level land near creeks or rivers may have poor surface runoff. Water may stand on the land and interfere with its use.

LAND CLASSES

Eight land classes are used. These are based on the factors in land capability. The eight classes are divided into those classes that can be cultivated and those unsuited for cultivation.

Land Suited for Cultivation

Four of the eight classes are suited for cultivation. Some of the classes require conservation practices to retain fertility. The classes are:

17-21. This Texas land is being prepared for sugar beets—a good use based on suitability. (Courtesy, Texas Department of Agriculture)

- **Class I: Very Good Land**—Class I land has few limitations. It is nearly level, has deep soil and good internal drainage, and the surface drains well. It is suitable land for most crops if the climate is satisfactory.
- **Class II: Good Land**—This class is not quite as good as Class I land but is very good. The slope may be up to 8%. Some practices may be needed to conserve the soil. The land can be used to grow many crops.
- **Class III: Moderately Good Land**—This land has slopes up to 10% and may have slow internal drainage. Special conservation practices should be used. The land can be used for certain crops and for pasture.
- **Class IV: Fairly Good Land**—This land has severe limitations. Slope and erosion as well as internal drainage may restrict the use of this land. It is best suited for permanent pasture, tree farming, hay, and wildlife habitat.

Land Unsuited for Cultivation

Four classes of land are unsuited for cultivation. This land tends to have a high percentage of slope or be very low and wet.

- **Class V: Unsuited for Cultivation**—This land can be used for pasture, hay crops, and tree farming. It may be wet, eroded, and have a slope that causes problems.
- **Class VI: Not Suited for Row Crops**—This land is too steep to use in growing row crops. It may easily erode and have a rugged, hilly surface. Class VI land is best suited for tree farming and pasture.
- **Class VII: Highly Unsuited for Cultivation**—This land should not be cultivated. Erosion and other problems make this land best suited for tree farming and wildlife habitat. Sometimes it can be used for pasture.

17-22. This land is excellent for fish ponds. It isn't flooded with overflow from a nearby stream and has a soil high in clay that will hold water. The land would not be good for row crops because of poor internal drainage and slope.

- **Class VIII: Unsuited for Plant Production**—This land may be very wet and in marshes or bayous. It is used for wildlife and recreation. Some of it may have a use in aquaculture. Areas known as "wet lands" may be in this class.

MANAGEMENT PRACTICES

Producers can use management practices that help conserve the soil. These will assure fertile soil in the future for producing food, fiber, and shelter. These practices also assure that the land will produce the highest possible yields.

Major practices include preventing erosion and draining wet land.

EROSION

Erosion is the loss of soil. It may be worn or carried away by water, wind, or other means. Plowing makes soil more easily lost. Soil covered with grass or other plants is less likely to erode.

The kind of erosion varies with the climate. Large, dry fields that aren't protected by a crop cover are most likely to be eroded by the wind. Excessive rain or other precipitation can wash soil away, especially if the land is sloped.

Kinds of Erosion

Erosion is a natural process in land that is undisturbed by plowing or other activity. The rate of natural erosion is very slow. Some say that erosion is a part of the weathering process.

17-23. Gully erosion can be easily seen in this field that has not been properly managed.

Artificial erosion is caused by people when they disturb the natural cover on the land. Plowing, cutting trees, excavating for a road or building, or other activities make the soil erode easily.

In agricultural land, erosion may occur in sheets or as gullies. Sheet erosion involves layers of the soil being washed or blown away. It may happen and go unnoticed. Gully erosion is more easily seen. It involves crevices or small ditches developing in the land. These can grow larger and larger so that any use of the land is impossible. Gully erosion is most likely to occur on hilly land that has been plowed.

17-24. This road builder is trying to control erosion on the highway site during construction. A sheet of plastic material has been stretched between posts to keep soil from washing into a stream and causing pollution.

Controlling Erosion

Controlling erosion involves keeping the soil in place. Protecting the soil from wind and water runoff will help slow down erosion. Several practices are available to control erosion.

- **Minimum and no tillage cropping**—Soil that isn't plowed is less likely to be lost. ***Minimum tillage*** cropping involves planting seed in land with very little plowing. No tillage cropping means that the land isn't plowed at all to prepare a seedbed. The seed are planted in the

17-25. Erosion in this soybean field has resulted in the loss of valuable top soil.

crop stubble from the previous year. Chemicals are used to control weeds. The crops aren't cultivated. Plowed land is not exposed to wind and water.

- **Mulching—*Mulching*** involves covering the soil with a layer of protective material. Sometimes crop stubble and other debris are left on the field as a natural mulch. Small gardens and flower beds may be covered with sheets of paper or plastic or layers of tree bark or sawdust. In addition to protecting the soil, mulch also prevents the evaporation of water.
- **Terracing—*Terracing*** involves building earthen ridges or embankments that slow the rate of water runoff. Terraces are made with gradual slopes and divert flowing water to grassy areas that won't wash like a plowed field. Terraces should be covered with grass to keep them from washing away. Also, terraces should be built so that equipment can cross them easily.
- **Strip cropping—*Strip cropping*** involves planting crops in strips. Plowed crops may be planted in alternating strips with crops that don't require plowing. The strips slow down the speed of wind and rate of water runoff.

17-26. Strip cropping is used with corn and hay on this Virginia farm to prevent soil loss.

- **Contour plowing**—***Contour plowing*** involves plowing around or across a slope rather than "with" the slope. The plowing is done so that the water flows gradually. Rows of crops should not go up and down hills but in a direction that slows down runoff. The rows are at angles (known as contours) that allow water runoff but slow its speed and retain the soil.
- **Crop rotation**—Planting different crops over the years on land helps reduce erosion. ***Crop rotation*** is alternately planting crops that use different nutrient amounts and conserve soil fertility. The amount of organic matter in the soil is increased because different plants make stems, leaves, and roots that hold the soil in different ways. For example, planting corn one year and wheat the next involves two different kinds of vegetation being left on the soil.
- **Diversion ditches and levees**—Water can sometimes be diverted around a field. Small ditches and levees can be built to do this.

DRAINAGE

Agricultural drainage involves moving excess water from the land. Open ditches and underground drain tiles or tubes may be used. Land with poor surface runoff and internal drainage can be made more productive.

Drainage systems need to be carefully designed and installed. Ditches and tiles that are in the wrong place won't work properly. In addition, care must be taken so that flow of water into streams does not change the natural flow of the stream.

LAND FORMING

Land forming involves shaping the surface of the land so that it is more productive. Low places in unlevel fields may be filled and high places planed off.

17-27. The land in this large field has been formed to allow good drainage and make it easy to use large equipment, such as this eight-row planter. The planter applies fertilizer and herbicide as well as plants the seed.

AgriScience Connection

USING PLANTS TO CLEAN POLLUTED SOIL

Soil is sometimes polluted by hazardous wastes, including some metals. Zinc, lead, cadmium, nickel, and other metals may get into soil and damage its ability to produce. This is particularly true near old refineries, dumps, and storage areas.

Scientists have found that some plants help cleanse the soil. When grown on polluted land, the root systems take up the metals and transport them to their stems and leaves. The plants can be cut like hay and hauled away, leaving the soil less polluted. Metal-scavenging plants are known as hyperaccumulators, with *Thlaspi* plant shown here with a massive root system being an example. (Courtesy, Agricultural Research Service, USDA)

Land forming is particularly necessary for some crops and irrigation systems. Rice requires land that will allow water to evenly cover the surface of the field. Crops that have irrigation water flooded onto the surface need to be shaped so that the water covers the field.

Land forming is also used to help remove excess water from land. Low places are filled so that puddles of water don't stand after the rest of the land has dried. Water puddles can slow field preparation and planting.

Large land planes and laser systems are used in leveling or forming land.

REVIEWING

MAIN IDEAS

Soil is the earth's most important resource. It is made of solid materials, water, air, and living organisms. The solid materials include minerals and organic matter. Most minerals have weathered from rocks and other materials. The major minerals are sand, silt, and clay. Organic matter is the decomposed remains of plants and animals, often known as humus. Soil also contains many elements that plants need, such as nitrogen, phosphorus, and potassium.

Soil has physical, chemical, and biological characteristics. Physical characteristics refer to the texture and structure of the soil. Chemical characteristics refer to the presence of chemical elements and general conditions that they cause. Biological characteristics refer to all of the living plants and animals found in soil.

Soil varies considerably in composition. The proportions of the minerals and organic matter can be quite different. Soil texture is based on the amounts of the major ingredients that are present. Texture also has a lot to do with how the soil is managed. Clay soils hold water better, while sandy soils don't. There are advantages and disadvantages to all kinds of soils.

The soil profile is a vertical view of the structure of the top few feet of the earth's crust. It shows soil horizons—topsoil, subsoil, parent material, and bedrock.

Most of the nutrients in soil are in the top few inches, known as topsoil. Conservation practices need to be used to keep the topsoil from eroding.

Land is classified on the basis of how it is best used. Classification includes soil as well as other features, such as slope and water supply.

Management of the soil can help retain its fertility for future generations. Controlling erosion and properly draining and forming land can help maintain productivity.

QUESTIONS

Answer the following questions using complete sentences and correct spelling.

1. What is soil? Where is it found?
2. What are the ingredients in soil? How do these ingredients determine the nature of the soil?
3. What is organic matter? How is it formed?
4. What is the soil triangle? Why is it useful?
5. Why is biological activity important in the soil?
6. How is soil formed? What processes are involved?
7. What is a soil profile? What horizons are found?
8. Explain water table.
9. What forms of water are found in soil? Distinguish between the forms.

10. How is water lost from the soil? What can be done to reduce the loss?
11. What is internal drainage? Why is it important?
12. What is pH? What pH ranges are found? Why is pH important?
13. How is pH changed?
14. What is land classification? What factors are used in determining the class of land?
15. Distinguish between the two major classes of land.
16. What is erosion? How is it prevented?
17. What other management practices may be used to make land more productive?

EXPLORING

1. Determine if your state or school participates in land judging. If so, get a copy of the regulations and form a land judging team.

2. Collect a sample of soil and test it for pH. Run the test in your school lab or send it to a soil testing laboratory. Determine the change needed in the pH, if any, for selected plants.

3. Using a transit, lay out a contour line on the school grounds or nearby land. Your teacher or a soil conservation technician can help with this activity.

4. Study the profile of the soil in your community. Determine the depth of the horizons and general texture of the materials. Also, determine the parent material for the soil.

18

APPLYING PRINCIPLES OF CHEMISTRY

Everything is made out of something! Living plants and animals as well as non-living materials are made out of tiny blocks of material that can be broken into simple forms.

Chemistry is a study of the characteristics of the substances that form these building blocks. It deals with what things are made of and how they are put together.

The tiny blocks (known as chemical elements) join together to form larger, more complex things. Small plants and animals as well as large mountains and open plains are made of these blocks. Most things in our lives are made of only a few different kinds of blocks!

How the blocks are put together determines what is made. This is much like using lumber in building. The structure is a result of how the pieces of material are cut and fit together. The same is true with the blocks that build the earth—some are best for certain uses.

Knowing about the building blocks of the earth's resources helps in achieving a better understanding of agriscience. They provide explanations about events that happen.

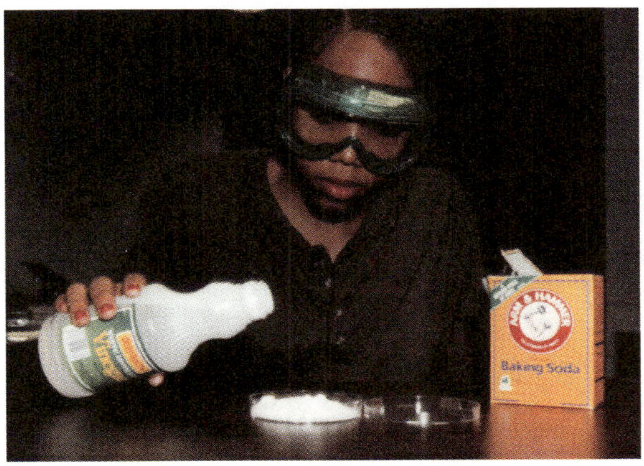

18-1. Creating a chemical reaction with vinegar and baking soda is easy to do.

OBJECTIVES

This chapter introduces important principles of chemistry. It has the following objectives:

1. Explain the concept and properties of matter.
2. Describe the major kinds of matter.
3. Distinguish between elements, atoms, and molecules.
4. Describe how new compounds are formed.
5. Explain the importance of carbon.
6. Distinguish between solutions and suspensions.
7. Explain acids, bases, and salts.
8. Describe common chemical processes in agriscience.
9. Describe the importance of measurements in chemistry.

TERMS

acid
alcohol
alkane
atom
atomic number
antifreeze
base
bonding
chemical reaction
colloidal suspension
corrosion
chemical properties
compound
diffusion
electron valence
elements
emulsion
hydrocarbon
ion
Law of Conservation of Mass
Law of Definite Composition
mass
metal
mixture
molecule
nonmetals
organic chemistry
oxidation
Periodic Law
Periodic Table of the Elements
phase change
physical properties
polarity
salt
solute
solution
solvent

MATTER: WHAT THE EARTH IS MADE OF

Matter is anything that has volume and mass.

Volume is often referred to as "taking up space." Volume is usually obvious: A tree in the forest takes up space, just as the milk in a jar does. Most of the things that take up space can be seen. Some things, such as the gases in the air, are not readily visible. Whether we see it or not, matter has volume.

Mass is the amount of matter that an object contains. Most often, mass is referred to as weight, but it is different from weight. The properties of the matter must be considered. Mass is usually measured in kilograms (kg) but grams (g) and milligrams (mg) are also used for amounts less than a kilogram.

STATES OF MATTER

Matter is found in three states: solid, liquid, and gas. Matter can be changed from one form to another but it cannot be created or destroyed. This is the ***Law of Conservation of Mass***.

Water is commonly used as an example of all three states: liquid, solid (ice), and gas (water vapor). Water is the only matter that occurs in nature in all three forms.

AgriScience Connection

SAVING OUR ENVIRONMENT FROM CHEMICAL WASTES

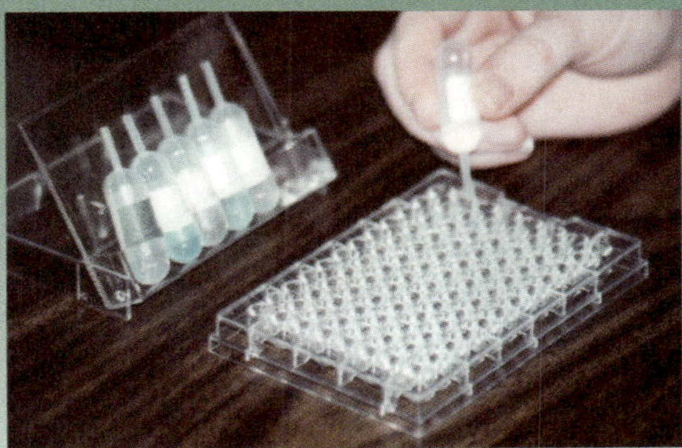

Practices in chemistry are changing to prevent excess pollution in the environment. After running an experiment, waste disposal may be a problem. What do you do with a big test tube filled with acid?

Cut back on waste by changing the way you go about some reactions. Use very small quantities of chemicals to reduce the amount of waste. Some tests and experiments can be done using microchemistry. Microplates with tiny wells can be used in place of test tubes and beakers for many experiments. All you often need is a couple of drops as compared to a big test tube filled with acid.

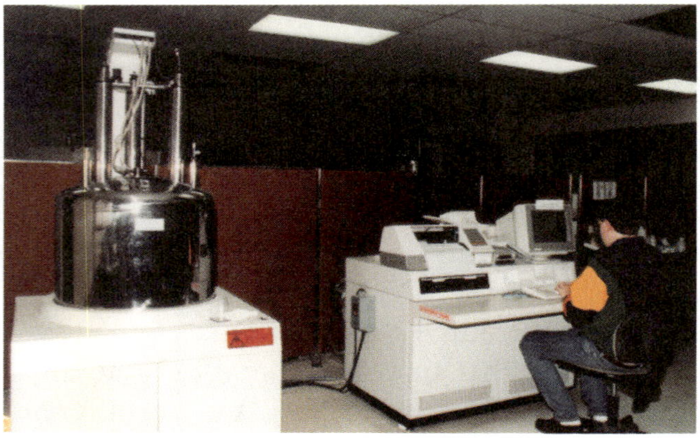

18-2. Using a nuclear magnetic resonance (NMR) machine to identify a compound. (Courtesy, Chemistry Department, Emory University, Atlanta)

When Changes Occur

When matter changes from one state to another, it has gone through a ***phase change***. Two conditions are involved when a phase change occurs: temperature and pressure.

Temperature. Temperature is measured with a thermometer and refers to the presence of heat. It is the activity of the molecules in matter. As heat increases, molecules move more rapidly. At a certain temperature, they are so active that they no longer stay together. At this point they become a gas. The liquid state occurs at a temperature range between a gas and a solid. When molecules move so slowly that they stop being a liquid, they become a solid.

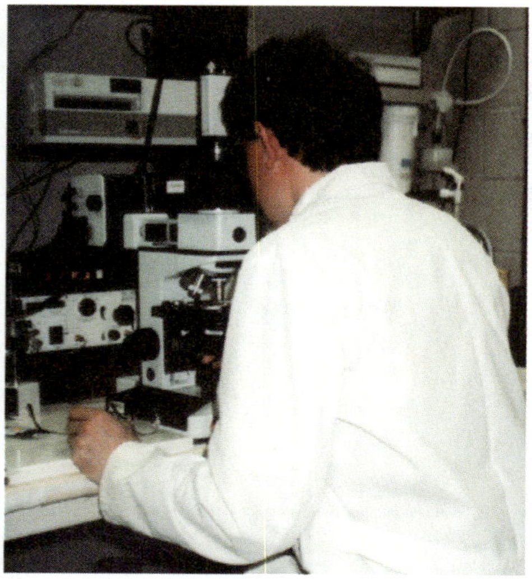

18-3. Complex equipment and laboratory procedures are sometimes needed in chemistry. This photograph shows the use of light microscopy to study surfactant aggregation. (Courtesy, Chemistry Department, Emory University, Atlanta)

Matter varies in the amount of heat required for the different phases to occur. Water becomes a solid at 32°F (0°C) and a gas at 212°F (100°C) at sea level. When water becomes a solid, it expands (increases in volume or gets bigger). The force of expansion can crack rocks, burst water pipes, or break the iron parts of engines.

Substances mixed with water change the temperatures at which water freezes and boils. This is why ***antifreeze*** is added to the water in engines that operate in cold weather or to the water that fills tractor tires. Most of the antifreeze is ethylene glycol. It lowers the freezing point of water. The higher the concentration of antifreeze, the lower the freezing point. This means that more antifreeze should be put in a cooling system during very cold weather.

Other kinds of matter change phases at different temperatures. Iron becomes a liquid at about 2,800°F (1,500°C). It returns to a solid below this temperature. Making steel involves heating iron to a liquid form and cooling it in various shapes and forms. Other matter may be added to the iron while it is a liquid (molten) to produce different kinds of steel, such as stainless steel.

Pressure. Pressure often refers to atmospheric pressure but can refer to other kinds of pressure, such as that in a fuel tank or can of food. Pressure affects the temperature at which substances change from solids to liquids and gases. A good example is water. As the pressure goes down, so does the temperature needed for it to boil and become a vapor.

Gases that are kept in storage tanks are often compressed. Examples include fuels, such as butane and liquefied petroleum, and refrigerants that help make coolers in flower shops cool. Solids and liquids can be compressed very little. Some gases will become liquids when compressed, such as liquefied petroleum.

Gases tend to occupy all of the space that is available to them. A tiny can of fumigant can be used in a large greenhouse to control white flies or other pests. This is because the fumigant gas spreads throughout all of the area of the greenhouse by ***diffusion***. Diffusion is when a substance moves from an area of greater concentration to an area of lesser concentration.

Solids will turn into liquid forms at lower temperatures when under pressure. Ice will become a liquid below the normal freezing point if pressure is applied to it. An example is under the blades of ice skates.

Properties of Matter

Matter has two types of properties: physical and chemical. Each type of matter has a set of identifying properties. These properties are important when learning to use chemical substances.

Physical properties can be determined without changing the identity or composition of matter. Physical properties include color, odor, solubility

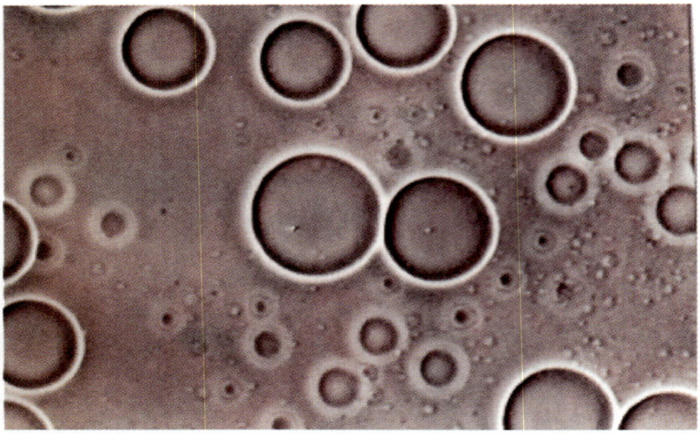

18-4. Studying the physical properties of matter may require considerable magnification of the compounds involved. This photograph shows greatly magnified soap vesicles used in surfactants. The compound is didodecyldimethyl ammonium bromide. (Courtesy, Chemistry Department, Emory University, Atlanta)

(how much dissolves), melting and boiling points, hardness, density, and crystal formation. Differences in matter can be determined by observing the physical properties. All matter doesn't possess all of the types of physical properties. Examples are a rock that may be odorless and the fact that liquid forms don't have hardness.

Chemical properties describe how matter changes when it is combined with other matter. Some matter reacts violently to being combined, such as when fuel burns. Other matter is relatively chemically inactive, such as gold or the noble gases. Information about how materials react is helpful when using them. Most all materials have labels that describe their dangers, such as the fact that ammonium nitrate fertilizer can be an explosive.

KINDS OF MATTER

Matter is classed into three general groups: elements, compounds, and mixtures.

Elements

Elements are substances that cannot be broken down into simpler materials by ordinary means. All known matter is made up of 109 elements; 92 of them are natural elements and the other 17 are artificial. Most of the artificial elements are radioactive and were discovered after the creation of the atomic bomb.

Each element has a name as well as a symbol. The major elements in the crust of the earth are oxygen, silicon, iron, aluminum, calcium,

Table 18-1
Common Elements and Their Symbols

Element	Symbol	Element	Symbol
Aluminum	Al	Magnesium	Mg
Arsenic	As	Manganese	MN
Calcium	Ca	Mercury	HG
Carbon	C	Nickel	Ni
Chlorine	Cl	Nitrogen	N
Chromium	Cr	Oxygen	O
Cobalt	Co	Phosphorus	P
Copper	Cu	Potassium	K
Fluorine	F	Silicon	Si
Gold	Au	Silver	Ag
Hydrogen	H	Sodium	Na
Iodine	I	Sulfur	S
Iron	Fe	Tin	Sn
Lead	Pb	Zinc	Zn

sodium, potassium, and magnesium. Oxygen is present in the largest amount. Oxygen makes up about 46% of the earth's crust.

The Periodic Table. The ***Periodic Table of the Elements*** is used to group and organize the elements for study. The table shows relationships and differences among the elements.

The properties of elements can be predicted from the table. The table includes the symbol for each element, atomic numbers, and atomic masses.

Most periodic tables are in columns and rows. The rows across are known as periods. The vertical columns are called groups or families. These groupings are based on similarities of the elements.

Atomic Numbers. All elements have ***atomic numbers***. This number is based on the number of protons in the nucleus of the atom. An ***atom*** is the smallest unit in which an element can exist. Some atoms can exist alone, such as gold (Au); others exist in pairs or different combinations, such as oxygen (O_2). The nucleus is the central part of an atom, has a positive charge, and contains most of the mass of the atom. The positive charge is caused by the presence of protons.

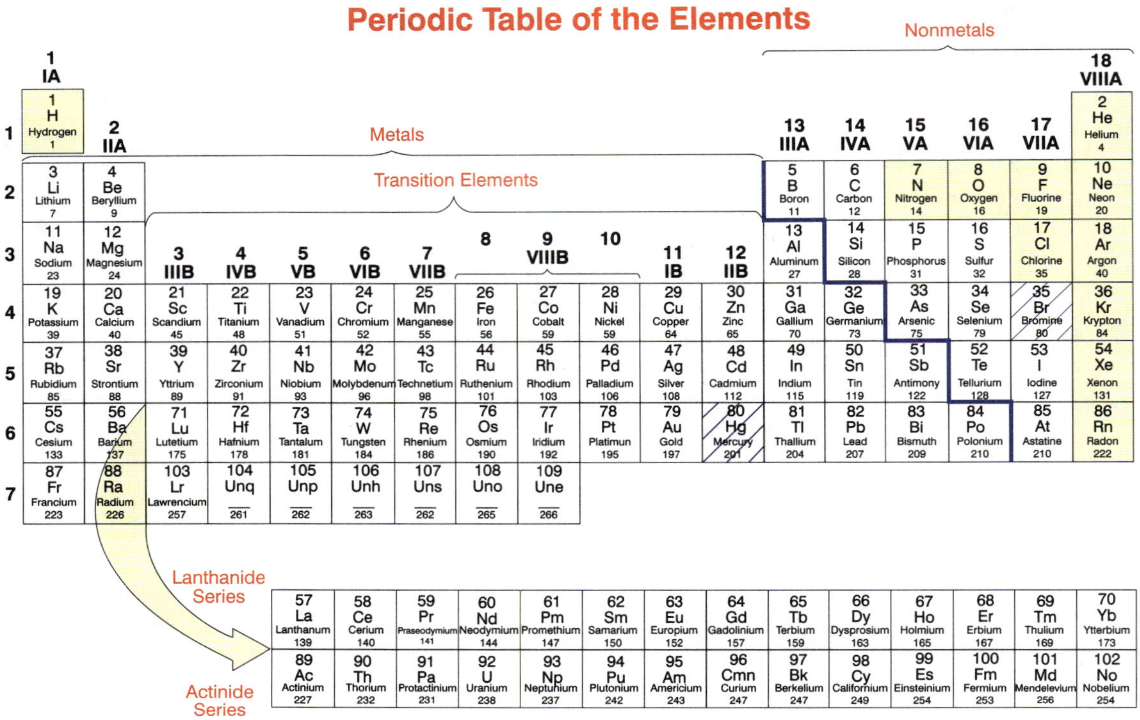

18-5. A Periodic Table of Elements.

The properties of elements are dictated by their atomic numbers. This is stated as the ***Periodic Law***: The chemical and physical properties of the elements are periodic functions of their atomic numbers.

The atomic mass (weight) of an element is the total of the number of protons and neutrons. Neutrons are in the nucleus along with protons. Neutrons are particles that have a neutral electrical charge and contain mass.

Electron Valence. Atoms have electrons that move in electron clouds like swarms of bees around the nucleus. Electrons have a negative charge and are important in the combination of elements to form compounds. The outer orbital may have an incomplete number of electrons. Trying to obtain a stable number of electrons causes matter to give electrons to or take electrons from other elements. The electrons involved in bonding are called valence electrons. The valence of an element determines how it will react chemically—with which elements and with how many other elements it will join to form a compound.

Two Categories of Elements

Each element has its own unique characteristics. However, there are some similarities among the elements. On the basis of the similarities, the elements can be grouped into two categories: metals and nonmetals.

18-6. Using a large Periodic Table of the Elements to study the elements.

Metals. ***Metals*** are easily identified by their brilliant appearance. Some have a metallic luster or shine. A few metals also have distinct colors, such as the color of gold (Au) and copper (Cu). Metals are good conductors of heat and can be formed into many shapes, such as iron into plows. Metals also transfer electrons when they react. Most metals are mined from nature as ore.

Most metals are solids at normal temperatures. One exception is mercury (Hg)—a material formerly used in gauges to measure temperature and pressure. The use of mercury has declined because it is highly poisonous.

Nonmetals. ***Nonmetals*** are poor conductors of heat. They are brittle and not easily made into other shapes. Nonmetals gain electrons when they react. Most are solids at room temperature, such as iodine (I) and phosphorus (P). Some nonmetals are gases at room temperature, such as

18-7. Three minerals used as sources of copper are chalcopyrite (top), azurite (bottom left), and malachite (bottom right).

neon (Ne) and radon (Rn). Chlorine (Cl), hydrogen (H), nitrogen (N), and oxygen (O) are common nonmetal gases.

Compounds

A ***compound*** is a substance that is made of two or more elements that have been chemically combined. Compounds can be broken into two or more simpler substances by ordinary chemical means. While combined, the two elements lose their separate chemical identity.

Pure elements rarely occur in nature. They are usually found in compounds. For example, pure sodium doesn't exist in the earth's crust. It is usually combined with other elements to form compounds, like table salt (sodium chloride—$NaCl$).

A compound is always made up of the same elements in a definite relationship. Thus, the ***Law of Definite Composition***: all compounds have a definite composition by mass.

The Law of Definite Composition is used to make many agricultural products. Since the same proportions of raw material are always used to produce a specific product, the amounts of ingredients necessary for a compound stay the same.

18-8. Table sugar is a compound with a definite composition by mass.

Mixtures

A ***mixture*** is a substance that has parts with different properties. Air is a mixture because it is made of several gases. Each gas in the air is a different substance and could be separated from the others usually according to their different boiling points.

Complete fertilizers are mixtures. Compounds that provide nitrogen, phosphorous, and potassium are blended together. The amounts of each substance to use in the blend are based on the intended analysis of the fertilizer. None of the fertilizer ingredients are used in elemental form; they are used as compounds. The element separates and becomes available to plants in the soil. Of course, availability depends on the pH and other characteristics of the soil.

CHEMICAL BONDS: NEW COMPOUNDS

Atoms of elements combine to form compounds. The same compound is always made up of the same atoms in the same ratio. Compounds are expressed in chemical formulas. Examples of common compounds are shown in Table 18-2.

Compounds can be formed as molecules. A ***molecule*** is the smallest group of atoms that acts together to form a stable, independent substance.

The simplest molecules are made of only two atoms. An example is O_2 (free oxygen). Oxygen free in nature is stable only when it bonds with another oxygen atom.

Many molecules are made of more than two atoms. An example of a molecule with three atoms is H_2O (water). One molecule of water is made

Table 18-2
Common Compounds Used in AgriScience

Compound	Formula	State as Found in Earth Environment
Acetic acid (vinegar)	$HC_2H_30_2$	liquid
Ammonium nitrate (nitrogen fertilizer)	NH_4NO_3	solid
Calcium carbonate (limestone)	$CaCO_3$	solid
Carbon dioxide	CO_2	gas
Citric acid (lemon juice)	$C_3H_4(OH)(COOH)_3$	liquid
Hydrochloric acid (muriatic acid)	HCl	liquid
Sodium bicarbonate (baking soda)	$NaHCO_3$	solid
Sodium chloride (table salt)	$NaCl$	solid
Sodium hydroxide (lye)	$NaOH$	liquid
Sodium nitrate (nitrogen fertilizer)	$NaNO_3$	solid
Sulfuric acid (oil of vitriole)	H_2SO_4	liquid
Water	H_20	liquid, gas, and solid

of two atoms of hydrogen and one atom of oxygen. A molecule of glucose sugar, as made by plants, has six atoms of carbon, twelve atoms of hydrogen and six atoms of oxygen ($C_6H_{12}O_6$).

IONS

As mentioned earlier, an atom has space in which electrons move around the nucleus. The outer space can be incomplete. It can gain or lose one or more valence electrons. When an atom gains or loses a valence electron, it develops a charge. This charged particle is known as an ***ion***. For example, a chlorine atom can gain one electron to become the chloride ion and have a negative charge. The process of gaining or losing an electron is known as ionization.

BONDING

Bonding is when two or more elements join together to form a compound. Valence electrons are shared or transferred by the atoms forming chemical bonds in the compound. Ionic bonds are formed when electrons are transferred. Sodium chloride ($NaCl$) is an example of ionic bonding. Covalent bonding occurs when electrons are shared. Natural gas (methane, or CH_4) is an example of covalent bonding.

Bonding usually results because two or more atoms with an unstable number of outer shell electrons form a stable compound. For example, water is a stable compound under many conditions. The elements oxygen and hydrogen aren't as stable as the compound water.

18-9. Set up for a reaction in a vented hood.

CHEMICAL REACTIONS

Four types of chemical reactions are used in agriscience, as follows:

- **Composition**—Composition occurs when two or more substances join to form a more complex substance. For example, $2Na + Cl_2 \rightarrow 2NaCl$.
- **Decomposition**—Decomposition occurs when one substance breaks down to form two or more other substances. For example, $2NaCl \rightarrow 2Na + Cl_2$.
- **Single replacement**—Single replacement results when one element replaces another element in a compound. For example, $Zn + 2HCl \rightarrow ZnCl_2 + H_2$.
- **Double replacement** (ionic)—Double replacement reactions occur when two compounds exchange elements. For example, $NaOH_{aq} + HCl_{aq} \rightarrow NaCl_{aq} + H_2O$. (Note: The symbol $_{aq}$ means aqueous or "in water.")

Some chemical reactions involve oxidation and reduction, which causes them to depart from these exact patterns. (Oxidation and reduction are covered later in this chapter.)

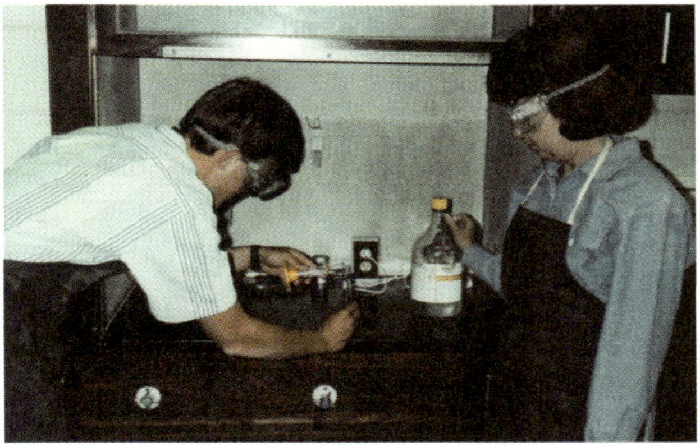

18-10. Testing a reaction. Note that the test is being made in an agriscience chemical laboratory ventilation hood. Eye safety is important.

CARBON: DIFFERENT AND NECESSARY

Carbon is an unusual and important element. It is found in all living things and in the foods they use. Carbon is different from other elements because it can bond itself in almost endless chains, designs, and shapes.

The study of carbon is known as ***organic chemistry***. It includes living organisms and the use of carbon in their structures. All organic compounds contain carbon and involve covalent bonding. Carbon ranks 17th among all elements in terms of volume found on the earth.

18-11. One molecule of aspirin is depicted by the CPK model. (Aspirin is chemically acetylsalicylic acid. The colors in the model represent these elements: red is O, black is C, and white is H.)

In addition to being found in all living things and in food, carbon is also important in other compounds. Coal, oil, natural gas, diamonds, graphite, and many other products contain carbon. Of course, wood and paper contain carbon.

Important gases contain carbon. Carbon dioxide is used by plants and given off by animals. Burning wood and fuel produces gases with carbon, such as carbon dioxide and carbon monoxide.

HYDROCARBONS

Many of the important resources used in agriscience are ***hydrocarbons***. The common hydrocarbons are used for fuel. They are found in crude oil and natural gas. Gasoline, lubricating oil, plastic, artificial fibers, and rubber are mixtures of hydrocarbons.

Hydrocarbons are fairly simple organic compounds made of only two elements: hydrogen and carbon. The difference is based on how the carbon bonds are arranged and the type of bonding. The number of atoms varies with the kind of hydrocarbon compound.

Natural gas is mostly methane (CH_4). It is found in porous formations below the surface of the earth. Other petroleum hydrocarbons are often located near natural gas. Natural gas and petroleum were formed millions of years ago by plants and animals that decayed under pressure.

Most petroleum undergoes refining to make it into more usable forms. Gasoline, diesel fuel, engine oil, and grease are produced at different stages in the refining process.

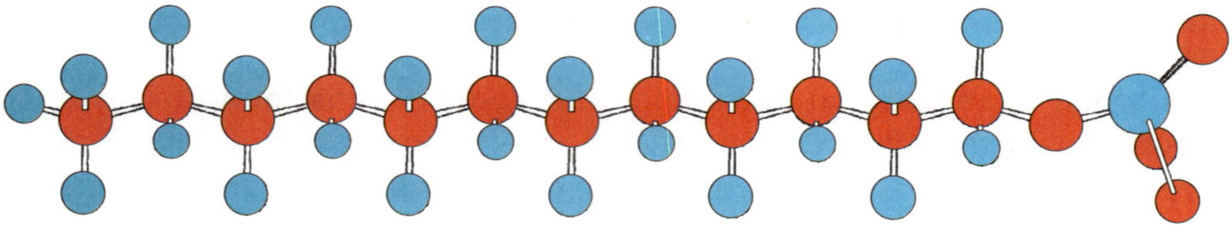

18-12. A computer-generated hydrocarbon molecule. (This shows lauryl sulfate [$CH_3(CH_2)_{10}CH_2OSO_{3-}$], a substance used in detergents and toothpaste.) (Courtesy, Chemistry Department, Emory University, Atlanta)

AgriScience Connection

WASHING OFF DANGEROUS CHEMICALS

Some chemical substances are extremely hazardous. Labs where they are used should provide appropriate equipment in case of an accident. A few substances can cause quick damage to the skin and body.

A first step with many substances is washing with plenty of water. Labs often have emergency showers and eye wash stations. If acids, bases, or other liquids get spilled or splashed on you, quickly go to the emergency station and turn the water on.

SUBSTITUTION COMPOUNDS

Alkanes are hydrocarbons that have single covalent bonds only. One of the hydrogen atoms can be replaced with an atom of another element. The halogen elements, chlorine, fluorine, bromine, and iodine, can combine easily with hydrocarbons to form halide compounds. Minor changes in the carbon bonds result in different products that are similar in some respects.

The compounds known as ***alcohols*** are made from organic halides. Several kinds of alcohol are used in agriscience. Methanol and ethanol are used as fuel or as additives to fuel. Some of these are made from grains and may be combined with other fuels. An example is gasohol, which is a combination of 10% ethanol and 90% nonleaded gasoline. Fuels made from grain tend to be more expensive than those made from petroleum.

Other forms of alcohol include ethylene glycol (a form of alcohol that is used as antifreeze) and glycerol, or glycerin (a form of alcohol used as an ingredient in many products, including medicines, cosmetics, and soap as well as in cellophane manufacture).

SOLUTIONS AND SUSPENSIONS

Solid materials are sometimes dissolved in water or other fluids. Dissolving breaks apart a solid into particles that are loosely attached to the solvent. Agriscientists often use liquids containing other materials, such as pesticides and biologicals (kinds of animal medicines).

The solid materials may be diffused (spread) throughout the liquid in two forms: solution and suspension. Sometimes the materials don't stay diffused. Steps need to be taken to get them to stay in a particular form.

Solutions and suspensions don't create new chemical compounds. They are mixtures of two or more materials that could be separated back into their original forms. In some cases, however, separating them is fairly difficult or impractical.

SOLUTIONS

A ***solution*** is a mixture of two or more substances. The amounts that can be mixed together vary. There is usually a limit to the amounts that will go into solution.

The liquid material is known as a ***solvent***. It dissolves the other substance. The substance that is dissolved is known as a ***solute***. Molecules of the solute are distributed evenly throughout the solvent.

An easy-to-see example of a solution involves placing a couple of drops of food coloring in a glass of water. The food coloring evenly spreads throughout the water by diffusion (moving from higher to lower concentration). After a period of time all of the water will have an equal amount of color.

Salt and sugar are also commonly mixed into solutions. In both cases, the water will dissolve only a certain amount of the salt or sugar. Beyond this amount, the salt or sugar will remain as crystals in the bottom or accumulate around the top of the container.

In a solution, the material passes with the liquid through various filters. A salt or sugar solution will pass through the filters with the water.

SUSPENSIONS

Solutes and solvents don't always form good solutions. The solutes may be of particles too large to dissolve and remain suspended by the movement of the solvent molecules. The solution will have a cloudy appearance and the materials may settle to the bottom of the liquid.

Colloids

Colloidal suspensions involve materials in the liquid that can be fairly easily filtered out with parchment or other semipermeable membranes (semipermeable membranes let some substances through but not others). For example, clay soil particles will be filtered out even though they may appear to be thoroughly mixed in water.

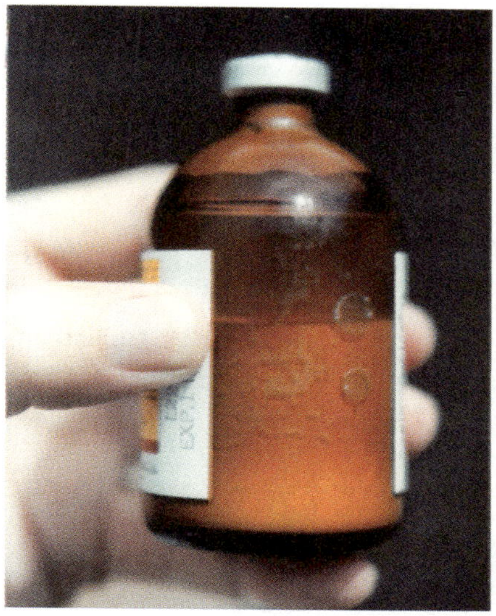

18-13. This bacterin is a colloidal suspension. Note that the level of the liquid is near the top of the bottle. The solid materials have settled beginning about half-way down the bottle. The bacterin must be shaken before use to keep the particles in suspension.

Emulsions

Emulsions are colloidal suspensions made of two solutions. One liquid is suspended as tiny droplets in the other liquid. Oil and water don't normally mix but when placed in a container and shaken, they will break into tiny droplets that will stay suspended. After a few minutes, the water will begin settling to the bottom and the oil will rise to the top.

Oil and water don't mix because of ***polarity***. Polarity is when different substances have different negative and positive charges. Water molecules are strongly attracted to each other. Oil molecules have a weak attraction for each other. Even though oil and water are liquids, they don't mix because water has a higher polarity. Water is denser than oil; when the two are mixed, the water rapidly separates from the oil and goes to the bottom.

Milk is a good example of an emulsion. The butter fat (which is oil) will rise to the top as cream. Milk processing plants homogenize milk by breaking the fat particles into smaller pieces that stay suspended in the liquid milk.

In fuel storage tanks, water will settle to the bottom of the tank and not be pumped into the fuel tanks of the engines being filled. If more fuel is added to the storage tank, the motion causes the water to be mixed with the fuel. It could get into engine fuel tanks if pumped before it has a chance to settle. This is why some people never buy gasoline at a station that is having fuel added to its storage tank.

AGRISCIENCE APPLICATIONS

Many solutions and suspensions are used in agriculture. Knowing the characteristics of each will help in their use.

Removing Suspensions: Filtration

Many kinds of filters are used in agriscience. Some are needed to protect human health; others are used to protect engines, water systems, and other systems involving liquids that could be contaminated. A few examples are presented here.

18-14. A glass bowl fuel filter separates water from fuel by collecting water at the bottom. Any water is easy to see and is removed by a small valve underneath.

- **Fuel filters**—Fuel for diesel and gasoline engines should pass through a filter before it reaches the engine. If it doesn't, delicate parts of the engine could become clogged. For example, the diesel injector nozzles could become clogged with small particles of rust, soil, bacteria, or other solids. Filtering will remove materials that are suspended in the fuel. Filters must be regularly replaced because they become clogged with particulate.
- **Dust**—Some people work in dusty environments. The dust is made of tiny particles of substances (soil, pollen, fertilizer, grain, etc.). Since dust is a suspension, masks over the nose and mouth (also known as respirators) can filter these materials from the air that people breathe.
- **Chemicals**—People applying chemicals in fields, greenhouses, or other places need to wear masks that protect their respiratory systems from particles suspended in the air. Tiny droplets of water will be filtered out of the air and not reach the lungs. Solutions of gas, however, may not be removed by filtration, such as a gaseous fumigant in a greenhouse. Masks with filters containing charcoal are more likely to filter out particles because they stick to the charcoal.

Having a Good Mix: Pesticides

Pesticide materials that are sprayed on crops or livestock often use water as a carrier or solvent. In some cases these materials are oil-based and do not mix well with water. Here are two examples.

- **Surfactants**—Surfactants are known as wetting agents and make a solution "wetter." Surfactants help oil and water mix evenly. Even the leaves of plants have a waxy coating that resists the even flow of a water-based pesticide over them. A surfactant helps overcome the failure of oil and water to mix. Small amounts of surfactants are often added to pesticides as they are being mixed. Some people use soap as a surfactant.
- **Tank agitation**—Pesticide application equipment should keep the solution well mixed as it is applied. Materials that don't mix well should be used in tanks with agitators. The agitators move the liquid around and keep it mixed well. If not mixed sufficiently, the pesticide will settle to the bottom and not be evenly applied to the crop.

18-15. Active ingredient must be mixed throughout a pesticide. This shows the use of a chemical "soaked" into ground corn to control ants.

Giving a Good Dose: Animal Medicines

Animal medicines may be suspensions of materials. Two examples are presented here.

- **Biologicals**—Liquid medicines given to animals must provide the recommended uniform dosage. For example, an injectable antibiotic may be an aqueous (water) suspension. If the bottle sits in the refrigerator for a few days, the antibiotic material will settle to the bottom of the bottle. The top of the bottle's contents will be clear and the bottom will be white. The bottle must be shaken to get all of the antibiotic in suspension. If not, an animal will not get the intended dosage.

- **Orally-administered liquids**—Animal medicines may be given by mouth using a dose syringe. A large number of animals may be given medicine from the same container. An example is cattle wormer. The medicine should be thoroughly mixed before the first dose is given and kept mixed thereafter. If not mixed properly, some animals may not get enough of the wormer to rid them of worms. Others may get too much.
- **External therapeutants**—Medicines are sometimes mixed with water or other liquids before being applied to an animal. A good example is formalin used to treat fish in a tank. The formalin must be mixed throughout the water in all parts of the tank. If it isn't, some of the fish won't be treated.

ACIDS, BASES, AND SALTS

Groups of substances that form compounds and solutions with electrical charge are known as acids, bases, and salts. The solutions they form contain ions and are sometimes known as electrolytes.

Acids, bases, and salts can be very dangerous to people, other animals, property, and the environment. When used properly, some of these substances are useful in food preparation or in industrial processes. Regardless, extreme care is essential. These materials can often cause severe burns and injury to the body.

Measuring the strength of these materials involves using a numeric scale known as pH.

pH

pH is a measurement that represents the number of hydrogen ions in a solution. It ranges from 0 to 14; 7.0 is neutral. Below 7.0 is acidic and above 7.0 is basic (alkaline). A neutral solution, such as pure water, has a pH of 7.0. (Soil pH was discussed in the previous chapter.)

Acid materials have a number below 7.0. The lower the number, the greater the acidity. For example, a pH of 4.5 is far more acidic than a pH of 6.0.

Basic materials have numbers above 7.0. The higher the number, the greater the basicity or alkalinity. A pH of 8.5 is more basic (alkaline) than a pH of 7.5.

Acid materials can be made less acidic by adding basic materials to them. This is the way soil pH management works. Lime or crushed limestone, with a pH of about 8.2, is added to soil with a pH of 4.8 to raise the pH and lower the acidity.

pH is often measured with a pH meter or with paper strips that have color indicators. Using the specially treated paper to measure pH is known as colorimetric pH determination.

Most plants and animals have preferred pH ranges where they attain their best growth and health.

18-16. Liquid compounds must be carefully tested to determine pH and the presence of acids, bases, or salts. (Courtesy, McCarty Foods, Jackson, Mississippi)

ACIDS

Acids are compounds that give up protons to water molecules to form hydronium ions. Protons are designated as H^+ in water. It is a hydrogen atom that has transferred an electron to another atom. The hydrogen ion does not have a negative electron and is thus a positive ion. This makes the H^+ ion have strong attraction to the polar water molecule forming the hydronium (H_3O^+). An acid is a proton donor.

Acids vary in strength. Vinegar (acetic acid) is a weak acid used in food preservation. Lemon juice (citric acid) is often used in foods and beverages for flavor. Other acids have agricultural uses, such as hydrochloric acid (also known as muriatic acid). The formula for hydrochloric acid is HCl. In addition, some fertilizers will form acids in the soil, such as sulfur which is used to increase the acidity of soil for certain crops. The sulfur forms sulfuric acid (H_2SO_4), which lowers the pH of the soil. (Acids taste sour like grapefruit. Grapefruit contain citric acid.)

BASES

Bases are compounds that produce hydroxide ions (OH^-) in water. Household ammonia (NH_4OH), lye (NaOH), and slaked lime [$Ca(OH)_2$]

are three examples of bases. A base is a proton acceptor. Bases taste bitter like soap.

Water-containing bases, such as ammonia, are used to clean and control the growth of microorganisms. Compounds that are basic (alkaline) are added to the soil to raise the pH (make it less acid). Slaked lime or crushed limestone are often used on soil.

SALTS

Salts are formed when strong acids and bases are combined. This most often occurs when an acid is neutralized with a base. Their combined pH adds to 7.0. For example, when sodium hydroxide (lye) and hydrochloric acid are combined, sodium chloride is formed along with water. Sodium chloride has neither an acid nor a basic taste—its tastes salty! However, there are many other kinds of salt.

Salts tend to have a neutral pH, but the pH of some salts varies. Mixing an acid and a base doesn't necessarily produce a neutral (7.0) pH. The pH of the salt depends on the strength of the acid and base that are combined. A strong acid and a weak base will form a salt that has an acid pH. A strong base and a weak acid will form a salt that has a basic pH.

The operation of some kinds of batteries (known as dry cells) uses salt-forming processes to make electrical energy. A paste of ammonium chloride (NH_4Cl) is placed inside a zinc casing. Turning on the switch (of a flashlight or radio, for example) causes contact to be made so that the electrical charges move and conduct electrical current.

Concentrations of different kinds of salt can build up in soil. This most often happens when irrigation water contains a salt or has basic materials that react with acid materials. Good quality irrigation water is free of salt and salt-forming substances. Water should be tested before it is used for irrigation.

CHEMICAL CHANGES

Chemical changes are important in agriscience. A chemical change occurs when a substance becomes another substance with different characteristics. ***Chemical reactions*** are involved in this chemical change process. Compounds different from the elemental (element) forms result. Sometimes, a compound reacts with another compound to form different substances. However, some compounds and elements don't react with each other.

Some chemical changes are a natural part of life processes, others are created by people. In many cases, chemical changes in living things can be encouraged or slowed down by changing the temperature, amount of nutrients, and other conditions.

PLANT AND ANIMAL FOOD

Important chemical changes are carried out by plants and animals. The world's food supply depends on creating conditions for plants and animals to complete chemical processes.

Photosynthesis

Photosynthesis is the process whereby plants make the sugar known as glucose. This sugar serves as the major food for the plant and for the animals that eat plants. By-products of photosynthesis include non-food fiber and shelter for humans.

The photosynthetic process involves plants using carbon dioxide from the air and water from the soil to produce sugar and elemental oxygen. Chlorophyll and sunlight must be present for the process to occur. The sugar is stored in various forms in the plant, including starch in tubers and seed.

This process is probably the most important chemical change that takes place on the earth! Agriscience involves encouraging the process. Plants can be given fertilizer and water. Good conditions help plants be more productive in their sugar making.

18-17. Growing plants in sand tubes allows careful testing of needed chemical elements in research laboratories.

Respiration

Animals reverse the process of photosynthesis. They eat the sugars made by plants and use the oxygen in respiration. The sugars are broken down into products the body can use by the process of digestion. The animals breathe in oxygen and give off carbon dioxide and other substances. Plants in turn use the materials given off by animals for the process of photosynthesis.

Agriscience involves helping animals have the right nutrients and keeping them healthy so that the processes can occur.

CORROSION: LOSS OF PROPERTY

Chemical reactions can destroy and weaken property. This is known as ***corrosion*** or rusting. It nearly always involves changes in a metal of some type, such as iron or silver. In most cases, two processes are involved: oxidation and reduction.

Oxidation is when the atom of the metal gives up electrons. Most of the time oxygen is involved, hence the name oxidation. The metal becomes more positive in its oxidation state.

18-18. The brown color of this new water tank is a sign that it is beginning to rust. The rust must be removed before painting.

Reduction is when the electrons given up in oxidation are captured by atoms of the same or another metal. The metal attains a more negative oxidation state.

Oxidation and reduction are illustrated by the rusting process.

Rusting

Rusting is primarily associated with iron and some of its alloys (mixtures). As iron rusts, it weakens and loses its strength. Rusting occurs when iron combines with oxygen in the presence of moisture. Unprotected structures made out of iron will rust rapidly if left out in the rain.

Rust is a brownish material formed when iron or steel is exposed to damp air. The chemical reaction that occurs produces hydrated iron oxide ($Fe_2O_3 \cdot 3H_2O$). Oxygen from the air combines with the iron in the presence of water.

Farm machinery should be painted and stored under a shed. Shiny plow points can be coated with oil to protect them from rusting.

Galvanized metal products are sometimes used because they don't rust. Examples of galvanized metals include zinc-coated metal roofing and zinc-coated barbed fencing wire. Zinc doesn't oxidize and it protects the iron underneath from oxygen and water. Over the years, zinc may wear thin or be weakened by acid rain. When this occurs, the iron is then exposed to oxygen and will likely rust.

Rust should be removed before re-painting machinery or a metal roof. This is done by brushing with a wire brush to get down to the bright metal or grinding in other ways. Acids are sometimes used to remove rust, but are impractical on most machinery. Once the rust has been removed, the metal should be painted as soon as possible. If not repainted, the metal will quickly rust again.

Acid Reactions

Acids react with metals, weakening and destroying them in the process. Some contact may be unavoidable; however, steps can be taken to minimize damage.

Fertilizer materials often form acids, particularly those containing nitrogen. As with rusting, moisture speeds the process along. Improperly stored fertilizer can cause metal buildings and equipment to corrode. Equipment that applies fertilizer should be cleaned after use to remove any residues that could cause corrosion. Avoiding spills and cleaning are two important steps to take in preventing corrosion. In addition, plastic-lined storage areas and equipment can help prevent corrosion.

18-19. The battery stores and generates electrical energy for an engine. Chemical reactions inside the battery generate the electrical power needed to turn the starter motor when cranking the engine on this tractor.

Batteries are used in many machines to provide electrical energy. A good example is the battery in a pick-up truck that starts the engine. Almost everyone is familiar with corrosion around a battery ter-

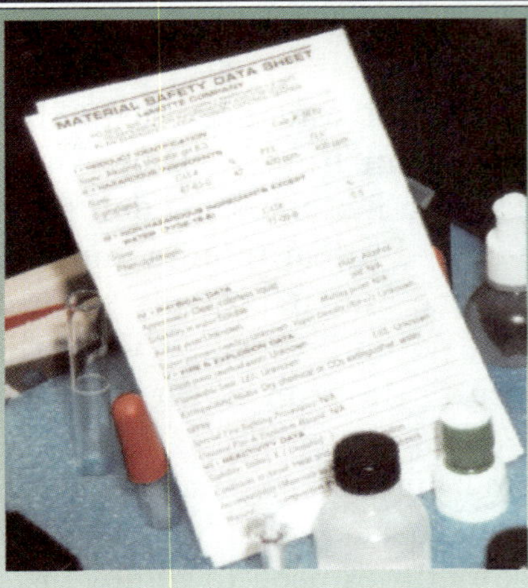

AgriScience Connection

BEING SAFE: USING MSDS

People need to know about the chemicals they are using. All chemical substances must have a Material Safety Data Sheet (MSDS) from the manufacturer. Federal regulations require that the sheets provide information about the substance and practices for safely using it. Information on fire and explosion hazards is also included.

MSDS may accompany individual containers of substances or be a part of a kit containing several substances. Kits with more than one chemical substance will likely have a separate sheet for each. These sheets should be kept available in case of an emergency or the need for information.

minal. A battery operates on the basis of chemical reactions. Sometimes this reaction gets to places where it is damaging. A small amount of the sulfuric acid in the battery may get on the outside of the battery. It will react with the lead and copper, forming a large amount of corrosion. The corrosion should be washed away with water containing baking soda. The baking soda neutralizes the acid. A thin layer of oil should be applied to the battery post and cable when it is dry. (Care should be taken to avoid contact with the acid or corrosion. They can burn skin and make holes in clothes.)

PREVENTING SPOILAGE: FOOD PRESERVATION

Various chemicals are used to keep food safe and improve its taste. Acids, bases, and salts are often added to food. Mild forms of certain chemical compounds improve the flavor of food. Some compounds prevent the growth of microorganisms that cause spoilage.

Pickling, salting, and using bases are examples of ways chemicals are used to preserve food. More information on these is presented in Chapter 21.

MEASUREMENTS IN CHEMISTRY

Chemistry often involves making measurements. Very small quantities are sometimes involved. These measurements must be accurate and carefully made. Errors can result in inaccurate results and cost agricultural producers a lot of money.

Tests must often be performed on soil, water, feed, food, and other substances. As discussed in the previous chapter, testing soil for nutrients involves a process of collecting a representative sample and using appropriate analyses. The findings must be interpreted in terms of what a crop needs.

18-20. Laboratory tests of soil cultures.

Measurements in chemistry are made using the metric system. Most materials are equated with water. For example, 1 milliliter (mL) of water is 1 cubic centimeter (cm^3) and has a mass of 1 gram (g). Another common example is the larger amount based on a liter: 1 liter (L) of water is equal to 1,000 cubic centimeters and has a mass of 1 kilogram (kg). (Refer to Chapter 3 for a detailed explanation of the metric system. Three convenient tables are presented in Appendices A, B, and C.)

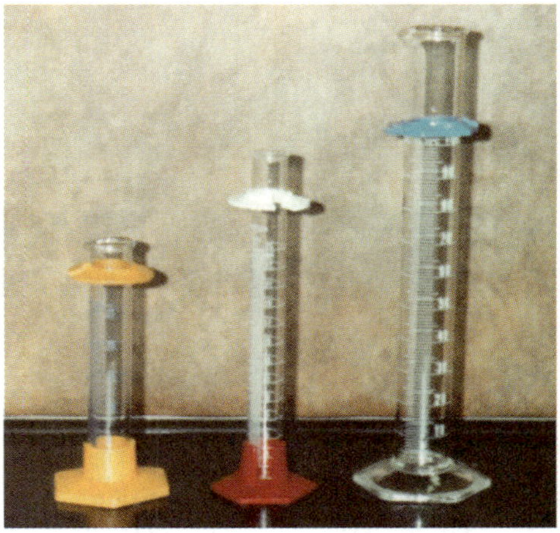

18-21. Graduated cylinders are used to accurately measure liquid chemical compounds.

18-22. Beakers graduated in milliliters are often used in chemistry laboratories.

AgriScience Connection

CHEMICAL PROCESSES ARE USED TO CLEAN WATER

Hazardous heavy metals sometimes pollute water. Sophisticated water cleaning facilities are used to remove the metal substances from the water. This shows a treatment plant for removing mercury, lead, and cadmium from water. Without cleaning, the water is hazardous to plants and animals.

A good knowledge of chemistry is needed to design and operate plants to remove metals from water. This plant is operated by the Unipure Environmental Company and is processing water from a site that has been polluted. (Courtesy, Unipure Environmental, Fullerton, California)

REVIEWING

MAIN IDEAS

Chemistry is the study of the substances that make-up living and nonliving things on the earth. Scientists have identified 109 different elements; 92 of these occur naturally.

Matter is anything that has volume and mass. This means that it occupies space and has mass. Mass is the amount of matter that something contains. Temperature and pressure can cause matter to change from solid to liquid, liquid to gas, etc.

Chemical elements can be placed in two categories: metals and nonmetals. Elements form compounds with different characteristics by chemically combining. Mixtures are different in that they do not chemically interact. Mixtures are only physically combined and keep their different chemical properties.

Carbon is a very important element in plant and animal life. Even though it is not one of the most abundant elements, it is found in many compounds. Hydrocarbons are found in petroleum. Alcohols are related to the hydrocarbons.

Acids, bases, and salts are substances with electrical charges. Salts are formed when bases and acids are mixed. Most salts have a near-neutral pH. Acids have pH ranges below 7.0, while bases (alkalines) have pH ranges above 7.0.

Solutions and suspensions are important in agriscience. Some solutions are made when something is dissolved in a liquid substance. The liquid is known as a solvent, with the substance dissolved known as a solute. Water is the most common solvent. Suspensions involve solid materials (colloids). They are mixed in a liquid, but often settle out. Emulsions involve mixing two solutions, such as water and oil. Sometimes they don't mix very well.

Many applications of chemistry are found in agriscience. These range from soil fertility to disease prevention, preservation of food, and protection of property from rust and corrosion.

QUESTIONS

Answer the following questions using complete sentences and correct spelling.

1. What is the law of conservation of mass?
2. What two conditions are involved in phase changes?
3. What are the two properties of matter?
4. What is an element? How many elements are known to exist naturally? Artificially? Name five common elements and give their symbols.
5. What is the Periodic Table of the Elements? How is it used?
6. What is the periodic law?

7. Name and distinguish between two categories of elements.
8. What is a compound? How is a compound related to the law of definite composition?
9. What is a mixture?
10. What is organic chemistry? What two major areas are involved?
11. Explain the differences between solutions, solvents, and solutes.
12. How do suspensions, colloidal suspensions, and emulsions differ?
13. Explain the following as related to agriscience: filtration, surfactants, and agitation.
14. What are acids, bases, and salts? How do they relate to each other?
15. What important agriscience chemical reactions occur in oxidation and food preservation?

EXPLORING

1. Test the effects of antifreeze on water. Add antifreeze to one of two containers with equal amounts of water. Place both containers in a freezer. Regularly check to see which freezes first. Be careful to use containers that won't easily break upon freezing.

2. Visit a supermarket and make a list of products that are solutions, colloidal suspensions, and emulsions.

3. Examine the label on a pesticide (herbicide or insecticide) container. Write down the chemical formula for the material. Make a list of the elements found in the active ingredient.

4. With the assistance of your teacher, test in the laboratory a compound to determine the elements that are in it. Use chemistry books and other materials for references.

19

APPLYING PRINCIPLES OF PHYSICS

The ways food, fiber, and shelter are produced have changed, but the fact that humans need these things hasn't changed. As the world's population increases, the needs of more people must be met.

People have developed ways of getting more done with less physical effort. Just think of the number of acres one person can plow with a large tractor in a day—10, 20, 30, or more acres. Compare this to digging in the soil with a hand shovel. Think further: suppose the shovel didn't have a handle! It would take a long time to get the job done!

People use machines, energy, and many other aids to help provide food, fiber, and shelter. Masses of people require mass production. If not produced in mass, very few people will get the things they want.

Agriscience involves using many principles of physics. People who work in certain areas of the industry need a superior knowledge of physics.

19-1. Building construction involves many principles of physics. Using a hammer to drive a nail is just one example.

OBJECTIVES

This chapter introduces the use of physics in agriscience. It has the following objectives:

1. Explain areas of physics used in agriscience.
2. Explain work and power.
3. Name and explain simple machines.
4. Describe mechanical advantage.
5. Explain the use of thermal energy in agriscience.
6. Explain the use of electrical energy in agriscience.
7. Describe the use of compression in agriscience.

TERMS

acceleration
alternating current
ampere
circuit
conductor
current electricity
deceleration
direct current
draft
electricity
force
ground fault circuit interrupter
heat engine
horsepower
hydraulic power
inclined plane
insulator
kinetic energy
lever
machine
mechanical advantage
mechanics
motion
optics
physics
pneumatic power
power
pulley
screw
sound
speed
thermal energy
volt
watt
wedge
work

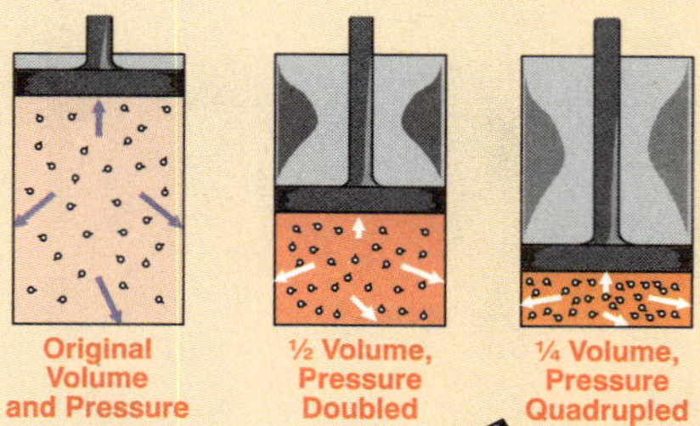

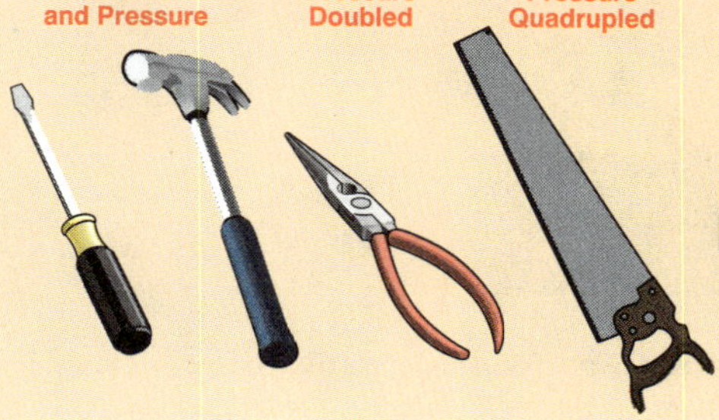

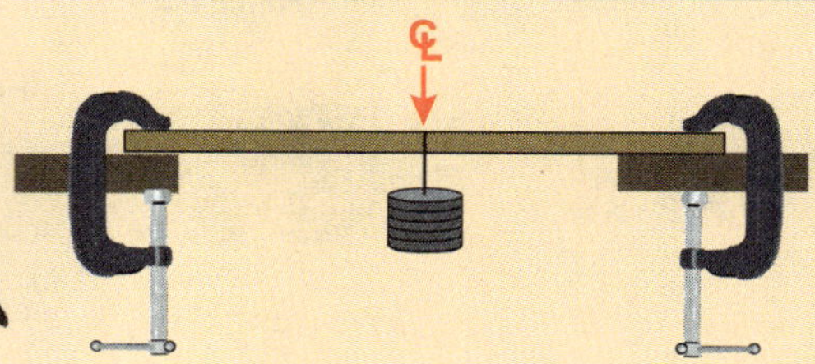

PHYSICS: WHY THINGS DO WHAT THEY DO

Physics deals with matter and energy and how the two relate to each other. The focus is on material things in the world and how people can use them to their advantage. Physics involves studying and using energy in various forms. The application of physics in agriscience is often referred to as mechanical technology. (The major emphasis in this chapter will be on principles of mechanical technology.) Another important area is computer technology. Physics involves the use of mathematical science, with major emphasis on mathematical relationships. Mathematics is used to express principles or laws of physics.

AREAS OF PHYSICS

Physics is often divided into several areas of study. Some of these have considerable applications in agriscience.

Mechanics

Mechanics deals with forces exerted by objects when at rest or moving. Objects that are in motion are known as dynamic, while those that are at rest are known as static. Energy is required for an object to move.

AgriScience Connection

PHYSICS AND JACKET LETTERING

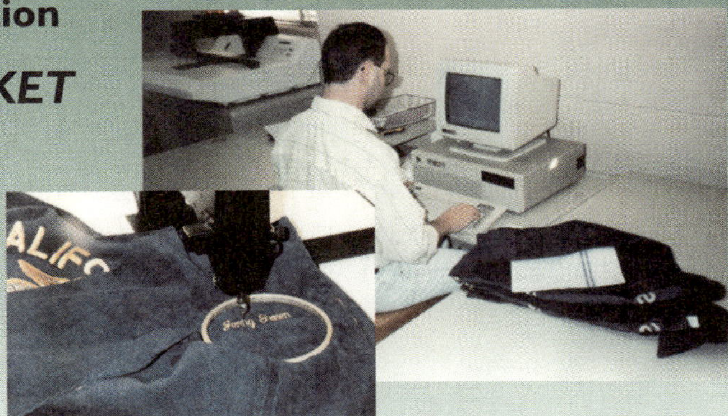

Many students in agricultural education are FFA members and wear official FFA jackets. The jackets usually have their name, chapter name, and state carefully lettered on the corduroy material.

Today jackets are lettered using a computer-controlled sewing machine. The lettering to be put on the jacket is keyed into the computer. The jacket is placed under the machine where it is automatically lettered. All letters are precise and the same on all jackets. This is a big change from a few years ago when all lettering was done by hand!

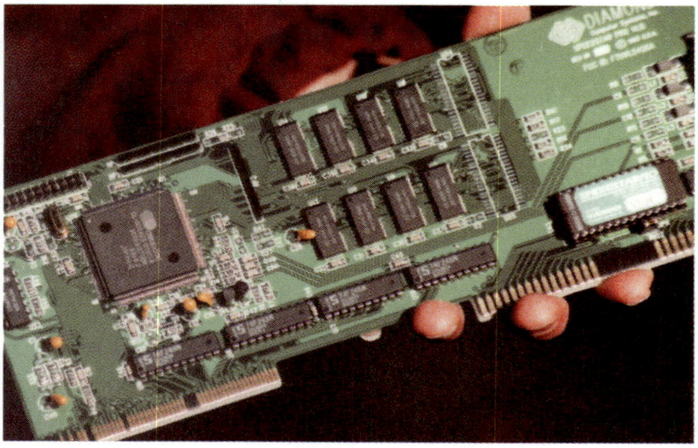

19-2. Transistors, integrated circuits, and other uses of physics have made the modern computer possible.

Many mechanical devices are used in agriscience. These apply various principles of physics. They also use other energy in their operation, such as a tractor that uses diesel fuel in its engine.

Particular areas of physics that relate to agriscience include hydraulics and compression. Hydraulics includes the movement of fluids to perform useful functions. Compression uses increased or decreased air pressure to perform work.

Heat

Heat is a form of energy that results from the motion of atoms and molecules. ***Kinetic energy*** is a result of an object's mass and motion. Heat refers to the total kinetic energy of an object. This can be compared with temperature, which is the average kinetic energy. Heat is the flow of energy caused by temperature differences. Temperature measures the change in heat. Temperature and heat are related but are not the same thing.

Heat is important in many agriscience activities. It is a factor in internal combustion engines, crop drying equipment and keeping young animals warm in cold weather.

Sound

Sound results when an object vibrates or moves. Waves are sent by vibration through a medium, such as air or water. Human ears pick up these vibrations and translate them into sound.

Sound is managed in different ways. A muffler is put on an engine to reduce the sound given off as the fuel suddenly burns. People who work

19-3. Electric motors are used in many areas of agriscience. This cut-away shows the coils of wire, magnets, and other parts of a motor.

near loud noises should wear hearing protection devices to keep the sound from damaging their hearing.

Ultrasonics are sounds that make vibrations too high for the human ear to hear. Various noise-making devices are used to frighten away unwanted animals. A good example is the use of ultrasonics to frighten cormorants away from fish hatcheries and growing ponds. Cormorants are birds that can eat many pounds of fish over a growing season. It is illegal to shoot a cormorant with a gun.

Magnetism and Electricity

Together, magnetism, and electricity are often known as electromagnetism. Electricity can produce magnetism. Magnetic forces can be used to produce electrical fields, such as the fields around the coils in engines or speakers in radios.

Magnetism can also be used to produce electricity. Electric power plants use the mechanical energy of falling water or steam to operate generators that produce electricity (electrical energy). Electric motors produce electromagnetic fields. Large windings of wire are used around magnets in building electric motors.

Light

Light makes vision possible. It is more than vision of the human eye; it includes the use of microscopes, lens and mirrors, and laser beams. Studying the use of light is known as ***optics***. The use of laser equipment in agriculture is a practical application of optics.

Electron and Nuclear Physics

Electron and nuclear physics are often treated as two separate areas. Electron physics deals with understanding the structure of atoms and molecules. Nuclear physics is concerned with the atomic nucleus, particle acceleration, and radioactivity. These processes are used to produce energy, preserve food, treat diseases, and for other uses.

USING PHYSICS

Much of what is done in agriscience involves using physics. It is not always necessary to understand the processes that are involved, but often knowledge of the physical principles involved allows people to do a better job. Almost everything people do involves the use of physics.

WORK AND POWER

Work and power are often needed to accomplish a task in agriculture. As used in physics, work is a little different from the duties of a job.

WORK

Work is moving an object through a distance when there is some resistance to its movement. Resistance (also known as inertia) is opposition to movement. The more mass an object has, the greater the inertia. A heavy bag of feed doesn't slide easily on the warehouse floor—it offers resistance.

A good example of resistance is what happens when a plow moves through the soil. The plow must turn the soil as well as break roots of

19-4. Pushing a pickup truck produces work only if the truck is moved.

plants that may be growing in it. This resistance to movement of a plow is often called ***draft***. Several things can happen if the plow gets too deep and the soil is hard to turn. The engine of the tractor will labor under the load and more fuel will be used. The plow or some part might break if the load is too great. If the draft is excessive, the tractor could stall and the engine quit running.

Two things are necessary for work to occur. First, work requires that force be applied to the object and, second, the object must be moved a distance. The object must move in order for there to be work. If an object is pushed very hard but doesn't move, no work has been accomplished.

Force

Force is the push or pull on an object. The object is either moved or damaged by force. This is what happens when an automobile crashes into a tree. The force of the impact damages the automobile and may break or damage the tree, depending on the size and force of the car.

Formula

Just as with most principles of physics, work can be expressed as a mathematical formula. Work is the force applied multiplied by the distance through which the object is moved. The formula is:

$$\text{work} = \text{force} \times \text{distance} \; (W = F \times d)$$

Both metric (SI) and English measures of work are used. In the SI, force is measured in newtons (N) and distance is measured in meters (m). This means that work is newton-meters. One newton-meter is equal to one joule (j). Work is calculated as joules.

In the English system, force is measured in pounds (lb) and distance in feet (ft). Work is calculated as foot-pounds. The same formula is used for both systems.

Here is an example of how work is calculated: Suppose a person pushes a small pickup truck 10 m using 200 newtons of exerted force. How much work was done?

$$\begin{aligned} \text{work pushing the pickup} &= 200 \text{ N} \times 10 \text{ m} \\ &= 2{,}000 \text{ newton-meters (joules)} \end{aligned}$$

Here are three useful conversions from SI to English:

$$
\begin{aligned}
1 \text{ meter} &= 3.28 \text{ feet} \\
1 \text{ newton} &= .225 \text{ pounds} \\
1 \text{ foot-pound} &= 1.356 \text{ joules}
\end{aligned}
$$

(In this example of work, the distance the pickup is moved is 10 m × 3.28 ft/m or 32.8 feet and the force is .225 lbs/N × 200 N = 45 pounds.)

$$
\begin{aligned}
\text{work pushing the pickup} &= 45 \text{ lb} \times 32.8 \text{ ft} \\
&= 1{,}476 \text{ foot-pounds (ft-lb)}
\end{aligned}
$$

Factor-Label Method. Many times in scientific calculations, a factor-label method is used. The advantage of the method is to think through the concepts of the labels prior to the calculations. This helps to logically organize the problem. Problems are set up so labels cancel out in the denominator and numerator leaving the label of the answer.

Here is how the problem in pushing the pickup truck would be worked using the factor-label method:

$$\text{force in lb} \quad \frac{200\ \cancel{\text{N}}}{} \left| \frac{.225 \text{ lb}}{1\ \cancel{\text{N}}} \right. = 45 \text{ lb}$$

$$\text{distance in ft} \quad \frac{10\ \cancel{\text{m}}}{} \left| \frac{3.28 \text{ ft}}{1\ \cancel{\text{m}}} \right. = 32.8 \text{ lb}$$

$$
\begin{aligned}
W &= F \times d \\
W &= 45 \text{ lb} \times 32.8 \text{ ft} \\
W &= 1{,}476 \text{ ft-lb}
\end{aligned}
$$

Energy

The ability to do work is known as energy. Fuels are frequently used to create energy so that work can be done. For example, gasoline is used in internal combustion engines to provide energy through heat. Heat is created when the fuel burns.

POWER

Power is the rate at which work is done. Time is a factor in doing the work. A force moves an object a distance in a certain amount of time.

Work is the same if an object is moved slowly or quickly. Power varies with the speed of movement.

Power is stated as units of work divided by units of time. In SI, power is stated in watts. A watt is the power needed to do one joule of work in one second.

Formula

The formula for power is:

$$\text{power} = \text{work} \div \text{time or } \frac{w}{t} = \frac{F \times d}{t}$$

In the above example of pushing the pickup truck, assume that 20 seconds (s) were needed to move the truck the distance of 10 meters. The amount of power in SI was:

power to push the pickup = 2,000 N·m ÷ 20 s = 100 watts.
(A watt is a joule per second.)

In the English system, power is measured in foot-pounds per second. The amount of power to move the pickup in the English system was:

power to push the pickup = 1,476 ft-lb ÷ 20 s = 73.8 ft-lb/s

Horsepower

Power is often stated as ***horsepower***. Engines used in tractors, pickup trucks, and other power equipment are given horsepower ratings.

19-5. A lot of horsepower is needed for some equipment, such as this hay cutter and conditioner. (Courtesy, KMN Modern Farm Equipment, Inc., West Memphis, Arkansas)

Horsepower is the amount of power used to do 550 foot-pounds of work per second. (It is 746 watts in SI.) The term, horsepower, originated with the steam engine. The power of a steam engine was compared to the power of a typical draft horse by James Watt, inventor of the steam engine.

The formula for horsepower is:

$$\text{horsepower} = \text{power} \div 550 \text{ ft-lb/s}$$

In the example of the pickup truck, the horsepower used to move it in the English system was (remember, the power calculated above was 73.8 foot pounds per second):

$$\begin{aligned}\text{horsepower to movepickup} &= 73.8 \text{ ft-lb/s} \div 550 \text{ ft-lb/s/horsepower} \\ &= 0.134 \text{ horsepower}\end{aligned}$$

The horsepower to move the pickup in the SI system was (remember, power was 100 watts):

$$\begin{aligned}\text{horsepower to move pickup} &= 100 \text{ watts} \div 746 \text{ watts/horsepower} \\ &= 0.134 \text{ horsepower}\end{aligned}$$

Both SI and English calculations result in the same horsepower required to move the pickup.

Since the pickup was on a smooth street and pushed by one person, the amount of horsepower was very small—only a little over 10% of one horsepower. If the pickup was traveling 50 miles per hour down a highway, the horsepower would be much greater.

The factor-label method of working this problem in the English system is:

7.38 ~~ft-lb/s~~	1 horsepower	= 0.134 horsepower
	550 ~~ft-lb/s~~	

The factor-label method of working this problem in the metric system is:

100 ~~watts~~	1 horsepower	= 0.134 horsepower
	746 ~~watts~~	

19-6. Automated systems in poultry production reduce the need for human labor.

MOTION

Motion is a part of work and power. ***Motion*** involves moving or changing position. Most of the time, motion is thought of as being in a straight (rectilinear) direction. It can also be in a circle, known as curvilinear motion.

Examples of rectilinear motion include a tractor clipping pasture, a truck going down the highway, and a handsaw cutting a board into two pieces.

Examples of curvilinear motion include a wheel turning on an axle, a fan that ventilates a chicken house, a screw-on lid on a jar, and the crankshaft of an engine.

Everything on the earth is in motion all of the time. The motion is often undetectable in everyday life, but it is on-going. The earth is rotating and revolving at a high speed.

The motion that people can detect is of one object compared to another. A tractor can move through a field cultivating the rows of crop plants. The plants appear to stay in the same place and the tractor passes over them.

Speed

Speed is the rate of motion. Some things move fast; others move slow. A race horse runs fast and must run faster than the other horses to win the race. A horse race is usually timed to see how long it takes the horse to run the race.

Tractors and other powered equipment are capable of certain speeds in a field. They can go faster when doing certain jobs than they can others. Tractors pulling heavy plows may go slower than those raking hay.

Speed is measured in miles per hour or kilometers per hour. Most motor vehicles have a speedometer that indicates the speed of the vehicle. Many jobs performed on crops are done with tractors moving at speeds of 5 to 10 miles per hour (mph). Speed must be adjusted to the job. Tractors cultivating tiny plants may need to travel slower than those cultivating larger plants. Faster speed tends to move the soil more, and the soil could be moved on top of tiny plants and destroy them. Larger plants stand taller and are stronger. Some soil against their stems will not likely cause a problem.

Speed is important in determining how long it takes to do a certain job. For example, if a 50-acre pasture is being clipped with a 6-foot cutter, how long will it take to clip the pasture at a speed of 8 mph? An acre is approximately 210 feet long and 210 feet wide. The cutter will have to make 35 passes over an acre. Each pass will be 210 rectilinear feet long, or a total of 7,350 rectilinear feet in an acre using 6-foot swaths. Fifty acres would contain 367,500 rectilinear feet (50 × 7,350). The total rectilinear distance for the pasture is 69.6 miles (367,500 ÷ 5,280 [feet in a mile]). At a speed of 8 mph, nearly 8.7 hours would be required to clip the pasture. (This doesn't include time spent turning around, servicing the tractor, and performing other off-task duties.)

The factor-label method of calculating this problem would be as follows:

50 ~~acres~~	35 ~~passes~~	210 ~~ft~~	1 ~~mi~~	1 hr	= 8.7 hr
	1 ~~acre~~	1 ~~pass~~	5,280 ~~ft~~	8 ~~mi~~	

Acceleration and Deceleration

Speed may change. ***Acceleration*** is increasing the speed of an object over time. ***Deceleration*** is when it slows down or decreases speed over time.

Speed is an important factor in carrying out many agricultural jobs. Tractor speed can be varied to suit the situation. Fuel consumption and power are related to speed. Labor requirements are also related to speed.

Using faster speeds, less time may be required to do the work. Of course, too much speed can cause many problems, including unsafe operation of the tractor.

Refer to the previous example of using a tractor to clip 50 acres of pasture. If the speed of the tractor was increased to 12 mph, how much time would be required to clip the pasture? At 12 mph, 5.8 hours would be needed to clip the pasture.

Using the factor-label method, the calculations would be as follows:

$$\frac{50\ \cancel{\text{acres}}}{} \left| \frac{35\ \cancel{\text{passes}}}{1\ \cancel{\text{acre}}} \right| \frac{210\ \cancel{\text{ft}}}{1\ \cancel{\text{pass}}} \left| \frac{1\ \cancel{\text{mi}}}{5{,}280\ \cancel{\text{ft}}} \right| \frac{1\ \text{hr}}{12\ \cancel{\text{mi}}} = 5.8\ \text{hr}$$

MACHINES AND MACHINERY

A ***machine*** is a device that helps transmit power, force, and motion. Machines do work! They may be very simple things that we use in our daily lives. Regardless, they help by increasing power, making more force, or increasing or decreasing motion.

Several simple machines can be joined to make a complex machine. The large machines are sometimes known as machinery. This is because they involve many different uses of simple machines. In agriculture, large hay balers, planters, plows, and harvesters involve many simple machines.

Machines have reduced the amount of physical effort humans must use to do work, such as to till the soil. With a machine, one human can do far more work. Modern agriscience couldn't have developed without the use of machines.

SIMPLE MACHINES

All of the machinery used is a combination of six simple machines. These six machines provide a lot of work with less human effort. The

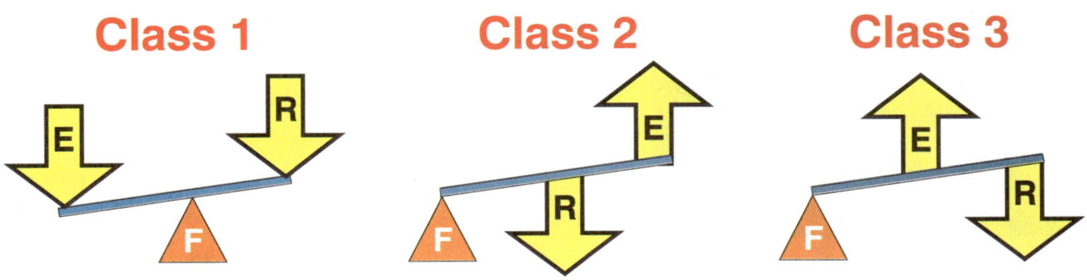

19-7. Three classes of levers. Note: In the figure, F is fulcrum, R is resistance, and E is effort.

19-8. The pry bar is an example of a first-class lever. Note the locations of the ffort, fulcrum, and resistance.

19-9. A hammer is an example of a third-class lever.

simple machines are: lever, wheel and axle, pulley, inclined plane, wedge, and screw.

Lever

A ***lever*** is a rod or bar that rotates on a fulcrum, or pivot point. Force is applied at one end to lift a load at the other end. Levers help lift loads with less effort. For example, a person can lift 1,000 pounds with a lever.

A lever has two parts: effort arm and load arm. The effort arm is the distance between the force and the fulcrum. The load arm is the distance between the fulcrum and the load.

Levers are classified by the positions of the fulcrum. With first-class levers, the effort arm is the long part of the lever and the load arm is the short part of the lever. A good example of a first-class lever is a crowbar. This is a simple machine that is used to pry boards apart and pull nails out of boards. The effort arm between the fulcrum and force is the longest part of the lever. This makes it much easier to take boards apart and remove nails.

A second-class lever involves having the fulcrum at one end and the effort arm at the other end. The load arm is between the fulcrum and the effort arm. An example is the wheelbarrow.

A third-class lever has the fulcrum at one end and the resistance arm at the other end. The effort arm is between the two. A good example is the hammer used in driving nails. The action of the arm is a part of the lever.

Wheel and Axle

Wheels and axles are important parts of all moving vehicles. Wheels are usually thought of as being round. Examples include the wheels on pickup trucks and tractors. Other uses of wheels and axles involve handles that make round motions. Handles attached to shafts are wheel and axle machines, such as the crank on a hand-operated grinder or the bit brace used in drilling holes. Another example is the pedal system on a bicycle. The pedals go around like a wheel and are attached to an axle.

The use of wheels and axles can increase power or speed, but not both. Turning a big wheel on a small axle increases power. Turning a smaller wheel on a larger axle increases speed.

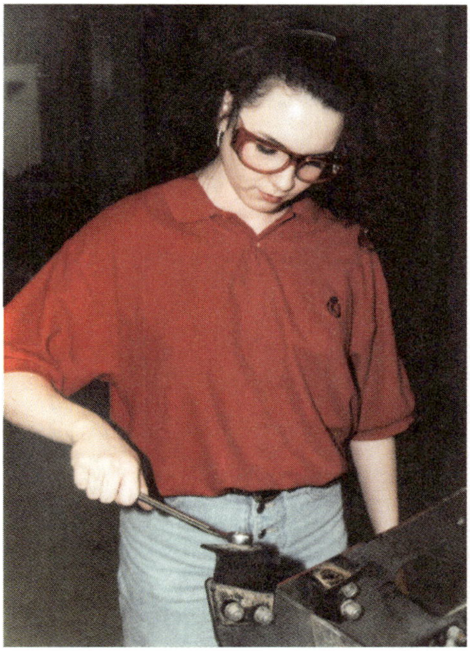

19-10. A wrench provides power not available with the bare hand. It is much like a lever but is a wheel and axle. (The nut on the bolt is a wheel.)

Pulley

A ***pulley*** is a special kind of wheel. It changes the direction of force using a belt or rope. The wheel is free to spin on an axle. The wheel usually has a groove in the middle for holding the belt in position.

Pulleys are used to provide a mechanical advantage in lifting similar to that of a lever. Pulleys with belts that provide the power to turn the parts of an engine change the direction of force of the crankshaft. Several pulleys used together result in much less effort being necessary to lift a heavy object.

Pulleys arranged as a block and tackle can be used to lift heavy loads with less effort. A good example is the block and

19-11. A brace with bit used to bore holes is an example of a wheel and axle.

19-12. Belt on a pulley of a gasoline engine is used to transfer power.

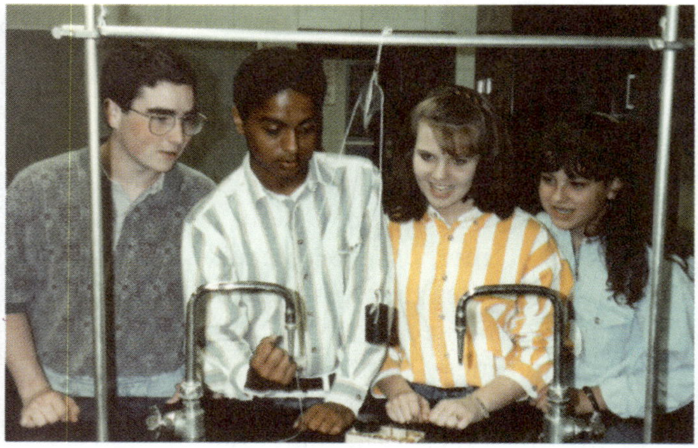

19-13. A fixed pulley is used here to lift a weight.

tackle used to remove a heavy engine from a piece of machinery.

Pulleys may be fixed or movable. A fixed pulley is mounted in one place, while a movable pulley rises and falls with its load. Movable pulleys are used to increase force. Pulling on one end of the rope can lift an object much heavier in relation to the force that is applied.

Inclined Plane

Inclined planes are straight, sloping surfaces. A ramp in a feed mill or a handicapped access ramp to a building are inclined planes. People

19-14. A cattle loading chute is an example of an inclined plane.

can move much heavier weights pushing them up a ramp than they can by lifting them straight up. Inclined planes also help load products. Cattle are loaded by driving them up ramps into trucks. It would be nearly impossible to load them without a ramp.

Wedge

A ***wedge*** is much like an inclined plane. Wedges have sloping surfaces that taper to a point. They are used to lift heavy loads short distances or to separate products. For example, wedges are used to split wood. Wedges may be driven into the wood, forcing it to split apart. Another example of a wedge is the opener on a planter. The opener pushes the soil apart so that the seed can be dropped into the ground. A covering wheel closes the drill opening made by the opener.

Wedges vary in angle. Angle refers to the amount that the wedge thickens away from the point. Wedges that are long and gradually taper are easier to drive into wood but split the wood more slowly. Wedges that get thick faster are more difficult to drive but split the wood more quickly.

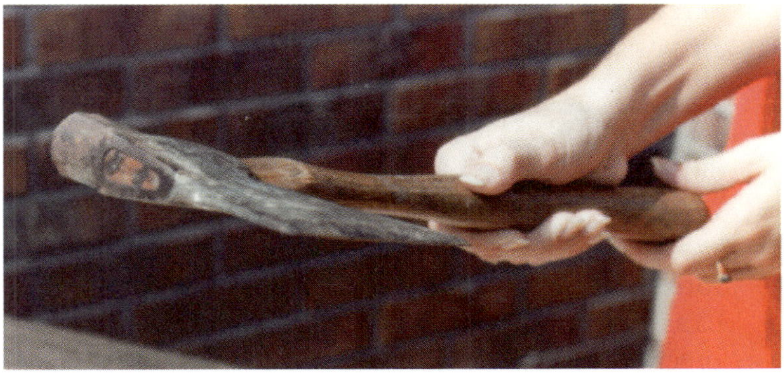

19-15. The head of a hatchet is an example of a wedge.

Screw

Like the wedge, the ***screw*** is a modified inclined plane. The incline plane on a screw is long and circular. It goes around the shaft (middle part) of the screw. Screws are used in many ways: as fasteners to hold materials together, as augers to move grain through a tube to a bin, and as the lifting mechanisms of jacks that lift a pickup truck with a flat tire.

Screws involve using rotating motions. The pitch of the screw is the angle of the thread (incline) around the shaft. Screwdrivers use rotating

19-16. Two examples of a screw machine are shown here. On the left is a square-head wood screw. A C-clamp is shown on the right.

19-17. An auger is a kind of screw. (This stainless steel auger at a food processing plant is used to move raw products inside for processing.)

force to cause the screw to move into the wood. Wrenches may be used to put bolts and nuts together. With effort, the screw can be made very tight so that it will hold heavy loads securely. Sometimes lock washers are put on screws to keep them from working loose. A lock washer holds the screw in position and doesn't allow it to turn backward.

ADVANTAGES OF MACHINES

Machines cannot create energy but they can control and multiply the force. A machine never does more work than the energy put into it. They can convert energy from one form to another, such as diesel fuel into mechanical power. This makes them very useful in many agricultural jobs.

MECHANICAL ADVANTAGE

Mechanical advantage explains why machines are so useful and important. Mechanical advantage is how machines multiply the force that they receive. With mechanical advantage, a small force can become a larger force—often much larger. A good example is a long lever. A little force on the end of a long lever can lift a heavy weight on the other end.

With mechanical advantage, the lever must rest on a fulcrum. A fulcrum is the support for a lever that allows it to turn. The closer the fulcrum is to the load, the less effort required to move the load.

19-18. A hoist has the mechnical advantage of a pulley system to lift heavy loads.

The formula for mechanical advantage (MA) calculates the ratio of input and output power. Here is the formula:

$$MA = \text{output force} \div \text{input force}$$

Here is an example: Suppose a lever is used to lift one side of a farm tractor. The side to be lifted weighs 1,000 pounds, known as the output force. The effort put on the lever is 20 pounds, known as the input force. The mechanical advantage is:

$$\begin{aligned} MA &= 1{,}000 \text{ lbs} \div 20 \text{ lbs} \\ &= 50 \text{ lbs} \end{aligned}$$

The same formula can be used with SI by substituting metric weights.

All machines have mechanical advantage. If they didn't, who would use them? Formulas for mechanical advantage for all simple machines are found in introductory physics books.

19-19. Using electric-powered trimmers allows people to do more with less effort.

WORK PRODUCED

Work is a product of force and distance. A machine will not increase both the force and the distance moved. The above lever is a good example. The person lifting the tractor must move the lever a much greater distance than the tractor will be lifted. The person uses the long part of the lever (known as the effort arm) to gain the advantage in lifting the tractor a short distance.

The Law of Conservation of Energy states that the amount of work produced by a machine is the same as the work put into the machine. Of course, some of the energy is lost because of friction in the machine. Some energy is required to rotate the lever around its fulcrum! No machine is 100% efficient. Efficiency is sacrificed to make jobs easier.

The length of a lever and where the fulcrum is located are important in mechanical advantage and work. Distance is measured from the fulcrum to each end of the lever. Here is how it is done:

output force = F_2 (weight lifted)
output distance = D_2 (distance the object is moved)
input force = F_1 (weight exerted on lever)
input distance = D_1 (distance the long end [effort arm] of the lever is moved)

The formula is:

$$F_1 \times D_1 = F_2 \times D_2$$

In the example of lifting one side of a tractor, how far would the effort arm (D_1) need to be moved if the tractor was lifted .2 foot (D_2)? Here is how it is calculated: (Remember, 20 pounds of force was applied to the long end of the lever, and this is known as D_1. The weight lifted was 1,000 pounds, known as F_2.)

$$\begin{aligned} 20 \text{ lb} \times D_1 &= 1{,}000 \text{ lb} \times .2 \text{ ft} \\ 20D_1 &= 200 \text{ ft-lb} \\ D_1 &= 200 \text{ ft-lb} \div 20 \text{ lb} \\ D_1 &= 10 \text{ ft} \end{aligned}$$

THERMAL ENERGY

Thermal energy is energy produced by heat. It is produced when fuel burns, such as gasoline in an engine. This process converts fuel into work.

Heat causes gases to expand rapidly. When in a strong container (such as an engine cylinder), this rapid expansion can be used to move mechanical parts. This converts energy to mechanical power in heat engines.

AgriScience Connection

CHECKING ENGINE OIL

Internal combustion engines must be properly lubricated. If not, the metal parts that turn and rub against each other will get hot and burn in a short time. Engines should be regularly checked to see that the crankcase contains sufficient oil.

The oil is checked by removing the dipstick from the crankcase. All dipsticks have marks and writing that indicate if the oil level is okay or if oil is needed. Always add the proper oil if some is needed.

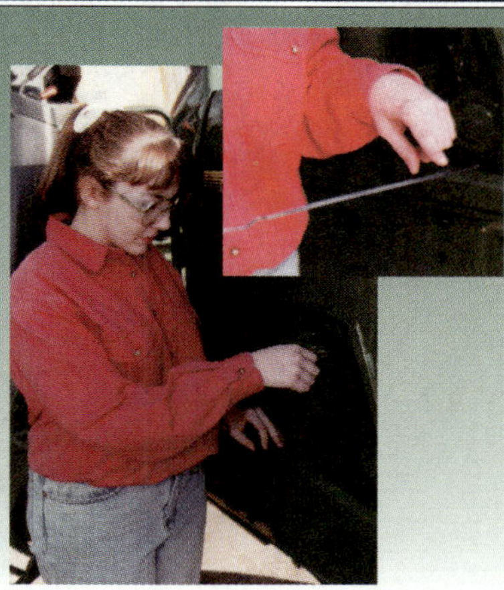

INTERNAL COMBUSTION ENGINES

The internal combustion engines on lawn mowers, tractors, and pickup trucks are good examples of ***heat engines***. The burning of fuel (gasoline, diesel, etc.) produces hot gases.

Fuel is placed in one end of a strong cylinder and ignited (burned). The burning fuel makes gases that rapidly expand inside the cylinder. One end of the cylinder is closed with a piston that can be moved. The rapid expansion pushes the piston creating motion. The piston is connected by a lever to the crankshaft that is rotated by the moving piston. The crankshaft is connected to axles and wheels, pulleys, and other machines that can do work.

19-20. Adjusting a one-cylinder gasoline engine.

FUELS

Three kinds of fuels are commonly used in internal combustion engines in agriscience: gasoline, diesel, and liquefied petroleum gas.

Gasoline

Many of the engines used in agriculture use gasoline as fuel. These engines are often

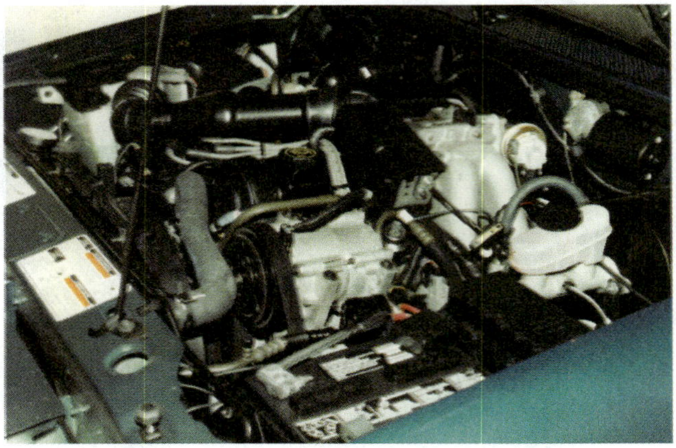

19-21. Four-cylinder gasoline engine in a pickup truck.

found in pickup trucks and used to run small equipment. A gasoline and air mixture is compressed in the cylinder and burned with a spark from a spark plug.

The ratio of air to gasoline in the fuel mixture in a cylinder is 9,000 to 1. This means that 9,000 gallons of air are needed for 1 gallon of gasoline. That is a lot of air!

Gasoline is made by refining petroleum (also known as crude oil). Most gasoline is a blend of many kinds of hydrocarbons. Each of the hydrocarbons burns differently in an engine.

Diesel

Diesel fuel is used in more powerful engines. The engines are heavier than gasoline engines and are designed to run long hours and pull heavy loads.

Diesel engines do not have spark ignition systems. The pistons create high compression in the cylinders. The high pressure causes heat which ignites the fuel when it is injected into the cylinder. The heat causes the fuel to burn rapidly. The burning creates gas pressure that forces the piston backward in the cylinder.

Liquefied Petroleum Gas

Liquefied petroleum gas (known as LP gas) is made of butane and propane. It is stored as a liquid under pressure in bottles or strong tanks. Tractors, pickup trucks, and other lighter-weight engines can be run on LP gas.

Gasoline engines are sometimes converted to LP gas. All processes of the engine are the same except for the way LP gas is stored and gets into the engine. The fuel requires a spark to ignite. An engine fueled with LP gas costs about one-third less to run than one that uses gasoline.

Other Fuels

Many other fuels are used on a limited basis to power engines. Alcohol produced from grain may be used alone or mixed with gasoline. Biodiesel manufacture with oils from grain is gaining use. Methane produced from decaying organic matter is used on a small scale. Some experiments have been done using hydrogen gas, but this is not available on a wide scale. Electricity is used to power vehicles on a small scale, such as for golf carts, loaders, and low-power equipment.

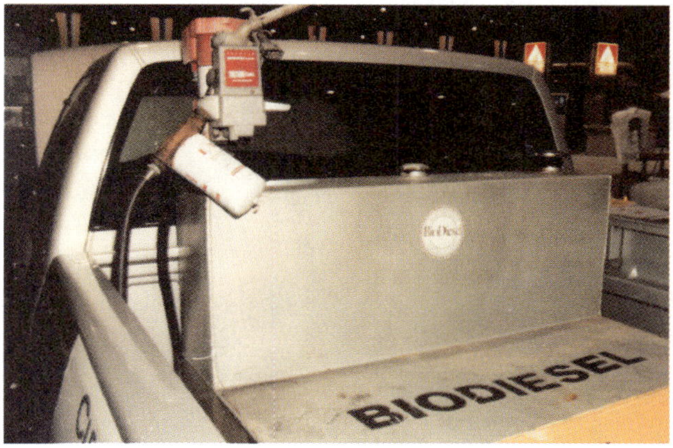

19-22. Pickup truck equipped to operate on biodiesel fuel.

ELECTRICAL ENERGY

Electrical energy comes from electricity. ***Electricity*** is the flow of electrons, which are small units of negative electrical charge. These electrons are too small to be seen with a microscope. All materials are made of protons and electrons that have electrical charge. The behavior of everything around us is due in part to these electrical charges.

CURRENT ELECTRICITY

Current electricity is the most important kind of electricity in agriscience. This is the kind of electricity that light bulbs, electric motors, and heating elements use. It is produced by electrical generators, batteries, and solar cells. Current electricity is electricity that is caused by flowing electrons.

Another form, static electricity, helps to explain electrical charges. Almost everyone is familiar with the static electricity that builds up when combing hair or walking on a carpet on a cold, dry day. The sparks produced are due to static electricity. Static electricity can be a source of danger. In grain elevators and flour mills with a lot of dust in the air, a single spark can result in a huge explosion.

Kinds of Current Electricity

Two kinds of current electricity are used in agriscience: direct and alternating. The kind of electric current depends on its source.

19-23. Alternating current (AC) electric motor.

Direct current (DC) is electricity that always flows in the same direction. It is produced by batteries and some kinds of electrical generators. DC electricity is used in the ignition systems of tractors, trucks, and other engines.

Alternating current (AC) regularly reverses its direction of flow—it alternates. AC goes through cycles; two changes in direction make up one cycle. The AC cycle is measured in hertz. One cycle per second is one hertz (Hz). Most of the current electricity in North America is 60 Hz. AC electricity is used in homes, on farms, and in agribusinesses to provide light, heat, cooling, and other purposes. Many farms and agribusinesses have stand-by electrical generators in case the power goes off.

Current can be changed from one form to another. A rectifier is used to convert AC to DC.

How Electrical Current Is Measured

Electrical current is measured to determine its rate of flow, pressure, and the amount available. Most people are familiar with the electric meter that the power company uses to measure the amount of electricity used. The amount a customer is charged is based on this measurement.

Ampere. An ***ampere***, or amp, is a measure of the rate of flow of electrical current. One ampere is made of a huge number of electrons that flow every second. Amperes are measured with an ammeter. Most electrical devices require wires that can conduct a certain flow of amperes. Larger wires will conduct more amperes of electricity, much like larger pipes will allow more water to flow.

Voltage. ***Volts*** are used to measure electrical potential energy. It is the difference in electrical potential between two points in an electrical circuit. The difference in potential that pushes electrical charges through a conductor is known as voltage. Most light electrical equipment in North America runs on 110 volts. Heavier equipment, such as welders and electric heaters, may need 220 volts. Some electrical power lines carry very high voltage. The voltage is reduced with transformers to the amount used in agriscience. Big voltages and big wires make it efficient for the power company to send electricity. It is "stepped down" when sent to the customer.

Wattage. The amount of electrical energy supplied or used is measured in ***watts***. Most electrical appliances, including light bulbs, have wattage ratings. For example, household light bulbs are usually 60- or 100-watt bulbs.

Combinations. Amperes, watts, and volts are used in various combinations to calculate electrical use and demand. The name plates of most electrical devices provide information about the amperes, watts, and volts needed to operate them. Proper installation of wiring systems requires knowing the electrical demands on the system.

19-24. The name plate on this small electric drill gives information about the electrical current requirements.

Conductors and Insulators

All electrical current must have a ***conductor***. A conductor is a substance that allows the electrons to flow freely. Copper, gold, and aluminum are good conductors of electricity.

For safety purposes, all electrical wires and other devices must be properly insulated. An ***insulator*** is a material that does not conduct electricity. Rubber, plastic, and glass are commonly used as insulation. Electric wires are usually covered with a layer of plastic insulation. Without insulation, the wire is dangerous and could cause electrocution as well as fires and other damage.

Electrical Circuits

A ***circuit*** is the path electricity follows. Most circuits are made of wires and other materials that are good conductors.

A circuit must have two wires. One wire carries the current from its source to where it is used. The other wire carries the current back. In addition, most electrical circuits have a third wire known as the ground wire. Ground wires prevent shock in case the other wires fail, a short occurs, or excess charge builds up. Lightning rods must have a wire to the ground (known as the ground wire) to prevent shock and fire.

Two kinds of circuits are commonly found: parallel and series. Parallel circuits have more than one path the electricity can follow. The electricity can flow to several output devices (lights, appliances, computers, etc.) along the path of the circuit. Any device on a circuit connected in parallel can be on or off without affecting the others. Household and business wiring is connected in parallel.

A series circuit has only one path that electricity can follow. It must flow through each output device located on the circuit. Any break in the circuit prevents all the output devices from working. This kind of circuit has limited uses. If used to connect a string of lights, for instance, when one bulb burns out or becomes loose, the entire string goes out.

Electrical Controls

Many kinds of devices are used to regulate and control activities in agriscience. These often involve using or controlling the use of electrical power.

Switches turn electrical power on and off. These range from simple toggle switches that operate a light to complex switches that control other devices.

19-25. Toggle switches are used to turn electric lights on and off.

19-26. Simple timer used for home appliances.

Thermostats are used to regulate the amount of heat and cooling. Their use depends on the result that is wanted. To keep something cool, a thermostat is set to the desired temperature. If it gets too warm, the thermostat operates a switch that provides power to operate the cooler. When it gets cool enough, the thermostat turns the power off.

Timers may be used to operate lights and other equipment. For example, a timer may be used to turn the lights on and off in a greenhouse so that the plants get more light than the day is providing. The timer is set to turn the lights on or off at a certain hour of the day or night. Night-time lighting is also used in poultry houses.

19-27. Control used on a dairy farm to regulate milk storage tank temperature and washing.

Security lights may have light sensors that turn a light on when it gets dark at night and back off in the morning in response to increasing sunlight.

REGULATIONS AND SAFETY

Electricity can be dangerous to humans and other animals. It can damage property and cause injury when used improperly. Buildings can burn down if a

19-28. Controls in a large agriculture product processing plant.

19-29. Ground fault circuit interrupter on the end of a power supply cord.

short circuit creates heat that sets wood or other material on fire.

Various regulations about the use of electricity have been established, known as codes. The National Electrical Code serves as the basis for electrical work. Local cities and power companies may also have codes. Always follow the codes and observe safety practices in using electricity.

All electrical devices should be grounded to protect the people who use them. Avoid using electrical appliances in wet conditions.

Many small appliances have a ***ground fault circuit interrupter***, known as GFCI. The GFCI is designed to protect people from shock when using electricity in areas that are wet or have water. The GFCI may be built into a breaker box or made a part of the cord where the appliance is plugged into a receptacle.

COMPRESSION POWER

Liquids and gases may be compressed to create power. The machines that do the compression are known as compressors. They are usually pumps with tanks or reservoirs filled with liquids and gases that can be used to create compression.

HYDRAULIC POWER

Hydraulics is the study of liquids at rest and in motion. In some cases, gases may be a part of hydraulics. Agriscience operations often involve using ***hydraulic power***—an important kind of compression power.

19-30. The wheels on a grain drill are raised and lowered by hydraulics. The cylinder is easily seen in this photo.

Pressure can be used to transfer or increase a force. Liquids, such as water and oil, can be placed under pressure that allows them to lift heavy weights, operate hydraulic motors, and perform other work.

The pressure in a hydraulic system is often measured as pounds per square inch (psi). Leaks in the system reduce the pressure and may result in the loss of liquid. Also, leaks can be dangerous and injure people. Hoses and connections should be kept tight and in good condition.

Many tractors have hydraulic systems that lift the equipment pulled by the tractor. Also, equipment may need power at a place that can't be easily reached with levers, belts, or axles. Hydraulic pressure can be transferred in oil through hoses to places where hydraulic motors can covert the pressure into movement.

Hydraulic systems operate using a pump powered by a heat engine or electric motor that creates pressure on the oil in the system. The oil may be moved to other locations where pistons in cylinders are moved. Movement of the pistons causes a lever to move. The lever is connected to wheels, axles, or other levers. This series of actions causes lifting.

An example is a disc harrow. A disc harrow may have wheels that can be raised or lowered. When the wheels are lowered, the blades of the disc harrow don't touch the soil. The weight of the harrow rests on the wheels. This is done when moving the harrow from one field to another. When the wheels are raised, the blades go into the soil. The depth of tilling can also be regulated by the position of the wheels.

COMPRESSED AIR

Air can be used to do work. Air is compressed when it is put under pressure. Using compressed air to do work is known as ***pneumatic power***.

19-31. The tire pump is an example of using a piston inside a cylinder to create air pressure.

Lowering air pressure creates a vacuum. A vacuum operates on the basis of pressure inside a container being lower than the pressure of the air in the surrounding area. Suction is created.

Air pressure is measured in pounds per square inch (psi). Most things that are filled with air have a maximum pressure and will burst like a balloon if the pressure is excessive. Air pressure is measured with an air gauge.

Air is compressed to create pressure. Air under pressure is used to fill tires that support heavy weights. It can also be used for other purposes, such as operating a paint sprayer. Compressed air is used in the tanks of some water supply systems to create pressure that causes the water to flow.

Pneumatic tools are powered by compressed air. Examples are wrenches that remove lugs from wheels, riveters, and air hammers that break concrete. Some pneumatic tools are used to drive nails and other fasteners in woodworking.

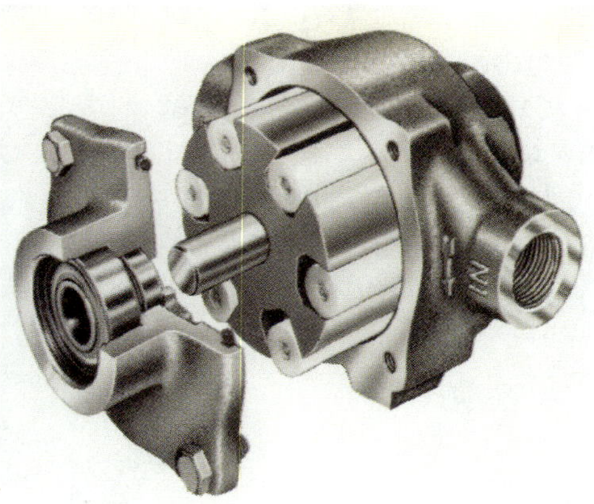

19-32. Cut-away view of a roller pump. (Courtesy, Hypro Corporation, New Brighton, Minnesota)

The suction of air (lowered pressure) is used to move some products. Air suction is used to load and unload cotton or move products through a processing plant. Suction may be used to remove unwanted parts of animals while they are being processed, such as the internal organs of chickens and fish.

Air pressure and suction can be created with piston or rotary pumps. The hand-operated tire pump is an example of a piston pump. When the piston in the cylinder is pushed down, the pressure in the cylinder increases and air is forced out through a small tube connected to a tire. The pressure in the cylinder causes the air to move into the tire.

Rotary pumps have rollers that turn inside a pump housing. Rollers create a steady pressure whereas a piston pump may create pressure in spurts.

REVIEWING

MAIN IDEAS

Many areas of physics are applied in agriscience. Most of the applications involve mathematical formulas and calculations.

Physics is used to increase the ability of people to do work. Mechanical power has replaced human and animal power. One person can do a lot more work with a tractor than could be done without one.

Work is moving an object through a distance. Force is applied to make it move. It the object doesn't move, no work has been done.

Power is the rate at which work is done. Time is an important factor in computing the amount of work that can be done. Horsepower is a measure of power that involves doing 550 foot-pounds of work a second.

Motion is an important part of work and power. Speed is the rate of motion. The rate can be increased or decreased.

Simple machines are used to accomplish work. These machines can be put together in complex combinations to form machinery that does a lot of work. Machines provide mechanical advantage. This means that the machine requires less input for the output it will provide.

Thermal and electrical energy are used in agriscience. Thermal energy involves changing heat into mechanical energy. This is why internal combustion engines are so important in agriculture. Electrical current can be used for many purposes, including light, power for equipment and heat.

Compression power includes hydraulics and pneumatics. Hydraulics is the compression of liquids. Pneumatics is the compression of air. Various tasks can be accomplished using hydraulics and pneumatics.

QUESTIONS

Answer the following questions using complete sentences and correct spelling.

1. What is physics? What areas are included?
2. What is work? Power? Distinguish between the two.

3. Calculate the amount of work if five pounds of force are used to push a lawn mower 500 feet.
4. Calculate the amount of power if it takes six minutes to push the lawn mower in question three.
5. What is horsepower?
6. What is motion? Speed?
7. How does speed relate to the time required to do a job?
8. What are the six simple machines? Name examples of the use of each.
9. What is mechanical advantage?
10. What is a heat engine? Why is heat important in the engine?
11. What is current electricity? What are the two kinds and how do they differ?
12. How is electric current measured?
13. Distinguish between an electric current conductor and insulator.
14. Why is a GFCI important?
15. What is hydraulic power? How is it used in agriscience?
16. What is pneumatic power? How is it used in agriscience?

EXPLORING

1. Arrange to visit a building construction site where electrical wiring and controls are being installed. Observe the procedures used in putting the wires, switches, and other devices in place. Prepare a report on your observations.

2. Observe machinery that is used in some area of agricultural industry. Make a list of the different simple machines that you see. Write a brief overall description of what the machinery does.

3. Construct a simple electric motor. Refer to the Activity Guide for *Introduction to World AgriScience and Technology* for information on how to build a motor.

4. Use a gauge to determine the air pressure in the tires on a vehicle. Determine the recommended pressure for the tires and add or reduce the amount of air so that the tires are properly inflated.

20

MARKETING TECHNOLOGY IN AGRISCIENCE

Almost everyone has had the experience of going to a store to purchase an item that the store didn't have. This leaves a consumer with the choice of buying the item at another store, or going without it.

Although the reason for the shortage is not always obvious, if it's food, fiber, or shelter, several things could have happened. It could be out of season or not available because of a crop failure. Today, these two situations are unlikely to occur for most things in North America.

The absence of a particular item in a store is usually due to problems other than production. Chances are good that the item has been produced and could be available. But, it isn't!

It is more common for problems to arise in transporting an item to the right place, in the right form, and at the right time. This is marketing!

20-1. Marketing is to meet the demands of consumers for food and other goods and services. Just think of all of the activities needed to have fast foods!

OBJECTIVES

This chapter introduces important areas of agricultural marketing technology. It has the following objectives:

1. Describe the importance of agricultural marketing technology.
2. Explain ways that agricultural produce is marketed.
3. List and explain the major functions in agricultural marketing.
4. Describe the role of marketing infrastructure.
5. Explain the role of communication in agricultural marketing.

TERMS

advertising
agricultural marketing technology
assembling
blemish
communication technology
custom harvesting
custom order marketing
demand
distribution
grading
grain elevator
harvesting
marketing
marketing infrastructure
niche market
labeling
packaging
packing shed
processing
promotion
retail market
roadside market
selling
transporting
vertical integration

MARKETING: PRODUCING A GOOD CROP ISN'T ENOUGH

Marketing is getting the goods and services that people want to them. It links the producer of the goods and services with consumers. Marketing agricultural products is different from other items, such as furniture and automobiles.

AGRICULTURAL MARKETING TECHNOLOGY

Agricultural marketing technology is all of the processes involved in providing people with what they want, in the form they want it, and when they want it.

Agricultural marketing is the movement of food, fiber, and shelter materials from the producer to the consumer—the person who uses them. It is a complex process that involves transporting a wide range of products, including food crops, animal and animal products, fiber crops, and wood. Hogs, tomatoes, sheep, apples, eggs, milk, wool, pulpwood, and all of the other things that are produced are marketed.

The focus of agricultural marketing technology is on meeting the demands of consumers. Many things have to be done to get food ready for

AgriScience Connection

GETTING A HIGHER PRICE FOR PRODUCTS

People who want to get pearls from oysters can do so with this creative marketing approach. The small tub contains several oysters with cultured pearls. Customers can buy the oyster, open it, and have one or more pearls. The price received for the oyster is much greater than would be received through regular pearl markets. The oyster is not eaten!

Creative entrepreneurs can find ways to meet market opportunity. All you need to do is identify what consumers are willing to pay for and find a way to meet the demand.

the consumer. In North America today, very few people want to do their own processing. Not many people buy live chickens and take them home to be plucked! They want to get their food in ready-to-cook form, their clothes ready to wear, and their shelter products ready to use.

When Marketing Begins

Marketing begins with a decision to produce something. The producer must study the situation. Only products that can be sold should be produced. Further, only those that can be sold at a price high enough to make a profit should be produced.

Producers use technology to help them make decisions. They get the best information they can from experts on agricultural marketing. They read news articles on market trends. They assess what exists in their community that is needed to make marketing possible. They use computer

20-2. Studying market information helps in making decisions about what to produce.

models to help make a good assessment. They may also sign agreements with buyers to assure that their products can be sold after they are produced.

A demand for the product must exist. This means that there is someone who wants what is produced.

Consumers Are Important

Consumers are the people who buy and use what is produced. Everyone is a consumer. Consumers create demand.

Buying a hamburger or blue jeans is consuming. In North America, millions of people buy hamburgers and jeans every day. This creates the need for agriscience to produce the materials needed to make the hamburgers and jeans. Due to the large amounts consumed, systems of providing for these needs have been developed.

Hundreds of millions of people will consume a huge amount of food, fiber, and shelter. And, they have preferences about what they want. They won't buy just any product; they want specific things. Agricultural marketing involves studying consumer preferences and producing what they want.

Demand is what people want. Further, they must want it enough to spend their money to get it. They also change their wants from time to time. Styles and notions about foods change. Consumers will reflect these changes in what they demand.

20-3. Marketing specialists gather information world wide. (Courtesy, Progressive Farmer Radio Network)

Producers Are Commercial

Today's farmers and ranchers are commercial producers. They produce plants and animals, and the products of plants and animals, that can be sold. If it can't be sold, it shouldn't be grown!

If something that can't be sold is produced, the producer is stuck with it. The product can't be sold and the producer can't recover the money spent to produce it. Why would a producer grow something that couldn't be sold?

RELATED AREAS OF MARKETING TECHNOLOGY

In addition to food, fiber, and shelter, many other products are marketed in similar ways. Some of these are marketed from the grower; others are marketed to agricultural producers.

Horticulture

20-4. Preparing flower arrangements and keeping them fresh is important to a florist.

The products of ornamental plant growers, such as flowers and shrubs, must reach consumers in good condition. Flowers must arrive in fresh condition. No one wants dead flowers! The demand for many flowers is seasonal—people want roses for Valentine's Day and poinsettias at Christmas time. Growers must target the major portion of their crop for the times of the year when demand is greatest.

Supplies and Services

Producers of plants and animals use many inputs in growing them. In growing crops, they need seed, fertilizer, equipment, fuel, and other inputs. In raising animals and producing their products, producers need feed, animal medicines, equipment, and other

20-5. Relating to customers helps a horticulture business prosper.

things found on livestock farms and ranches. These are the supplies and services that are used to grow crops and raise livestock. They are marketed to producers by agribusinesses.

TECHNOLOGICAL METHODS IN AGRICULTURAL MARKETING

Marketing technology must provide a channel or link between the producer and the consumer. Different marketing channels are used depending on the producer's preferences, opportunities, and kind of product. Some producers only want to produce their product and get it to a processor. Other producers like direct marketing and dealing with customers. The amount produced is another important part of selecting a marketing channel.

ROADSIDE AND RETAIL MARKETS

In roadside and retail markets, the producer sells directly to the consumer. This adds to the tasks involved in producing plants and animals. The producer harvests, grades, packages, and weighs the product(s), and runs the market.

Some kinds of products lend themselves to roadside and retail markets; others don't. Examples of produce sold through these markets include: fresh vegetables and fruit, eggs, melons, and nuts. Except for fish, meat products are not usually marketed to the consumer by the producer. This is due to the amount of processing required and the regulations that must be followed.

20-6. An attractive roadside market outlet operated by Bay View Farm in Hawaii.

20-7. Roadside stand on a California strawberry farm.

Roadside markets are located on the farm. They work best for farms that are on major roads or state highways. The producer operates the retail market in addition to producing what is sold.

Retail markets are usually located in nearby towns or cities and sell to consumers. Sometimes the markets are a part of a planned farmers' market where a number of different growers have small stalls or booths for selling their produce. Retail markets can also be separate stores operated by the producer.

CENTRAL LOCATIONS— PACKING SHEDS AND ELEVATORS

Producers may deliver their harvested produce to packing sheds and grain elevators. These places buy the produce from many growers. All of the produce is assembled together for shipment to processing plants. Before shipping, the produce is graded and the products are put in uniform lots.

The packing sheds and elevators negotiate with processing plants to get the best possible prices. In some cases, several growers may cooperatively own these facilities.

Packing sheds receive, sort, and combine produce from a number of different farms. This provides an advantage to growers—combining what they have produced separately into a larger volume. Larger volumes are more appealing to processing plants.

Livestock auction sales have been popular with some producers. These have declined in recent years as more producers deal directly with packing plants or order buyers. (Order buyers are people who buy certain kinds of animals or plants for other people. They are paid a fee for their work.)

Grain elevators receive, grade, dry, store, and sell grain to mills. They have long been important in the marketing of corn, wheat, and other grains. They are usually located on railroads and highways. These facilities receive the grain and prepare it for further marketing. The elevators may dry grain that has too much moisture. Elevators carefully store grain until it is sold to mills. The mills may make food for people or feed for animals.

20-8. Grain elevators are central locations where farmers take soybeans. This photograph shows a truck loaded with soybeans being emptied. (Courtesy, American Soybean Association, St. Louis, Missouri)

PROCESSING PLANTS

Some producers market directly to processing plants. These facilities convert the raw product into a form that consumers will buy. In some cases, they convert it to a form that may require additional processing at a retail store, such as fresh meat that will be cut into portions for customers.

In most cases, the processing plants want large volumes of produce. They also want produce that meets their needs and standards. Some processors may contract with growers to produce certain kinds of crops. These growers know that they will be able to sell what they produce before they plant it.

Processing plants alter produce in a variety of ways. Food is preserved and packaged for supermarkets, restaurants, and consumers. Fiber is woven

20-9. The auction sale is important in some locations as a part of the marketing process.

into cloth that is then made into clothing at garment plants. Pulpwood is used to make paper and other related products.

CUSTOM ORDERS AND VERTICAL INTEGRATION

Custom orders and vertical integration have some similar marketing technology, but are different. Both are based on knowing ahead of time that an outlet for the product exists.

Custom order marketing is when buyers (processing plants and others) contract with producers to grow a certain plant or animal for them. For example, a pickle company may sign growers to grow a certain kind of cucumber. In addition, the pickle plant may specify growing methods and the exact size the cucumbers must be when harvested.

Vertical integration is when a non-farm agribusiness owns the processing plant and all of the sources of supplies for producing the plant or animal. The farmer or rancher merely provides land and labor. This is often used for broiler production. A broiler processor will sign up growers and specify exactly how the birds are to be raised. The grower is paid a set amount for each bird that is raised. The processor provides the baby chicks, furnishes the feed, sends a field representative to supervise the operation, and hauls the chickens to the plant for processing. People who grow chickens on their own might not be able to sell them to a processor.

NICHE MARKETING

Some growers are able to develop special markets to fit their situations. These growers are said to have a ***niche market***. Here are a few examples:

20-10. Organic vegetables are niche marketed at this street-side stand.

- **Lambs**—In areas with a number of ethnic groups, lambs can sometimes be grown and marketed live. This is especially true for certain religious celebrations.
- **Goats**—A few goat growers produce for the barbecue season of the year. Young goats are sold live and the buyer does the slaughtering. Some buyers prefer to take goats to a custom slaughter house.
- **Recreational fish**—A few aquafarmers operate fee lakes on their farms. The public can fish in a pond by paying a fee. The fee varies but usually involves a flat rate for admission plus a per pound charge for the fish that are caught. Some fish farms get additional money by providing fish dressing services that are done for a fee.
- **Exotic animals and plants**—Some producers find growing unusual animals and plants profitable. They often market directly to consumers but may market to retail stores. Examples include organic vegetables, "run-about" eggs (laid by chickens that aren't penned up), game birds, tropical pet fish, and other pets, such as dogs, cats, and snakes.

MARKETING INVOLVES MANY FUNCTIONS

Getting products to consumers the way they want them involves many processes. These processes are sometimes known as marketing functions. These functions convert the product into a form consumers want and transport it to them. The functions are the links between the producer and the consumer.

Marketing is more than selling what has been produced. Many people only view marketing as changing ownership of products or services but it is far more than this. Further, the ownership of a product may change several times between the producer and the consumer.

Nine marketing functions are presented in this chapter.

HARVESTING AND ASSEMBLING

Harvesting and assembling begin at the point of production. They are the first steps in getting what has been produced ready for the consumer.

Different crops require different methods of ***harvesting***. Some products are harvested with machines; others must be harvested by hand. Harvesting includes picking apples, digging potatoes, and cutting sugarcane. It also includes seining fish, herding cattle into a catch pen, shearing sheep, and

20-11. Cauliflower being harvested in Salinas Valley, California.

milking cows. Logging, tapping trees for sap, and cutting Christmas trees are also forms of harvesting. Digging ornamental plants and cutting long-stem roses are two examples of harvesting horticulture crops.

Care must be used to assure a good product. Foods must be kept free of impurities and harmful organisms. Fibers must be kept dry and free of trash so that their color isn't affected. Fruits and vegetables must be protected from wilting and injury.

Assembling begins after harvesting. The products are delivered to a central location for the next functions in marketing. Assembling is important because of a concept known as "economy of scale." This means that marketing people want to deal with large volumes. Many farms don't have large enough volumes unless they combine with other farms.

Assembling involves massing large quantities together to make marketing more efficient. Larger farms may do their own assembling. Smaller farms may take harvested products to packing sheds or processing plants.

20-12. Sugar beets are assembled in large piles at this North Dakota piling station.

Examples of assembling include cattle sale barns, grain elevators, and cotton gins.

As with harvesting, assembling varies according to the product. In some cases, route trucks may pick up products, such as milk. In other cases, the product may be packaged on the farm ready for the consumer, such as apples that are put in boxes.

GRADING

Grading is sorting products for uniformity. It often occurs during or after assembling. People want products that are of uniform size and quality. They don't want to buy a carton of eggs and have a few that are small, some middle size, and others that are large.

Various standards are used in assessing products. Often these standards have been set by the government or by an association of producers. Government standards are designed to protect consumers from unsafe products and assure uniformity. Processors, supermarkets, and others may have their own standards. Large restaurant chains have definite standards that dictate how processors will prepare products for them.

Grading often involves sorting on the basis of several factors: size, color, blemishes, and varieties. Size is often a factor in the price a producer receives. For example, small cucumbers sell for more money than large cucumbers and large watermelons sell for more than small melons. Every product has standards of its own that are used in grading for size.

Color is a factor in some products. People want lettuce leaves that are green and eggs that have yellow yolks. They want cotton that is white and not yellowed by rain. They want oranges that are orange and limes that are green.

20-13. Eggs are being graded to provide uniform quality for the consumer. (Courtesy, Marco Nicovich, Mississippi State University)

20-14. A quality cauliflower showing no defects or early signs of decay.

Blemishes are defects in products. Potatoes with rotten spots are blemished. Hogs with injuries are discounted. Eggs that have dirty shells aren't as desirable as those that are clean. Of course, no one wants products that have worm holes.

Products must be of uniform varieties or kinds. Fish that are mixed, such as tilapia and bass, aren't desired and must be sorted. Red and white potatoes can't be mixed. Yellow and white corn must be kept separate.

TRANSPORTING

Transporting is moving the produce from one place to another. Milk must be moved from the dairy farm to a processing plant. Freshly picked beans must be moved to the cannery (or packing shed for the fresh market). Sweet corn cannot be kept long without loss of sugar content in kernels.

Refrigeration will slow down deterioration. Some products must be immediately refrigerated on the farm, such as milk. Milk that isn't cooled and kept cool will spoil quickly. The bacteria count will increase and to a point where it can't be sold.

Products are transported in different kinds of trucks, barges, airplanes, wagons, and box cars. Some are hauled in tanks, such as milk; others are hauled on open trailers, such as pulpwood. Some must be refrigerated or

20-15. Trucks and trailers are used to haul tomatoes to market.

frozen while being hauled; although this usually applies to products that have been processed. Some must be hauled to keep them alive and in good condition, such as hogs, cattle, and some fish.

PROCESSING

Processing is all of the changes made in a product that prepare it for consumption. Products are processed in many different ways. Some of the processing is done to keep them from spoiling; other processing makes them easier to prepare. (Chapter 21 gives more information on processing.)

Here are a few examples of processing of different products.

- **Logs**—Logs may be made into a number of different products. We typically think of them being sawed into boards (lumber). Plywood, veneers, and other wood products are made from logs. In addition to sawing, processing lumber involves planing the boards to make them smooth, drying to remove the sap and, in some cases, adding a preservative to help them resist rot and insect damage.
- **Cattle**—Beef cattle are typically slaughtered and made into various cuts and products. Some of the meat will be sold fresh and some will be canned or frozen for later consumption.
- **Wheat**—The grain kernels of wheat are milled to remove the husk and made into flour. The flour is used to bake bread. Flour may be prepared in different ways; some may have baking additives to make it easier to use in cooking.
- **Tomatoes**—Tomatoes are popular food items that are processed in many ways. Many tomatoes are sold as fresh tomatoes—vine ripe. Some are dried and used in that form. Other tomatoes are canned and made into products such as catsup, soup, and sauce. Most pizza sauce started as tomatoes!

PACKAGING

Packaging is placing products in containers. Many different kinds of containers are used. All containers are designed to keep a good quality product and make it easier to handle in various quantities. Some products are packaged several ways.

Packaging materials include plastic, wood, paper, metal, and glass. Eggs and milk are often put in plastic cartons. Strawberries and tomatoes may be put in paper packages. Canned asparagus and snap beans are put in

20-16. Packaging for this fresh asparagus in a supermarket involves using two rubber bands and a label.

metal cans. Pickles and apple sauce may be put in glass jars. Fruit may be put in wooden crates or pasteboard boxes.

Portioning

Most containers divide the product in portions or sizes that consumers want. This is known as portioning. Some containers are large, such as the can of beans used in a school cafeteria. Others are small, such as the jar of pimento peppers used at home when making pimento and cheese spread.

Shipping Crates

In addition to the individual containers, the cans and jars may be placed in large pasteboard boxes for shipping. This makes it easier to haul

20-17. Harvested lettuce is being hauled from the field in large shipping crates.

and move the products into and out of trucks, warehouses, and stores. Packaging filler material may be added to keep the products from bumping together while being moved.

Modern facilities involve using wooden pallets stacked with boxes. The boxes on the pallets may be held together with plastic wrap and lifted with a fork lift. All of this saves human labor and keeps down the cost of food and other products.

Labeling

Labeling is a step in packaging that is used to identify the product. The name of the product as well as other information, such as weight and nutrition, are given on the label. Labels may also describe how the consumer is to fix the product, such as a microwavable dinner.

Labels also provide information about where the product was processed and the date by which it should be used.

20-18. Apples may be individually labelled.

20-19. This loaf of bread is attractively and conveniently packaged.

STORING

Most products have to be stored, which means keeping them until they are needed. Some can be kept for a long time with little loss of quality if stored properly. Others will deteriorate fairly rapidly even under the best conditions. The way the product has been processed will influence storage life.

Most marketing involves keeping the amount of products on hand as small as possible to meet the demand. This requires businesses to order and buy carefully. It also reduces the amount of money a business must invest in warehouses and other facilities.

Products must be stored properly. Some must be in refrigerated coolers, such as fresh vegetables, eggs, meat, and milk. Others can be stored without refrigeration as long as the temperature is kept reasonable, such as in a well ventilated warehouse where canned foods are kept.

How a product has been processed has a lot to do with how it can be stored. Frozen foods must be kept frozen. Dried foods must be protected from moisture, insects, and rodents.

ADVERTISING AND PROMOTING PRODUCTS

Advertising and promotion are used to get more people to use a product. Agricultural marketing companies want to sell as much of a product as they can. They want people to buy what they have to sell.

Many different technologies are used in advertising and promoting products. Some people study these areas to develop skills in advertising and promoting products.

Advertising

Advertising involves communicating information to people that is intended to persuade them to buy a product. Advertising may involve radio and television, newspapers and magazines, billboards, and other means. The purpose is to make people aware of products and encourage them to buy them.

Promoting

Promotion includes advertising, but it goes further. Promotion may involve providing free samples of food in a supermarket. Prices are reduced to encourage people to buy a product who might not otherwise buy it. Reduced prices attract buyers and increase the amount sold.

Special events may be used. June is known as "dairy month." A variety of approaches are used to encourage the use of dairy products. Some places have strawberry festivals, apple harvest carnivals, and other events that promote products.

20-20. Activity at the New York Cotton Exchange shows bidding underway on cotton. (Courtesy, Ann P. Bruch, New York Cotton Exchange, New York)

SELLING

A sale occurs when the ownership of a product changes. ***Selling*** is the process of making a transaction between a buyer and a seller. Most people use money to buy products; however, sometimes things of equal value are exchanged.

Agricultural products may be sold several times during the marketing process. The producer may sell them to a processing plant. The processing plant may sell them to a warehouse company (known as a jobber). The warehouse may sell them to a retail store, which sells them to the consumer. All the people involved try to make a profit when they sell. If they don't, they won't be able to stay in business and help with marketing.

DISTRIBUTING

Once a product has been prepared, it must be moved to a place where the consumer can get it. This is much more than hauling, but hauling is an important activity.

Distribution is getting what is needed to the right place at the right time. The right amounts must arrive when needed. A good example is the hamburger. All of the things needed to make a hamburger must get to the restaurant. If everything gets there but the bun, the restaurant can't make and sell hamburgers to its customers.

Distribution requires careful planning. All activities must be carried out as planned. Orders must be placed and processed. Trucks must make their runs and deliver what was ordered. All along the way, the product must be kept in a good, wholesome condition.

AgriScience Connection

MARKETING ORANGES REQUIRES SCIENCE AND TECHNOLOGY

The oranges in this grove were carefully selected to produce citrus to meet consumer demand. In the grove, the latest cultural practices are followed to assure quality fruit. Once grown, the oranges are picked, placed in crates, shipped to a packing shed, and graded and placed in boxes. Trucks deliver them to warehouses and supermarkets throughout North America. Of course, some are further processed into juice and other products. (Courtesy, Florida Department of Citrus)

Getting products to the right place on time is a key to success. Restaurants and supermarkets without food have nothing to sell!

MARKETING INFRASTRUCTURE

Infrastructure—a big and important word in agricultural marketing. It refers to all of the things that must be in place for marketing to occur. It includes production agriculture as well as the supplies and services producers need. Infrastructure definitely includes the marketing facilities for products.

Marketing infrastructure is those things that support and make marketing possible. For example, a road is needed to haul a product to a processing plant. A truck with special equipment may be needed to do the hauling. The processing plant is infrastructure because the product couldn't be processed without it.

20-21. Considerable equipment and labor are needed to harvest large amounts of broccoli.

HARVESTING INFRASTRUCTURE

Once plant or animal products have been produced on the farm, they must be harvested. A way to harvest them must be available.

Some products require very expensive harvesting machinery. These machines may cost more than one producer can afford. An individual in a particular area may buy one of the machines and use it to harvest crops for several farmers. Wheat is a good example. Only the largest farming operations can afford to own combines. ***Custom harvesting*** is when one person buys a combine and harvests the wheat of other farmers for a fee. Custom harvesting is widely used with wheat, hay, cotton, and several other crops.

Products shouldn't be grown if there isn't a way to harvest them. A good example is fish. Many farmers have built fish ponds, stocked them with fish and fed out the fish without arranging for harvesting equipment. The equipment includes a seine, seine hauler, a small boat to help in setting the seine, and haul trucks to take the fish to a processing plant. The crop shouldn't be started until these are available.

Another good example is dairy farming. Cows have to be milked twice every day all year long. Milking equipment must be on the farm and people must be available to do the milking. In addition, storage facilities must also be present to hold the milk until it is taken to a processing plant. If a processing plant isn't reasonably close by, dairy production may not be a good choice.

TRANSPORTATION INFRASTRUCTURE

Transportation is needed to move products through the marketing process. Roads, railroads, water ports, and airports are all used in marketing.

20-22. Railroad box cars are being loaded with canned milk products at a processing plant.

Bridges and road surfaces must be strong enough to support the heavy weights of trucks.

The right kinds of vehicles must be available. Trucks that are properly equipped to haul produce are needed. If timber is being harvested, pulpwood, pole, or log trucks will be needed. If the wood is chipped, trucks with chip trailers are needed. Milk must be hauled in refrigerated tank trucks; cattle hauled in a cattle truck; and cotton hauled in cotton wagons. Almost every plant or animal produced has unique transportation requirements.

If the needed transportation is not available, marketing will be more costly and could be impossible.

ASSEMBLY AND PROCESSING FACILITIES

Packing sheds and processing plants need to be available. Hauling products long distances adds to the cost of production. Further, harvested products can deteriorate in quality while being hauled.

20-23. Onions are being unloaded at a processing plant.

A good example is grain. Grain elevators or other outlets need to be convenient. Hauling grain long distances costs money. As costs of hauling go up, profit goes down.

For many animals and plants, assembly and processing facilities are essential. They must be present before production is begun. If not, careful thought should be given to producing some other product.

NEEDED FINANCES

Producers need money to finance the costs of crops and animals. Banks and other lending agencies are needed. If unavailable or unwilling to make a loan for a particular crop, the infrastructure of finances doesn't support marketing.

PEOPLE SKILLS

People who know how to perform certain work must be available. If the labor supply is short, the infrastructure won't support the production of certain crops. This is especially true if the people are required to have certain skills.

Increasingly, shortages of people to work in various areas of agriculture have developed. In some cases, workers are brought in from other countries to do the work.

Labor applies not only to the production of the plants or animals, but also includes all of the marketing functions. People must be available to work in processing plants.

The work is useful and contributes to the overall benefit of the people. The notion that some work is not good is unfortunate. If the work produces needed food, fiber, and shelter, it is worthwhile!

20-24. Individuals with many different skills are needed in agricultural marketing, such as this pasta machine operator. (Courtesy, ARA Services, Philadelphia, Pennsylvania)

COMMUNICATION TECHNOLOGY IN AGRICULTURAL MARKETING

Marketing involves the transfer of information. People who don't stay up-to-date are likely to lose money because they aren't aware of what is needed in marketing.

Regular reports of market information are provided on many plant and animal crops. ***Communication technology*** includes all of the ways people exchange information about marketing.

Some plants and animals are traded on commodity exchanges. These are located in central cities where buyers and sellers can get together. Most of the people involved are traders; however, a few growers market through these facilities.

MEDIA REPORTS

Radio, television, newspapers, and specialized marketing publications provide important information about marketing. Daily market quotes are broadcast or printed. Producers can keep up with the price trends and make decisions about buying and selling.

Some of the media emphasize reporting the news behind the market. This often includes worldwide events that affect the prices that are paid. For example, heavy rain in South America can influence the price of soybeans in North America and vice versa.

20-25. Computers are often used to get and manage marketing information.

COMPUTER NETWORKS AND THE WORLD WIDE WEB

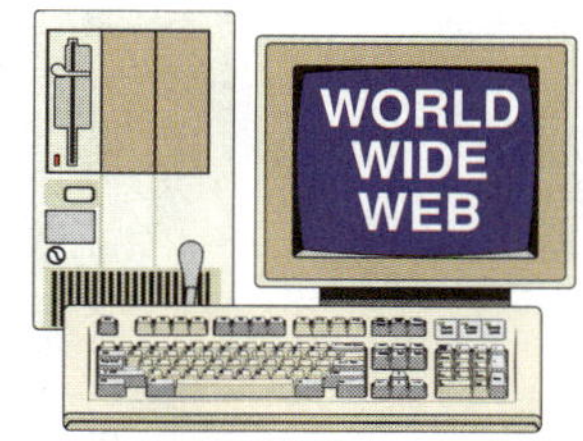

20-26. The world wide web (www) which is accessed with a web browser through an internet provider is being used as a part of communication technology.

Several services in agricultural marketing are available through computer networks. To use the networks, a modem and access to a telephone line are needed. These provide information about marketing as well as other areas of agriculture. Some allow people to interact with the system to buy, sell, or get more information.

The world wide web (www) is accessed through an internet provider. Many sources of agricultural marketing information can be found in www. Some producers have home pages that provide detailed information about the status of their crops and current selling prices.

Some training is needed to use the networks and world wide web effectively. This can usually be learned in a short time. Of course, the producer must have a computer with a modem and an internet service provider.

BROKERS AND CONSULTANTS

Brokers and consultants are people who keep up with the market and help others make marketing decisions. They are experts on marketing. Staying informed about the marketing situation is their only business. They help producers make marketing decisions. They charge a fee for their work.

Consultants are often able to help producers get higher prices than the producers would get otherwise. Some consultants work with producers

20-27. A marketing manager is preparing a television spot for promoting Springwater brand cultured fish.

20-28. AgriScience teachers have classes that teach marketing processes.

throughout the year. They help with soil testing, variety selection, fertilizer use, insect control, harvest management, and other production activities.

EDUCATORS AND SERVICE SPECIALISTS

Various agencies offer free marketing assistance to producers. Some of these can be useful sources of information for making marketing decisions.

Vocational and technical education programs at high schools and community colleges may have classes and adult groups that are taught marketing processes. Some schools have marketing specialists to help with educational programs. Students may be taught marketing processes through competitive activities, such as the Marketing Plan Project of the National FFA Organization.

State agricultural agencies may have marketing specialists. They help link producers with possible market outlets. In some cases, they promote the development of new markets.

Associations of growers may get involved in providing education and services. Producers of crops join together to learn and gain more power in the marketing process. A good example is almond growers in California. Extensive marketing is carried out by a growers' association.

Cooperative Extension Service agents often have marketing information. Some have college degrees in areas of agricultural marketing. Most local communities have an extension service office.

20-29. Television is used to provide agricultural marketing information.

20-30. Agricultural marketing provides an abundance of food for North America. (Courtesy, ARA Services, Philadelphia, Pennsylvania)

REVIEWING

MAIN IDEAS

Marketing is linking the producer with the consumer. Only plants and animals for which there is a market should be produced. A producer should never begin production without having a market.

Many marketing technologies are available and used. Each has advantages and disadvantages. The approaches used are often based on the preferences of the producer. The marketing channels are: (1) roadside and retail markets, (2) central locations—packing sheds and elevators, (3) processing plants, (4) custom orders and vertical integration, and (5) niche markets.

Getting produce to consumers involves several marketing functions. These convert the raw produce of the farm or ranch into the product that consumers want. The functions are: (1) harvesting and assembling, (2) grading, (3) transporting, (4) processing, (5) packaging, (6) storing, (7) advertising and promotion, (8) selling, and (9) distributing.

In order to market, certain facilities must be available. These are known as marketing infrastructure. Roads, vehicles, assembling and processing plants, finances, and labor are all necessary.

Communication of marketing information is essential. Without good information, it is difficult to be good at marketing. People involved with marketing must keep up to date.

QUESTIONS

Answer the following questions using complete sentences and correct spelling.

1. What are marketing and agricultural marketing technology?
2. Why are consumers important?
3. What methods are used in agricultural marketing? Briefly describe each method. Give an example of a plant or animal product that could be marketed using each method.
4. What functions are involved in marketing? Briefly describe each function. Give examples of what may be done to different products during the functions.
5. What is marketing infrastructure? What are the major areas of agricultural marketing infrastructure?
6. What is communication technology in agricultural marketing? What are examples of communication technology?
7. Why is information about marketing important?
8. What are the sources of marketing education and services in a local community?

EXPLORING

1. Survey your community to determine how producers market crops and livestock. This might involve interviewing a few producers or agricultural professionals. The class may want to develop the questions to ask as a group activity.

2. Trace the marketing process of one of your favorite foods. If the food has more than one ingredient, trace all of the ingredients. For example, if you like spaghetti with meat sauce, use reference books and other materials to help trace the origins of the raw materials and the processes involved.

3. Make a collection of the agricultural marketing information in the local newspaper. Collect the information for at least a week. Analyze price changes and try to forecast future price levels.

4. Arrange to tour a local marketing facility, such as a packing shed, processing plant, or grain elevator. Have the manager explain how the facility operates and what it expects of producers.

21

PROCESSING TECHNOLOGY IN AGRISCIENCE

The way people live has changed dramatically. They have come to expect things that are ready and easy to use. They don't want to spend a lot of time making foods, clothing, and shelter. People want things that are convenient.

Many of the things people use today would be difficult to make at home starting from the beginning. Imagine the amount of work it would take to produce all of the ingredients needed for a pizza.

All plants and animals and their products must be prepared for use. Many years ago people did this in their homes. Today, people rely on others to prepare their food, fiber, and shelter. People also want them to be of good quality at an economical price. A huge network of activities to meet the needs of consumers has developed. Many of these involve careful and important processes.

21-1. Broilers are processed on an automated line. One processing plant will prepare thousands of broilers in one day. (Courtesy, James Lytle, Mississippi State University)

OBJECTIVES

This chapter introduces important areas of agricultural processing technology. It has the following objectives:

1. Explain the meaning and importance of processing.
2. List and describe methods of food preservation.
3. Explain spoilage in food.
4. Describe methods used in processing fiber products.
5. Describe methods used in processing wood products.
6. Describe safety and regulations that apply to processing.

TERMS

- aseptic packaging
- botulism
- canning
- classing
- convenience preparation
- cotton seed meal
- curing
- drying
- fabricating
- fermenting
- fleece
- food processing
- freezing
- ginning
- hardwood
- irradiation
- log
- pickling
- planing
- plywood
- portioning
- preserving
- pulp
- pulpwood
- refrigeration
- saw mill
- saw timber
- seasoning
- softwood
- spoilage
- synthetic food
- textile mill
- veneer

PROCESSING: MEETING THE NEEDS OF PEOPLE

Processing is preparing food, fiber, and wood for people to use. Several areas of the marketing process overlap with processing. Processing is grading, fabricating, preserving, portioning, convenience preparation, and packaging. Consumers are willing to pay more for their food, clothing, and shelter if it is easy to use.

Most people only think of food when they think about processing the produce of farms. Processing also includes fiber and wood products.

21-2. Cooking hamburgers on a grill is popular with many people.

PROCESSING FOOD

Food processing is all of the steps that are done to prepare food for people. What is done varies according to the food product and how it is prepared. Some items don't require a lot of processing; others must have considerable preparation.

Processing also includes preparing foods for convenient preparation. Consumers want pre-cooked packaged food that can easily be made into a meal. This may involve putting a small can of tuna with crackers, pickle relish, and mayonnaise in one convenient package. At other times, pre-cooked frozen meals that only have to be heated for a few minutes may be prepared.

Procedures in Processing

Food processing generally includes grading, fabricating, preserving, portioning, preparing convenient forms, and packaging. Some of these overlap with marketing technology. After all, agricultural marketing technology involves getting food, fiber, and shelter to people!

- **Grading**—Grading is sorting items for uniformity and quality. It is done in processing plants if the products weren't graded as a part of marketing.

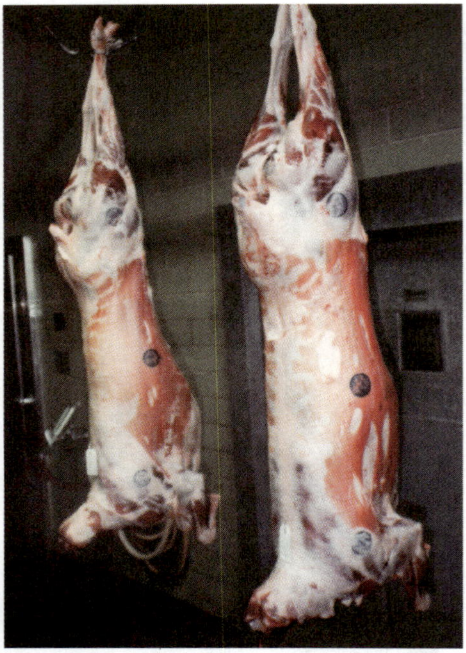

21-3. Lamb carcasses hanging in a processing plant.

- **Fabricating**—***Fabricating*** is making the product into forms that can be used. This step varies considerably, depending on the product. For example, livestock must be slaughtered and the carcasses cut into pieces that can be easily used. Humane procedures must be used with live animals.
- **Preserving—*Preserving*** is treating food to keep it from spoiling. Many different procedures are involved. These will be described in more detail later in this chapter.
- **Portioning**—***Portioning*** is preparing foods in the right quantities for consumers. Examples include putting two steaks in a package, putting 16 ounces of vegetables in a can, and having the right combinations of vegetables, meat, and other items in a frozen dinner.
- **Preparing in convenient forms—*Convenience preparation*** is making foods easy for consumers to use. Sometimes the foods are pre-cooked and only need to be heated. Some even come in containers that can be used to heat and serve the food.
- **Packaging**—Packaging is using containers that protect food from spoilage and keep it in the right amounts. Packaging materials include glass, metal, plastic, mesh fabric, and paper and other materials used in making containers.

21-4. Macademia nuts may be bought unshelled or shelled. The shells are very hard to crack. Consumers appreciate the convenience of the shelled nuts.

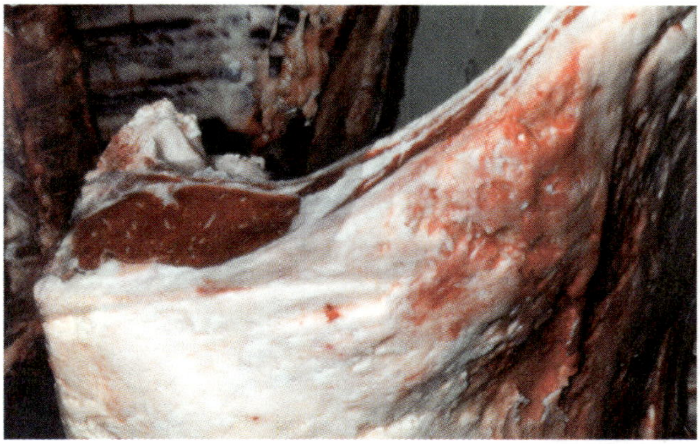

21-5. Beef loin is a tender and valuable part of the beef carcass.

Highly Processed Food Products

Foods may be made into a wide range of products. Some require a lot of processing. Other products allow processors to use low quality materials to make higher quality products. In all cases, products that consumers want must be produced.

Here are a few examples:

- **Wieners**—Wieners are made from meat pieces that are of little value by themselves, such as scraps from bones and organ tissues of the animals. Beef, pork, and poultry meats may be used in making wieners. The meat is finely ground, thoroughly cooked, and formed into the shapes that are desired.

- **Boneless products**—Boneless products include hams, fish fillets, turkey breasts, beef steaks, and sandwich meats. In some cases, the bones are carefully removed and the meat is pressed backed together. Making a boneless ham involves removing the bone from the ham muscles and reforming it into a shape resembling the original ham. Many of these products are thoroughly cooked during processing. To be safe, always read the label to see if the product requires cooking before it can be eaten.

- **Oils**—Many oils used in cooking and food preparation are made from grains. Corn, soybeans, safflower, and canola grains are often used to make oil, known as vegetable oil. These substitute for animal fat, such as lard from pork and tallow from beef. The grains and seeds are processed in oil mills. In addition to oil, a by-product known as meal usually results. The meal is often used as a source of high protein animal feed.

21-6. Soybean oil in a beaker and soybean meal in a scooop are surrounded by soybeans that haven't been processed. (Courtesy, American Soybean Association, St. Louis, Missouri)

21-7. Cotton fibers are made into durable, comfortable clothing. (Courtesy, National Cotton Council of America, Memphis, Tennessee)

PROCESSING FIBER

Processing fiber is converting raw material into cloth and other fabrics that can be used. Cotton, wool, and linen are three important natural fibers. Many artificial fibers are made.

The fibers must be cleaned and dried to remove excess moisture. They are woven into fabrics that are dyed or left their natural colors. The fabrics are usually transported and sold as large rolls or bolts.

Fabrics are used in making clothing, furniture upholstery, carpet, towels, bed sheets, and curtains for the home and office. These products are made in cutting and sewing plants. Products that are knitted, such as socks and sweaters, are made in knitting mills.

The major use of fiber is in clothing manufacture. Clothing is made in garment plants. Automated cutting and sewing machines are used as much as possible in clothing construction.

More information is presented on processing fiber later in this chapter.

WOOD PROCESSING

Wood is produced and used to make many different products. When and how the wood is harvested is related to how it is later used.

21-8. Logs are sawed into lumber at a saw mill.

There are two kinds of wood: hardwood and softwood. ***Hardwood*** is from oak, sweet gum, hickory, maple, and cherry trees. ***Softwood*** is from pines, firs, and spruces. Softwood is more likely to be used for making lumber, paper, plywood, and similar products. Hardwood is more likely to be used for furniture, flooring, and cabinets.

Pulpwood is made into paper, particle board, and other materials. For all of these, wood fibers are broken apart and pressed back together in a new form. Trees for pulpwood are cut when smaller and younger than trees used for logs.

Plywood is made by forming short logs into long, thin sheets that are glued together. Plywood is a strong, useful wood product.

Saw timber is the largest and oldest trees. It is fabricated into lumber at a saw mill. Logs are cut into boards of different dimensions, such as 1, 2, or 4 inches thick and 4, 6, 8, 10, or 12 inches wide. Lumber has many uses in constructing homes and other buildings.

Specialty products are made from different kinds of wood. Examples include handles for hand tools made from hickory wood, black walnut wood used in making fine furniture, various hardwoods used to make railroad cross ties, and balsa wood used to make very light objects, such as floats for fishing lines.

SYNTHETIC FOOD MANUFACTURING

A ***synthetic food*** is an edible product made from non-food raw materials or altered food materials. Food processing companies are developing synthetic foods to achieve certain purposes in the human diet, such as no fat. One of the first developed is ***olestra***—a fat substitute. (The trade name for olestra is Olean.)

In 1996, the U.S. Food and Drug Administration approved limited use of olestra. The first uses are in making snack foods such as potato chips, crackers, and tortilla chips. Since olestra provides no calories as do the oils now used in cooking potato chips, for example, the calories in potato chips are reduced for those cooked in olestra.

Olestra is a synthetic mixture of sugar and vegetable oil. It passes through the body without digestion or absorption. Some people who eat foods cooked in olestra may have side effects such as upset stomachs and stomach pain. (Some natural foods also have these side effects with people.) It may interfere with the absorption of nutrients from other foods. The use of "fake fat" and other edible synthetic products will likely increase in the future.

Synthetic foods should meet standards of wholesomeness just as natural foods. They should be prepared in a clean environment and be free of contamination.

FOOD PRESERVATION

The main purpose of food preservation is to keep food from spoiling. Some foods will spoil quickly if not preserved; however, these same foods can often be kept for a long time without spoiling if properly preserved.

People want safe food. They don't want to eat food that will make them sick. They want food that is wholesome and has healthful nutrients.

21-9. A bulging can is a sign of spoiled food. Never eat food from such a can.

SPOILAGE

Food can be unfit for people to eat. Good, wholesome food can spoil (go bad). ***Spoilage*** is when the food is unsafe to eat and the flavor has changed. For example, sour milk has a definite change in flavor. Spoiled food may make people sick; sometimes it even causes death.

Spoilage is caused by several different things. Most of these can be controlled.

Growth of Microorganisms

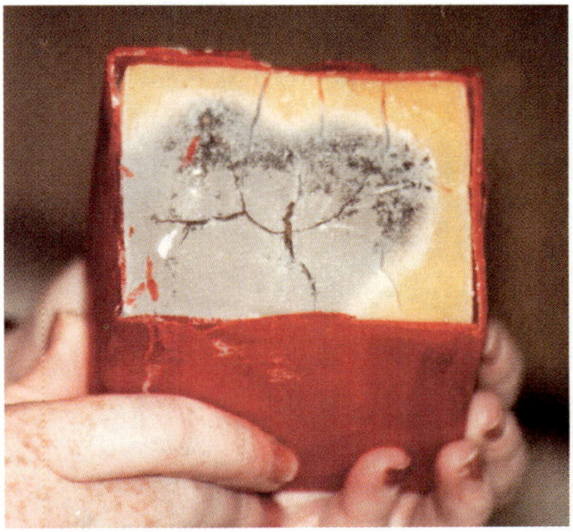

21-10. Cheese with mold is an example of food that has spoiled.

Several kinds of microorganisms can cause spoilage. Most spoilage is caused by bacteria and fungi, such as molds and yeast. Molds grow best at about 90°F (32°C) in places that are moist. The mold on bread develops when the bread is kept at this temperature.

Fresh vegetables and fruit will often begin to decay within a few days. This is due to the presence of microorganisms. Storing fresh foods properly will slow the growth of these organisms.

Botulism is probably the best known food poisoning. It is caused by the bacterium *Clostridium botulinum*. Food contaminated by botulism may be fatal to people.

Most of these microorganisms are killed at high temperatures. They don't survive in extreme cold, high acid conditions, salt and sugar concentrations, and conditions without moisture.

All preserved food uses at least one method of controlling the growth of microorganisms. Once the organism is controlled, the food can be sealed in air-tight containers. Microorganisms must have oxygen to live..

Rodents

Rodents are small animals, such as rats and mice, that damage food and other products. They are more likely to cause problems with grain and similar products that are stored in bins and warehouses. Not only do the rodents eat some of the stored food but they also contaminate the food with feces.

Insects

Insects can get into grain, fruit, vegetables, and other food products. Fruit flies may lay eggs on the fruit and these eggs can later hatch. Weevils get into grain and eat the nutritious germ and starch. The presence of insects and insect parts in human food is controlled by law. Most food products have a tolerance level for insects and insect parts that is very low. No one wants flour that contains insects!

AgriScience Connection

FROM BERRY TO BAG

Coffee berries are picked when they are turning red. Each berry contains one or two beans—most have two. Berries with one bean are more valuable and are known as "pea berries." Coffee berries are hand picked and delivered to a warehouse or processor. Here, drying, removing the hull of the berry, and bagging take place. In some cases, the beans may be roasted and prepared for sale rather than put in large bags.

This shows a coffee plantation in the South Kona District of Hawaii. The plantation is located on a mountain that rises above the Pacific Ocean. Drying takes place in the large greenhouse-type structures. The beans are regularly raked by hand to allow even drying. Moisture, dust, insects, and other contaminants are kept out.

Undesired Growth

Some food products will begin to grow after harvest if not properly stored. A good example is the onion that sprouts in warm storage. Beans and peas as well as the seed in grapefruit and oranges may also sprout. Sprouting is a particular problem with potatoes.

The presence of sprouts is a sign that the food is spoiling. Most people don't want to eat sprouts; however, some kinds of sprouts are used for food such as alfalfa and bean sprouts.

Chemical Contamination

Food products can be contaminated with chemical residues, such as those used to control weeds in the crop. Harvested food crops should never be stored or hauled in areas where pesticides were stored. This also applies to feed crops for animals, such as feed for chickens and cattle.

Fuel that gets on food is also a source of chemical contamination. Gasoline and diesel fuel should not be allowed to come into contact with grain, fruit, vegetables, and other foods.

Foreign Objects

Undesireable objects in food products make them unfit for human consumption. Metal and glass pieces should be kept out of food. Small rocks may sometimes get into crops when they are harvested. For example, a bean harvester may accidentally pick up a small rock about the same size as a bean. Rocks are difficult to remove from some food crops. Pieces of wood and stem are also foreign materials, but may not pose a serious danger to consumers. Pieces of rubber and plastic also contaminate food. A good example is foods tied in bunches with rubber bands. Food preparers must be careful to keep the rubber bands out of the food.

Purity

Food products should be of uniform species. Beans and peas can't be mixed together. No one wants to open a can of tomatoes and find a kernel of corn! Beef and pork should be kept separate and not mixed. Of course, various mixtures of foods are made when preparing soup, stew, and other products. Proper labeling is a must so that people know what they are getting.

Improper Preservation

Foods will spoil if not properly preserved. Cans of food must be properly heated and sealed. Bulges in cans are signs of spoilage. Discoloration in food is also a sign that some spoilage is under way. Using the wrong additives can also make food unsafe to eat.

21-11. Discoloration on the orange is a sign of decay likely due to improper storage.

METHODS OF PRESERVATION

Food can be preserved in different ways. Different foods require different methods of preservation. Further, some of the methods of preservation help make products that people want. Some foods have better tastes after preserving than they did fresh.

Food preservation and creating products that people like often go together. For example, people like cucumber pickles. Pickling is used to preserve the cucumbers. Various spices are added to pickles to make them more appealing to people.

The acid levels of food products are important in food preservation. For example, tomatoes have higher acid than beans and corn. Microorganisms are less likely to survive in the high acid foods.

The methods of food preservation are presented in the following sections. Table 21-1 shows recommended methods of preserving selected food products.

Table 21-1
Recommended Methods of Preserving Different Foods

Food	Method of Preserving		
	Canning	Freezing	Storing[1]
Vegetables			
asparagus	XXXX	XXXX	
lima beans	XXX	XXXX	
snap beans	XXXX	XXX	
beets	XXX	XX	
broccoli		XXX	
carrots	XXXX	XXX	XXX
sweet corn	XXX	XXXX	
Irish potatoes			XXXX
sweet potatoes	X		XXXX
pumpkins	XX		XXX
squash	X	X	X
tomatoes	XXXX		
Fruit:			
apples	XX	XX	XXXX
berries	XXXX	XXXX	
grapes	XXXX	XX	
peaches	XXXX	XXXX	
Meats:			
beef	XX	XXXX	
pork	XX	XX	
chicken	XX	XXXX	

[1]Most all foods can be stored temporarily in refrigeration.

Some of the foods listed here can be stored without refrigeration if the proper conditions are maintained.

Note: The number of X's indicates the rating of the method of preservation for the food, with more X's meaning that the method is better.

Canning

21-12. Boiler used to provide steam for heating food during canning.

Canning is placing food in a container and heating it to kill all microorganisms. The food containers are held at the right temperature for a period of time and then cooled. Products vary in the times and procedures that are used. Heating kills the microorganisms that cause spoilage. An airtight seal on the container keeps other microorganisms from getting into the container after it cools.

Most foods are canned in glass jars or metal cans. The cans are typically steel coated with a very thin layer of tin. Some cans are coated with a layer of enamel.

The seal is the final closure of the can or jar after the food has been put in it. It is essential that the lid be air tight. Seals that leak will result in food spoilage. Spoiled food can cause people to get sick and, possibly, die.

Before food is placed in a jar or can, it should be carefully prepared. Only good food free of decay and contamination should be canned. Some food will need trimming, cutting, or shelling. The food should be carefully washed before it is placed in the jar or can.

Canning has been used for a long time to preserve food. It is an economical way to keep food from spoiling. After the canning process,

21-13. Canned food section in a supermarket.

AgriScience Connection

POTATOES INTO FRENCH FRIES

Huge amounts of potatoes are made into cut potato products each day. French fries are favorites with many people. You can be sure that you have a wholesome food from this processing plant.

Potatoes are quickly washed, peeled, inspected, cut, frozen, and packaged for shipment. The top photo shows peeled potatoes being released by the peeler onto a conveyor belt for inspection. Following inspection, they are cut into the desired sizes and quick frozen. Very low temperatures are used in quick freezing—often in blasts of air minus 25°F. After freezing, the potato products go into bags inside boxes where they are placed on a pallet and wrapped with plastic. (Courtesy, RDO Frozen, Park Rapids, Minnesota)

21-14. Large freezing facility that quick freezes individual portions of fish.

most jars or cans can be stored at room temperature. Canned foods can be stored for months and, in some cases, years before use.

Proper equipment must be used in canning. Large pressure cookers known as retorts are used to heat the cans of food. Heating sealed cans outside a pressurized container will result in the cans exploding. Improperly heating jars can result in the jars bursting and glass fragments cutting the people around them.

Freezing

Freezing is using low temperatures that freeze the water in food products. This slows and prevents the growth of microorganisms. Freezing is the method of food preservation that best retains the nutrients

21-15. Frozen food section in a supermarket.

in foods. However, some foods taste better if preserved other ways, such as cucumbers.

Just as with canning, foods should be carefully prepared for freezing. Only food free of decay should be frozen. Most freezing involves blanching vegetables and fruit to stop natural processes in the foods. Blanching is putting the food in hot water or steam for a few minutes which destroys the microorganisms that cause spoilage.

Fresh food is quickly frozen at temperatures well below freezing, often about –30°F (–34°C). After freezing, the food must be kept frozen and not allowed to thaw until it is used.

Meats, vegetables, fruit, juices, and other foods are preserved by freezing.

Fermenting

Fermenting is a method of preservation that involves using certain bacteria, molds, and yeasts. It is used in making cheese, bread, vinegar,

21-16. Milk goes through several steps in the fermentation process for making cheese.

and other products. The action of the microorganisms must be carefully controlled to get the desired product.

Drying

Drying involves removing the moisture from food products. Some drying is a natural process, such as making raisins by sun-drying grapes. Other drying is artificial and is sometimes known as dehydration.

Many dehydrated foods have had considerable processing to remove the moisture from them. They are in forms that can be made ready to eat by adding water. Sometimes the product must be heated after adding water or hot water must be added to the food. Examples include dried milk, soup, instant coffee, macaroni, and gelatin. Beef jerky is a dried food that doesn't need water added to it.

Curing

Curing is adding substances to foods to prevent spoilage. Salt, sugar, sodium nitrate, and other materials are used. Smoke is sometimes used in curing. Meat may be hung in a smokehouse or an artificial, liquid smoke flavoring may be added to the meat.

The natural circulatory system in meat may be used in curing. The ham is a good example. The curing solution may be injected under pressure into the blood vessels of the ham. This assures that the curing solution reaches all parts of the ham.

Curing creates conditions where spoilage organisms can't grow; salt and sugar prevent their growth.

Refrigeration

Refrigeration is keeping food products at temperatures slightly above freezing. Refrigeration is sometimes known as cold storage. Most fresh meats, milk, fruit, vegetables, and other fresh products high in moisture can be kept for several days with refrigeration. Refrigeration slows the growth of microorganisms that cause food to spoil.

Products that are being refrigerated should not be allowed to freeze. Freezing might make them unacceptable by changing their texture and flavor. In some cases, freezing would spoil the food.

21-17. Fresh refrigerated food section in a supermarket.

Nearly all homes have refrigerators. They are used to keep foods before cooking as well as to store foods that have been cooked. Most foods can be refrigerated for only a few days.

Irradiation

Irradiation is treating food with electrically charged particles such as X-rays, electron beams, or gamma rays. Low amounts of radiation will kill spoilage organisms. It is used to keep potatoes from sprouting, control insects in vegetables and grain, control parasites in meat, and slow down ripening processes in fruit and berries. The latter use helps keep foods fresh longer.

Aseptic Packaging

Aseptic packaging is one of the newer methods of food preservation used in North America. It was approved for use in the United States in 1981. Few foods are now available that have been preserved by this method, though its use is increasing.

Aseptic packaging involves sterilizing the food and the container it is stored in, such as plastic and foil. The amount of time the foods are heated while being processed is kept to a minimum. Long heating causes foods to lose their flavor.

Foods in aseptic packaging can be stored for weeks or months at room temperature without spoilage. Milk is a good example. Milk spoils in several days in paper or plastic cartons under refrigeration but will keep for weeks if preserved using aseptic methods.

Pickling

Pickling is placing foods in acid solutions to prevent the growth of organisms that cause spoilage. Vinegar is commonly used. Cucumbers, cauliflower, okra, peppers, and other vegetables are pickled. Some meat products are pickled, such as pigs feet and eggs. Many pickling solutions have spices and other flavoring materials added to enhance the taste of the food product.

Other Methods of Preservation

Several other methods are used to protect food from spoilage. Here are five examples:

- **Salting**—Salting involves using concentrations of salt to prevent the growth of spoilage organisms. Meats, especially pork, have been preserved by salting for a long time.
- **Fumigants**—Fumigants may be used in grain bins and potato warehouses to kill insects and rodents.
- **Chemical additives**—Chemicals may be added to foods to prevent spoilage, such as sulfur dioxide and citric acid.
- **Waxing**—Some foods are coated with a thin a layer of parafin (wax). The wax holds moisture in and keeps decay organisms out.
- **Pasteurization**—Pasteurization involves heating a food product to a certain temperature for a period of time to kill microorganisms that may be present. Milk is the most widely used pasteurized food. All fresh milk should be pasteurized. In addition, dairy products should only be made with pasteurized milk.

21-18. These rutabagas have been coated with wax to prevent spoilage.

21-19. A modern dairy processing plant.

- **Jelly and jam**—Fruit and berries are often preserved with sugar to make jelly, jam, and preserves. These products are often used as spreads on breads.
- **Cooking**—Cooking stops spoilage processes. Cooked foods can be kept under refrigeration for several days.
- **Using bases**—Bases are chemical compounds with a high pH. They are the opposite of acids. An example of using a base in food preservation is in making hominy. Corn kernels are soaked in a lye (sodium hydroxide) solution. Another example is using lime to preserve potatoes.

PROCESSING FIBER PRODUCTS

Fibers can be classified as two types: natural and artificial. Natural fibers are the most important in agriscience.

Natural fibers come from plants and animals. The major fibers from plants include cotton, linen, and jute. The major fibers from animals are wool, silk, and mohair. One natural fiber is made from rocks—asbestos. In North America, cotton and wool are most widely used.

Artificial fibers may be manufactured into products similar to natural fibers. Nylon and rayon are common examples.

COTTON PROCESSING

Cotton processing begins with ***ginning***, which is done at cotton gins. Gins do several things with the raw cotton, known as seed cotton. First, the gin separates the fiber and seed. Secondly, the gin removes trash, then dries and packages the lint cotton into bales. Lint cotton is the fiber without the seed. Cotton-producing areas have gins, which are large processing plants.

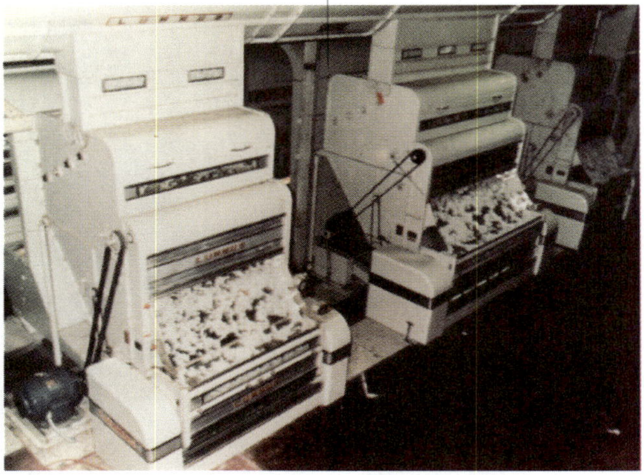

21-20. Gin stands in a cotton gin separate the lint and seed.

Classing

After ginning, the lint cotton is classed. ***Classing*** involves judging the cotton on the basis of length of fiber (staple), thickness of the fiber (micronaire) and grade, which is the presence of trash and color. Cotton should be white. Rain on open cotton bolls causes it to change to off-white.

Weaving

Cotton is woven into cloth at a ***textile mill***. Additional cleaning, spinning, and finishing are steps in the process. Spinning involves separating and twisting the fibers into tiny threads. After weaving, the cloth may be dyed and treated with other chemicals to give it a particular finish. Some cloth goes through a machine that prints colored designs on it.

21-21. Bale of lint cotton ready to go to a textile mill.

21-22. Manufacturing in a garment plant. (Courtesy, National Cotton Council of America, Memphis, Tennessee)

Manufacturing

After the textile mill, the cotton fabric goes to a plant for manufacture into various products. Most fabric goes to garment plants for manufacture into clothing. The fabric is cut into pieces, sewed, labeled, inspected, folded, packaged, and otherwise prepared for shipment.

Some cotton is made into yarn. This goes to a knitting mill for making knitted products, such as socks.

Use of Cotton Seed

Seed removed from the lint could be used to plant next year's cotton crop. This isn't widely done because most cotton producers use improved seed. The seed aren't thrown away, but are used for making food and feed products.

Cotton seed usually go to oil mills where they are made into vegetable oil and other products. A good example is oleomargarine—a butter substitute. The oil processing leaves a meal product known as ***cotton seed meal***. The meal is used as a high protein concentrate to feed animals.

WOOL PROCESSING

Wool is used to make clothing, rugs, blankets, upholstery, and other fabric products. Various kinds of wool are produced by different species of animals. Most wool in North America is from medium-wool sheep. (The kinds of wool were covered in Chapter 13.)

21-23. A truck loaded with large sacks of wool ready for grading and weaving.

Shearing

The wool on a sheep is known as ***fleece***. Shearing is removing the fleece with power clippers so that the sheep isn't injured. New procedures are being developed to chemically remove wool, but these aren't widely used. Most sheep are sheared once a year—usually in the spring. This keeps the sheep cooler in the hot summer weather. A new coat is grown before winter.

Grading

After shearing, the wool is sorted to remove trash and off-color fleece. It is then graded to determine its fineness, length, crimp, and color. Crimp refers to the straightness of the wool fibers. Fineness is determined by the diameter of the fibers.

Weaving

At the weaving plant, the wool is cleaned with a detergent to remove excess oil and prepare it for weaving. The fibers are arranged in parallel and woven into yarn or fabric. Different kinds of wool are used to make different fabrics. The wool is dyed to give it the desired color.

Manufacturing

After weaving, the fabric is made into garments and other products. This involves cutting the fabric according to the pattern, sewing the pieces together, and adding buttons, zippers, and decorative trim to the final garment. Wool yarn is knitted into sweaters, hats, and other products that are known for warmth and durability.

PROCESSING WOOD AND TIMBER PRODUCTS

Forestry products are made into many kinds of lumber, plywood, particle board, veneer, and paper. These products are used in many ways—shelter being the most important.

Lumber and plywood are important in building houses and other small buildings. In recent years, particle board has become more important in building construction. Veneers are used to make fine furniture.

PRODUCING LUMBER

Lumber is made by sawing logs into boards. ***Logs*** are the large stems of trees that require many years to grow. The trees are felled (cut down) and cut into log-lengths in the forest. Most logs are 12 to 20 feet (3.66

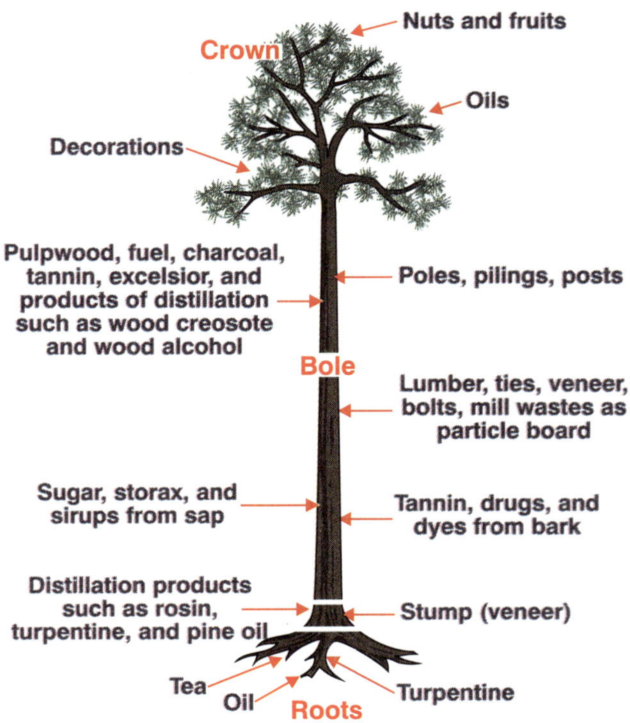

21-24. Examples of products made from different parts of trees.

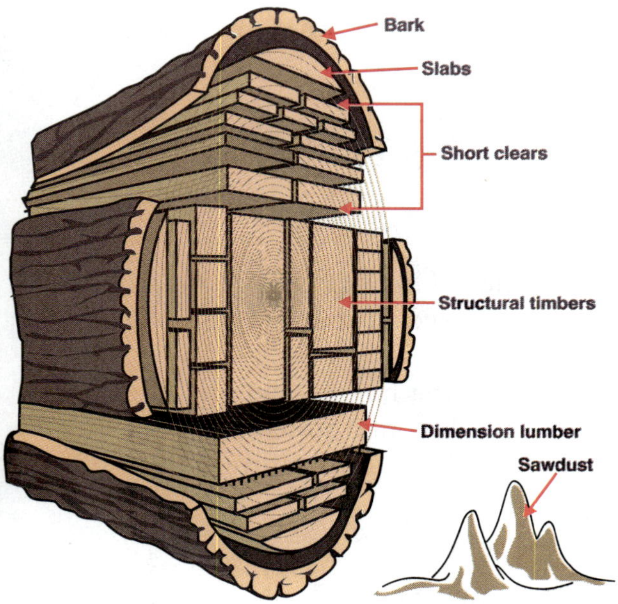

21-25. Products from a sawlog.

to 6.1 m) or longer. Logs must also have a minimum diameter that makes them useful in making wood products. The diameter varies with the kind of wood and the product to be made. It is obvious that a board 10 inches (25.4 cm) wide requires a log that is over 10 inches in diameter at the smallest end.

Saw Milling

Logs are cut into boards at a ***saw mill***. Large saws rip the logs open, creating sawdust where the cut is made. The boards are cut to specific sizes for lumber. As sawed, logs have a rough finish and are called rough lumber. After sawing the lumber is graded. Important factors in grading are the presence of knots (where a limb grew on the stem), size, and presence of bark or defects, such as worm holes. Boards that grade high (no. 1) have few or no knots and all the bark has been removed.

Seasoning

Most logs are sawed into lumber while still green (sappy). The lumber can't be used until the moisture has been reduced. This involves ***seasoning*** or drying. The lumber may be naturally or artificially dried.

Natural drying is also known as air drying. This involves stacking the lumber so that air moves between it in the lumber yard. Since this is a slow process, artificial drying is more common.

Artificial drying involves using a dry kiln. This is a specially constructed building that can be heated to speed up the drying process. Wood that isn't dry will warp and cause problems in building construction.

Planing

After drying, the lumber is planed. ***Planing*** is smoothing and sizing lumber. Planing removes the roughness resulting from sawing and carefully sizes the lumber. All boards of the same dimension should be exactly the size specified with no variation. After planing the lumber is ready for shipment and use.

Treating

Some wood products are treated to make them resistant to rot and insect damage. Only approved chemicals should be used in treating wood. Wood products treated with some preservatives can't be used where people or animals will touch them. The chemicals are considered hazardous to health.

PRODUCING PAPER

Smaller trees are used to make paper. They may be cut and hauled as long poles or made into pulpwood lengths of about 6 feet (2 m). Most go to a chipping mill where the bark is removed and the wood is chopped. The chips go to the paper mill.

Paper is made by breaking wood into ***pulp***. The wood may be broken apart with chemicals and by a grinding action. Pulp is screened and washed to remove impurities. Large machines take the pulp and rapidly form it into continuous sheets of paper. The thin mixture of pulp and water is spread over a meshed wire that lets the water drain away. The remaining mat of fiber moves between large rollers that squeeze out more water. The paper goes through other rollers and dryers. The fibers bond together during the drying process.

OTHER WOOD PRODUCTS

Wood is used to make many products. These products often require the wood of certain kinds of trees.

Furniture

Furniture requires special kinds of woods that are cut, milled, and seasoned properly. Most furniture is made from hardwoods but some pines and cedars are used. Wood serves as the frame over which upholstery and cushions are fitted. Wood also makes beautiful table tops and other furniture with fine wood finishes.

Plywood

Both hardwood and softwood trees are used to make plywood. The softwood plywood is more likely to be used in building construction. Hardwood plywood is used in cabinet and furniture manufacturing.

A special kind of hardwood material is ***veneer***. This is made with a thin layer of valuable wood over thicker layers of lower cost wood. For example, walnut may be put over gum or low-grade oak. Some veneers are used as paneling on the walls of homes and offices.

Plywood is made by slicing logs into thin sheets and gluing them together. The cutting process is somewhat like unrolling the log along its growth rings. The wood is "unwound" in a long continuous piece. The wood is used while still green and is dried as a part of the process of making plywood. Plywood is made in various thicknesses, such as $1/4$ or $1/2$ inch (0.63 or 1.27 cm). It is cut into sheets that are usually 4 feet (1.22 m) wide and 8 feet (2.44 m) long.

21-26. A quality log can make a large amount of veneer. This compares the thickness of a quarter with veneer. (The veneer shown here is from yellow poplar and will be used in making doors.)

SAFETY AND REGULATIONS IN PROCESSING

Processing food, fiber, and wood products is done using certain standards. Many times regulations have been established by government agencies to protect consumers. Food processing is highly regulated.

SAFETY

Processing often involves using machinery and equipment that could cause injury. People should know how to properly use equipment. Carelessness can result in injury.

Protective clothing should be worn. Boots and eye protection should be used. In noisy places, devices to protect hearing are needed. In some plants, gloves that protect workers from cuts or hazardous substances are worn.

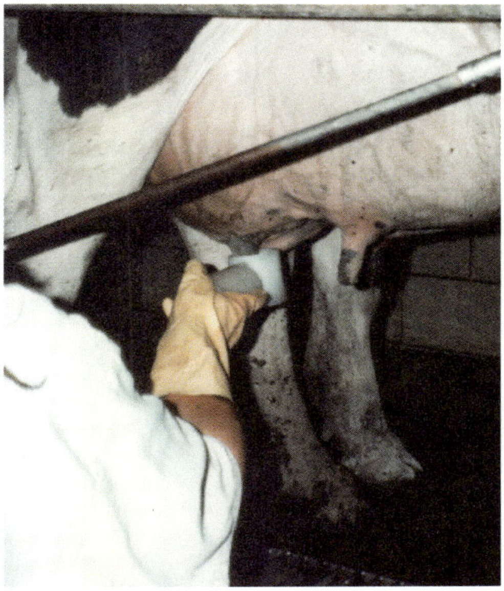

21-27. Teats of cows are dipped to sanitize them before milking.

SANITATION

Sanitation is keeping the places where food is processed clean. Regular washing and cleaning are needed to prevent the growth of microorganisms. Equipment, tables, floors, and other parts of the facility should be cleaned. Anything that the food touches must be clean.

WASTE DISPOSAL

Most processing creates waste materials. These must be disposed of properly. Water from processing cannot be run into streams without treatment. The general rule is that water run into a stream cannot change the general condition of the stream.

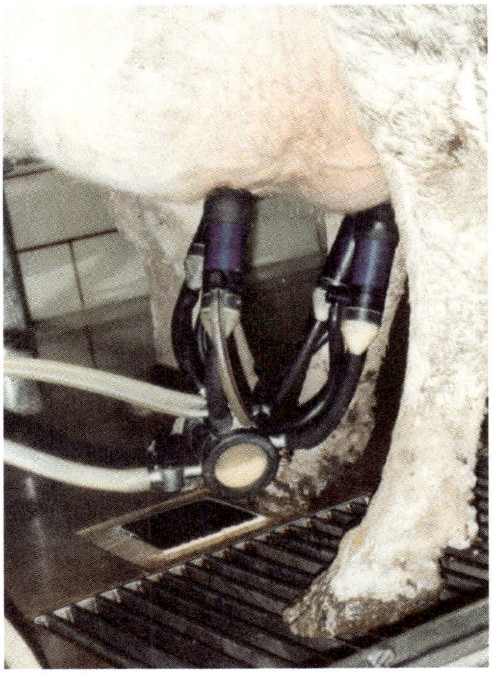

21-28. Milking equipment must be kept clean to keep the bacteria count low in milk.

Some processing creates solid waste materials. These must be disposed of by moving them to landfills or processing them into by-products. In most cases, very little is actually thrown away at a processing plant. Most things can be used to produce other products. For example, some of the waste from meat processing is used to make protein concentrates for animal feed.

CONTROL OF RODENTS, INSECTS, AND OTHER ANIMALS

Food processing plants must control rats, mice, roaches, birds, and other animals. These should not be allowed in places where food is being processed. Birds should not fly around in canneries or other facilities.

INSPECTION

Processing plants that produce food are regularly inspected for sanitation. The regulations must be met if the plant is to continue operating.

Inspectors may be hired by state or federal government agencies or by the processing plant. Their role is to assure wholesome food for people.

21-29. U.S. Department of Agriculture inspectors are evaluating a potato in a processing plant. (Courtesy, RDO Frozen, Park Rapids, Minnesota)

REVIEWING

MAIN IDEAS

Processing is preparing food, fiber, and shelter products for use by people. The procedures involved vary with the kind of product and the way it is prepared.

Food and fiber processing is a part of the agricultural marketing process. The steps overlap and are used to add value to what is produced. People are willing to pay more because products that have been processed are more convenient to use.

Processing includes several functions: grading, fabricating, preserving, portioning, preparing in convenient forms, and packaging. All food products go through one or more of the steps of processing

Food preservation is an important area of processing. It involves preparing foods to keep them from spoiling. The methods of food preservation are: canning, freezing, fermenting, drying, refrigerating, aseptic packaging, irradiating, and pickling.

Fiber processing is converting the fiber into forms that people want. It includes weaving and spinning as well as manufacturing.

Wood products are made from softwood and hardwood timber. Lumber is made from larger trees, while paper is made from smaller trees. The manufacturing processes may be complex.

Safety practices and regulations must be followed in processing. Sanitation and other procedures are musts in food processing to assure the consumer receives wholesome food products.

QUESTIONS

Answer the following questions using complete sentences and correct spelling.

1. What is processing? Why is it important?
2. What are the five major steps in processing food?
3. What is food preservation? Why is it important?
4. What is spoilage? What causes food to spoil?
5. What major methods of food preservation are used? List and briefly describe each.
6. How are fibers processed? List and briefly describe the steps involved for cotton and wool.

7. How is lumber processed? What steps are involved?
8. How is paper made?
9. What wood products are made in addition to lumber and paper?
10. What safety and sanitation procedures should be followed in processing?

EXPLORING

1. Arrange a tour to a local lumber yard. Note how the wood has been cut and the products that are available. Prepare a report on your findings.

2. Make a display of different kinds of food packaging materials. Use empty packages from your home, school cafeteria, or other places.

3. Preserve a food product using one of the methods of preservation. Consult with a home economics teacher and use reference materials on food preservation.

4. Visit a clothing store and observe the fabrics that are used. Make a list of the different articles of clothing and the kinds of fabrics used to make them. Determine if the fabrics are natural or artificial.

Appendixes

Appendix A
Common Weight Conversions and Equivalents Used in AgriScience

Customary System	
1 tablespoon (tbs)	= 3 teaspoons (tsp) or 0.5 fluid ounce (fl. oz)
1 cup	= 8 fl. oz
1 pint (pt)	= 2 cups of 16 fl. oz
1 quart (qt)	= 2 pt or 4 cups or 32 fl. oz
1 gallon (gal)	= 4 qt
1 bushel (bu)	= 8 gal or 32 qt

Metric System	
1 milliliter (ml)	= 1 cubic centimeter (cc) or 1 gram (g)
1 centiliter (cl)	= 10 ml
1 deciliter (dl)	= 10 cl or 100 ml
1 liter (L)	= 10 dl or 1,000 ml or 1 kilogram (kg)
1 dekaliter (dal)	= 10 L
1 hectoliter (hl)	= 10 dal or 100 L
1 kiloliter (kl)	= 100 hl or 1,000 L or 1,000 kg

Equivalents	
1 ounce	= 28.3495 grams
1 pound	= 453.59 grams or 0.45359 kilograms
1 kilogram	= 35.27 ounces or 2.2046 pounds
1 metric ton	= 2,204.6 pounds
1 hectare	= 2.47 acres
1 liter (L)	= 1.057 quarts
1 liter (L)	= 0.2642 gallons
1 gallon	= 3.79 L
1 meter (m)	= 39.37 inches
1 acre	= 0.40468 hectares
1 inch	= 2.54 centimeters (cm)
1 centimeter (cm)	= 0.4 inch

Appendix B
Common Volume Measures Used in AgriScience

Customary Fluid Measures

4 gills (gi)	= 1 pint (pt)	= 28.88 cubic inches
2 pints (pts)	= 1 quart (qt)	= 57.75 cubic inches
4 quarts (qts)	= 1 gallon (gal)	= 231 cubic inches
31.5 gallons (gals)	= 1 barrel (bbl)	= 4.21 cubic feet
2 barrels (bbls)	= 1 hogshead (hhd)	= 8.42 cubic feet
7.5 gallons	= 1 cubic foot	= 1,728 cubic inches

Customary Dry Measures

2 pints	= 1 quart	= 67.2 cubic inches
8 quarts	= 1 peck (pk)	= 537.6 cubic inches
4 pecks (pks)	= 1 bushel (bu)	= 2,150.4 cubic inches or 1.244 cubic feet

Appendix C
Weight Measures Commonly Used in AgriScience

Customary Weights

Avoirdupois Weight:	
437.5 grains (gr)	= 1 counce (oz)
16 ounces	= 1 pound (lb)
100 pounds	= 1 hundredweight (cwt)
2,000 pounds	= 1 ton (T)
2,240 pounds	= 1 long ton
Troy Weight:	
24 grains	= 1 pennyweight (pwt)
20 pennyweight	= 1 ounce
12 ounces	= 1 pound
Apothecaries Weight:	
20 grains	= 1 scruple (sc)
3 scruples	= 1 dram (dr)
8 drams	= 1 ounce
12 ounces	= 1 pound

Metric Weights

1 centigram (cg)	= 10 milligrams (mg)
1 decigram (dg)	= 10 centigrams and 100 mg
1 gram (g)	= 10 dg, 100 cg and 1,000 mg
1 kilogram (kg)	= 1,000 grams
1 metric ton (t)	= 1,000 kg

Glossary

Abiotic factor—nonliving things in an ecosystem.

Acceleration—the increasing change in the speed of an object over time.

Acid—a compound that gives up protons to water molecules to form hydronium ions.

Acquired immunity—overcoming a disease causing the animal to develop immunity to getting the disease again in the future.

Acute disease—a disease that afflicts an animal for a short period of time and may be quite severe.

Additive—an addition placed in feed while it is being manufactured to preserve it and enhance the growth of animals.

Advertising—involves communicating information to people that is intended to persuade them to buy a product or service.

Aerial stem—that part of the plant stem that grows above ground.

Agricultural drainage—involves moving excess water from the land through open ditches and underground drain tiles or tubes.

Agricultural industry—all of the processes involved in getting food, fiber, and shelter to the consumer.

Agricultural marketing technology—all of the processes involved in providing people with what they want, in the form they want it and when they want it.

Agriculture—the science of growing crops and raising livestock.

Agriculture pollution—any pollution caused by agriculture activities.

Agriscience—the use of science in producing food, fiber, and shelter.

Agronomy—the study of plants and how they relate to the soil.

Air pressure—the force caused by the weight of the air on an object due to the earth's gravitational attraction.

Alcohol—the compound made from organic halides.

Algae—plant-like organisms that live as single cells or in groups.

Alkane—a hydrocarbon that has single covalent bonds only.

Allele—each different form of a gene.

Alternating current—electricity that regularly reverses its direction of flow.

Alveoli—found in the mammary glands, they use nutrients from the blood to make milk.

Ampere—a measure of the rate of flow of electrical current.

Anatomy—the study of the form, shape and appearance of animals.

Animal science—the area of agriscience dealing with the production of animals for food.

Annual—a plant that has a complete life cycle within one year.

Annual growth rings—the layer of cells formed for each year a tree grows.

Antibody—a secondary defender of disease produced by organisms in response to an invasion of pathogens.

Antifreeze—a substance mixed with water that changes the temperature at which water freezes.

Aquaculture—the science of water farming.

Aquatic—an animal that grows in water.

Area—the measurement of surfaces; based on the length of the sides of an object or a space.

Artificial erosion—loss of soil caused by people when they disturb the natural cover on the land.

Artificial insemination—involves collecting the semen from a male and depositing it in the reproductive tract of the female.

Aseptic packaging—involves sterilizing the food and the container it is stored in.

Asexual reproduction—involves reproducing plants by using vegetative parts of the plant.

Assembling—involves massing large quantities together to market more efficiently.

Atmosphere—the air that surrounds the earth.

Atom—the smallest unit in which an element can exist.

Atomic number—the number assigned to elements based on the number of protons in the nucleus of the atom.

Auxin—plant hormone that controls stem growth and regulates fruit development.

Avoirdupois weight system—a weight system based on the pound weight 16 ounces and the ounce weighing 16 drams.

Bacteria—one-cell organisms in the monera kingdom.

Bactericide—chemical used to control bacteria.

Balanced ration—one which provides all of the nutrients needed by animal in the right amount and proportion.

Barometer—the device used to measure air pressure.

Base—a compound that produces hydroxide ions in water.

Basic human needs—things that support human life; include food, fiber, and shelter.

Beef-type—beef animal with higher amounts of lean muscle; indicated by good muscling in the loin and hind quarters.

Beneficial insect—helps plants to grow by improving the environment.

Biennial—a plant that lives two seasons.

Binomial nomenclature—a two-part scientific name.

Biodegradable—the term used to describe materials that will break down from the action of bacteria in sunlight.

Biogenesis—means that life comes from life.

Biological pest control—involves using living organisms to control pests.

Biology—the study of living things, life science.

Biosphere—the area of the earth that supports life.

Biotechnology—involves using biology to develop new products.

Biotic community—the group of plants and animals that live together.

Biotic factors—plants, animals, and other living organisms in an ecosystem.

Bivalve mollusk—a class of mollusk with two-sided shells that are hinged together.

Blemish—defects in products.

Blood—liquid in the circulatory system of animals; 90 percent water.

Bloodline—a group within a breed, all members of a bloodline tend to have one common ancestor.

Board foot—a piece of wood that is 1 foot square and 1 inch thick.

Body system disease—a disease that attacks only certain systems of the body.

Bonding—joining two or more elements together to form a compound.

Bone—any of the separate parts of the hard tissue forming the skeleton of most full grown vertebrate animals.

Bony fish—fish in the class Osteichthyes.

Botany—about plants.

Botulism—a food poisoning caused by *Clostridium botulinum*.

Brain—the mass of nerve tissue in the cranium of vertebrate animals that is the main part of the nervous system, the center of thought, and the organ that receives sensory impulses and transmits motor impulses.

Breathing—respiration; involves air entering and leaving the lungs.

Breed—animals of the same species that have a definite identifying characteristic and a common origin.

Breeding—the process of helping animals reproduce.

Broiler—a chicken 6 to 12 weeks old weighing 2½ pounds or more.

Browse—leaves, twigs and young shoots of trees or shrubs which animals feed on.

Budding—taking a bud from one plant and moving it to another.

Bulb—underground stem with layers similar to leaves, including onion, tulip, and jonquil.

Caliper—a measuring device with jaws or legs that are moved to determine the thickness, diameter or distance between things.

Canning—placing food in a container and heating it to kill all microorganisms.

Capillary water—the coating of thin layers of water on soil particles.

Carbohydrate—any of certain organic compounds, including the sugars, starches, and celluloses.

Career opportunities—employment alternatives.

Cartilage—a tough, elastic, whitish animal tissue; the skeletons of embryos and young animals are composed largely of cartilage, most of which later turns to bone.

Castration—removing the testicles from a male.

Cell—tiny building blocks that have a definite structure, the basic structure of life.

Cell cycle—sequence of events in a cell from one mitosis to another.

Cell division—the process of cells splitting.

Cell specialization—variations in cells so they can perform specific functions.

Cell structure—refers to the general pattern of organization and relationship in a cell.

Celsius scale—the metric system of temperature measurement, in which water boils at 100°C and freezes at 0°C.

Cereal grains—wheat, rice and corn.

Chain—typically used to measure land.

Chemical properties—those properties that describe how it changes when it is combined with others.

Chemical reaction—what occurs when a substance becomes another substance with different characteristics because of a chemical change.

Chemical weathering—when new substances are formed by chemical reaction that occurs when rocks and minerals interact with carbon dioxide and oxygen.

Chewing mouth—bite off, chew and swallow parts.

Chlorophyll—the green pigment found in the chlorplasts of plant cells which must be present for photosynthesis to take place.

Chromosome—that part of a cell that controls activity, made up of proteins and nucleic acids.

Chronic disease—a disease that afflicts an animal for a long time.

Circle—a plane figure bounded by a single curved line every point of which is equally distant from the point at the center.

Circuit—the path electricity follows.

Circulation—the moving of nutrients and other materials throughout an organism.

Circulatory system—the complex system in an animal for the moving of food nutrients, digested food and other materials containing a fluid known as blood.

Classing—involves judging the cotton on the basis of length of fiber (staple), thickness of the fiber (micronare) and grade, which is the presence of trash and color.

Cleavage—getting the new DNA to join with the DNA that has been cut.

Climate—an average of all the weather conditions for an area over a period of time, usually a year or more.

Climate zone—one of the three areas the earth is divided into based on temperature.

Clone—producing offspring genetically identical to their parent, such as by tissue culture.

Cloning—a process of asexually reproducing organisms.

Cloud—condensed water vapor.

Colloidal suspension—involves material in the liquid that can be fairly easily filtered out with parchment or other semipermeable membranes.

Colostrum—after the birth, the first milk given which is high in antibodies and other substances that help the new animal survive.

Common names—the names in everyday use given to plants or animals.

Communication technology—includes all of the ways people exchange information about marketing.

Complete flower—flowers with all four principal parts; sepals, petals, stamens, and pistil.

Compound—a substance that is made of two or more elements that have been chemically combined.

Compound leaf—one leaf divided into two or more leaflets.

Computational science—involves using computers to do mathematics and science problems.

Computer simulation—a computer emulating a process based on the data it is given in order to gather information to answer questions.

Concentrate—a feed that is low in fiber (less than 18%) and high in energy.

Conductor—a substance that allows electricity electrons to flow freely.

Cone—a cylinder that tapers to a point at one end.

Consumer—a person with certain needs that must be met.

Contact poison—poison that must come into contact and is absorbed through the skin.

Contagious disease—one that can be spread by direct or indirect contact.

Continental crust—the crust that is made up of inorganic materials as well as materials from plants and animals.

Contour plowing—involves plowing around or across a slope rather than with the slope.

Controlled breeding—involves breeding plants or animals to achieve certain characteristics in the offspring.

Convenience preparation—making foods easy for consumers to use.

Cord—a stack of wood that measures 4 feet wide by 4 feet high by 8 feet long.

Core—center.

Corm—an underground stem with thin leaves, such as garlic and gladiolus.

Corrosion—the chemical reaction, rusting, when a change in a metal occurs through oxidation and reduction.

Cotton seed meal—meal product remaining after oil has been extracted from cotton seed.

Cow and calf production—keeping cows to produce calves that will primarily be used for meat.

Cracking and rolling—a method used to break the hard outer coatings on grain.

Crawfish—a crustacean found in freshwater all over North America and similar to a shrimp.

Cropping model—using a computer program and information about a crop for assistance in making decisions about the cultural practices to use.

Crop rotation—planting crops in alternating growing seasons that use different nutrients and conserve soil fertility.

Crossbreeding—mating animals of different breeds but definitely of the same species.

Crust—surface.

Crustacean—any of a class of arthropods, including shrimps, crabs, barnacles, and lobsters, that usually live in the water and breathe through gills.

Cubic units—a volume measurement based on a cube having six equal sides.

Cultivar—a crop variety that is cultivated and retains its features when reproduced.

Cultural pest control practices—those things that can be done to keep pests out.

Curing—adding substances to foods to prevent spoilage.

Current electricity—electricity that is caused by flowing electrons.

Customary measurement system—a system of measurement used in every day life in North America, including inches, feet, pounds, and other measures.

Custom harvesting—harvesting work for a fee by the owner of equipment and supplies for a producer.

Custom order marketing—when buyers contract with producers to provide a specific crop or animal.

Cutting—using a short section of the plant stem for propagation.

Cylinder—a circle with height.

Dairy-type—cow with the capacity to produce a large amount of milk.

Debeaking—the removal of the tip of the beak of a chick.

Deceleration—when an object slows down or decreases speed over time.

Demand—what people want.

Deoxyribonucleic acid—(DNA) the control substance in the nucleus of cells.

Dew—form of liquid precipitation that forms on the leaves of plants and other materials when the temperature of air cools so that it cannot hold as much moisture as the air contains.

Dicot—plants with two seed leaves.

Dicot seed—a plant seed embryo with two seed leaves.

Diet—the type and amount of feed and water an animal eats.

Diffusion—a gas occupying all space available to it.

Diploid cell—a cell with all of the chromosomes characteristic of a species.

Digestible nutrient—part of a feedstuff that can be digested or broken down.

Digestion—the process that changes food into simpler forms that can be absorbed.

Direct current—electricity that always flows in the same direction.

Disease—something that causes pain, injury or inability to do normal things.

Distribution—getting what is needed to the right place at the right time.

Domestication—to tame and breed for many purposes of man.

Dominant—traits that are caused by alleles that cover up or mask the recessive traits of other alleles.

Donor—provider.

Donor cells—cells that provide the new material.

Double helix—the structure known by scientists which DNA forms.

Draft—resistance to movement in machines, such as the resistance a plow has in the soil.

Draft horse—a horse weighing upwards of 1,400 pounds developed to pull heavy loads.

Dry fertilizer—fertilizer produced as pellets or granules.

Dry fruit—one that develops as a pod or in a hull.

Drying—involves removing the moisture from food products.

Earth science—study of the environment in which plants and animals grow.

Ecosystem—all of the parts of a particular environment.

Ectotherm—an animal that adjusts body temperature to its environment.

Effluent—discharged water from factories or farms.

Electricity—flow of electrons.

Electron valence—the electrons that move around the nucleus of an atom.

Elements—substances that cannot be broken down into simpler materials by ordinary means.

Embryo—an immature plant or animal that will grow if properly cared for.

Embryo splitting—to cut an embryo cell into two parts which each develop into a complete animal.

Embryo transfer—involves removing an embryo from its mother and placing it in another female to develop.

Emulsion—a colloidal suspension made of two solutions.

Endosperm—area in a seed where food is stored for the embryo to use.

Endotherm—an animal that maintains a certain body temperature.

Energy—provided by food to support life processes of living organisms.

Entomology—the branch of zoology that deals with insects and related small animals.

Environment—all of the factors that effect the life of a living thing.

Environmental disease—caused by elements in the environment that aren't right.

Environmental pollution—made up of all of the ways that people contaminate their environment.

Environmental science—involves ways of wisely using and protecting the things around us.

Erosion—the washing or wearing away of soil.

Estrus cycle—the cycle in the female reproductive system that prepares it for reproduction.

Estrus synchronization—involves using hormones to change the estrus cycle to get several females to come in heat at the same time.

Exoskeleton—an animal with a hard outer shell or covering to protect and support the body.

Experiment—a trial or test; used to discover something unknown or verify a procedure.

External parasite—a parasite that lives on the outside parts of an animal (*see* parasite).

External respiration—the exchange of gases in the lungs between the blood and the atmosphere.

Fabricating—making the product into forms that can be used.

Fahrenheit scale—the customary measurement of temperature, in which water boils at 212°F and freezes at 32°F.

Family—the fifth division in the scientific classification system of plants and animals.

Fat—what excess energy is changed to in the body.

Feed—to provide as food something necessary for growth, development or existence.

Feeder pig—a pig sold at a weight of 30 to 60 pounds to other producers to grow to a meat-size hog.

Feedlot—a confinement finishing system whose aim is to get animals to grow and put on body fat.

Feedstuff—feed ingredients.

Fermenting—a method of preservation that involves using certain bacteria, molds and yeasts.

Fertilization—the union of the pollen cell with the ovule in plants; union of male and female sex cells.

Fertilizer—any substance used to provide plant nutrients.

Fertilizer analysis—tells the nutrients that are in the fertilizer.

Fetus—a developing animal in the reproductive tract of a female.

Fiber—primarily material used to make clothing and shelter; what remains of feed after the sugar, protein and other substances are removed.

Fibrous root system—a root system made of many small roots and spread out through the soil.

Fleece—the wool on a sheep.

Fleshy fruit—those that have large fibrous structures surrounding the seed.

Flower—the part of a flowering plant that contains the reproductive organs.

Flushing—involves using a saline solution to remove developing embryos.

Food—the solid and liquid material humans eat.

Food chain—the sequence in which living things in an ecosystem get food.

Food processing—all of the activities to prepare food for human consumption.

Force—the push or pull on an object.

Forestry—the science of growing and using forests.

Free access—animals have access to the feed when they want it.

Freezing—using low temperatures that freeze the water in food products.

Fruit—a mature ovary of a flowering plant.

Fuel—any material that provides energy; may be in a solid, liquid, or gas form.

Fumigant—pesticides in gas form and enter the respiratory system of organisms.

Fungi—the classification ranging from tiny yeasts to large mushrooms.

Fungicide—chemical used to control disease caused by fungi.

Galaxy—a segment of the universe which is made up by clusters of dust, gases and stars.

Game management—procedures followed to protect and improve wildlife and fish.

Gene—the specific determiner of heredity.

Gene expression—a method used to isolate a gene that does certain things.

Genetic code—the information in an organism that causes it to function.

Genetic engineering—an advanced form of biotechnology.

Genetic manipulation—artificial means of producing desirable traits in plants and animals.

Genetic pest control—using genetics to develop crops that are resistant to pests.

Genetics—the study of heredity in plants and animals.

Genome—the heredity material in a cell.

Genotype—the genetic makeup of an organism.

Genus—the sixth division of the scientific classification system of plants and animals.

Geotropism—how plants respond to gravity.

Germination—the sprouting of a seed to grow a new plant.

Gestation—the period between fertilization and full development of the fetus.

Gibberellic acid—affects stem and leaf growth, fruit development, flowering, cell division, and other plant activities.

Ginning—the process of separating cotton lint (fiber) and seed.

Glucose—sugar made by photosynthesis in plants.

Glycolysis—the process of releasing energy to help plants grow and reproduce.

Gonadotropin—a hormone injected into a cow that causes the cow to release 8 to 20 eggs during estrus (see superovulation).

Good seed—seed that will germinate, grow and become the intended plant.

Grading—sorting products for uniformity.

Grafting—placing a section of a stem of one plant onto another plant.

Grain elevator—a facility that receives, grades, dries, stores and sells grain to mills.

Gravitational water—the water that fills cracks and air spaces between soil particles.

Grinding—involves making the particle size of feed smaller.

Ground fault circuit interrupter—an electrical circuit breaker designed to protect people from shock when using electricity in areas that are wet or have water.

Growing season—the period of time after the last frost in the Spring and before the first frost in the Fall.

Growth—an organism increasing in size by adding cells or by the cells getting larger.

Growth chamber—a structure that controls and monitors the environment.

Habitat—the place a plant or animal lives under natural conditions.

Haploid cell—cells produced by meiosis that have single chromosomes rather than homologous pairs of chromosomes.

Hardwood—wood with fine grain and more difficult to cut and shape than softwood; most hardwood trees have broad leaves, such as oak and hickory.

Harvesting—the first step after production in getting what has been produced ready for the consumer.

Hay—made of the leaves and stems of plants that have been cut and dried for feed.

Health—the condition of the body and a measure of how the functions of life are being performed.

Heartwood—the innermost xylem of a tree.

Heat engine—internal combustion engine.

Herbaceous stems—soft and green stems which only contain a small amount of xylem.

Herbicide—chemical used to control weeds.

Heredity—the passing of the traits of parents to offspring.

Heterozygous—when the alleles are different in an organism.

Homozygous—when the alleles are the same in an organism.

Horizon—one of the four layers of materials from the soil surface down several feet.

Horsepower—a way of stating power in engines and other power equipment.

Horticulture—the science of growing certain food and ornamental plants.

Humidity—the amount of moisture in the air.

Humus—well decomposed organic matter.

Hybrid—improved plant or animal developed through a breeding program.

Hydraulic power—using the force of a liquid or gas placed under pressure to do work.

Hydrocarbon—fairly simple organic compounds made of only two elements, hydrogen and carbon, which are commonly used for fuel.

Hydroscopic water—a thin layer that sticks to soil particles.

Hydrosphere—all of the water on earth.

Immunity—resistant to a certain disease.

Immunochemistry—new ways of making medical products for humans using animals.

Imperfect flower—a flower that lacks a stamen or pistil.

Implant—involves placing a solid material under the skin.

Inbreeding—mating animals of the same breed with closely related animals.

Inclined plane—a simple machine made of a straight sloping surface.

Incomplete flower—a flower that does not have all four principal parts.

Infectious disease—a disease caused by organisms that get inside the body.

Injection—involves using hypodermic needles and syringes to get a substance into the body system.

Insect—any of a large class of small arthropod animals characterized in the adult stage by division of the body into head, thorax, and abdomen with three pair of legs.

Insecticide—chemical that controls insects.

Insulator—a material that does not conduct electricity.

Integrated pest management—a complex process for controlling pests that involves using the control methods that have the best outcomes for the well-being of society.

Interferon—protein substances being developed through immunochemistry.

Internal drainage—how readily water moves through the soil.

Internal parasite—a parasite that lives inside an animal (see parasite).

Internal respiration—the exchange of gases between the cells and the blood within the body.

Invertebrate—an animal that does not have a backbone.

In vitro—in glass (usually a test tube).

Involuntary muscle—a muscle that is automatically controlled by a lower part of the brain.

Ion—the charged particle when an atom gains or loses a valence electron.

Irradiation—treating food with electrically charged particles such as X-rays, electron beams or gamma rays.

Isolation—involves physically separating from others.

Issue—a point in question that has more than one side.

Kelp—the common name of brown seaweed in Europe and North America.

Kinetic energy—a result of an object's mass and motion.

Kingdom—the first division in the scientific classification system of plants and animals.

Label—a document used to provide information.

Labeling—a step in packaging used to identify the product.

Lactation—milk production by female mammals.

Lamb—a young sheep.

Land capability classes—agricultural land classified on the basis of its productive capability.

Landfill—large earthen pits for waste disposal.

Land forming—shaping the surface of the land so that it is more productive.

Laser—a device that produces a special kind of light known as coherent light.

Law of Conservation of Mass—matter is found in three states: solid, liquid and gas; and can be changed from one form to another but it cannot be created or destroyed.

Law of Definite Composition—all compounds have a definite composition by mass.

Layering—getting roots to grow from the stems of plants.

Leaf—the part of a plant that makes food through photosynthesis.

Legumes—plants that have the ability to take nitrogen from the air and put it into nodules that grow on their roots.

Lever—simple machine consisting of a bar or rod that rotates on a fulcrum.

Life cycle—the length of the plant's or animal's life.

Life science—the study of living things.

Life span—the period of life.

Ligation—involves uniting or attaching the two DNA fragments.

Light horse—a horse weighing 900-1,400 pounds used for riding, driving, racing and other purposes.

Linear measurement—the distance between two points.

Liquid fertilizer—fertilizer nutrients dissolved in water and applied by spraying on the soil.

Living condition—all of the life processes that are essential for something.

Loam—a soil made of nearly equal amounts of sand, silt, and clay.

Log—large main stems of trees used for making lumber and other wood products.

Locomotion—movement.

Machine—a device that helps transmit power, force and motion.

Macromineral—minerals needed by animals in larger amounts than other minerals which are trace or microminerals.

Maintenance—amount of energy, provided by feed, needed to keep an animal from losing weight when not producing milk or other products.

Major elements—elements that are needed in large amounts for growth.

Mammals—a class of warm-blooded, usually hairy vertebrates whose offspring are fed with milk secreted by the female mammary glands.

Mammary glands—female glands that produce milk for the newborn young.

Mantle—the layer between the core and crust of the earth.

Market hog—a 220 pound hog.

Marketing—providing the goods and services people want.

Marketing and processing—the part of the agricultural industry that moves crops and livestock from the farm to the people who use them.

Marketing infrastructure—those things that support and make marketing possible.

Mass—amount of matter in an object; properties of matter determine mass.

Meal—ground feed.

Measurement—involve s determining the number of units in something.

Measuring stick—a graduated device typically made of wood, plastic or metal often the length of one unit and marked in smaller units.

Mechanical advantage—how machines multiply the force that they receive.

Mechanical pest control—using tools or equipment for control of pests.

Mechanical technology—involves the use of machines and equipment to do work.

Mechanical weathering—when larger rocks break into smaller rocks.

Mechanics—deals with forces exerted by objects when at rest or moving.

Meiosis—division process in sex cells.

Metals—category of elements easily identified by their brilliant appearance.

Metamorphosis—the stages an insect goes through from egg to adult.

Metric system—the measurement system used by all scientists around the world.

Mineral—substance needed for good health (*see* minerals).

Mineral block—minerals and other materials compressed into a large cube.

Minerals—important in the maintenance, growth, reproduction and other functions.

Mineral soil—soil that is higher in minerals than other materials.

Minimum tillage—involves planting seed in land with very little plowing.

Mitosis—a type of cell division for growth and repair, a sequential process of cell division.

Mixed feed—feed made of a variety of ingredients.

Mixture—a substance that has parts with different properties.

Molecular biotechnology—involves changing the structure and parts of cells.

Molecule—the smallest group of atoms that acts together to form a stable, independent substance.

Mollusk—an animal with a soft body often enclosed wholly or in part in a mantle and a calcareous shell of one or more pieces and usually having gills and a foot such as oysters, snails, clams, mussels, abalone, slugs and octopuses.

Monocot—plants with one seed leaf.

Monocot seed—a plant seed embryo with one seed leaf.

Monogastric—an animal that digests food without making improvements to it.

Motion—involves moving or changing position.

Mulch—cover on soil to prevent moisture loss and weed growth; examples include straw, paper, and plastic sheets.

Mulching—involves covering the soil with material.

Multicellular organism—organisms comprised of billions of cells.

Mutant—an offspring that has traits genetically different from its parents.

Mutation—change that naturally occurs in the genetic material of an organism.

Mutton—meat of an older sheep.

Natural immunity—when an animal is immune to a disease.

Natural insemination—involves the male depositing semen in the vagina of the female during copulation.

Natural resources—the naturally occurring things that support life, provide fuel or are used in other ways by humans.

Needle teeth—eight sharp teeth of newborn pigs.

Nematocide—chemical used to control nematodes.

Nematode—tiny worm-like organism that lives in the soil and attacks the roots and stems of plants.

Niche—its own special way of living.

Niche market—producers able to develop special markets to fit their situations.

Nitrogen—among the most important elements in the growth of plants.

Non-contagious disease—a disease caused by conditions or substances that aren't transferred from one to another.

Nonmetals—a category of elements that are poor conductors of heat, brittle and not easily made into other shapes.

Nonpoint source pollution—pollution from sources that cannot be readily identified, such as all engines that release exhaust.

Non-renewable natural resources—resources once used that cannot be renewed.

Non-ruminant—an animal with a stomach with one chamber that does not further refine what it ingests.

Non-selective herbicide—chemical which will kill all plants upon contact, contact herbicide.

Nori—red seaweed; one form of algae used as food.

No-till cropping—growing crops without plowing the land.

Nucleus—central part of a cell that controls cell activity.

Nutrient—anything nutritious for health and growth.

Nutrients—chemical substances in feed that support the life processes (see nutrient).

Olestra—synthetic non-digestible cooking oil; trade name is Olean.

Optics—studying the use of light.

Organ—collections of tissues that work together to perform certain functions.

Organic chemistry—the study of carbon.

Organic matter—rotting plants and animals.

Organic soil—soil that is higher in organic matter than other materials.

Organism—a living thing.

Organismic biotechnology—dealing with intact or complete organisms.

Organ system—several organs which work together to perform an activity.

Osmosis—water moving from a greater concentration of water to a lower concentration.

Outcrossing—mating of the same breed but of different families in the breed.

Ovule—small egg; structure in plants that becomes a seed when fertilized.

Ovum—egg or female sex cell.

Oxidation—a chemical process between oxygen and other materials, such as when an atom of a metal gives up electrons; a process in animal respiration that produces chemical energy from digested food.

Packaging—placing products in containers.

Packing shed—a facility that receives, sorts and combines produce from a number of different farms.

Palatability—has to do with how well an animal likes a feed.

Parasite—usually multi-celled animal organisms that live in or on other animals.

Parasitic disease—caused by organisms on the host disrupting its normal life.

Parent material—the mass of material from which something develops.

Particulate—small solid particles in the air.

Parts per billion—(ppb) measurement stated as parts per unit.

Parts per million—(ppm) measurement stated as parts per unit.

Parts per thousand—(ppt) measurement stated as parts per unit.

Parts per unit—a system of measurement based on sets of units, such as parts per thousand.

Parturition—the process of giving birth.

Pasture—land where grasses and other plants grow for animals to graze.

Pathogen—the microorganism that gets into the body and causes infectious disease.

Pedigree—the names of the ancestors of an animal.

Pellet—feed made by forcing ground forage through holes in a heavy steel plate to give them shape.

Percent germination—the percentage of plant seeds expected to grow.

Perennial—a plant that lives for more than two seasons.

Perfect flower—the stamen and pistil in the same flower.

Periodic law—the properties of elements are dictated by their atomic numbers.

Periodic Table of the Elements—the table used to group and organize the elements for study.

Permanent pasture—pastures planted with grasses and legumes that will live and grow for years.

Pesticide—chemicals used to control pests.

Phagocyte—cells that surround and destroy disease-causing microorganisms.

Phase change—when matter changes from one state to another.

Phenotype—the appearance of an organism.

Phloem—moves food from where it is manufactured to other parts of a plant.

Phosphorus—what plants need to store and transfer energy and to grow.

Photosynthesis—the process by which plants make food.

Phototropism—how plants respond to light.

Phylum—the second level, division, in the scientific classification system of plants and animals.

Physical properties—those properties which can be determined without changing the identity or composition.

Physical science—the study of nonliving matter around us.

Physics—deals with matter and energy and how the two relate to each other.

Physiology—focuses on the functions of the cells, tissue, organs and systems of the body.

Phytoplankton—form of algae that produces oxygen in water; used as food by some aquatic organisms.

Pickling—placing foods in salt solutions (brine) to prevent the growth of organisms that cause spoilage.

Placenta—the bag structure in the uterus of the female which nourishes and protects the developing animal.

Planimeter—a device that measures area by tracing the boundaries shown on a photograph.

Planing—smoothing and sizing lumber.

Plant disease—abnormal conditions in living plants.

Plant hormones—substances that directly or indirectly cause plants to grow.

Planting depth—how deep a seed is planted in its growing medium or soil.

Plant pest—anything that causes injury or loss to a plant.

Plant tissue culture—propagating plants using single cells or smaller groups of cells.

Plasma—the liquid part of blood which is 92% water.

Platelet—not complete cells in blood which are essential for blood to clot.

Plywood—sheets of wood product made by gluing several layers of thin wood together; most plywood is in 4 x 8 ft. sheets.

Pneumatic power—using compressed air to do work.

Point source pollution—pollution from a readily identifiable source such as a factory.

Polarity—the positive or negative charge of a substance.

Polar climate zone—those areas on earth where the average temperature for a year is below 50°F (10°C).

Polled—naturally without horns.

Pollen—the male sex cell in plants.

Pollination—the transfer of pollen from an anther to a stigma of a flower of the same species.

Pony—a small horse weighing under 900 pounds.

Portioning—preparing foods in the right quantities for consumers.

Potassium—what plants need for photosynthesis, moving sugar and other functions.

Poultry science—concerned with raising chickens and related fowl.

Power—the rate at which work is done.

Precipitation—any form of water that falls to the earth.

Preconditioning—performing routine management well ahead of the time of stress.

Prefix—what is added before the unit (measure) being used to increase or decrease numbers in the metric system.

Pregnancy testing—one of several procedures used to determine if females are pregnant.

Pregnant—the period while a female has a developing animal in her reproductive tract.

Presentation—the fetal position when giving birth.

Preserving—treating food to keep it from spoiling.

Primary defenders—the main defense of disease, the skin and mucous membranes of the body.

Primary growth—linear growth.

Primary root—the main root which grows from the seed.

Primary succession—the development of a biotic community where none existed.

Processing—all of the changes made in a product that prepare it for consumption.

Production agriculture—the farming part of the agricultural industry.

Production system—focus on the outcome of production—the kind of product produced.

Promotion—includes advertising, but goes further to attract new customers, reinforce the current customer base or increase the amount sold.

Propagation—involves increasing the number of plants.

Protective treatment—using chemicals on disease-free plants to prevent disease from developing.

Protein—food substance important to growth, maintenance, reproduction and other functions.

Protein supplement—high in protein and TDN.

Protoplasm—the unique chemical organization in an organism.

Puberty—the age at which an animal is capable of reproduction.

Pulley—a simple machine consisting of a kind of grooved wheel and a rope or belt that provides mechanical advantage in doing work.

Pulp—wood broken apart with chemicals and by a grinding action.

Pulpwood—young and usually softwoods that are made into pulp for manufacturing paper, particle board, and similar products.

Pulse rate—movement in arteries of animals due to heart beats.

Purebred—animals registered in a breed.

Purebred breeding—the mating of purebred animals.

Radiation—a kind of energy that travels as waves.

Ration—the total amount of feed an animal gets in a 24-hour period.

Receptors—various parts of the nervous system that collect information from the environment.

Recessive—traits that are covered up by dominant alleles.

Recipient—the female animal that receives the transfer of the split embryo.

Recombinant DNA—taking a tiny amount of DNA from one chromosome and moving it to another.

Rectangle—a surface where the two opposite sides are of equal length and the sides that touch are of unequal length.

Recycling—involves recovering and reusing materials instead of throwing them away.

Red blood cell—erythrocytes which contain a protein called hemoglobin which carries oxygen in the blood.

Refrigeration—keeping food products at temperatures slightly above freezing.

Relative humidity—the amount of water in the air compared with the air's ability to hold water vapor.

Remote sensing—involves gathering and recording data from a great distance.

Renewable natural resources—those resources which can be replaced when they are used.

Repair—what occurs when parts of an organism wear or are damaged; injuries are healed, worn cells may be replaced, etc..

Reproduction—the process of organisms giving rise to new individuals of the same kind.

Research—careful and diligent search for answers to questions or to solve problems.

Respiration—the process an organism uses to provide its cells with oxygen so that energy can be released from digest food.

Respiration rate—number of breaths taken in a minute by an animal; frequency of inhalation and exhalation of air.

Retail market—a market usually located near or in towns or cities and selling to consumers.

Revolution—one complete rotation.

Rhizome—underground stems that send up shoots to start new plants, such as Johnson grass.

Roadside market—a producer operating a retail market located on the farm in addition to producing what is sold.

Rodent—small gnawing animal with four continuously growing incisors, including rats, mice, squirrels, and beaver.

Roguing—removing infected plants from a field, forest or orchard.

Root cap—the tip of the primary root that protects the root as it grows into the soil.

Root hairs—very small roots off of primary and secondary roots that grow between soil particles.

Roots—the parts of a plant that usually grow underground.

Rotation—turning, revolving.

Roughage—feedstuff that is high in fiber and low in energy.

Ruminant—an animal with a stomach divided into three or four chambers for further refining what was ingested.

Salt—what is formed when a strong acid and base are combined.

Sanitation—keeping areas clean.

Sapwood—the xylem in a tree that is two to four years old.

Saw mill—the place where logs are cut into boards.

Saw timber—the largest trees harvested for making lumber.

Scheduled feeding—fed only at certain times of the day.

Science—knowledge of the world around us arranged as facts or truths to show relationships.

Scientific method—the five important steps in studying science.

Scientific names—the two-word names of a species.

Scion—section of stem or other plant part that is grafted to another plant.

Screw—modified incline plane; a simple machine.

Seasoning—drying of lumber to reduce the moisture content.

Seaweed—algae forms cultivated in water for food or food ingredients.

Secondary defenders—the natural responses to many diseases which try to keep the animal healthy.

Secondary growth—increases in the diameter of stems and roots in plants.

Secondary root—a branch off of the primary root.

Secondary succession—what occurs when an area is destroyed by a natural disaster or by people and the plant and animal life sequentially develop.

Seed coat—outer layer of a seed that protects the embryo and holds the seed parts together.
Selling—process of making a transaction between a buyer and a seller.
Seed—container of new plant life.
Seed-borne pest—pest in the seed and goes where the seed goes.
Seed certification—a process to assure quality and is used in the improvement of plant varieties and to assure genetic purity.
Seed-germ—plant embryo.
Selective herbicide—a chemical which will kill only certain kinds of plants.
Sexual reproduction—using seed to propagate plants or animals.
Shelter—housing humans use.
Side-dressing—applying along the sides of rows or growing plants.
Silage—made from cut green plants by chopping them into small pieces and placing the "chop" into a silo.
Simple leaf—one with only one blade.
Single woody trunks—tree trunks of many cells with strong walls.
Softwood—wood that is light and easy to cut; from pine, fir, spruce, and similar trees.
Soil—the outer layer of the earth's surface.
Soil aeration—tiny cracks and other breaks in the soil allowing air to enter.
Soil analysis—test used to determine the nutrients present in soil.
Soil pH—a chemical way of expressing the acidity and alkalinity of soil.
Soil profile—a vertical section of the soil at a particular place.
Soil sampling—involves collecting samples of soil for analysis.
Soil structure—the arrangement of the soil particles into shapes or pieces.
Soil texture—the proportion of sand, silt and clay in the soil.
Soil tilth—the physical condition of the soil.
Solar system—the group of bodies that are held together or served by a sun.
Solute—the substance that is dissolved in a solution.
Solution—a mixture of two or more substances.
Solvent—what the liquid material of a solution is known as.
Sound—the result when an object vibrates or moves.
Spanish goat—a class of goat kept for meat and to control unwanted vegetation.
Spay—remove the ovaries of a female.
Species—the seventh division in the scientific classification system of plants and animals where all organisms are very similar.
Speed—the rate of motion.
Sperm—male sex cell.
Spinal cord—the thick cord of nerve tissue of the central nervous system, extending down the spinal canal in the backbone.
Spoilage—when food is unsafe to eat and the flavor has changed.
Square—a surface in which all four sides are of equal length.
Square units—how area is reported in measurement.
Standard of living—the income of the people and the products available to them.
Steer—a calf that is castrated before sexual maturity.
Stem—that part of a plant that supports leaves and flowers, transports water and other material, and stores food.
Stocker calf—a calf that is put in a feedlot or on pasture for additional growth and fattening.
Stomach poison—toxic material that is eaten.
Straightbreeding—the mating of animals of the same breed.
Stratosphere—the second layer of the atmosphere above the earth extending out about 30 miles (50 km).
Strip cropping—involves planting crops in strips which slow down the speed of the wind and rate of water runoff.
Subterranean stem—that part of a plant stem that grows below the ground.
Succession—the process of growing and changing, involves the natural continual replacement of the organisms that live in a place.
Sucking mouth—suck only without capability of chewing.

Superovulation—a procedure where the reproductive potential of a female is increased through the number of eggs produced during their estrus cycle.

Supplies and services—the many things required by farming that are manufactured off of the farm.

Surfactant—a material that helps solutions do what they are supposed to do.

Sustainable agriculture—the term used to describe all of the things done in agriscience to maintain the ability to produce food, fiber and shelter.

Symbiosis—biotic factors forming a relationship that has benefit to at least one of them.

Symptom—the way a disease shows itself in an animal.

Synthetic biology—creating life-like characteristics with chemicals and non-living substances.

Synthetic food—an edible product made from non-food raw materials or altered food materials.

Systemic medicine—medications absorbed into the bloodstream.

Systemic poison—poison inside the plant that had been applied to the soil and taken up by the roots.

Tape measure—a measuring device usually made of metal or fabric that the tape rolls when not in use.

Taproot—plants with one main root that grows downward.

Taxonomy—the classification of living things.

Technology—the practical use of science.

Temperate climate zone—bands around the earth between the tropical and polar zones where the average temperature is above 65°F (18°C) in the summer and below 50°F (10°C) in the winter.

Temperature—refers to how hot or cold something is; a measure of the heat energy of molecules as they move about.

Temporary pasture—pastures planted for winter, summer or semi-permanent grazing.

Terracing—involves building earthen ridges or embankments that slow the rate of water runoff.

Terrestrial—an animal that grows on land.

Textile mill—the facility that weaves cotton into cloth.

Therapeutant—chemical used with disease present, therapeutic chemical.

Thermal energy—energy produced by heat.

Thermometer—a device used to measure temperature.

Thickness scale—used to measure very small distances and make settings.

Thigmotropism—how plants grow over and around solid objects such as rocks.

Tissue—groups of cells that are alike in activity and structure.

Tissue analysis—involves testing samples of tissue.

Tissue culture—taking a group of cells or a single cell and growing it.

Top-dressing—applied over the ground where plants are growing.

Topical medicine—therapeutant applied to the skin or outer surface.

Toxic products—substances that contain poison or have the potential to poison animals and plants.

Trace elements—nutrients that are needed in small amounts for growth.

Transformation—process of changing the DNA in a cell.

Transformed cells—cells that receive new material.

Transgenic animals—animals whose genes were changed when they were embryos.

Translocated herbicide—herbicide absorbed into the plant and moved all through its ports.

Transpiration—the process of the leaves of plants giving off water.

Transporting—moving the product from one place to another.

Trap cropping—involves planting a small plot of the crop near the field about two weeks before the main crop is planted in order to trap pests.

Trapezoid—a shape with four sides, two of which are parallel.

Triangle—a surface that has three sides.

Tropical climate zone—a wide band around the earth on both sides of the equator where the average temperature is above 65°F (18°C).

Tropism—plant movement.

Troposphere—the layer of the atmosphere closest to the earth.

Tuber—enlarged underground branch of a plant, such as the potato.

Ultrasonic transducer—a device that produces ultrasound which are high frequency waves that can be used to indicate movement or development in organisms.

Universe—all that exists—both known and unknown.

Upgrading—a purebred male is mated with a grade female with the intent to improve the quality of the offspring.

Variability—the difference between each animal in a species.

Variety—a group of related organisms in a species.

Vascular system—circulatory system of a plant.

Vector—what carries new DNA into a cell.

Vegetative propagation—reproducing plants asexually; getting a plant part to grow as another plant.

Veneer—thin layer of valuable or high quality wood placed over low quality wood.

Vertebrate—animals with internal skeletons and backbones.

Vertical integration—when a non-farm agribusiness owns the processing plant and all of the sources of supplies for producing the plant or animal.

Vesicle—tiny sphere-shaped structures with cell-like characteristics; used in synthetic biology research.

Viability—the growing quality of seeds.

Vital signs—signs of life; breathing, pulse rate and body temperature.

Vitamins—substances needed for good health; important in regulating body functions, keeping the body health, and developing resistance to disease.

Volt—used to measure electrical potential energy.

Volume—the total size of an object.

Voluntary muscle—a muscle that is controlled by the thinking part of the brain.

Wakame—brown seaweed used for food.

Waste disposal—the process of getting rid of wastes.

Water—a colorless, transparent liquid; probably the most important of all nutrients.

Water cycle—the process of water circulating from one part of the hydrosphere to another.

Water table—the point in the earth where all of the spaces are filled and no more water can be held; natural level of free water below the surface of the earth.

Watt—used to measure the amount of electrical energy supplied or used.

Weather—the conditions that exist in the atmosphere including temperature, air pressure, wind, moisture, clouds and precipitation.

Weathering—how materials that are the crust of the earth are changed.

Wedge—sloping surface that tapers to a point; similar to an inclined plane.

Weed—a plant that is growing where it isn't wanted.

Weighing scale—a machine that measures weight.

Weight—the heaviness of something.

White blood cell—leukocytes which help fight off disease.

Wholesome—promoting or conductive to good health.

Wilting—the stems and leaves of a plant are no longer rigid and erect.

Wind—moving air.

Woody stems—the stems of shrubs and many crop plants with rigid tissue.

Work—moving an object through a distance when there is some resistance to its movement.

Working animal—one used for power, pleasure and other purposes.

Xylem—conducts water and nutrients throughout a plant.

Zoology—the science of animal life.

Bibliography

Books

Baker, MeeCee and Robert E. Mikesell. *Animal Science Biology and Technology.* Danville, Illinois: Interstate Publishers, Inc., 1996.

Bienz, D.R. *The Why and How of Home Horticulture*. New York: W.H. Freeman and Company, 1993.

Boone, Lester V. *Producing Farm Crops.* Danville, Illinois: Interstate Publishers, Inc., 1991.

Boyd, James S. (Revised by Carl L. Reynolds). *Practical Farm Buildings*. Danville, Illinois: Interstate Publishers, 1993.

Buriak, Philip and Edward W. Osborne. *Physical Science Applications in Agriculture*. Danville, Illinois: Interstate Publishers, Inc., 1996.

Burke, Stanley R. and T.J. Wakeman. *Modern Agricultural Mechanics,* 2nd ed. Danville, Illinois: Interstate Publishers, 1992.

Burton, L. DeVere. *Agriscience and Technology*. Albany, New York: Delmar Publishers, 1992.

California Fertilizer Association. *Western Fertilizer Handbook*. Danville, Illinois: Interstate Publishers, 1995.

Camp, William G. and Thomas B. Daugherty. *Managing Our Natural Resouces*. Albany, New York: Delmar Publishers, Inc., 1991.

Causey, Ann S. *Environmental Action Guide: Action for a Sustainable Future.* New York: The Benjamin/Cummings Publishing Co., Inc., 1991.

Chiras, Daniel D. *Environmental Science*. New York: Benjamin/Cummings Publishing Co., 1991.

Ensminger, M.E. *Animal Science*, 9th ed. Danville, Illinois: Interstate Publishers, Inc., 1991.

Ensminger, M.E. *Beef Cattle Science*. Danville, Illinois: Interstate Publishers, Inc., 1997.

Ensminger, M.E. *Horses and Horsemanship*, 6th ed. Danville, Illinois: Interstate Publishers, Inc., 1990.

Ensminger, M.E. *The Stockman's Handbook*, 7th ed. Danville, Illinois: Interstate Publishers, Inc., 1992.

Ensminger, M.E. and R.O. Parker. *Swine Science*, 5th ed. Danville, Illinois: Interstate Publishers, Inc., 1984.

Fenske, Theodore H., Richard M. Drake, and Allen W. Edson. *Arithmetic in Agriculture*. Danville, Illinois: Interstate Publishers, Inc., 1951.

Foster, Albert B. *Approved Practices in Soil Conservation*. Danville, Illinois: Interstate Publishers, Inc., 1982.

Greer, William J. *Animal Health*. Danville, Illinois: Interstate Publishers, 1992.

Hamilton, William H., Donald F. Connelly, and D. Howard Doster. *Agribusiness: An Entrepreneurial Approach*. Albany, New York: Delmar Publishers, 1992.

Hawkins, Del I., Roger J. Best, and Kenneth A. Coney. *Consumer Behavior: Implications for Marketing Strategy*, 5th ed. Homewood, Illinois: Irwin, 1992.

Herren, Ray V. and Roy L. Donahue. *The Agriculture Dictionary*. Albany, New York: Delmar Publishers, 1991.

Hokanson, Clifford M. *Applied Problems in Mathematics for Agriculture*. Danville, Illinois: Interstate Publishers, Inc., 1984.

Holland, I.I. and G.L. Rolfe. *Forests and Forestry*. Danville, Illinois: Interstate Publishers, Inc., 1997.

Hunsley, Roger E. and W. Malcolm Beeson. *Livestock Judging, Selection and Evaluation*. Danville, Illinois: Interstate Publishers, Inc., 1991.

Ingels, Jack E. *Landscaping: Principles and Practices*, 4th ed. Albany, New York: Delmar Publishers, 1992.

Janick, Jules and James E. Simon, ed. *New Crops*. New York: John Wiley and Sons, Inc., 1993.

Kircher, Harry B., Donald L. Wallace, and Dorothy J. Gore. *Our Natural Resources and Their Conservation*. Danville, Illinois: Interstate Publishers, Inc., 1992.

Kotler, Philip. *Marketing Management*, 7th ed. Englewood Cliffs, New Jersey: Prentice Hall, Inc., 1991.

Krebs, Alfred H. (Revised by Michael E. Newman). *AgriScience in Our Lives*. Danville, Illinois: Interstate Publishers, Inc., 1994.

Lee, Daniel O'C. and John F. Wickins. *Crustacean Farming*. New York: Halsted Press, 1992.

Lee, Jasper S. *Commercial Catfish Farming*. Danville, Illinois: Interstate Publishers, Inc., 1991.

Lee, Jasper S. and Michael E. Newman. *Aquaculture — An Introduction*. Danville, Illinois: Interstate Publishers, Inc., 1997.

Lee, Jasper S., Chris Embry, Jim Hutter, Rick Rudd, Jody Pollok, Lyle Westrom, and Austin Bull. *Introduction to Livestock and Poultry Production*. Danville, Illinois: Interstate Publishers, Inc., 1996.

Luening, Robert A., Richard M. Klemme, and William P. Mortenson. *The Farm Management Handbook.* Danville, Illinois: Interstate Publishers, Inc., 1991.

Newman, Michael and Walter J. Wills. *Agribusiness Management and Entrepreneurship.* Danville, Illinois: Interstate Publishers, Inc., 1994.

Newton, David E. *Environmental Chemistry.* Portland, Maine: J. Weston Walch, 1991.

Oppenheimer, Harold L. *Cowboy Arithmetic — Cattle as an Investment.* Danville, Illinois: Interstate Publishers, Inc., 1985.

Petrides, George A. *Peterson First Guides, Trees.* New York: Houghton Miflin Company, 1993.

Phipps, Lloyd J. and Carl L. Reynolds. *Mechanics in Agriculture.* Danville, Illinois: Interstate Publishers, Inc., 1992.

Plaster, Edward J. *Soil Science and Management.* Albany, New York: Delmar Publishers, Inc., 1992.

Porter, Lynn, Jasper S. Lee, Diana L. Turner, and Malcolm Hillan. *Environmental Science and Technology.* Danville, Illinois: Interstate Publishers, Inc., 1997.

Reiley, H. Edward and Carroll L. Shry, Jr. *Introductory Horticulture.* Albany, New York: Delmar Publishers, 1991.

Romans, John R., William J. Costello, Kevin W. Jones, C. Wendell Carlson, and Marion L. Greaser. *The Meat We Eat.* Danville, Illinois: Interstate Publishers, Inc., 1994.

Rushing, Felder. *Gardening Southern Style.* Jackson, Mississippi: University Press of Mississippi, 1987.

Russell, N.J. and G.W. Gould (edited). *Food Preservatives.* New York: Van Nostrand Reinhold, 1991.

Sager, Robert J., David M. Helgren, and Saul Israel. *World Geography Today.* Chicago, Illinois: Holt, Rinehart and Winston, 1992.

Saland, Leonard. *Contemporary Chemistry: A Practical Approach.* Portland, Maine: J. Weston Walch, 1991.

Schoell, William F. and Joseph P. Guiltinan. *Marketing: Contemporary Concepts and Practices,* 5th ed. Boston: Allyn and Bacon, 1992.

Schraer, William D. and Herbert J. Stoltze. *Biology: The Study of Life.* Englewood Cliffs, New Jersey: Prentice Hall, 1995.

Schroeder, Charles B., Eddie Dean Seagle, Lorie M. Felton, John M. Ruter, William Terry Kelley, and Gerard Krewer. *Introduction to Horticulture: Science and Technology.* Danville, Illinois: Interstate Publishers, Inc., 1997.

Seiden, Rudolph. *Livestock Health Encyclopedia.* New York: Springer Publishing Co., Inc., 1968.

Smith, Marcella, Jean M. Underwood, and Mark Bultmann. *Careers in Agribusiness and Industry.* Danville, Illinois: Interstate Publishers, Inc., 1991.

Swiader, John M., George W. Ware, and J. P. McCollum. *Producing Vegetable Crops*, 4th ed. Danville, Illinois: Interstate Publishers, Inc., 1992.

Towle, Albert. *Modern Biology*. Chicago, Illinois: Holt, Rinehart and Winston, 1991.

U.S. Department of Agriculture. *Agricultural Statistics 1991*. Washington, D.C., 1991.

Warner, Linda A., Loretta Kett Bierer, Sherry A. Lawson, and Tracey L. Cohen. *Life Science, The Challenge of Discovery*. Lexington, Massachusetts: D. C. Heath and Company, 1991.

Witt, Steven C. *Brief Book — Biotechnology, Microbes and the Environment*. San Francisco: Center for Science Information, 1990.

Wood, Clair G. *Safety in School Science Labs*. Portland, Maine: J. Weston Walch, 1991.

Wright, John W., ed. *The Universal Almanac 1993*. New York: Andrews and McMeel, 1993.

Zim, Herbert S. and Alexander C. Martin. *Trees*. New York: Golden Press, 1987.

Monographs

Lee, Jasper S. "Biotechnology: Science at Work in Agricultural Industry." Washington, D.C.: United States Department of Agriculture, May 1988.

Lee, Jasper S. *Food Preservation Manual*. Blacksburg, Virginia: Agricultural Education Program, 1975.

Lee, Jasper S. and James N. Butler, Jr. "The Importance of Selected Mathematics Concepts/Skills and Applications as Perceived by Mississippi High School Agriscience and Mathematics Teachers for the Introduction to Agriscience Curriculum." Miss. State: Department of Agriculture Education and Experimental Station.

Lee, Jasper S. "A Reference Unit on Vegetable Production." Miss. State: Agricultural Education Department, 1969.

Sattelle, David B. "Biotechnology in Perspective." Hobsons Publishing, 1990.

"Vocational Applied Mathematics — Agriculture: Crop Production." Atlanta, Georgia: Georgia Department of Education, 1991.

"Vocational Applied Mathematics — Agriculture: Livestock." Atlanta, Georgia: Office of Vocational Education, 1986.

"1996 Weed Control Guidelines." Miss. State: Mississippi Cooperative Extension Service, 1996.

Index

A

B

C

D

E

O

P

R

S